W0037259

Periodic Table of the Elements with the Gmelin System Numbers

1 H 2																1 H 2	2 He 1
3 Li 20	4 Be 26											5 B 13	6 C 14	7 N 4	8 O 3	9 F 5	10 Ne 1
11 Na 21	12 Mg 27											13 Al 35	14 Si 15	15 P 16	16 S 9	17 Cl 6	18 Ar 1
19 * K 22	20 Ca 28	21 Sc 39	22 Ti 41	23 V 48	24 Cr 52	25 Mn 56	26 Fe 59	27 Co 58	28 Ni 57	29 Cu 60	30 Zn 32	31 Ga 36	32 Ge 45	33 As 17	34 Se 10	35 Br 7	36 Kr 1
37 Rb 24	38 Sr 29	39 Y 39	40 Zr 42	41 Nb 49	42 Mo 53	43 Tc 69	44 Ru 63	45 Rh 64	46 Pd 65	47 Ag 61	48 Cd 33	49 In 37	50 Sn 46	51 Sb 18	52 Te 11	53 I 8	54 Xe 1
55 Cs 25	56 Ba 30	57** La 39	72 Hf 43	73 Ta 50	74 W 54	75 Re 70	76 Os 66	77 Ir 67	78 Pt 68	79 Au 62	80 Hg 34	81 Tl 38	82 Pb 47	83 Bi 19	84 Po 12	85 At	86 Rn 1
87 Fr	88 Ra 31	89*** Ac 40	104 71	105 71													* NH_4 23

**Lanthanides 39	58 Ce	59 Pr	60 Nd	61 Pm	62 Sm	63 Eu	64 Gd	65 Tb	66 Dy	67 Ho	68 Er	69 Tm	70 Yb	71 Lu

***Actinides	90 Th 44	91 Pa 51	92 U 55	93 Np 71	94 Pu 71	95 Am 71	96 Cm 71	97 Bk 71	98 Cf 71	99 Es 71	100 Fm 71	101 Md 71	102 No 71	103 Lr 71

A Key to the Gmelin System is given on the Inside Back Cover

Gmelin Handbook of Inorganic Chemistry

8th Edition

Organometallic Compounds in the Gmelin Handbook

The following listing indicates in which volumes these compounds are discussed or are referred to:

Ag	Silber B5 (1975)
Au	Organogold Compounds (1980)
Bi	Bismut-Organische Verbindungen (1977)
Co	Kobalt-Organische Verbindungen 1 (1973), 2 (1973), Kobalt Erg.-Bd. A (1961), B1 (1963), B2 (1964)
Cr	Chrom-Organische Verbindungen (1971)
Cu	Organocopper Compounds 1 (1985), 2 (1983)
Fe	Eisen-Organische Verbindungen A1 (1974), A2 (1977), A3 (1978), A4 (1980), A5 (1981), A6 (1977), A7 (1980), A8 (1985), B1 (partly in English; 1976), Organoiron Compounds B2 (1978), Eisen-Organische Verbindungen B3 (partly in English; 1979), B4 (1978), B5 (1978), Organoiron Compounds B6 (1981), B7 (1981), B8 to B10 (1985), B11 (1983), B12 (1984), Eisen-Organische Verbindungen C1 (1979), C2 (1979), Organoiron Compounds C3 (1980), C4 (1981), C5 (1981), C7 (1985), and Eisen B (1929-1932)
Hf	Organohafnium Compounds (1973)
Nb	Niob B4 (1973)
Ni	Nickel-Organische Verbindungen 1 (1975), 2 (1974), Register (1975), Nickel B3 (1966), C1 (1968), C2 (1969)
Np, Pu	Transurane C (partly in English; 1972)
Pt	Platin C (1939) and D (1957)
Ru	Ruthenium Erg.-Bd. (1970)
Sb	Organoantimony Compounds 1 (1981), 2 (1981), 3 (1982)
Sc, Y, La to Lu	D6 (1983)
Sn	Zinn-Organische Verbindungen 1 (1975), 2 (1975), 3 (1976), 4 (1976), 5 (1978), 6 (1979), Organotin Compounds 7 (1980), 8 (1981), 9 (1982), 10 (1983), 11 (1984), 12 (1985)
Ta	Tantal B2 (1971)
Ti	Titan-Organische Verbindungen 1 (1977), 2 (1980) 3 (1984), 4 and Register (1984)
U	Uranium Suppl. Vol. E2 (1980)
V	Vanadium-Organische Verbindungen (1971), Vanadium B (1967)
Zr	Organozirconium Compounds (1973)

Gmelin Handbook of Inorganic Chemistry

8th Edition

Gmelin Handbuch der Anorganischen Chemie

Achte, völlig neu bearbeitete Auflage

Prepared and issued by

Gmelin-Institut für Anorganische Chemie
der Max-Planck-Gesellschaft
zur Förderung der Wissenschaften

Director: Ekkehard Fluck

Founded by — Leopold Gmelin

8th Edition — 8th Edition begun under the auspices of the Deutsche Chemische Gesellschaft by R. J. Meyer

Continued by — E.H.E. Pietsch and A. Kotowski, and by Margot Becke-Goehring

Springer-Verlag Berlin Heidelberg GmbH 1985

Gmelin-Institut für Anorganische Chemie
der Max-Planck-Gesellschaft zur Förderung der Wissenschaften

Gmelin Handbook of Inorganic Chemistry

8th Edition

Fe
Organoiron Compounds

Part B 9

Mononuclear Compounds 9

With 27 illustrations

AUTHOR Adolf Slawisch

EDITORS Jürgen Faust, Johannes Füssel, Marlis Mirbach

CHIEF EDITOR Adolf Slawisch

Springer-Verlag Berlin Heidelberg GmbH 1985

LITERATURE CLOSING DATE: 1983
IN SOME CASES MORE RECENT DATA HAVE BEEN CONSIDERED

Library of Congress Catalog Card Number: Agr 25-1383

ISBN 978-3-662-06929-5 ISBN 978-3-662-06927-1 (eBook)
DOI 10.1007/978-3-662-06927-1

Originally published by Springer-Verlag, Berlin · Heidelberg · New York · Tokyo in 1985
Softcover reprint of the hardcover 8th edition 1985

Preface

The present volume is a continuation of Series B on the mononuclear organoiron compounds and describes 545 compounds. It covers the literature completely to the end of 1983 and includes many references to the literature up to mid-1984.

This volume continues the description of mononuclear organoiron derivatives by treating compounds of the type $^4LFe(CO)_3$ where ^4L includes seven-, eight-, and nine-membered ring systems. Only a few larger rings of the $^4LFe(CO)_3$ type compounds are known, and they are also included. In the symbol ^mL_n, n is the number of organic ligands L in the compound, and m is the number of Fe-C bonds formed from an L ligand as already explained in the prefaces to "Kobalt-Organische Verbindungen" 1, New Suppl. Ser., Vol. 5, 1973, and "Nickel-Organische Verbindungen" 1, New Suppl. Ser., Vol. 16, 1975.

Series B so far comprises volumes B1 to B12, and a survey of this series has been given in the preface to B7 (1981).

Much of the data, particularly in tables, is given in abbreviated form without dimensions; for explanation see p. VIII. Additional remarks, if necessary, are given in the heading of the tables.

A formula index for volumes B8, B9, and B10 will be given in "Organoiron Compounds" B10.

Frankfurt
September 1985

Adolf Slawisch

Remarks on Abbreviations and Dimensions

Many compounds in this volume are presented in tables in which numerous abbreviations are used, and the dimensions are omitted for the sake of conciseness. This necessitates the following clarification:

The prefixes endo (same side of ring to that occupied by Fe) and exo (opposite side of ring) are used to denote stereochemistry of the complexes; see also pp. 44.

Temperatures are given in °C, otherwise K stands for Kelvin. Abbreviations used with temperatures are m.p. for melting point, b.p. for boiling point, dec. for decomposition, and subl. for sublimation.

NMR represents **nuclear magnetic resonance.** Chemical shifts are given as δ values in ppm; reference substances and signs for δ are shown in the scheme below:

		increasing field	→
		δ=0 for	
^{1}H	+	$Si(CH_3)_4$	−
^{11}B	+	$BF_3 \cdot O(C_2H_5)_2$	−
^{13}C	+	$Si(CH_3)_4$	−
^{19}F	−	$CFCl_3$	+
^{31}P	+	H_3PO_4	−

Coupling constants J, in Hz, are given as J(A, B) or as J(1,3) and refer to labelled structural formulas.

Multiplicities of the signals are abbreviated as s, d, t, q (singlet to quartet), quint, sext, sept, oct (quintet to octet), and m (multiplet); terms like dd (double doublet) and t's (triplets) are also used. Assignments referring to labelled structural formulas are given in the form C-4, H-3,5. The numbering deviates in some cases from official nomenclature so that corresponding values for compounds in the same chapter can be more easily compared.

Mössbauer spectra are represented by ^{57}Fe-γ and ^{119}Sn-γ: both the isomer shift δ (vs. $Na_2[Fe(CN)_5NO]$ or $BaSnO_3$ at room temperature) and the quadrupole splitting Δ are given in mm/s; the experimental error has generally been omitted. Other reference substances for δ are indicated after the numerical value, e.g., δ=0.23 (Fe).

Optical spectra are labelled as IR (infrared), R (Raman), and UV (electronic spectrum including the visible region): IR bands and Raman lines are given in cm^{-1}; the assigned bands are usually labelled with the symbols ν for stretching vibration and δ for deformation vibration whereas unlabelled bands belong to CO stretching vibrations. Intensities occur in parentheses either in the common qualitative terms (s, m, w, vs, br, etc.) or as numerical relative intensities; str., def., pol., and dp mean stretching, deformation, polarized, and depolarized, respectively. The CO stretching force constant and CO, CO interaction constant (in mydn/Å) are denoted as k and k′, respectively. The UV absorption maxima, λ_{max}, are given in nm followed by the extinction coefficient ε ($L \cdot cm^{-1} \cdot mol^{-1}$) or log ε in parentheses; sh means shoulder.

Solvents, or the **physical state** of the sample, and the temperature (in °C or K) are given in parentheses immediately after the spectral symbol, e.g., R(solid), ^{13}C NMR (C_6D_6, 50 °C). Common solvents are given by their formula (cyclo-C_6H_{12}=cyclohexane) except THF, which represents tetrahydrofuran.

Figures give only selected parameters. Bond lengths (in Å) or angles are mean values.

Table of Contents

Organoiron Compounds, Part B

Mononuclear Compounds 9

Page

Remarks on Abbreviations and Dimensions VIII

1.4.1.4.1.2.4 ^{4}L is a Seven-Membered Ring System 1
^{4}L is a Heterocyclic Ring System 2
^{4}L is an Isocyclic Ring 34
^{4}L is Cyclohepta-1,3-diene 34
^{4}L is a Substituted Cyclohepta-1,3-diene 37
^{4}L is Cycloheptatriene 102
$C_7H_8Fe(CO)_3$ 102
$C_7H_8Fe(CO)_3AlX_3$ 116
^{4}L is a Substituted Cycloheptatriene 116
^{4}L is a Monosubstituted Cycloheptatriene 116
^{4}L is a Disubstituted Cycloheptatriene 144
Other Substituted Cycloheptatrieneirontricarbonyl Complexes 179
^{4}L is a Cyclohepta-1,4-diene 197

1.4.1.4.1.2.5 ^{4}L is an Eight-Membered Ring System 207
^{4}L is a Cyclooctadiene 208
^{4}L is a Cyclooctatriene 219
^{4}L is a Cyclooctatetraene 230
$C_8H_8Fe(CO)_3$ and Its $AlBr_3$ Adduct 230
Preparation and Formation of $C_8H_8Fe(CO)_3$ 230
Physical Properties of $C_8H_8Fe(CO)_3$ 234
Chemical Behavior and Uses of $C_8H_8Fe(CO)_3$ 241
The Adduct $C_8H_8Fe(CO)_3AlBr_3$ 253
^{4}L is a Substituted Cyclooctatetraene 253

1.4.1.4.1.2.6 ^{4}L is a Nine-Membered Ring System 279

1.4.1.4.1.2.7 ^{4}L is a Ring System Greater than C_9 282

Table of Conversion Factors 285

Organoiron Compounds, Part B

Mononuclear Compounds 9

For general references see 1.4 in "Organoiron Compounds" B 6, 1981, pp. 1/2, and 1.4.1.4.1.2 in "Organoiron Compounds" B 7, 1981, pp. 1/2. Further references for the years 1973 to 1983 are given in the individual chapters.

1.4.1.4.1.2.4 ^{4}L is a Seven-Membered Ring System

General References:

T.-Y. Luh, Trimethylamine N-Oxide – A Versatile Reagent for Organometallic Chemistry, Coord. Chem. Rev. **60** [1984] 255/76.

M. Franck-Neumann, Synthetic Applications of Some Metal Carbonyl Complexes, Pure Appl. Chem. **55** [1983] 1715/32.

R.C. Kerber, Interaction of Organometallic Moieties with Carbanions and Other Electron-Rich Centers, J. Organometal. Chem. **254** [1983] 131/42.

A.J. Pearson, Natural Products Synthesis Using Organoiron Complexes, Pure Appl. Chem. **55** [1983] 1767/79.

S.G. Davies, Organotransition Metal Chemistry: Applications to Organic Synthesis, Pergamon Press, Oxford, Vol. 2, 1982, pp. 1/411.

B.E. Mann, Dynamic NMR Spectroscopy in Inorganic and Organometallic Chemistry, Ann. Rept. NMR Spectrosc. **12** [1982] 263/86.

J.C. Green, Gas Phase Photoelectron Spectra of d- and f-Block Organometallic Compounds, Struct. Bonding [Berlin] **43** [1981] 37/112.

R. Hoffmann, Theoretical Organometallic Chemistry, Science **211** [1981] 995/1002.

H. Behrens, Four Decades of Metal Carbonyl Chemistry in Liquid Ammonia: Aspects and Prospects, Advan. Organometal. Chem. **18** [1980] 1/53.

Št. Toma, Využitie niektorých zlúčenin prechodných kovov v organickej syntéze [Using Some Compounds of Transition Metals in Organic Synthesis], Chem. Listy **73** [1979] 1049/83.

R. Fields, Per- and Poly-Fluorinated Aliphatic Derivatives of the Transition Elements, Fluorocarbon Relat. Chem. **3** [1976] 308/55.

J.M. Kelly, Photochemistry of Inorganic and Organometallic Compounds, Photochemistry **7** [1976] 151/210.

B.L. Booth, Complexes Containing Metal-Carbon Sigma-Bonds, MTP [Med. Tech. Publ. Co.] Intern. Rev. Sci. Inorg. Chem. Ser. Two **6** [1975] 137/88.

H. Takaya, R. Noyori, Stereochemistry of Reactions Involving Transition Metal Complexes, Yuki Gosei Kagaku Kyokaishi **32** [1974] 2/19.

F. Pietra, Seven-Membered Conjugated Carbo- and Heterocyclic Compounds and Their Homoconjugated Analogs and Metal Complexes: Synthesis Biosynthesis Structure and Reactivity, Chem. Rev. **73** [1973] 293/364.

1.4.1.4.1.2.4.1 ⁴L Is a Heterocyclic Ring System

This section deals with $^4LFe(CO)_3$ compounds where ^4L is a seven-membered heterocyclic ring system containing B, N, or O as heteroatoms. Two types of N heterocycles, azepines and diazepines, are known (compare Table 1, pp. 5/20). Some of the compounds cited in Table 1 can be prepared by the following four methods:

Method I: $Fe(CO)_5$ and the organic ligand ^4L are irradiated for several hours in an inert solvent such as ether or tetrahydrofuran; the products are separated by chromatography on alumina or silica gel.

a. $Fe(CO)_5$ and the organic ligand ^4L are irradiated in tetrahydrofuran for 20 to 36 h (36 h for No. 4 [11], 30 h for No. 5 [1, 47], 24 h for Nos. 6 and 7 [11], and 20 h for No. 16 [1]). The reaction mixture is filtered, evaporated, and chromatographed on neutral alumina (Woelm, activity II) for No. 4 [11], on alumina containing 10% H_2O for No. 5 [1], or on basic alumina (Woelm) for Nos. 6, 7 [11], and 16 [1]. The eluent is hexane [1, 11].

b. $Fe(CO)_5$ and the B containing heterocycle ^4L are irradiated in ether in a Schlenk tube. The reaction mixture is filtered from $Fe_2(CO)_9$, the filtrate is evaporated, and the residue is chromatographed on silica gel (activity II to III) with pentane. The intensive yellow band is concentrated, filtered through silica gel, and crystallized at −78 °C to give No. 1. Three other bands are not identified [37]. Irradiation of $Fe(CO)_5$ with a mixture of I and II (R=H) in ether at −60 °C for 3 h, followed by distillation at 20 °C/15 Torr gives C_6H_6 (56%), and chromatography of the residue on neutral silica gel (Woelm, 5% H_2O) with petroleum ether gives $Fe_3(CO)_{12}$, No. 75, and C_6H_5OH (37%) [27]. Similar photolysis at 25 °C lowers the yield of No. 75 [4, 27]. Analogous photolysis with II(R=CH_3) at −60 °C gives $Fe_3(CO)_{12}$, No. 76 (5%), and 2-xylene (61%) [27]. Further elution with C_6H_6 gives III (0.7%) [27], possibly identical to $C_8H_{10}OFe_2(CO)_6$ (see 2.6.1.1 in "Organoiron Compounds" C5, 1981, p. 4) formulated with a closed ring [4], and 2,6-dimethylphenol (20%) [27]. Similar irradiation with II(R=CH_3) at room temperature in petroleum ether (b.p. 40 to 60 °C) for 22 h followed by chromatography on Al_2O_3 (activity III) with petroleum ether gives No. 76 and $C_8H_{10}OFe_2(CO)_6$ (see before) [4]. Analogous irradiation with II(R=CH_3) in tetrahydrofuran followed by chromatography on Al_2O_3 with petroleum ether gives eight different products; two of which are identical with those described before [4]. Analogous irradiation with 1-benzoxepine (^4L) for 15 h in petroleum ether gives No. 77 [4].

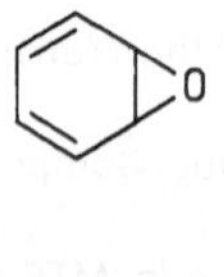

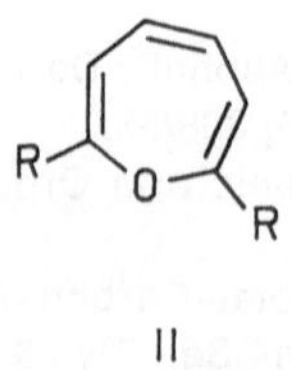

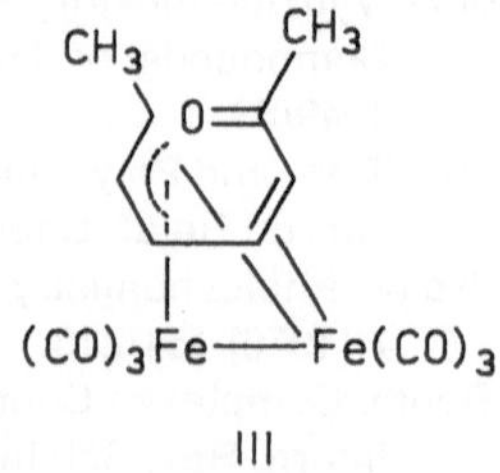

Method II: $Fe_2(CO)_9$ and the organic ligand ^4L react in an inert solvent. The reaction mixture is filtered, the solvent is removed from the filtrate, and the residue is chromatographed. This method is simpler and leads to cleaner products than Method I [11]. Cyclo-[3.3.3]azine derivatives do not give complexes with $Fe_2(CO)_9$ in ether at room temperature [41].

References on pp. 33/4

a. $Fe_2(CO)_9$ and ^4L react in hexane at room temperature for 12 h [36] or at reflux for 20 min. The reaction mixture is filtered [11, 36] through Celite, and evaporation of the filtrate gives No. 4 [11, 36]. Reaction of $Fe_2(CO)_9$ with ^4L in n-hexane at room temperature for 12 h, followed by chromatography on neutral Al_2O_3 with hexane as eluent, gives iron carbonyls. Subsequent elution with ether gives No. 58 [47].

b. $Fe_2(CO)_9$ and ^4L react in C_6H_6 at 20 °C, [9, 10, 20, 26, 40] (in the dark for No. 38 [26]) for 5 h (for Nos. 43, 44, 46 [9, 40], 56 [9], 58 [9, 10], 63, 64 [40], 68 [9]), for 8 to 10 h (for Nos. 44, 53, 56, 65, 66, and 69 [20]), or for 2 d (for Nos. 37 to 39) [26]. The reaction mixture is filtered, evaporated, and the residue is dissolved in C_6H_6 [9, 10, 20, 26, 40]. Chromatography on silica gel [9, 10, 20, 26, 40] with $C_6H_6/CH_3CO_2C_2H_5$ (1:1) gives Nos. 37 to 39 [26]. No. 38 is not obtained in its pure form due to its extreme instability [26]. The products No. 43, 44, 46, 56 [9], 58 [9, 10], and 68 [9] are eluted with cyclohexane/$CH_3CO_2C_2H_5$ (6:4). In a reinvestigation of this reaction, pyrroles V with R′=H, $R=SO_2C_6H_5$ (56%), $SO_2C_6H_4CH_3$-4 (45%), or with $R'=CH_3$, $R=SO_2C_6H_5$ (75%), $SO_2C_6H_4CH_3$-4 (73%) are found as the main products in the preparation of Nos. 43, 44, 63, and 64. Excess $Fe_2(CO)_9$, to form No. 44, does not lead to the formation of this compound [40]. The products No. 44, 53, 56, 65, 66, and 69 are eluted with $C_6H_6/CH_3CO_2C_2H_5$ (4:1) [20]. Refluxing $Fe_2(CO)_9$ and ^4L in C_6H_6, followed by work-up as in Method II a, gives No. 32 [44].

c. $Fe_2(CO)_9$ and ^4L (for ^4L for No. 21 see Formula IV) react in ether at room temperature for 12 h (Nos. 25, 33, 34, and 35) [42], for 24 h (No. 30) [44], for 65 h (No. 21) [42], or for 2d (No. 31) [44]. Concentration of the reaction mixture followed by chromatography on Al_2O_3 (activity II) with $C_6H_6/CH_3CO_2C_2H_5$ (5:1) for No. 33 or with benzine/$C_6H_6/CH_3CO_2C_2H_5$ (5:2:1) for Nos. 34 and 35 [42] or on silica gel with light petroleum ether/$C_6H_6/CH_3CO_2C_2H_5$ (1:1:1) for Nos. 25, 28 to 31 [44] gives the products. For the separation of No. 21 the reaction mixture is filtered, evaporated, and the residue is dissolved in C_6H_6 and cooled to +5 °C to separate the product from $Fe_3(CO)_{12}$. The solution is decanted, evaporated, and chromatographed on silica gel (Woelm, activity I). The yellow main fraction is separated on silica gel with hexane/ether (9:1) to give No. 21a and No. 21b (ca. 1:1) in 92% yield [39].

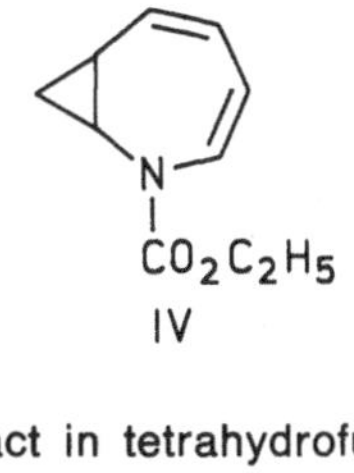

N
$CO_2C_2H_5$
IV

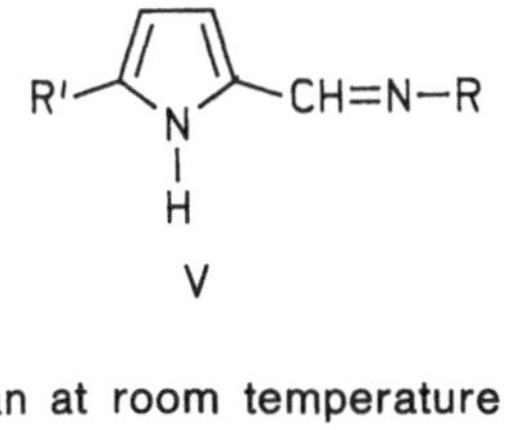

R′ N CH=N—R
H
V

d. $Fe_2(CO)_9$ and ^4L react in tetrahydrofuran at room temperature for 75 min. Filtration through Celite followed by evaporation gives Nos. 3 and 8 [11].

e. $Fe_2(CO)_9$ and ^4L react in petroleum ether at room temperature followed by chromatography on silica gel to give No. 58 [8].

Method III: $Fe_3(CO)_{12}$ and XXV (p. 27) are refluxed in C_6H_6 for 24 h. The reaction mixture is filtered through Hy-Flo Supercal. Evaporation leaves a brown glass, solidifying on trituration with CH_3OH to give No. 26 [15].

References on pp. 33/4

VI a VII b

According to SCF calculation, a complexation of VI with $Fe(CO)_5$ will take place at the seven-membered ring (see compounds No. 33 to 36) [24]; at room temperature, the onset of the equilibrium VII already begins, as seen by the broad coalesced peaks which characterize the ^{1}H NMR spectra of compounds No. 4 to 8. In the vicinity of 65 to 85 °C, the movement of the $Fe(CO)_3$ residue about the two possible positions becomes sufficiently rapid that a symmetrical spectrum of the AA′XX′YY′ type results. At 0 °C there is a well defined spectrum indicative of the fixed VIIIa (or alternatively VIIb). Further lowering of the probe temperature to −30 and −60 °C is seen to cause a new line broadening, coalescence, and finally the appearance of additional signals. Such observations can be best attributed to the freezing out of the normally rapid rotation of the CO_2R function about the C-N bond (compare XI, p. 23); when this rotational process is sufficiently retarded, the chemical shifts of the protons will understandably be different and unique and will give rise to the observed spectra (compare "Further information" for Nos. 4 to 8). This last consideration does not apply to No. 3 [11]. Molecular orbital calculations for $[C_7H_7Fe(CO)_3]^-$ predict a ^{3}L ligand and are of interesting consequences for $^4LFe(CO)_3$ compounds (^{4}L = azepine, Nos. 2 to 16, and oxepine, No. 75) [38].

Table 1
Compounds of the Type $^4LFe(CO)_3$ where ^{4}L is a Heterocyclic Seven-Membered Ring System. Further information on numbers preceded by an asterisk is given at the end of the table, pp. 21/32.
For abbreviations and dimensions see p. VIII.

No.	compound	method of preparation (yield in %), properties and remarks	Ref.
^{4}L is a B containing heterocyclic ring system:			
1	$C_7H_6B(C_6H_5)Fe(CO)_3$ (ring positions 2 to 7; B-C_6H_5)	Ib (58) m.p. 64.5 to 65.5° (from pentane at −78°), pale yellow, slightly air sensitive crystals ^{1}H NMR (benzene-d^6): 1.45 (m, H-4exo, 5exo), 1.95 (m, H-4endo, 5endo; J(H-3,4endo) <2), 3.90 (d, H-3,6; J(H-2,3) = 10.5), 4.67 (d, H-2,7), 7.47 (m, 3H in C_6H_5), 8.15 (m, 2H in C_6H_5) ^{11}B NMR (ether): 27.3 IR (n-hexane): 1994, 2051 mass spectrum: $[M-nCO]^+$ (n = 0 to 3), $[C_4H_4BC_6H_5Fe]^+$, $[C_6H_7BFe]^+$, $[C_6H_6Fe]^+$, $[C_5H_5BFe]^+$, $[Fe]^+$	[37]

References on pp. 33/4

Table 1 [continued]

No.	compound	method of preparation (yield in %), properties and remarks	Ref.
[4]L is an azepine:			
*2	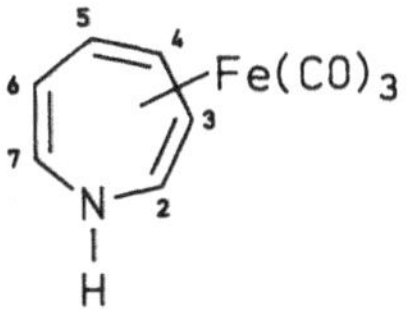 1H-azepine with positions 2–7 numbered, $Fe(CO)_3$ attached; N–H	see No. 5, p. 24 m.p. 41 to 42°, subl.p. 40°/high vacuum, orange colored, extremely air sensitive crystals IR (Nujol/Hostaflon): 1909(=NH); 1976, 1988, 2053(CO); 3460(NH)	[1]
3	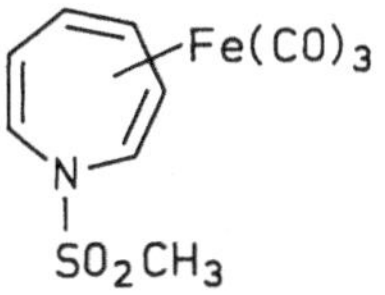 azepine with $Fe(CO)_3$ attached; N–SO_2CH_3	II d (69) m.p. 94.5 to 96° (from hexane), lustrous orange-yellow prisms ^{1}H NMR ($CDCl_3$): 2.92(s, CH_3), 3.5 to 6.5 (br, humps, 6 vinyl H, temperature dependent) IR ($CHCl_3$): 1170, 1350(SO_2N); 1635(C=C); 1990, 2070(CO) UV (C_2H_5OH): $\lambda_{max}(\varepsilon) = 232\,(14600)$ nm	[11]
*4	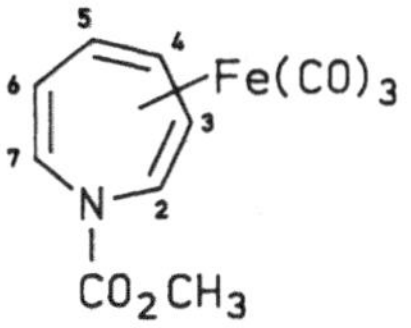 azepine with positions 2–7 numbered, $Fe(CO)_3$ attached; N–CO_2CH_3	I a (70.5), II a (53 to 69) m.p. 90 to 92°, m.p. 115 to 115.5°, m.p. 116 to 116.5° (from hexane), air stable highly crystalline yellow prisms, bright yellow crystals ^{1}H NMR ($CDCl_3$): 3.85(s, OCH_3); 4.6, 5.0, 6.3(br, H-2 to 7, temperature dependent) ^{1}H NMR ($CDCl_3$ at −10°): 3.5(H-5; J(H-4,5) = 8.0), 3.76(OCH_3), 4.44(H-3; J(H-3,5) = 1.0, J(H-3,4) = 4.0), 5.02(H-4,6; J(H-5,6) = 8.0), 5.98(H-2; J(H-2,3) = 6.0), 6.45(H-7; J(H-6,7) = 9.0) IR (Nujol): 1655(C=C); 1730, 1750(C=O); 1190(error in [11]?), 2070(CO) IR (hexane): 1735(C=O); 1980, 1993, 2053(CO) UV (C_2H_5OH): $\lambda_{max}(\varepsilon) = 250\,(17670)$, 297(5390) nm mass spectrum: $[M - nCO]^+$ (n = 0 to 3)	[3, 11, 36]
*5	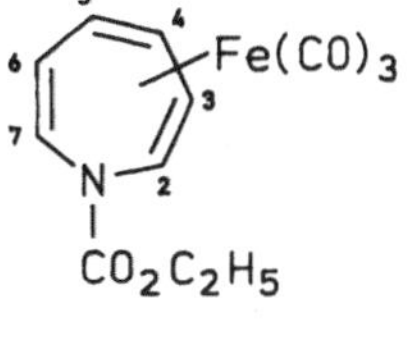 azepine with positions 2–7 numbered, $Fe(CO)_3$ attached; N–$CO_2C_2H_5$	I a (59 to 77), see also No. 2, p. 22 m.p. 86 to 88° (from hexane), m.p. 87°, subl.p. 60°/high vacuum, air stable orange-yellow crystals ^{1}H NMR ($CDCl_3$ at 0 °C): 1.32(t, CH_3; J = 7.2), 3.54(dd, H-5; J(H-5,6), ~J(H-4,5) = 7.8), 4.26(q, CH_2), ca. 4.47(H-3), 5.06(quasi t, H-4,6), 6.07(d, H-2; J(H-2,3) = 6.8), 6.52(d, H-7; J(H-6,7) = 9.0) IR (Nujol/Hostaflon): 1712; 1984, 2000, 2062(CO)	[1, 2, 14, 47]

References on pp. 33/4

Table 1 [continued]

No.	compound	method of preparation (yield in %), properties and remarks	Ref.
6	$(C_6H_6NCO_2CH_2C_6H_5)Fe(CO)_3$	I a (99) m.p. 127 to 128° (from hexane), yellow-orange needles ^{1}H NMR (CCl_4): 4.2 to 4.8 and 6.0 to 6.5 (br, 6 vinyl H, temperature dependent), 5.17 (s, OCH_2), 7.35 (s, C_6H_5) IR (Nujol): 1650 (C=C); 1750 (C=O); 2000, 2065 (CO)	[11]
7	$(C_6H_6NCO_2C(CH_3)_3)Fe(CO)_3$	I a (39.4) m.p. 70 to 70.5° (from hexane), yellow prisms ^{1}H NMR (CCl_4): 1.48 (s, $C(CH_3)_3$), 4.5 to 6.4 (br, humps, 6 vinyl H, temperature dependent) IR (Nujol): 1650, 1750 (C=O); 1990, 2070 (CO) UV (C_2H_5OH): $\lambda_{max}(\varepsilon) = 249.5 (18770)$, 298 (5825) nm	[11]
8	$(C_6H_6NCO_2C_6H_5)Fe(CO)_3$	II d (48.2) m.p. 108 to 109° (dec.), yellow prisms (from hexane) ^{1}H NMR ($CDCl_3$): 3.5, 4.6, 5.1, 6.1, 6.6 (br, peaks, 6 vinyl H, temperature dependent), 7.2 (m, C_6H_5) IR (CCl_4): 1704 (C=O); 2000, 2060 (CO)	[11]
*9	$(3\text{-}CH_3OC\text{-}C_6H_5NH)Fe(CO)_3$	see No. 12, p. 25	[14]
10	$(3\text{-}CH_3OC\text{-}C_6H_5NCH_3)Fe(CO)_3$	see No. 9, p. 25 ^{1}H NMR: 2.20 (s, $COCH_3$)	[14]
*11	$(3\text{-}OHC\text{-}C_6H_5NCO_2C_2H_5)Fe(CO)_3$	see No. 5, p. 24 m.p. 126 to 127° ^{1}H NMR: 1.41 (t, CH_3; J = 7.2), 10.09 (s, CHO)	[14]
*12	$(3\text{-}CH_3OC\text{-}C_6H_5NCO_2C_2H_5)Fe(CO)_3$	see No. 5, p. 24 m.p. 144 to 145°, yellow crystalline solid ^{1}H NMR: 1.43 (t, CH_3; J = 7.1), 2.32 (s, $COCH_3$)	[14]

References on pp. 33/4

Table 1 [continued]

No.	compound	method of preparation (yield in %), properties and remarks	Ref.
13	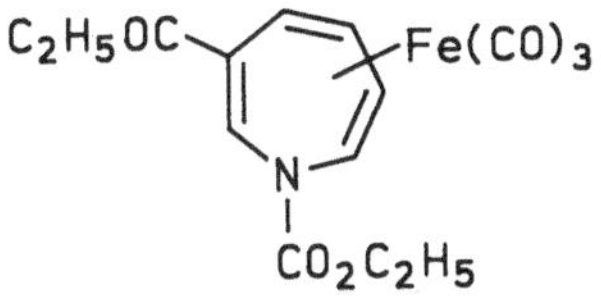 C_2H_5OC-azepine(N-$CO_2C_2H_5$)-$Fe(CO)_3$	see No. 5, p. 24	[14]
*14	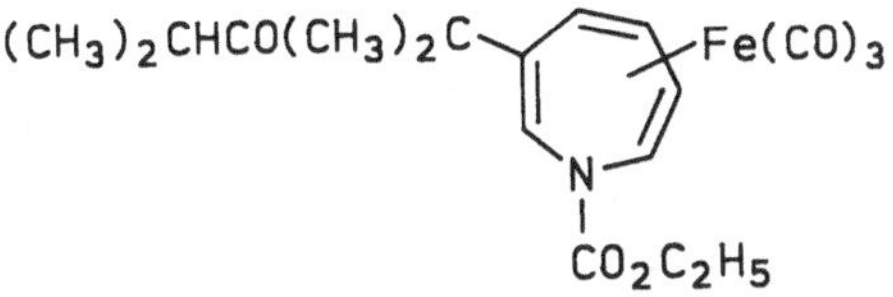 $(CH_3)_2CHCO(CH_3)_2C$-azepine(N-$CO_2C_2H_5$)-$Fe(CO)_3$	see No. 5, p. 24 yellow oil mass spectrum: $[M]^+$ (mixture with No. 22)	[43]
*15	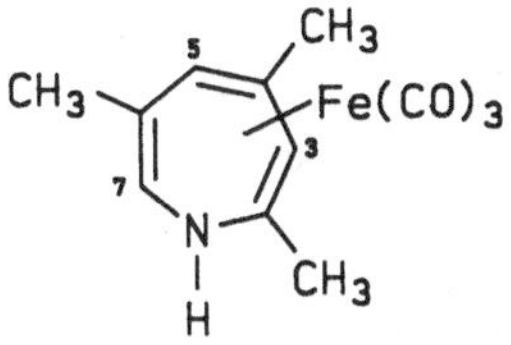 2,4,6-$(CH_3)_3$-1H-azepine(N-H)-$Fe(CO)_3$ (positions 3, 5, 7 marked)	see No. 9, p. 25 m.p. 60 to 63°, subl.p. 50°/high vacuum, air sensitive orange-red crystals IR (Nujol/Hostaflon): 1961, 1992, 2058	[1]
*16	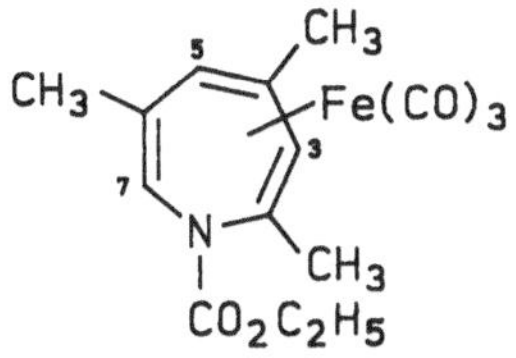 2,4,6-$(CH_3)_3$-azepine(N-$CO_2C_2H_5$)-$Fe(CO)_3$ (positions 3, 5, 7 marked)	I a (11, 6) m.p. 54°, yellow crystals IR (Nujol/Hostaflon): 1972, 1988, 2049	[1]
17	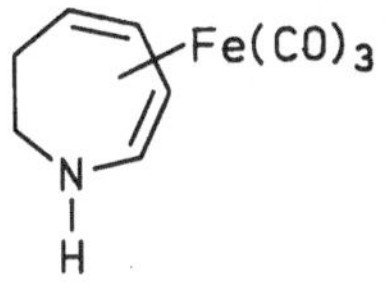 dihydroazepine(N-H)-$Fe(CO)_3$	see No. 2, p. 22 not isolated	[1]
18	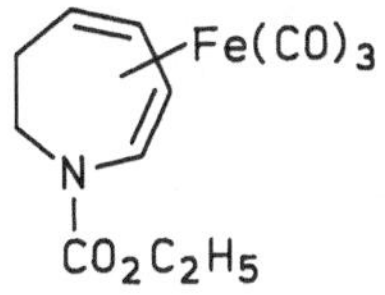 dihydroazepine(N-$CO_2C_2H_5$)-$Fe(CO)_3$	see No. 5, p. 24 not isolated	[1]
19	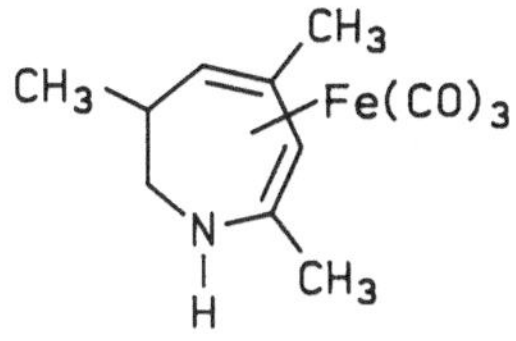 $(CH_3)_3$-dihydroazepine(N-H)-$Fe(CO)_3$	see No. 15, p. 25 not isolated	[1]

References on pp. 33/4

Table 1 [continued]

No.	compound	method of preparation (yield in %), properties and remarks	Ref.
20	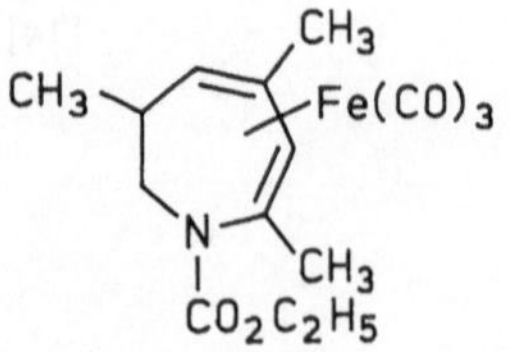	see No. 16, p. 26 not isolated	[1]
*21a	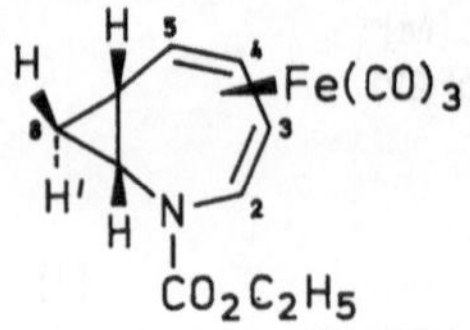	II c (46) m.p. 70 to 71° (from hexane), yellow crystals ^{1}H NMR (CS_2): −0.86 (H′-8; J(H′-8, H-8) = 5, J(H-6, H′-8) = 7.5), 0.72 (H-8; J(H-7,8) = 9, J(H-6,8) = 7), 1.52 (H-6; J(H-6,7) = 7.5), 2.95 (H-7; J(H-7, H′-8) = 5), 3.85 (H-5; J(H-5,6) = 8), 4.74 (H-3; J(H-3,4) = 5, J(H-3,5) = 1.5), 5.08 (H-4; J(H-2,4) = 2, J(H-4,5) = 8), 5.54 (H-2; J(H-2,3) = 7) ^{13}C NMR ($CDCl_3$): 14.6 ((CH_3), 14.9 (C-8), 15.8 (C-6), 34.1 (C-7), 62.0 (CH_2); 62.9, 72.9, 75.0, 89.6 (C-2 to 5); 156.8 (C=O), 209.3 (CO) IR (hexane): 1986, 1989, 2049 (CO) mass spectrum: $[M-nCO]^+$ (n = 0 to 3)	[39]
b	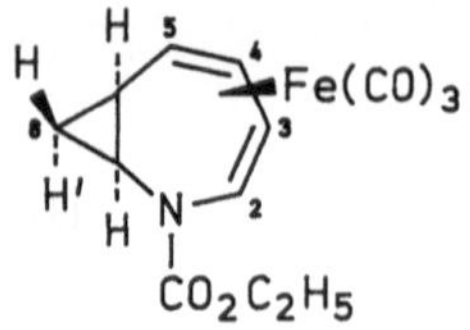	II c (46) m.p. 69 to 70° (from hexane) ^{1}H NMR (CS_2): −0.5 (H′-8; J(H-7, H′-8) = 8.5), 0.9 (H-8; J(H-7,8) = 4), 1.35 (H-6; J(H-4,6) = 1), 2.46 (H-7; J(H-6,7) = 7), 3.16 (H-5; J(H-5,6) = 4.5), 4.42 (H-3; J(H-3,4) = 5, J(H-3,5) = 2), 5.62 (H-4; J(H-4,5) = 6.5), 5.66 (H-2; J(H-2,3) = 7.5) ^{13}C NMR ($CDCl_3$): 14.0 (C-8), 14.6 (CH_3), 16.6 (C-6), 30.4 (C-7), 62.0 (CH_2); 53.3, 75.1, 77.3, 88.9 (C-2 to 5); 157.2 (C=O), 209.1 (CO) IR (hexane): 1987, 1993, 2051 (CO) mass spectrum: $[M-nCO]^+$ (n = 0 to 3)	[39]
22	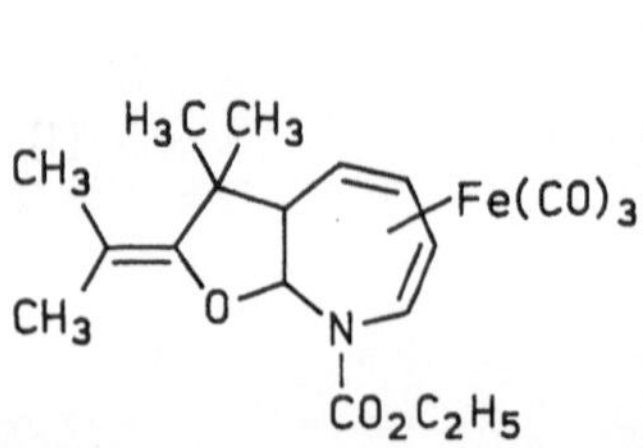	see No. 5, p. 24 yellow oil mass spectrum: $[M]^+$ (mixture with No. 14) for the oxidation with o-chloranil see No. 14, p. 25	[43]

References on pp. 33/4

Table 1 [continued]

No.	compound	method of preparation (yield in %), properties and remarks	Ref.
*23	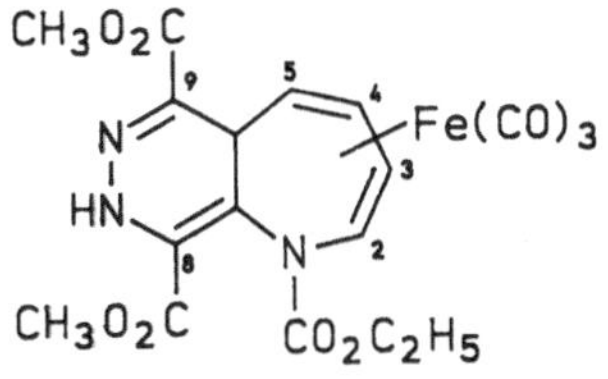 CH_3O_2C, N, HN, CH_3O_2C, N, $CO_2C_2H_5$, $Fe(CO)_3$ (positions 2, 3, 4, 5, 8, 9)	see No. 5, p. 24 m.p. 163 to 164° (dec.), yellow crystals (from CH_3OH) ^1H NMR (benzene-d_6): 0.85, 1.04 (t, CH_3); 3.17, 3.40 (s, 2 OCH_3); 3.30 (d, H-5), 3.90 (m, CH_2), 4.13 (dd, H-4), 4.32 (s, H-6), 4.71 (dd, H-3), 5.89 (d, H-2), 7.70 (s, NH, exchanges with D_2O) ^1H NMR ($CDCl_3$): 1.09 to 1.48 (m, CH_3), 3.22 (d, H-5; J (H-4,5) = 7.0); 3.74, 3.88 (s, 2 OCH_3); 4.00 to 4.20 (m, CH_2; J (CH_2, CH_3) = 7.5), 4.08 (s, H-6), 4.89 (dd, H-4; J (H-3,4) = 5.5), 5.56 (dd, H-3; J (H-2,3) = 5.0), 5.75 (d, H-2), 7.93 (s, NH, exchanges with D_2O) ^{13}C NMR ($CDCl_3$): 14.47 (q, C-6), 36.21 (d, CH_3), 52.38 (q, CH_3), 52.85 (q, CH_3), 62.34 (t, CH_2); 60.59 (d), 73.59 (d), 74.06 (d), 89.18 (d) (C-2 to 5); 119.47 (s), 121.70 (s), 132.07 (s) (C-7 to 9); 151.17 (s, C=O), 160.19 (s, C=O), 164.17 (s, C=O), 208.76 (s, CO) IR (Nujol): 1710 (urethane); 1730, 1740 (C=O); 1980, 2080 (CO); 3320 (NH) mass spectrum: $[M-nCO]^+$ (n = 2, 3)	[6]
24	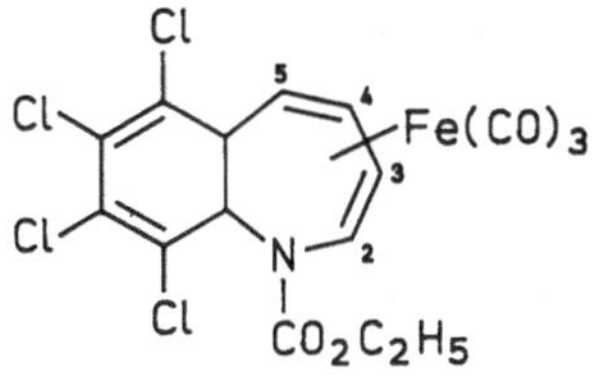 Cl, Cl, Cl, Cl, N, $CO_2C_2H_5$, $Fe(CO)_3$ (positions 2, 3, 4, 5)	see No. 5, p. 24 m.p. 126 to 128° (from ether), pale yellow crystals ^1H NMR ($CDCl_3$): 1.20 to 1.50 (m, CH_3), 2.80 (d, H-5), 2.96 (d, H-6), 4.10 to 4.40 (m, CH_2), 4.88 (m, H-3), 5.20 to 5.70 (m, H-2,4,7) IR (Nujol): 1710 (urethane); 1960, 1985, 2080 (CO) mass spectrum: $[M+2]^+$, $[M-CO+n]^+$ (n = 0, 2, 4)	[6]
25	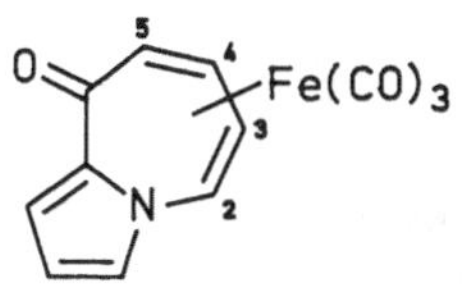 O, N, $Fe(CO)_3$ (positions 2, 3, 4, 5)	IIc (73) m.p. 134° (from acetone/petroleum ether), citrine needles ^1H NMR (acetone-d_6): 3.51 (H-5; J (H-4,5) = 8.19), 5.47 (H-2; J (H-2,3) = 6.48), 5.99 (H-3; J (H-3,4) = 5.13), 6.49 (H-4) IR (KBr): 1609, 1972, 1981, 2070 UV (C_2H_5OH): λ_{max} (log ε) = 227 (4.02), 258 (3.94), 320 (3.66) nm mass spectrum: $[M-nCO]^+$ (n = 1 to 3)	[44]

References on pp. 33/4

Table 1 [continued]

No.	compound	method of preparation (yield in %), properties and remarks	Ref.
*26	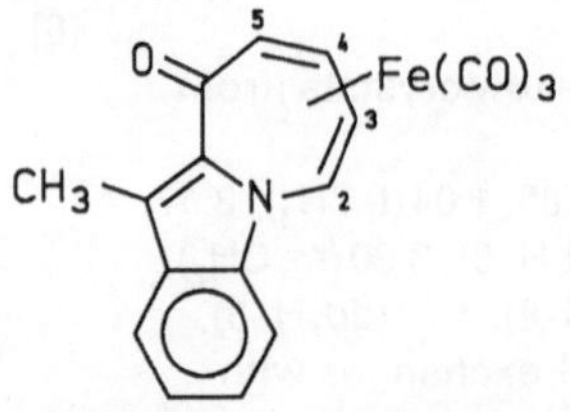 [structure: $Fe(CO)_3$ complex, CH_3-substituted oxo azepino-indole]	III (60) m.p. >196° (dec.), red-brown prisms (from CH_3OH) ^1H NMR ($CDCl_3$): 2.54 (s, CH_3), 3.65 to 3.95 (d, H-5; J=8), 5.4 to 5.9 (m, H-2,4), 6.1 to 6.4 (m, H-3), 7.1 to 8.0 (m, C_6H_4) IR ($CHCl_3$): 1616, 2016, 2072 UV (C_2H_5OH): λ_{max} (log ε) = 224 (4.50), 265, 334 (4.24) nm	[15]
27	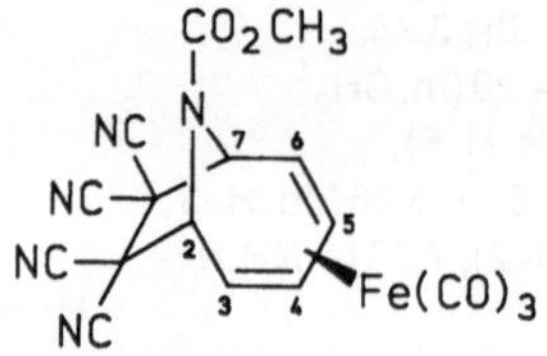 [structure: CO_2CH_3-N, tetracyano, $Fe(CO)_3$]	see No. 4, p. 23 pale yellow crystals (from CH_2Cl_2/hexane at 0°) ^1H NMR (acetone-d_6): 3.71 (OCH_3), 4.00 (H-3,6; J(H-3,4) = 7.5, J(H-3,5) = 1.6), 5.75 (H-2,7; J(H-2,4) = 8.8), 5.95 (H-4,5; J(H-4,5) = 4.5) IR (CH_2Cl_2): 1727 (C=O); 2015, 2019, 2078 (CO) mass spectrum: $[M-nCO]^+$ (n = 1 to 3)	[6, 36, 48, 49]
28	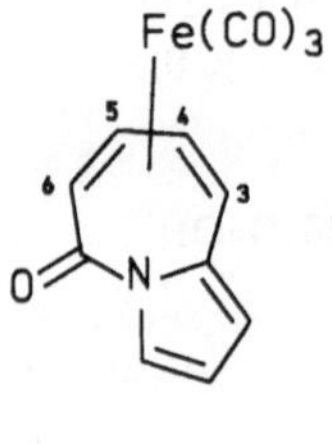 [structure: $Fe(CO)_3$, oxo pyrrolo-azepine]	IIc (64) m.p. 104° (from acetone/petroleum ether), orange-yellow needles ^1H NMR (acetone-d_6): 3.17 (H-6; J(H-5,6) = 7.65), 4.05 (H-3; J(H-3,4) = 7.78), 6.15 (H-4; J(H-4,5) = 4.95), 6.40 (H-5) IR (KBr): 1665, 1980, 1995, 2065 UV (C_2H_5OH): λ_{max} (log ε) = 231 (3.77), 272 (3.69), 332 (3.19) nm mass spectrum: $[M-nCO]^+$ (n = 0 to 3)	[44]
29	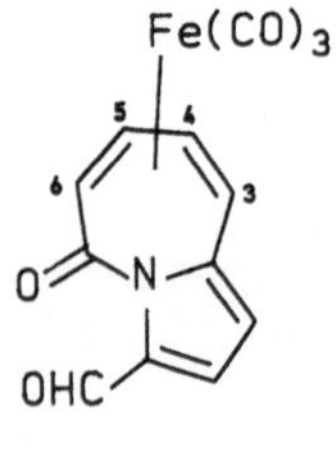 [structure: $Fe(CO)_3$, oxo pyrrolo-azepine, OHC]	IIc (86) m.p. 189° (from acetone/petroleum ether), red needles ^1H NMR (acetone-d_6): 3.24 (H-6; J(H-5,6) = 7.02), 3.97 (H-3; J(H-3,4) = 7.38), 6.36 (H-4; J(H-4,5) = 4.68), 6.54 (H-5) IR (KBr): 1650, 1982, 1995, 2065 UV (C_2H_5OH): λ_{max} (log ε) = 233 (4.53), 325 (4.08), 420 (3.48) nm mass spectrum: $[M-nCO]^+$ (n = 0 to 3)	[44]
30	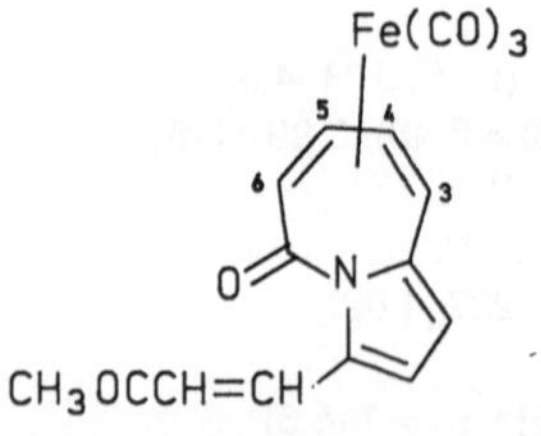 [structure: $Fe(CO)_3$, oxo pyrrolo-azepine, $CH_3OCCH=CH$]	IIc (48) m.p. 128° (from acetone/petroleum ether), orange-red needles ^1H NMR (acetone-d_6): 3.20 (H-6; J(H-5,6) = 7.53), 4.13 (H-3; J(H-3,4) = 7.49), 6.21 (H-4; J(H-4,5) = 4.35), 6.76 (H-5) IR (KBr): 1660 to 1665, 2065, 2990, 2980 mass spectrum: $[M-nCO]^+$ (n = 0 to 3)	[44]

References on pp. 33/4

Table 1 [continued]

No.	compound	method of preparation (yield in %), properties and remarks	Ref.
31	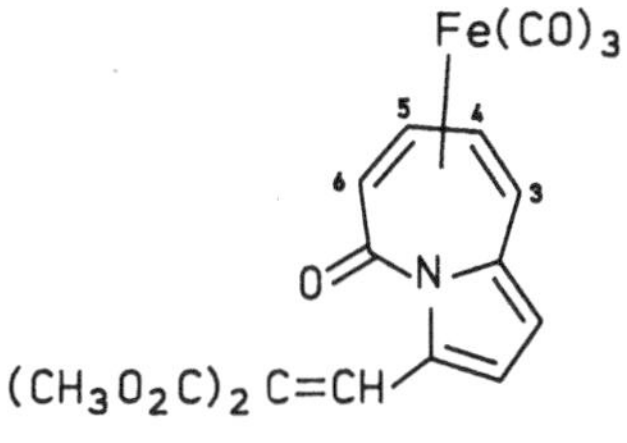 $Fe(CO)_3$ complex; $(CH_3O_2C)_2C{=}CH$	IIc (51) m.p. 168° (from CH_3OH), orange-yellow needles ^{1}H NMR (acetone-d_6): 3.24 (H-6; J(H-5,6) = 7.29), 4.00 (H-3; J(H-3,4) = 7.02), 6.17 (H-4; J(H-4,5) = 4.78), 6.84 (H-5) IR (KBr): 1650, 1700, 1985, 1995, 2062 UV (C_2H_5OH): λ_{max}(log ε) = 232 (4.36), 276 (4.16), 3.73 (4.20), 440 (3.82) nm mass spectrum: $[M-nCO]^+$ (n = 0 to 3)	[44]
32	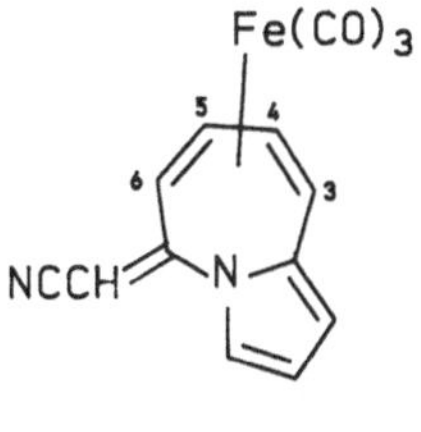 $Fe(CO)_3$ complex; NCCH=	IIb (28) m.p. 158° (from petroleum ether), ocher-yellow needles ^{1}H NMR (acetone-d_6): 3.99 (H-6; J(H-5,6) = 7.65), 4.03 (H-3; J(H-3,4) = 7.02), 6.23 (H-5) IR (KBr): 1955, 1996, 2061, 2210 UV (C_2H_5OH): λ_{max}(log ε) = 244 (4.55), 276 (4.52), 303 (4.41), 415 (3.98) nm mass spectrum: $[M-nCO]^+$ (n = 0 to 3)	[44]
33	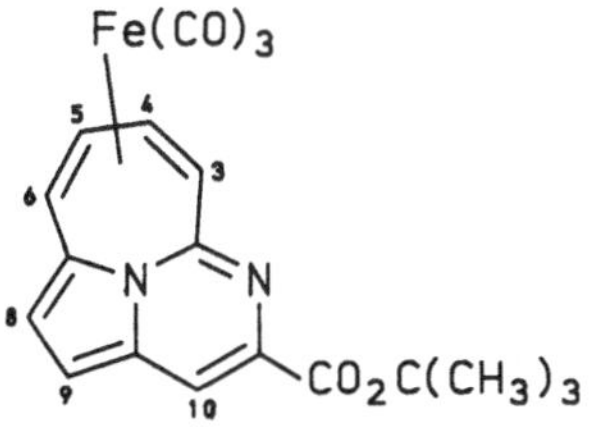 $Fe(CO)_3$ complex; $CO_2C(CH_3)_3$	IIc (66) m.p. 162° (from benzine), orange colored needles ^{1}H NMR ($CDCl_3$): 4.48 (H-6; J(H-5,6) = 7.9), 4.72 (H-3; J(H-3,4) = 7.6), 5.49 (H-5; J(H-4,5) = 5.5), 5.66 (H-4), 6.70 (H-8; J(H-8,9) = 4.0), 6.75 (H-9), 8.01 (H-10) IR (KBr): 1710 (C=O); 1980, 2005, 2040, 2060 (CO) mass spectrum: $[M-nCO]^+$ (n = 0 to 3)	[24, 42]
34	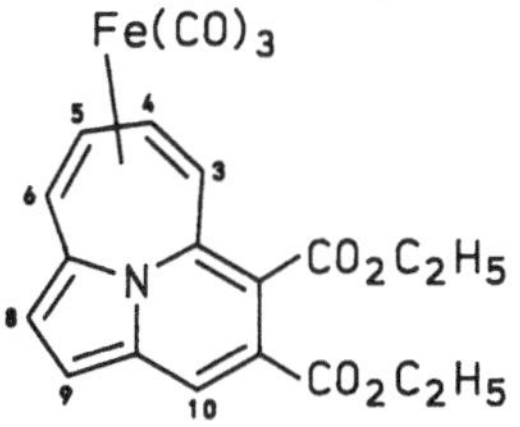 $Fe(CO)_3$ complex; $CO_2C_2H_5$, $CO_2C_2H_5$	IIc (84) m.p. 129° (from benzine), red needles ^{1}H NMR ($CDCl_3$): 4.45 (H-3,6), 5.1 (H-4,5), 6.78 (H-8; J(H-8,9) = 4.0), 6.90 (H-9), 8.10 (H-10) IR (KBr): 1700, 1720 (C=O); 1955, 1975, 2050 (CO); 2990 (CH) mass spectrum: $[M-nCO]^+$ (n = 0 to 3)	[24, 42]
*35	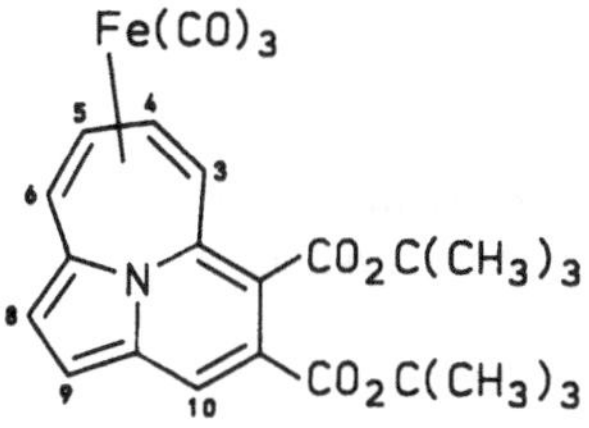 $Fe(CO)_3$ complex; $CO_2C(CH_3)_3$, $CO_2C(CH_3)_3$	IIc (84) m.p. 193° (from $CHCl_3$), orange-red needles ^{1}H NMR ($CDCl_3$): 4.46 (H-6; J(H-5,6) = 8.0), 4.55 (H-3; J(H-3,4) = 7.0), 5.02 (H-4,5; J(H-4,5) = 5.5), 6.70 (H-8; J(H-8,9) = 4.0), 6.86 (H-9), 7.93 (H-10) IR (KBr): 1705, 1725 (C=O); 1962, 1982, 2060 (CO)	[24, 42]

References on pp. 33/4

Table 1 [continued]

No.	compound	method of preparation (yield in %), properties and remarks	Ref.
		UV (C_2H_5OH): $\lambda_{max}(\log \varepsilon)$ = 207 (4.47), 240 (4.25), 306 (4.31), 430 (3.66) nm mass spectrum: $[M-nCO]^+$ (n = 0 to 3)	
36	$Fe(CO)_3$	see No. 35, p. 27 red powder (impure) ^{1}H NMR ($CDCl_3$): 4.56 (H-6; J(H-5,6) = 7.9), 5.22 (H-4; J(H-4,5) = 4.0), 5.27 (H-5), 5.48 (H-3; J(H-3,4) = 4.0), 6.99 (H-8), 7.11 (H-9), 7.95 (H-10) IR (KBr): 1750, 1810 (C=O); 1980, 2060 (CO) mass spectrum: $[M-nCO]^+$ (n = 0 to 3)	[24, 42]
^{4}L is a diazepine:			
*37	$Fe(CO)_3$, H–N, N–$COCH_3$	IIb, see also No. 53, p. 28 yellow ^{1}H NMR ($CDCl_3$): 2.12 (s, CH_3), 2.9 to 3.4 (m, H-6), 3.58 to 4.0 (br, m, H-3), 4.85 to 5.05 (q, H-5; J(H-4,5) = 7, J(H-5,6) = 5), 5.5 to 6.0 (m, H-2,4) IR ($CHCl_3$): 1600, 1635 (C=C); 1655 (C=O); 1950, 1970, 2040 (CO); 3420 (NH)	[26]
*38	$Fe(CO)_3$, H–N, N–$CO_2C_2H_5$	IIb, see also No. 58, p. 29 not obtained in pure form due to extreme instability	[26]
39	$Fe(CO)_3$, CH_3OC–N, N–$COCH_3$	IIb, see also No. 37, p. 27 ^{1}H NMR ($CDCl_3$): 1.98 (s, $COCH_3$-1), 2.0 (s, $COCH_3$-7), 2.85 to 3.33 (m, H-3), 2.9 (br, d, H-6exo; J(H-6exo, 6endo) = 16), 4.75 (br, q, H-6endo; J(H-5,6endo) = 5), 4.8 to 5.1 (m, H-5), 5.4 to 5.77 (m, H-2,4)	[26]
40	$Fe(CO)_3$, CH_3OC–N, N–$CO_2C_2H_5$	IIb, see also No. 38, p. 27	[26]
*41	$Fe(CO)_3$, N, N–H	see Nos. 53, 58, 70, pp. 28/31 m.p. 118 to 119° (from pentane), m.p. 121°, subl.p. 70°/10^{-4} Torr, yellow crystals, lemon-yellow plates ^{1}H NMR ($CDCl_3$ at room temperature): 3.88 (s, H-3,5, sharpens to a triplet at 70°),	[23, 32, 33]

References on pp. 33/4

Table 1 [continued]

No.	compound	method of preparation (yield in %), properties and remarks	Ref.
		5.80 (NH coalesces on raising the temperature), 6.86 (br, H-6) ^{1}H NMR ($CDCl_3$ at −31.5 °C): 3.64 (H-5; J(H-4,5) = 6.6, J(H-3,5) = 1.8), 4.05 (H-3; J(H-2,3) = 6.1), 4.67 (H-4; J(H-3,4) = 3.9), 5.60 (H-2; J(H-2,4) = 1.6), 6.89 (H-6; J(H-5,6) = 6.4, J(H-4,6) = 0.3), 7.26 (NH; J(NH, H-3) = 0.5, J(NH, H-2) = 4.5) ^{57}Fe-γ: δ = 0.28, Δ = 1.27 IR (Nujol): 1897, 1950, 2038 IR (n-hexane): 1976, 1990, 2052 (CO); 3275 (NH) IR (cyclohexane): 1976, 1990, 2052 mass spectrum: $[M-nCO]^+$ (n = 0 to 3), $[C_5H_6N_2]^+$	
*42	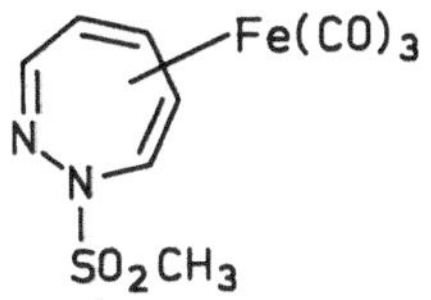 1-(methylsulfonyl)-1,2-diazepine $Fe(CO)_3$; N–SO_2CH_3	–	[13]
43	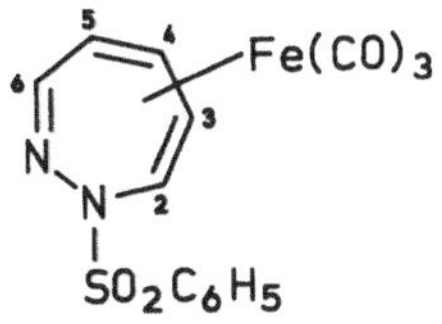 1-(phenylsulfonyl)-1,2-diazepine $Fe(CO)_3$; N–$SO_2C_6H_5$ (ring positions 2–6)	IIb (18) m.p. 139 to 140° (from acetone/hexane), yellow crystals ^{1}H NMR ($CDCl_3$): 3.34 (m, H-5; J = 7, J = 6, J = 1.65), 4.46 (m, H-3; J = 6.5, J = 4.4), 4.92 (m, H-4; J = 1.8, J = 0.5), 6.08 (q, H-2), 6.94 (q, H-6, 7.1 (C_6H_5)	[9, 40, 45]
44	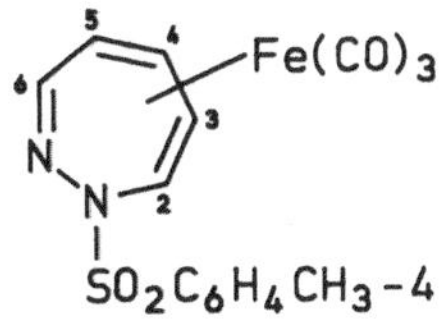 1-(tolylsulfonyl)-1,2-diazepine $Fe(CO)_3$; N–$SO_2C_6H_4CH_3$-4 (ring positions 2–6)	IIb (20) m.p. 160°, yellow crystals (from petroleum ether) ^{1}H NMR ($CDCl_3$): 3.46 (m, H-5; J = 7, J = 6, J = 1.65), 4.54 (m, H-3; J = 6.5, J = 4.4), 5.02 (m, H-4; J = 1.8, J = 0.5), 6.24 (q, H-2), 7.08 (q, H-6) ^{57}Fe-γ (78 K): δ = 0.31, Δ = 1.27 IR ($CHCl_3$/hexane): 1308, 1368, 1403, 1448 (C=C); 1496, 1596; 1999, 2020, 2080 (CO) mass spectrum: $[M-nCO]^+$ (n = 0 to 3), $[M-3CO-HCN]^+$, $[M-3CO-C_7H_7]^+$, $[M-3CO-Fe]^+$ and others	[9, 20, 40, 45]

References on pp. 33/4

Table 1 [continued]

No.	compound	method of preparation (yield in %), properties and remarks	Ref.
*45	1-methyl-1H-1,2-diazepine $Fe(CO)_3$ (N-CH_3)	see No. 41, p. 28, and No. 72, p. 31 m.p. 49 to 50° (from pentane at −78°), yellow-orange product ^{1}H NMR (liquid SO_2 at −10°): 3.09 (s, CH_3), 3.63 (m, H-5; J(H-3,5) = 1.9), 4.36 (m, H-3; J(H-2,3) = 4.6), 4.97 (m, H-4; J(H-4,5) = 6.9), 5.36 (dd, H-6; J(H-5,6) = 6.5), 6.95 (d, H-2) IR (cyclohexane): 1950, 1975, 2045 mass spectrum: $[M-nCO]^+$ (n = 0 to 3)	[32, 33, 35]
46	1-ethyl-1H-1,2-diazepine $Fe(CO)_3$ (N-C_2H_5)	IIb (95) m.p. 115°, subl.p. 60°/0.04 Torr, yellow crystals ^{1}H NMR ($CDCl_3$): 3.32 (m, H-5; J(H-4,5) = 7.0), 4.57 (m, H-3; J(H-3,5) = 1.65), 5.10 (m, H-4; J(H-4,6) = 0.5), 6.31 (q, H-2; J(H-2,3) = 6.5), 7.01 (d, H-6; J(H-5,6) = 6.0)	[9]
47	1H-1,2-diazepine $Fe(CO)_3$ (N-$CH_2CH{=}CH_2$)	see No. 41, p. 28	[32, 35]
*48	1H-1,2-diazepine $Fe(CO)_3$ (N-$CH_2C_6H_5$)	see No. 41, p. 28 yellow oil IR (hexane): 1974, 1987, 2050 (CO)	[23, 32, 35]
49	1H-1,2-diazepine $Fe(CO)_3$ (N-$C(CH_3){=}CH_2$)	see No. 41, p. 28	[35]
50	1H-1,2-diazepine $Fe(CO)_3$ (N-$COCF_3$)	see No. 41, p. 28	[32, 35]

Table 1 [continued]

No.	compound	method of preparation (yield in %), properties and remarks	Ref.
*51	1-(trichloroacetyl)-1H-1,2-diazepine $Fe(CO)_3$ ($COCCl_3$)	see No. 41, p. 28	[35]
*52	1-(chloroacetyl)-1H-1,2-diazepine $Fe(CO)_3$ ($COCH_2Cl$)	see No. 41, p. 28	[32, 35]
*53	1-acetyl-1H-1,2-diazepine $Fe(CO)_3$ ($COCH_3$)	IIb, see also No. 41, p. 28 yellow crystals (from petroleum ether) ^1H NMR ($CDCl_3$): 3.28 (H-5; J(H-4,5) = 7), 4.59 (H-3; J(H-3,4) = 4.3, J(H-3,5) = 1.6), 5.11 (H-4; J(H-2,4) = 1.7), 6.47 (H-2; J(H-2,3) = 6.7), 6.92 (H-6; J(H-5,6) = 6) ^{57}Fe-γ: $\delta = 0.31$, $\Delta = 1.24$ IR ($CHCl_3$/hexane): 1300, 1370, 1376, 1426, 1447 (C=C); 1603, 1680 (C=O); 1993, 2004, 2063 (CO) mass spectrum: $[M-nCO]^+$ (n = 0 to 3), $[M-3CO-HCN]^+$	[20, 23, 32, 35, 40]
54	1-acryloyl-1H-1,2-diazepine $Fe(CO)_3$ ($COCH=CH_2$)	see No. 41, p. 28	[35]
*55	1-cinnamoyl-1H-1,2-diazepine $Fe(CO)_3$ ($COCH=CHC_6H_5$)	see No. 41, p. 28	[32, 35]
*56	1-benzoyl-1H-1,2-diazepine $Fe(CO)_3$ (COC_6H_5)	IIb (84), see also No. 41, p. 28 m.p. 169 to 170° (from $CH_3CO_2C_2H_5$/hexane), yellow crystals ^1H NMR ($CDCl_3$): 3.2 (m, H-5; J = 7), 4.64 (m, H-3; J = 1.65), 5.07 (m, H-4; J = 4.4, J = 0.5), 6.5 (q, H-2; J = 1.8, J = 6.5), 6.79 (d, H-6; J = 6), 7.4 (C_6H_5) ^{57}Fe-γ (78 K): $\delta = 0.30$, $\Delta = 1.25$	[9, 20, 40, 45]

References on pp. 33/4

Table 1 [continued]

No.	compound	method of preparation (yield in %), properties and remarks	Ref.
		IR ($CHCl_3$/hexane): 1438, 1450 (C=C); 1497, 1582, 1604, 1670; 1994, 2003, 2064 (CO) mass spectrum: $[M-nCO]^+$ (n=0 to 4), $[M-3CO-HCN]^+$	
*57	1-($CO_2CH_2CCl_3$)-1H-1,2-diazepine $Fe(CO)_3$	see No. 41, p. 28	[32, 35]
*58	1-($CO_2C_2H_5$)-1H-1,2-diazepine $Fe(CO)_3$	IIa (8), IIb, IIe m.p. 115°, m.p. 115 to 116°, subl.p. 60°/0.04 Torr, yellow crystals, red-orange air stable crystals (from C_6H_6/hexane or ether/pentane (1:1)) ^{1}H NMR ($CDCl_3$): 3.32 (H-5; J(H-3,5)=1.65), 4.57 (H-3; J(H-2,3)=6.5, J(H-3,4)=4.4), 5.10 (H-4; J(H-4,6)=0.5), 6.31 (H-2; J(H-2,4)=1.8), 7.01 (H-6; J(H-5,6)=6); no temperature dependence observed IR ($CHCl_3$): 1770, 1990, 2000, 2060	[8, 10, 22, 45, 47]
*59	1-($CO_2C_3H_7$-i)-1H-1,2-diazepine $Fe(CO)_3$	IIb m.p. 104 to 105°, yellow crystals	[16]
60	1-CN-1H-1,2-diazepine $Fe(CO)_3$	see No. 41, p. 28	[32]
*61	7-CH_3-1H-1,2-diazepine $Fe(CO)_3$ (N–H)	see No. 65, p. 30, and No. 73, p. 32 m.p. 93 to 94°, m.p. 99°, yellow crystals or orange plates ^{1}H NMR ($CDCl_3$ at +35 °C): 2.16 (CH_3; J(H-4, CH_3) ~0.0), 3.60 (dd, H-5; J(H-5, CH_3)=0.25±0.1), 4.0 (m, H-3; J(H-3,5)=1.65±0.1, J(H-3,4)=4.5±0.1), 4.67 (m, H-4; J(H-4,5)=7.0±0.1), 5.75 (dd, H-2; J(H-2,4)=1.7±0.1, J(H-2,3)=6.1±0.1), 6.15 (br, s, NH; J(H-2, NH) ~0.0, J(H-3, NH ~0.0) IR (n-hexane): 1974, 1987, 2050 (CO)	[23, 33]

References on pp. 33/4

Table 1 [continued]

No.	compound	method of preparation (yield in %), properties and remarks	Ref.
		IR (cyclohexane): 1950, 1975, 2042 (CO); 3410 (NH) mass spectrum: $[M-nCO]^+$ (n = 0 to 3)	
*62	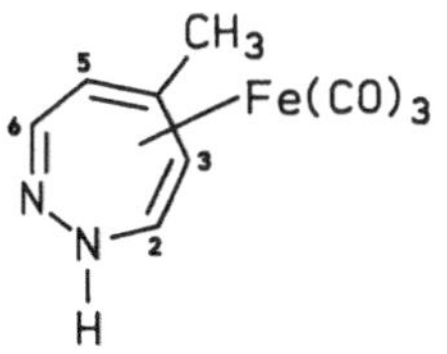 CH_3, $Fe(CO)_3$, N, N–H	see No. 67, p. 30 m.p. 97°, yellow crystals ^{1}H NMR ($CDCl_3$ at −40°): 1.79 (CH_3; J(H-5, CH_3) = 0.5), 4.0 (H-3; J(H-3,5) = 2.0 ± 0.2, J(H-3, CH_3) = 0.3), 5.28 (H-2; J(H-2,3) = 6.3, J(H-2, CH_3) = 0.2), 6.69 (NH; J(H-3, NH) = 0.4, J(H-2, NH) = 4.5), 6.78 (H-6; J(H-6, CH_3) ~ 0.0), 8.58 (H-5; J(H-5,6) = 6.2) IR (n-hexane): 1973, 1987, 2049 (CO)	[23]
63	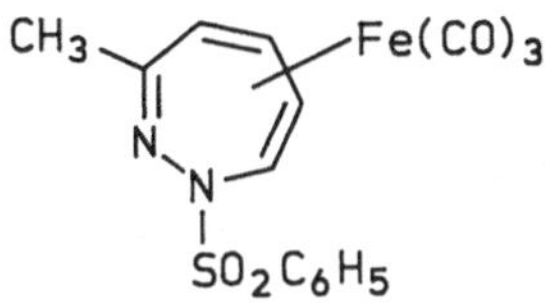 CH_3, $Fe(CO)_3$, N, N–$SO_2C_6H_5$	IIb (10)	[40]
64	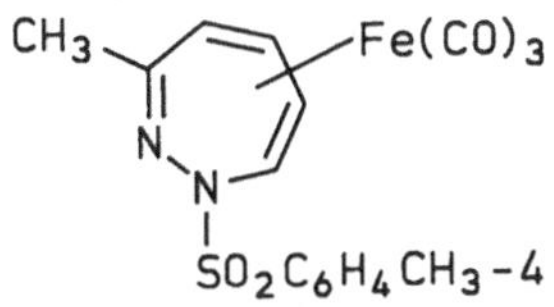 CH_3, $Fe(CO)_3$, N, N–$SO_2C_6H_4CH_3$-4	IIb (9)	[40]
*65	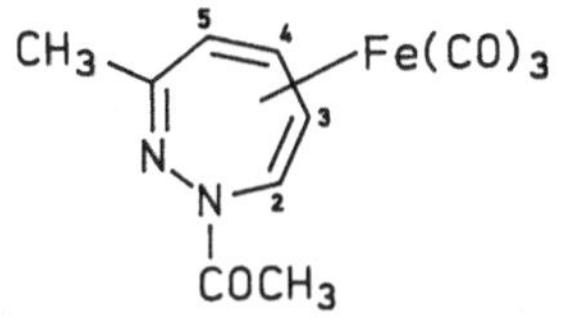 CH_3, $Fe(CO)_3$, N, N–$COCH_3$	IIb, see also No. 61, pp. 29/30 yellow ^{1}H NMR ($CDCl_3$): 2.11 (s, CH_3), 2.31 (s, $COCH_3$), 3.19 (dd, H-5; J(H-4,5) = 7.5, J(H-3,5) = 2), 4.59 (oct, H-3; J(H-2,3) = 7), 5.09 (oct, H-4; J(H-3,4) = 4.5), 6.39 (dd, H-2; J(H-2,4) = 2) ^{57}Fe-γ (78 K): δ = 0.31, Δ = 1.26 IR (hexane/$CHCl_3$): 1302, 1379, 1449, 1470, 1623, 1680; 1991, 2002, 2062 (CO) mass spectrum: $[M-nCO]^+$ (n = 0 to 3), $[M-3CO-HCN]^+$	[20, 30]
66	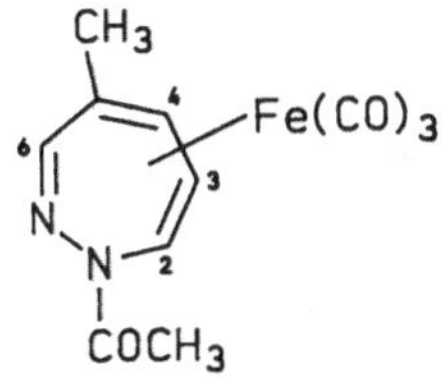 CH_3, $Fe(CO)_3$, N, N–$COCH_3$	IIb yellow ^{1}H NMR ($CDCl_3$): 1.91 (s, CH_3), 2.37 (s, $COCH_3$), 4.38 (dd, H-3; J(H-2,3) = 6.5), 4.97 (br, d, H-4; J(H-3,4) = 4.7), 6.34 (dd, H-2; J(H-2,4) = 2), 6.84 (s, H-6) ^{57}Fe-γ (78 K): δ = 0.32, Δ = 1.31	[20]

References on pp. 33/4

Table 1 [continued]

No.	compound	method of preparation (yield in %), properties and remarks	Ref.
		IR (hexane/$CHCl_3$): 1370, 1430, 1457 (C=C); 1504, 1520, 1605, 1679, 1690; 1987, 1998, 2058 (CO) mass spectrum: $[M-nCO]^+$ (n=0 to 3), $[M-3CO-HCN]^+$	
*67	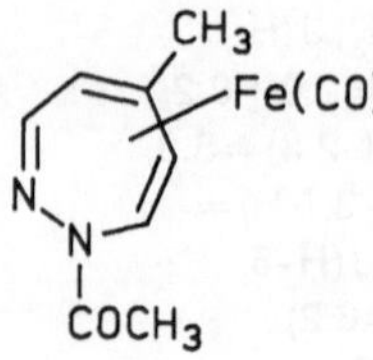 CH_3, $Fe(CO)_3$, N, N, $COCH_3$	–	[23]
68	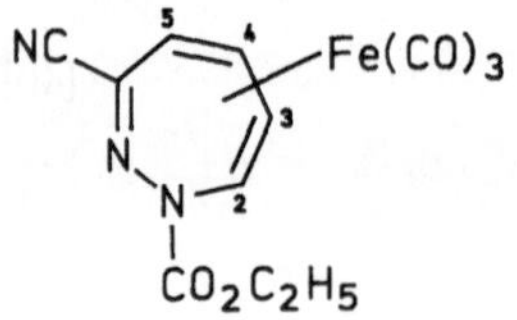 NC, $Fe(CO)_3$, N, N, $CO_2C_2H_5$; ring positions 2, 3, 4, 5	IIb m.p. 155 to 157°, subl.p. 60°/0.05 Torr, red crystals ^{1}H NMR ($COCl_3$): 3.55 (q, H-5; J=7), 4.59 (m, H-3; J=4.6), 5.1 (m, H-4; J=4, J=6, J=7.4), 6.37 (q, H-2; J=2) IR ($CHCl_3$): 1725 (C=O); 1990, 2000, 2060 (CO); 2235 (CN)	[9, 45]
69	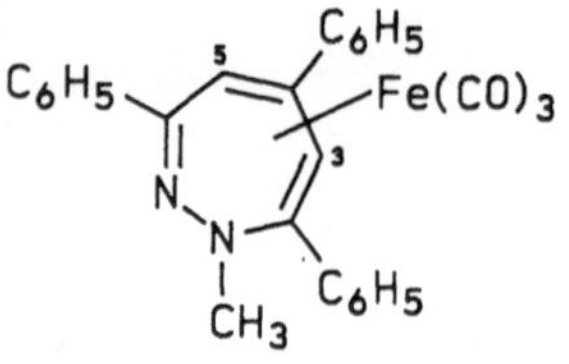 C_6H_5, C_6H_5, $Fe(CO)_3$, N, N, CH_3, C_6H_5; ring positions 3, 5	IIb red ^{1}H NMR ($CDCl_3$): 2.98 (s, CH_3), 4.74 (d, H-3; J(H-3,5)=2), 4.94 (d, H-5), 6.88 to 7.47 (m, 11H aromatic), 7.75 (m, 2H aromatic), 8.07 (m, 2H aromatic) ^{57}Fe-γ (78 K): $\delta=0.34$, $\Delta=1.25$ IR (hexane/$CHCl_3$): 1310, 1352, 1419, 1477 (C=C); 1483, 1551, 1581, 1599; 1956, 1974, 1980, 2041 (CO) mass spectrum: $[M-nCO]^+$ (n=0 to 3), $[M-3CO-CH_3]^+$, $[M-3CO-CH_2N_2]^+$, $[M-3CO-CH_3N]^+$, $[M-3CO-C_6H_5CN]^+$	[20]
*70	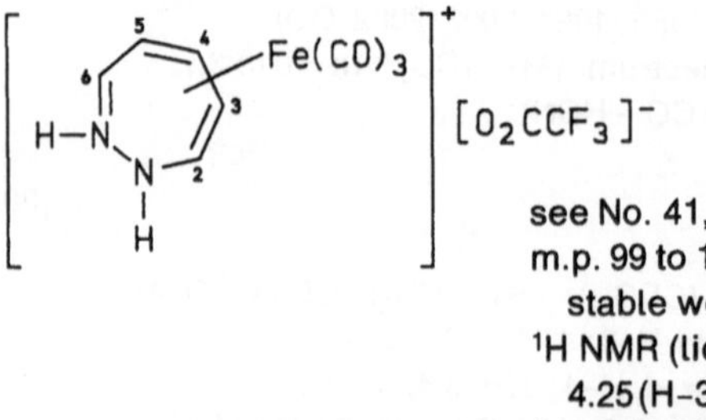 $[\ldots Fe(CO)_3]^+$ $[O_2CCF_3]^-$; H–N, N–H; ring positions 2, 3, 4, 5, 6	see No. 41, p. 27 m.p. 99 to 100° (from CH_2Cl_2 at −20°), air-stable well formed yellow-orange crystals ^{1}H NMR (liquid SO_2 at −10°): 10.6 (NH), 4.25 (H-3,5), 5.15 (H-4), 6.93 (H-2,6) ^{1}H NMR (95% CF_3CO_2D/SO_2 at 0°): 4.21 (H-3; J(H-3,4)=6.0), 4.25 (H-5; J(H-5,6)=6.7),	[29, 31, 33]

Table 1 [continued]

No.	compound	method of preparation (yield in %), properties and remarks	Ref.
		5.15 (H-4; J (H-4,5) = 6.0), 7.04 (H-2,6; J (H-2,3) = 6.7) ^{13}C NMR (80% CF_3CO_2H/H_2O at 0°): 64.7 (C-3,5), 94.7 (C-4), 116.5 (C-2,6), 213.4 (CO) IR (Nujol): 1650 (CO_2); 1985, 2060 (CO); 2300 to 3500 (several broad maxima)	
71	$[(H\text{–}N,\ N\text{–}COCH_3\ \text{diazepine})Fe(CO)_3]^+\ [O_2CCF_3]^-$	see No. 53, p. 28 ^{1}H NMR (CF_3CO_2H at −10°): 2.75 ($COCH_3$), 3.86 (H-5; J (H-4,5) = 7.8), 5.52 (H-3; J (H-2,3) = 5.5), 6.00 (H-4; J (H-3,4) = 5), 6.16 (H-2; J (H-2,4) = 1.8), 8.20 (H-6; J (H-5,6) = 7.8); NH is averaged with solvent signal there is a H bridge between NH and $COCH_3$	[33]
*72	$[(CH_3\text{–}N,\ N\text{–}H\ \text{diazepine})Fe(CO)_3]^+\ X^-$	X = OSO_2F: see No. 41, p. 28 m.p. 105° (dec.), air sensitive ^{1}H NMR (liquid SO_2 at −10°): 3.67 (CH_3), 4.02 (H-5; J (H-4,5) = 6.7), 4.51 (H-3; J (H-3,5) = 1.5, J (H-2,3) = 6.7), 5.23 (H-4; J (H-3,4) = 5.5), 6.21 (H-2; J (H-2,4) = 1.8), 7.28 (H-6; J (H-5,6) = 7.8), 8.29 (NH) IR (Nujol): 1975, 2071	[33]
		X = O_2CCF_3: see No. 45, p. 28 (not isolated) ^{1}H NMR (95% CF_3CO_2H at 0°): 3.93 (H-5; J (H-4,5) = 6.5), 4.71 (H-3; J (H-3,5) = 1.5, J (H-2,3) = 6.7), 5.33 (H-4; J (H-3,4) = 5.5), 5.95 (H-2; J (H-2,4) = 1.5), 7.10 (NH; J (H-2, NH) = 4.7), 7.65 (H-6; J (H-5,6) = 7.6), 9.73 (CH_3)	[20, 33]
*73	$[(6\text{-}CH_3,\ H\text{–}N,\ N\text{–}H\ \text{diazepine})Fe(CO)_3]^+\ [O_2CCF_3]^-$	see No. 61, p. 29 m.p. 91 to 92°, well formed air stable yellow-orange crystals ^{1}H NMR (liquid SO_2 at −10°): 2.39 (CH_3), 3.81 (H-5; J (H-4,5) = 7.3), 4.57 (H-3; J (H-3,5) = 1.5), 5.18 (H-4; J (H-3,4) = 4.5),	[33]

Table 1 [continued]

No.	compound	method of preparation (yield in %), properties and remarks	Ref.
		6.00 (H-2; J(H-2,3) = 6.6), 8.95 (NH), 14.80 (NH^+) IR (Nujol): 1980, 2065	
*74	$[CH_3C_4H_3N_2H_2Fe(CO)_3]^+$ $[O_2CCF_3]^-$	see No. 61, p. 29	[33]
	[4]L is an oxepine:		
*75	$C_6H_6OFe(CO)_3$	Ib (3) m.p. 51° (from pentane), subl.p. 20°/0.01 Torr, yellow needles, thermally stable up to 110° ^1H NMR (toluene-d_8 at ≧25°): 3.83 (H-4,5), 4.21 (H-3,6), 5.78 (H-2,7) ^1H NMR (toluene-d_8 at 2.5°): 2.89 (H-4; J(H-4,5) = 7.37, J(H-4,6) = 1.98), 3.51 (H-6; J(H-5,6) = 4.37), 4.24 (H-5; J(H-5,7) = 1.78), 4.55 (H-3; J(H-3,4) = 8.28), 5.64 (H-2; J(H-2,3) = 6.56), 5.75 (H-7; J(H-6,7) = 5.67) IR: 1987, 1996, 2054 mass spectrum: $[M-nCO]^+$ (n = 0 to 3), $[C_6H_6Fe]^+$	[4, 27]
*76	$2,7-(CH_3)_2C_6H_4OFe(CO)_3$	Ib (5) m.p. 46° (from petroleum ether), m.p. 84 to 86° (dec., from pentane), subl.p. 30°/0.01 Torr, sinters above 80°, air sensitive, splendid orange colored rhombs ^1H NMR (room temperature): 0.93 (CH_3), 1.25 (CH_3), 3.18 (2H), 5.58 (1H), 5.95 (1H) ^1H NMR (toluene-d_8 at −0.5°): 1.48 (CH_3-2; J(CH_3-2, H-3) = 0.7), 1.95 (CH_3-7; J(CH_3-7, H-6) = 0.2), 2.89 (H-4; J(H-4,5) = 7.4), 3.52 (H-6; J(H-4,6) = 1.7), 4.14 (H-5; J(H-5,6) = 4,6), 4.42 (H-3; J(H-3,4) = 8.3) IR (KBr): 6.95, 789, 806, 880, 892, 928, 956, 986, 1016, 1056, 1122, 1133, 1170, 1185, 1205, 1264, 1309, 1357, 1391, 1430, 1464, 1493; 1634 (C=C); 1919, 1942, 2028 (CO); 2825, 2874, 2941, 3039 IR: 1975, 1993, 2045 (CO) mass spectrum: $[M-nCO]^+$ (n = 0 to 3), $[C_8H_nFe]^+$ (n = 8, 10)	[4, 27]

References on pp. 33/4

Table 1 [continued]

No.	compound	method of preparation (yield in %), properties and remarks	Ref.
77	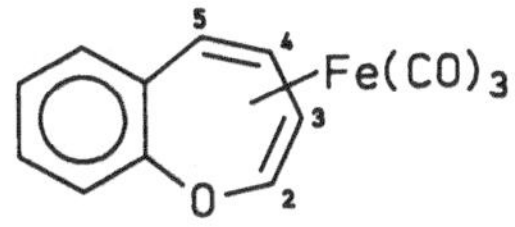	Ib (22.4) m.p. 116 to 117° (from petroleum ether (b.p. 40 to 60°)), subl.p. 80 to 90°/high vacuum, pale yellow needles, monomeric in C_6H_6 (by osmometry) ^1H NMR: 4.07 (H-2; J (H-2,3) = 7.5, J (H-2,4) = 1,5), 4.65 (H-4; J (H-3,4) = 5.0), 5.33 (H-3; J (H-3,5) = 1.9), 6.15 (H-5; J (H-4,5) = 5.5), 6.6 to 7.33 (m, 4H) IR (KBr): 678, 722, 734, 760, 813, 849, 867, 916, 948, 955, 972, 997, 1036, 1109, 1119, 1178, 1227, 1247, 1266, 1314, 1362, 1408, 1425, 1462, 1488, 1575, 1961, 2049, 2994 IR (n-hexane): 1986, 2000, 2058 (CO)	[4]
*78	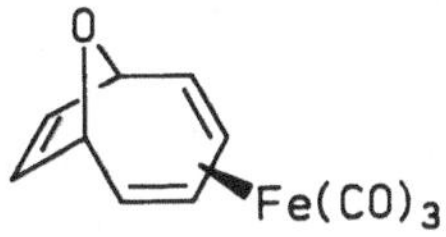	m.p. 67° (dec.), yellow crystals IR (hexane): 1991, 2001, 2059 (CO) mass spectrum: $[M-nCO]^+$ (n = 1 to 3)	[28]
supplements:			
*79	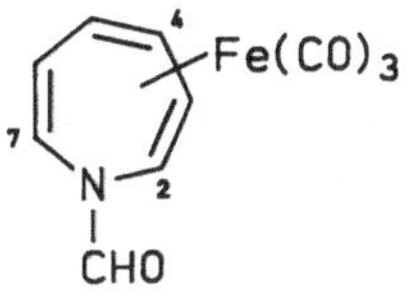	–	[49]
*80	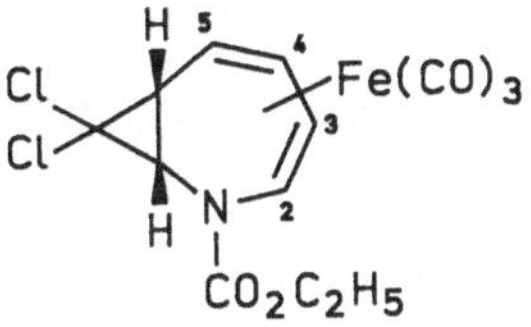	see No. 5, p. 24 m.p. 113 to 114°, pale yellow crystals (from n-hexane) ^1H NMR (benzene-d_6): 0.99 (t, CH_3; J (CH_3, CH_2) = 7.0), 2.03 (dd, H-6; J (H-6,7) = 11.5, J (H-5,6) = 7.5), 2.48 (dd, H-5; J (H-4,5) = 7.5), 3.14 (d, H-7), 3.7 to 4.5 (m, H-3,4, CH_2), 5.26 (br, d, H-2; J (H-2,3) = 6.5) IR ($CHCl_3$): 1705 (NCO); 1990, 2085 (CO) mass spectrum: $[M-nCO]^+$ (n = 0 to 3)	[51]

* Further information:

$HNC_6H_6Fe(CO)_3$ (Table **1**, No. **2**) is monomeric in C_6H_6 (isothermic distillation). The ^1H NMR spectrum in benzene-d_6 shows chemical shifts at 209 (H-1), 236 (H-3,4,5), 245 (H-2), 295 (H-6), and 301 (H-7) Hz [1]. The compound crystallizes in the orthorhombic space group Pbca-D_{2h}^{15}, V_h^{15} (No. 61) with the unit cell parameters a = 12.59(1), b = 24.07(2), and c = 12.67(1) Å; Z = 16 molecules per unit cell and 2 molecules in the asymmetric unit [5, 12, 21]. The measured density is 1.60 g/cm³ and the calculated 1.61 g/cm³ [21]. The $Fe(CO)_3$

References on pp. 33/4

group is bonded to a planar s-cis-butadiene fragment of the azepine ring. The complete azepine ring consists of two planes with a dihedral angle of 37° [12, 21]. The Fe-butadiene bond can be explained in a more classical way as a π-complex bond or as a Diels-Alder-type addition of Fe to the butadiene system. A careful discussion of bond length, bond angles, and hydrogen conformation indicates an intermediate state between the two possible models [12]. The main bond distances and angles are shown in **Fig. 1** [12, 21]. Calculations of the extended Hückel type give for C(2), C(4), and C(7) (for atom numbers see Table 1, p. 5) the HOMO coefficients +0.314, −0.186, and +0.10. This result predicts that 2,4-addition is the preferred cycloaddition mode for $(NC)_2C{=}C(CN)_2$ [49].

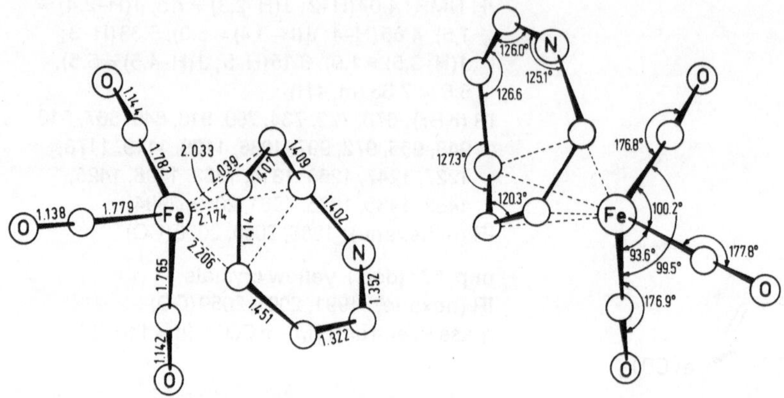

Fig. 1. The two molecules of the asymmetric unit in $HNC_6H_6Fe(CO)_3$ (No. 2) [12, 21].

The compound is very soluble in halogen free solvent and decomposes quickly in $CHCl_3$ or CCl_4. The compound takes up H_2 at room temperature in the presence of PtO_2/CH_3OH or CH_3CO_2H (possibly yielding No. 17). Reaction with $ClCO_2C_2H_5$ in the presence of a small amount of Na_2CO_3 in petroleum ether gives No. 5 (ca. 6%) [1].

$CH_3O_2CNC_6H_6Fe(CO)_3$ (Table **1**, No. **4**) crystallizes in the triclinic space group $P\bar{1}$-C_i^1, S_2^1 (No. 2) with the unit cell parameters a = 8.61(2), b = 6.49(3), c = 10.91(2) Å and α = 93.4(1)°, β = 93.0(1)°, γ = 104.5(1)°; Z = 2 molecules per unit cell. The measured density is 1.61 g/cm³, calculated 1.65 g/cm³. The main bond distances and angles are shown in **Fig. 2** [7, 18].

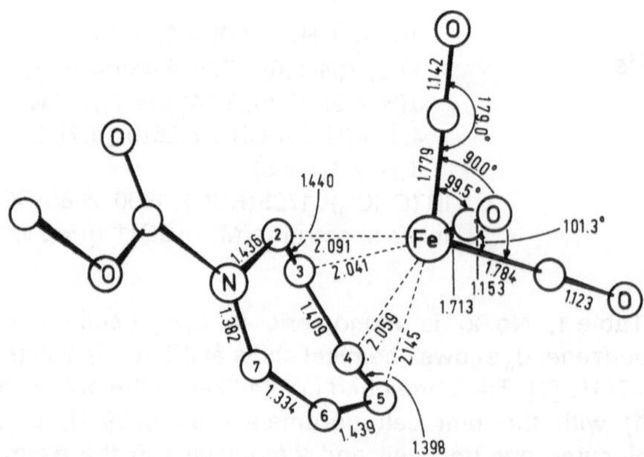

Fig. 2. Molecular structure of $CH_3O_2CNC_6H_6Fe(CO)_3$ (No. 4) [7, 18].

The Fe···C distances vary from 2.041(9) to 2.145(10) Å and the C–C distances in the diene units vary from 1.398(14) to 1.440(12) Å. The normals to the near-planar groups of atoms C-2 to 5 and N, C-2,5,6,7 are inclined at a dihedral angle of 42°45′ [18].

The compound reacts with $(NC)_2C{=}C(CN)_2$ (ratio 1:1) in C_6H_5CN or CH_2Cl_2 at room temperature. The products VIII and No. 27 (ratio 1:6) slowly precipitate after 0.5 h and are recrystallized from CH_2Cl_2/hexane at 0 °C (total yield 28%) [36, 48], cf. [6]. This reaction is repeated, and the ratio of products is confirmed by the ^{1}H NMR spectrum in acetone-d_6 (erroneously No. 4 is cited as No. 5 (?)). However, frontier orbital calculations predict that 2,4-addition (product VIII) is the preferred cycloaddition mode for $(NC)_2C{=}C(CN)_2$. Thus, No. 27 (2,7-addition) is produced by isomerization of VIII [49]. Similar reaction with $OC(CF_3)_2$ in a Carius tube in C_6H_6 at 80 °C for 24 h gives IX. With $(CF_3)_2C{=}C(CN)_2$, no reaction is observed under a variety of conditions [36].

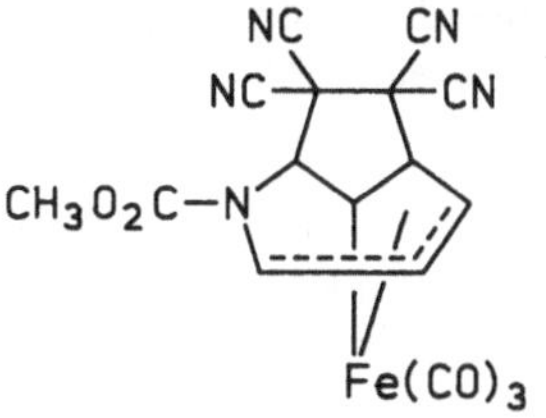

VIII 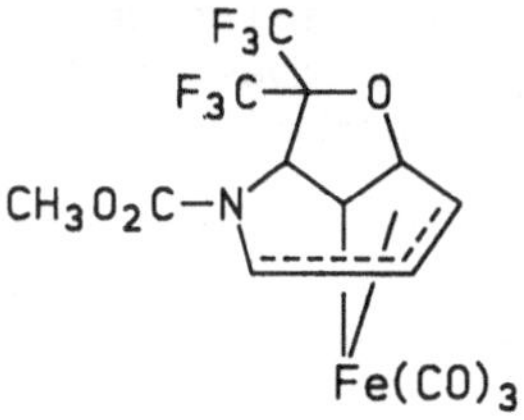 IX

$C_2H_5O_2CNC_6H_6Fe(CO)_3$ (Table **1**, No. **5**) is monomeric in C_6H_6 (by isothermic distillation) and is diamagnetic [1]. The compound equilibrates as shown in X and XI [2, 13, 27, 47, 50] and is proved by the examination of the temperature variable ^{1}H NMR spectrum [2]. Warming the $CDCl_3$ solution above 0 °C (see Table 1, p. 5) broadens the resonance lines, and, finally, pair-wise coalescence of H-2,7, H-3,6, and H-4,5 is observed. Further warming gives a symmetric average spectrum of the type AA′XX′YY′. The coalescence temperature at 34 °C gives, for the valence equilibrium X, $k \sim 60\ s^{-1}$ and the free activation energy ΔG^{*} = ca. 15.5 kcal/mol. A broadening of the resonance lines is also observed below 0 °C as well as a coalescence and the appearance of new signals. The rotation of the $CO_2C_2H_5$ group (see XI) freezes at −58 °C as shown by the chemical shifts at δ = 1.31 (t, CH_3), 1.40 (t, CH_3), 5.93 (d, H-2), 6.12 (d, H-2), 6.50 (d, H-7), and 6.66 (d, H-7) ppm. At the coalescence temperature of −28 °C, $\vec{k} \sim 10\ s^{-1}$ and ΔG^{*} = ca. 13.1 kcal/mol are calculated. From the integration ratio H-7:H-2, the equilibrium constant K is XI(cis):XI(trans) ~ 2.4, and ΔG at −58 °C is ca. 0.4 kcal/mol in favor of XI (cis) [2, 47], cf. [50].

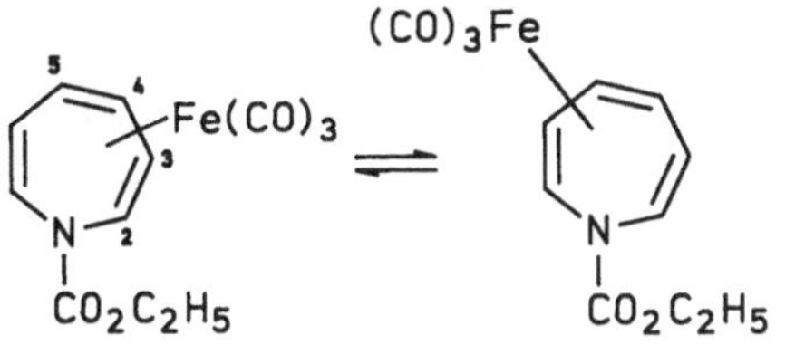

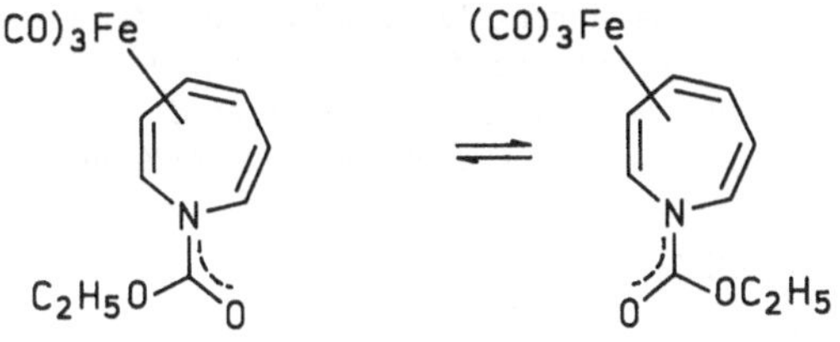

trans cis

X XI

Compound No. 5 is used as an example for the proposed "hapto" nomenclature in [46].

References on pp. 33/4

The compound No. 5 is easily soluble in organic solvents and takes-up H_2 at room temperature in the presence of PtO_2/CH_3OH or CH_3CO_2H (possibly yielding No. 18). Reaction with Na in CH_3OH at 40 °C for 5 h, followed by chromatography on neutral Al_2O_3 with n-hexane, gives compound No. 2 (60% yield). The compound No. 5 is soluble in conc. H_2SO_4 or CF_3CO_2H, but the solutions decompose quickly [1]. Reaction with HBF_4 in $O(COCH_3)_2$ at −15 °C gives a yellow BF_4 salt which decomposes with bases. Similar reaction at 0 °C gives a yellow BF_4 salt which on addition of a base gives No. 12 (42% yield). Analogous reaction with HBF_4 in $O(COC_2H_5)_2$ at 0 °C, followed by addition of a base, gives No. 13. Similar reaction with $SnCl_4/O(COCH_3)_2$ gives No. 12 in 22% yield, and reaction with $HCON(CH_3)_2/OPCl_3$ gives No. 11 in 19% yield [14]. Reaction of No. 5 (1.5 g) with a mixture of $Fe_2(CO)_9$ (6.3 g) and $OC(C(CH_3)_2Br)_2$ (4.1 g) in C_6H_6 at 60 °C for 8 h followed by filtration, evaporation of the filtrate, and chromatography of the residue on silica gel with C_6H_6 gives a mixture of No. 14 and No. 22 as a yellow oil [43]. Similar reaction with 3,6-bis(methoxycarbonyl)-1,2,4,5-tetrazine (see Formula XII) followed by work-up as before but eluting with n-hexane/$CH_3CO_2C_2H_5$ (3:2) gives No. 23 in 92% yield. The kinetic data for this reaction are shown in the following table:

temperature in °C	solvent	$\frac{\text{No. 5}}{\text{XII}}$	$k \cdot 10^3$ in s^{-1}	$\frac{k\text{(No. 5)}}{k\text{(XII)}}$
34.1	C_6H_6	100	1.22	—
34.1	ClC_6H_5	100	1.26	1
34.1	ClC_6H_5	50	1.36	—
34.1	ClC_6H_5	25	1.31	—
34.1	ClC_6H_5	10	1.26	—
30.0	ClC_6H_5	100	1.11	—
40.0	ClC_6H_5	100	1.67	—
45.0	ClC_6H_5	100	2.31	—
50.0	ClC_6H_5	100	2.84	—
34.1	$C_2H_4Cl_2$	100	2.84	—
34.1	CH_3CN	100	1.64	—

The lack of dependence of rate on change in solvent polarity, and the large entropy value of $\Delta S^* = -42.7 \pm 4.5$ e.u., rules out a dipolar intermediate and supports a concerted mechanism for this reaction. The activation energy is $E_a = 9.55 \pm 1.4$ kcal/mol. Refluxing compound No. 5 with XIII in C_6H_6 for 6 h, followed by work-up as before but elution with n-hexane/$CH_3CO_2C_2H_5$ (10:1), gives No. 24 in 66% yield [6]. Compound No. 5 reacts with the carbene CCl_2 to give No. 80. Thus, a mixture of 50% aqueous $NaOH/C_6H_6$, $[C_6H_5CH_2N(C_2H_5)_3]Cl$, and No. 5 is vigorously stirred at room temperature. $CHCl_3$ is added slowly during 30 min. The mixture is stirred for 1.5 h, then poured into H_2O, and extracted with CH_2Cl_2. The extract is dried over Na_2SO_4 and evaporated. Chromatography of the residue on silica gel with C_6H_6/n-hexane (3:1) gives No. 80 in 73% yield [51].

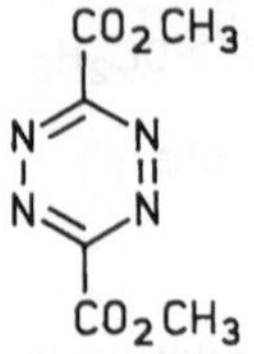

XII 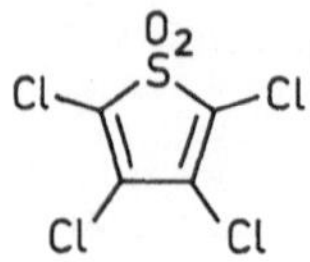 XIII 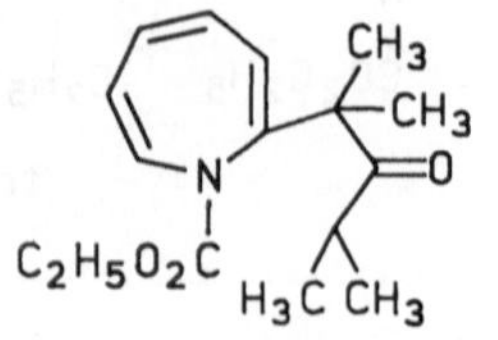 XIV 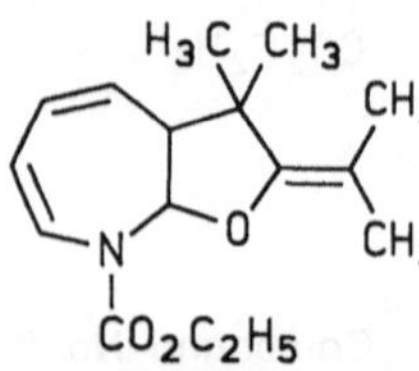 XV

$CH_3COC_6H_5NHFe(CO)_3$ (Table **1**, No. **9**) shows no line-broadening at room temperature, indicating the absence of tautomeric bond-switching (fluxional) behavior, and it is therefore probable that No. 9 has the same structure in solution and in the solid state [19]. The IR spectrum shows hydrogen bridges [14]. An X-ray structure determination shows that the crystals are triclinic, space group $P\bar{1}$-C_i^1, S_2^1 (No. 2), with the unit cell parameters a = 7.002(5), b = 9.006(6), c = 9.886(8) Å, α = 86°47(4)′, β = 96°11(4)′, γ = 114°19(3)′; Z = 2 molecules per unit cell. The measured density is 1.634 g/cm³, calculated 1.635 g/cm³. The main bond distances and angles are shown in **Fig. 3**. The dihedral angle between the plane of the butadiene fragment C-2 to 5 and the plane N, C-5 to 7 is 141°, the latter plane being inclined at 3.6° to that of the acetyl group [19]. Reaction of compound No. 9 with CH_3I gives No. 10 [14].

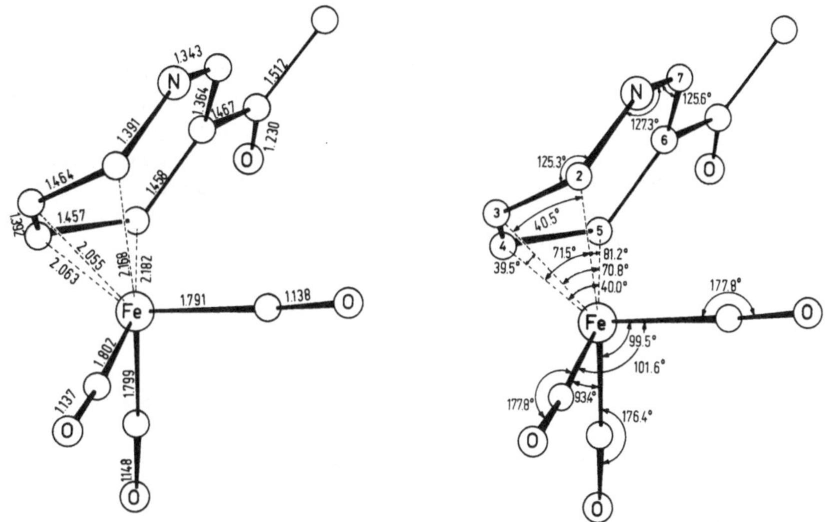

Fig. 3. Molecular structure of $CH_3COC_6H_5NHFe(CO)_3$ (No. 9) [19].

$OHCC_6H_5NCO_2C_2H_5Fe(CO)_3$ (Table **1**, No. **11**) crystallizes in the triclinic space group $P\bar{1}$-C_i^1, S_2^1 (No. 2) with the unit cell parameters a = 7.022(3), b = 10.550(4), c = 9.500(3) Å, α = 92°49(2)′, β = 94°32(2)′, γ = 96°8(2)′; Z = 2 molecules per unit cell. The measured density is 1.58 g/cm³, calculated 1.587 g/cm³. The main bond distances and angles are shown in **Fig. 4**, p. 26. The ring has two nearly planar groups of atoms C-2 to 4 and C-2,5 to 7, N. The dihedral angle between the planes is 138° [25].

$CH_3COC_6H_5NCO_2C_2H_5Fe(CO)_3$ (Table **1**, No. **12**) gives No. 9 with $NaOCH_3/CH_3OH$ [14].

$(CH_3)_2CHCO(CH_3)_2CC_6H_5NCO_2C_2H_5Fe(CO)_3$ (Table **1**, No. **14**). Dropwise addition of 2-chloranil in CH_2Cl_2 to the mixture of No. 14 and No. 22 in CH_2Cl_2 gives XIV (21%) and XV (24%) after 24 h at room temperature [43].

$(CH_3)_3C_6H_3NHFe(CO)_3$ (Table **1**, No. **15**) is monomeric in C_6H_6 (by isothermic distillation) and shows in the ^1H NMR spectrum in benzene-d_6 chemical shifts at 86 and 88 (CH_3-2,4), 111 (CH_3-6), 218 (NH; J(H-1,7) = 6.5 Hz), 237 (H-5), 260 (H-3), and 288 (H-7) Hz. The compound No. 15 is highly soluble in halogen free solvents but decomposes quickly in $CHCl_3$ and CCl_4. It takes up H_2 at room temperature in the presence of PtO_2/CH_3OH or CH_3CO_2H (possibly yielding No. 19) [1].

References on pp. 33/4

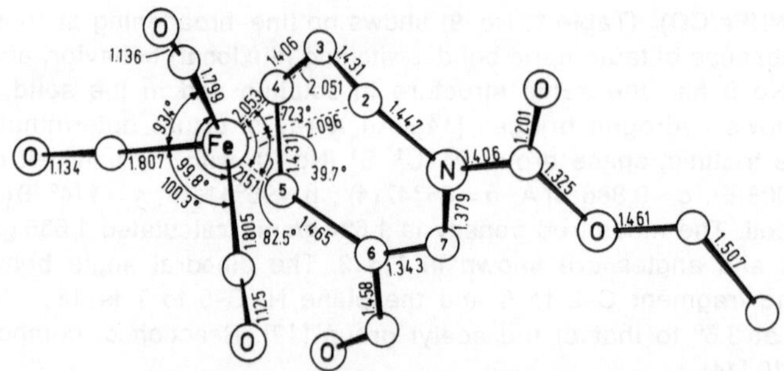

Fig. 4. Molecular structure of $OHCC_6H_5NCO_2C_2H_5Fe(CO)_3$ (No. 11) [25].

$(CH_3)_3C_6H_3NCO_2C_2H_5Fe(CO)_3$ (Table **1**, No. **16**) is diamagnetic and shows in the ^{1}H NMR spectrum in $CDCl_3$ chemical shifts at 81 (CH_3 in C_2H_5), 110 (CH_3), 119 (CH_3), 121 (CH_3), 257 (CH_2), 294 (H-5), 306 (H-3), and 342 (H-7) Hz. The compound is highly soluble in organic solvents, soluble in conc. H_2SO_4 and CF_3CO_2H, in which solutions quickly decompose. It takes-up H_2 at room temperature in the presence of PtO_2/CH_3OH or CH_3CO_2H (possibly forming No. 20). Reaction with Na in CH_3OH at 40 °C for 3 h followed by chromatography on neutral Al_2O_3 with n-hexane gives No. 20 (14%) [1].

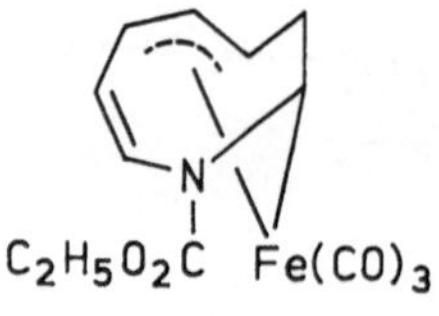

XVI 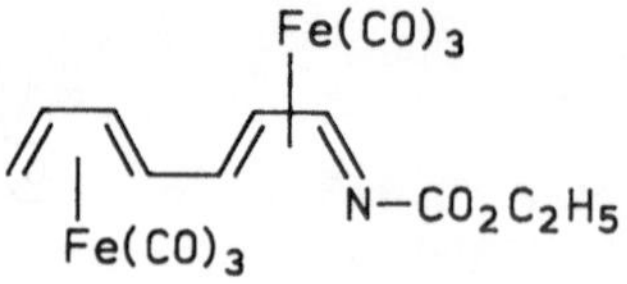 XVII 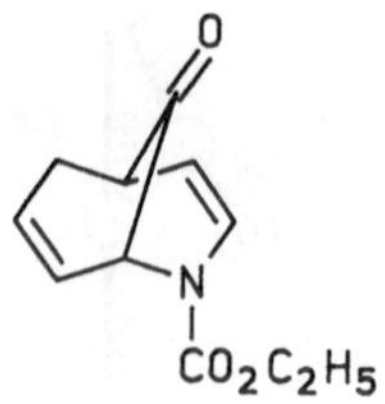 XVIII

$CH_2C_6H_6NCO_2C_2H_5Fe(CO)_3$ (Table **1**, Nos. **21a** and **b**) shows in the ^{1}H NMR spectra a distinct temperature dependence of the resonance lines due to the hindered rotation of the N-$CO_2C_2H_5$ bond (compare XI). The signals of H-2 and H-7 are markedly split at −10 °C (No. 21a) or −21 °C (No. 21b), but averaged in the limit spectrum at higher temperatures. Characteristic J(H-5,6) values are observed (see Table 1, p. 8). Thermal isomerization of the exo isomer No. 21a in benzene-d_6 at 90 °C for 30 h gives XVI, and similar isomerization of the endo isomer No. 21b in C_6H_6 gives XVII. Carbonylation of No. 21a in an autoclave in C_6H_6 with CO (160 atm, 80 °C, 13 h) gives XVI in 40% yield and XVIII (57%). Analogous carbonylation of 21b gives XVI in 25% yield, starting compound No. 21b (15%), XIX (7%), and XX (60% relative to reacted No. 21b) [39].

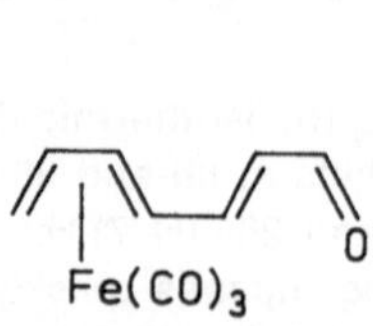

XIX 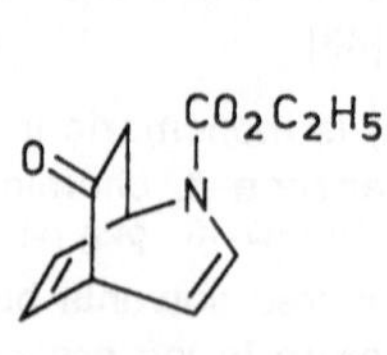 XX 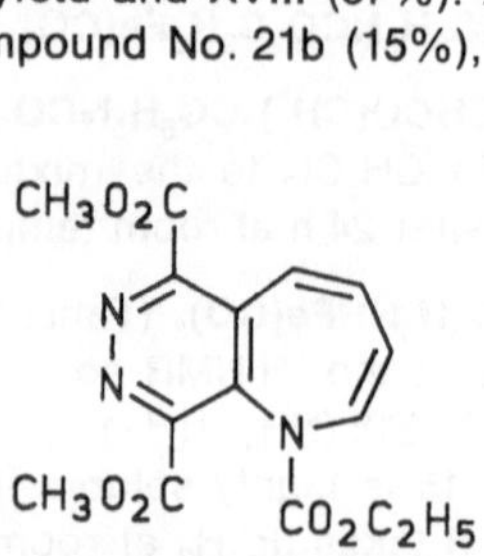 XXI

References on pp. 33/4

$(CH_3O_2C)_2C_2N_2HC_6H_5NCO_2C_2H_5Fe(CO)_3$ (Table **1**, No. **23**). Oxidative degradation with 2-chloranil in $CHCl_3$ at room temperature for 10 min liberates the organic ligand (71%). Refluxing with $ON(CH_3)_3$ in C_6H_6 for 2 h gives XXI in 32% yield [6].

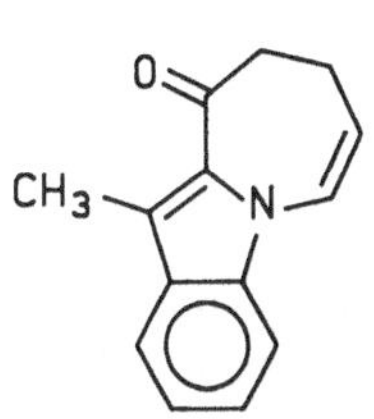

XXII 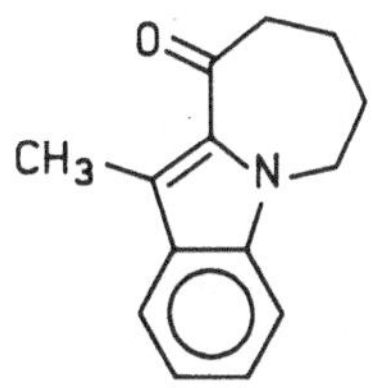 XXIII 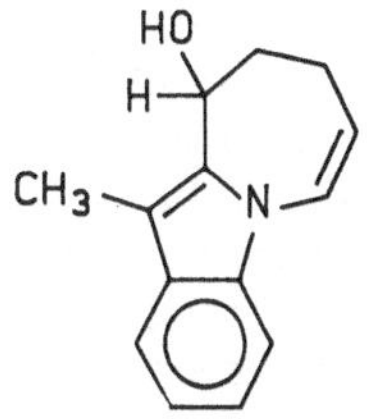 XXIV 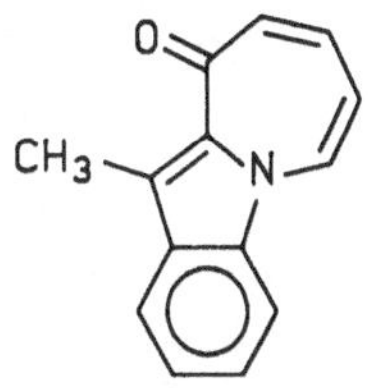 XXV

$C_{14}H_{11}NOFe(CO)_3$ (Table **1**, No. **26**) can still undergo reduction of the keto group [15, 17]. Thus, reduction with $NaBH_4$ in C_2H_5OH at room temperature for 5 h gives XXII, XXIII (57%), XXIV, and XXV. Reduction with $LiAlH_4$ in tetrahydrofuran at room temperature for 3 h gives XXVI (R=H), XXII, XXV, and XXVI (R=OH) [15].

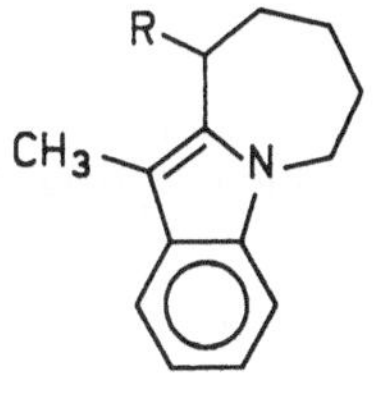

XXVI 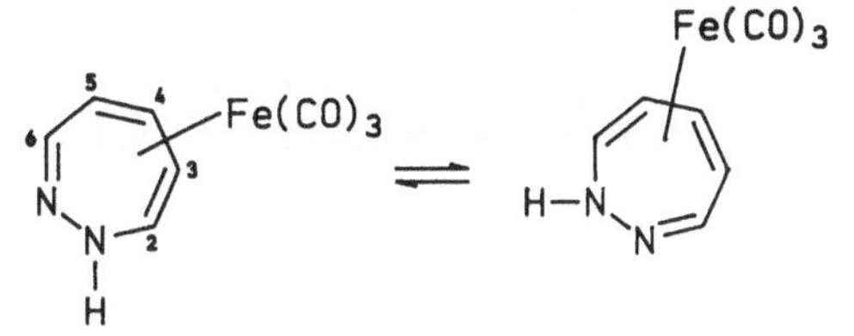 a XXVII b

$C_{22}H_{25}O_4NFe(CO)_3$ (Table **1**, No. **35**) is heated in a quartz tube for 3 min. The mixture is taken-up in C_6H_6 and chromatographed on silica gel with benzine/C_6H_6/$CH_3CO_2C_2H_5$ (5:2:1). Concentration of the solution gives partial decomposition. The isolated product No. 36 (50%) always contains insoluble impurities, possibly due to hydrolysis followed by prototropic isomerization to give an unstable (η^3-allyl)tricarbonylironketone. Refluxing No. 35 in quinoline causes decomplexation [42].

$RNNHC_5H_6Fe(CO)_3$ (Table **1**, Nos. **37** and **38** with R=$COCH_3$ and $CO_2C_2H_5$) react with $O(COCH_3)_2$ at room temperature within 12 to 16 h to give Nos. 39 and 40 (65%), respectively, after chromatography with C_6H_6/CH_2Cl_2 (1:1) as eluent. In fact, although No. 37 is found to be somewhat unstable under ordinary laboratory conditions and No. 38 is extremely unstable, both can be transformed in this way to yield Nos. 39 and 40 [26].

$HN_2C_5H_5Fe(CO)_3$ (Table **1**, No. **41**). The deuterated compound $DN_2C_5H_5Fe(CO)_3$ shows in the ^1H NMR spectrum ($CDCl_3$) at −39.5 °C chemical shifts at δ=3.67 (H-5; J(H-4,5)=6.6, J(H-3,4)=1.8), 4.09 (H-3; J(H-2,3)=6.0), 4.68 (H-4; J(H-3,4)=4.0), 5.61 (H-2; J(H-2,4)=1.6), 6.89 (H-6; J(H-5,6)=6.4, J(H-4,6)=0.3) ppm. The ^1H NMR spectral data can be ascribed to simultaneous tautomerism and fluxionality as shown in XXVII. Analysis of the ^1H NMR data for No. 41 yields at 25 °C the free energy $\Delta G^{*}=13.7\pm0.4$ kcal/mol and the entropy of activation $\Delta S^{*}=-6.2\pm4.0$ e.u., and for the deuterated derivative $DN_2C_5H_5Fe(CO)_3$ $\Delta G^{*}=13.4\pm0.4$ kcal/mol and $\Delta S^{*}=-6.9\pm4.0$ e.u., the same within experimental error [23]. Reaction with excess CF_3CO_2H [29, 31] or with equal amounts of CF_3CO_2H in CH_2Cl_2 at −20 °C overnight [33] gives No. 70 [29, 31, 33] in 50 to 60% yield [33]. Reaction with NaH/CH_3COCl

References on pp. 33/4

in tetrahydrofuran gives No. 53 after chromatography on alumina [23]. Acylation with 1 equivalent RCOCl, $NaHCO_3$ present, in C_6H_6 at room temperature for 1 to 2 h gives compounds of the type $RCON_2C_5H_5Fe(CO)_3$ with $R=CH_3$ (No. 53) [32, 35] in >95% yield [32], $R=C_6H_5$ (No. 56) [32, 35] in 86% yield, $R=CH{=}CHC_6H_5$ (No. 55) and $R=CH_2Cl_2$ (No. 52) [32, 35] in >95% yield, respectively [32], $R=CCl_3$ (No. 51), $R=CH{=}CH_2$ (No. 54) [35], and $Cl_3CCH_2O_2C/N_2C_5H_5Fe(CO)_3$ (No. 57) [32, 35] >95% yield [32] after chromatography on silica gel [32]. Acylation with $O(COCF_3)_2$ or BrCN under similar conditions gives $CF_3CON_2C_5H_5Fe(CO)_3$ (No. 50) [32, 35] in 94% yield [32] or $NCN_2C_5H_5Fe(CO)_3$ (No. 60) in 82% yield [32]. Similar reaction with $O(COCH_3)_2$ at room temperature gives $CH_3CON_2C_5H_5Fe(CO)_3$ (No. 53) [32]. Analogous reaction with CH_3OSO_2F gives No. 72 as the monohydrate [33]. Alkylation with CH_3Br in CH_3OH in a Carius tube at room temperature for 3d followed by 0.5 h at 75 °C gives $CH_3N_2C_5H_5Fe(CO)_3$ (No. 45). The reaction mixture is evaporated and the residue is taken up in ether. The organic layer is washed with 10% aqueous $NaHCO_3$, H_2O, and dried (K_2CO_3). Evaporation gives No. 45 [33]. Reaction with LiC_4H_9-n/$C_6H_5CH_2Br$ in ether gives $C_6H_5CH_2N_2C_5H_5Fe(CO)_3$ (No. 48) in 40% yield [23]. Alkylation with excess RX (X=Br or I), $NaHCO_3$ at room temperature for 20 h followed by chromatography on silica gel gives compounds of the type $RN_2C_5H_5Fe(CO)_3$ with $R=CH_3$ (No. 45) [32, 35] in 30% yield [32], $R=CH_2C_6H_5$ (No. 48) [32, 35] in 89% yield [32], $R=CH_2CH{=}CH_2$ (No. 47) [32, 35] in 75% yield [32], and $R=CH(CH_3){=}CH_2$ (No. 49) [35]. Similar reaction with CH_3OSO_2F [32, 35] in the presence of $C_2H_5N(C_3H_7\text{-i})_2$ in CH_2Cl_2 at 0 °C to room temperature gives $CH_3N_2C_5H_5Fe(CO)_3$ (No. 45) in 70% yield [32].

$CH_3O_2SN_2C_5H_5Fe(CO)_3$ (Table **1**, No. **42**) exhibits valence tautomerization involving the $Fe(CO)_3$ group, but shows no evidence of restricted rotation around the N-S bond (compare No. 5, p. 23) [13].

$CH_3N_2C_5H_5Fe(CO)_3$ (Table **1**, No. **45**). Protonation with CF_3CO_2H gives No. 72 with $X=O_2CCF_3$ [20].

$RN_2C_5H_5Fe(CO)_3$ (Table **1**, Nos. **48, 51, 52** with $R=CH_2C_6H_5$, $COCCl_3$, $COCH_2Cl$). Oxidation of Nos. 48 [32] or 52 [32, 35] with at least a 20-fold molar excess of freshly sublimed $ON(CH_3)_3$ in C_6H_6 at room temperature for 3 h liberates the organic ligand ^4L in 30 and 41% yield [32]. $ON(CH_3)_3 \cdot 2H_2O$ gives unsatisfactory results. Other reagents reported to be moderately successful in releasing nonnitrogen containing organic ligands from their $^4LFe(CO)_3$ complexes lead uniformly to extensive decomposition [32]. Similar oxidation of No. 51 liberates the organic ligand ^4L, too [35].

$CH_3CON_2C_5H_5Fe(CO)_3$ (Table **1**, No. **53**). The chemical shifts and coupling constants of the ^1H NMR spectrum are calculated in [20]. The ^1H NMR spectrum shows no change in the temperature range from −40 to +50 °C [20, 23], and attempts to induce fluxionality by raising the temperature to 100 °C are unsuccessful [23]. This result implies that the energy barrier due to hindered rotation of the N-CO function must be quite low [20]. Oxidation with at least a 20-fold molar excess of freshly sublimed $ON(CH_3)_3$ in C_6H_6 at room temperature for 3 h liberates the organic ligand ^4L in 91% yield (compare Nos. 48, 51, 52 before) [32]. Reduction with $NaBH_4$ in CH_3OH at room temperature for 5 to 6 h, evaporation at reduced pressure below 30 °C, treatment of the residue with ice water, extraction with CH_2Cl_2, and drying the organic layer with $MgSO_4$ followed by evaporation and chromatography on silica gel with $CH_3CO_2C_2H_5$ gives No. 37 in 45% yield [26]. Treatment with CF_3CO_2H gives No. 71 [33]. Reaction with $NaOC_2H_5$ in C_2H_5OH at room temperature for 1.5 h followed by quenching with H_2O, extraction with ether, drying the organic layer with K_2CO_3, and removal of the solvent gives No. 41 in 96% yield [33], cf. [23, 32].

$RN_2C_5H_5Fe(CO)_3$ (Table **1**, Nos. **55** and **56** with $R=COCH{=}CHC_6H_5$ and COC_6H_5) liberate the organic ligand ^4L on oxidation with at least a 20-fold molar excess of freshly sublimed

$ON(CH_3)_3$ in C_6H_6 at room temperature for 3 h (94 and 78% yield, respectively) [32, 35], compare Nos. 48, 51, 52.

$CCl_3CH_2O_2CN_2C_5H_5Fe(CO)_3$ (Table **1**, No. **57**) liberates the organic ligand ^4L in 86% yield with at least 20-fold molar excess of freshly sublimed $ON(CH_3)_3$ in C_6H_6 at room temperature with 3 h [32, 35], compare Nos. 48, 51, and 52.

$C_2H_5O_2CN_2C_5H_5Fe(CO)_3$ (Table **1**, No. **58**) is not dynamic at room temperature, and attempts to induce fluxionality by raising the temperature to 100 °C are unsuccessful [23]. The compound crystallizes in the triclinic space group $P\bar{1}$-C_2^1, S_2^1 (No. 2), with the unit cell parameters $a=12.410\pm0.10$, $b=17.530\pm0.15$, $c=6.696\pm0.008$ Å, $\alpha=99.20°\pm0.15°$, $\beta=116.58°\pm0.15°$, $\gamma=91.00°\pm0.15°$; Z=4 molecules per unit cell. The measured density is 1.59 ± 0.02 g/cm³, calculated 1.59 g/cm³. There are two molecules in the asymmetric unit. The main bond distances and angles are shown in **Fig. 5**. The dihedral angle between the planes of the complexed butadiene part and the rest of the ring is 18.5° in the one molecule and 17.3° in the other [22].

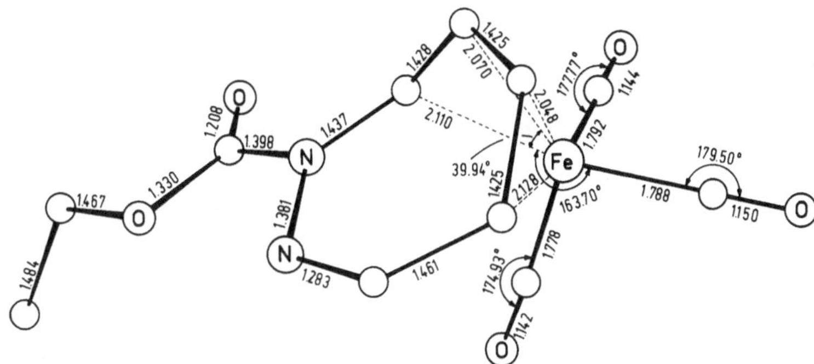

Fig. 5. Molecular structure of one of the molecules of $C_2H_5O_2CN_2C_5H_5Fe(CO)_3$ (No. 58) [22].

Reduction with $NaBH_4$ in CH_3OH at room temperature for 5 to 6 h, evaporation at reduced pressure below 30 °C, treating the residue with ice water, extraction with CH_2Cl_2, drying the organic layer ($MgSO_4$), and evaporation followed by chromatography on silica gel with $CH_3CO_2C_2H_5$ gives No. 38 [26]. Reaction with $NaOC_2H_5$ in C_2H_5OH at 0 °C for 1 h gives No. 41 in 60% yield after chromatography on alumina [23].

i-$C_3H_7O_2CN_2C_5H_5Fe(CO)_3$ (Table **1**, No. **59**) crystallizes in the monoclinic space group $P2_1/n$-C_{2h}^5 (No. 14) with the unit cell parameters a=12.277(3), b=17.241(5), c=6.639(2) Å, β=101.60(5)°; Z=4 molecules per unit cell. The measured density is 1.53 ± 0.02 g/cm³, calculated 1.545 g/cm³. The main bond distances and angles are shown in **Fig. 6**, p. 30. The dihedral angle between the planes of the complexed butadiene part and the rest of the ring is 40° [16].

6-$CH_3C_5H_4N_2HFe(CO)_3$ (Table **1**, No. **61**) shows by the well resolved nonfluxional temperature invariant ^{1}H NMR spectrum that the one tautomer, shown in the Table 1, p. 16, is stabilized [23]. Protonation of No. 61, which itself is incapable of achieving a time-averaged symmetry plane, cannot lead to a fluxional ion. With CF_3CO_2H in liquid SO_2 solution, four nonequivalent ring protons with an essentially identical coupling pattern remain. At low temperature (< -60 °C) two NH environments are observed corresponding to the imminium and the amine site of No. 73. At temperatures above 10 °C a single average signal appears. NH site exchange must then occur via a small equilibrium concentration of No. 74 which

References on pp. 33/4

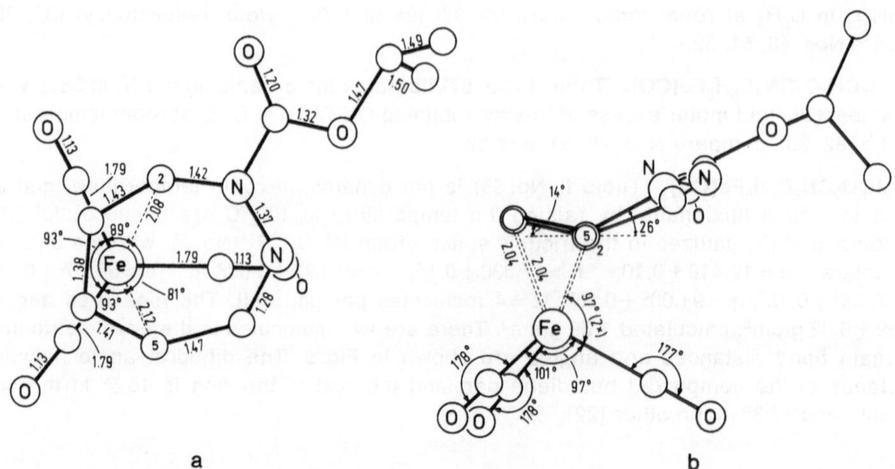

Fig. 6. Molecular structure of i-$C_3H_7O_2CN_2C_5H_5Fe(CO)_3$ (No. 59).
a) Projection on the plain of the C atoms in the $Fe(CO)_3$ moiety.
b) Projection in the direction of C-2,5, vertical to the projection in a [16].

is not observable in the spectrum [33], compare XXX. The nucleophilic properties of the NH lone pair are reflected in the isolation of only No. 65 from No. 61 and CH_3COCl [30].

4-$CH_3C_5H_4N_2HFe(CO)_3$ (Table **1**, No. **62**) is a dynamic molecule with $\Delta G^* = 15 \pm 0.4$ kcal/mol [23].

6-$CH_3C_5H_4N_2COCH_3Fe(CO)_3$ and **4-$CH_3C_5H_4N_2COCH_3Fe(CO)_3$** (Table **1**, Nos. **65** and **67**) are not dynamic at room temperature, and attempts to induce fluxionality by raising the temperature to 100 °C are unsuccessful. The compounds No. 65 and 67 give No. 61 in 85% or No. 62 in 62% yield in the reaction with $NaOC_2H_5$ in C_2H_5OH at 0 °C within 1 h after chromatography on alumina with ether as eluant [23, 33].

$[H_2N_2C_5H_5Fe(CO)_3]^+$ (a) ⇌ $[H_2N_2C_5H_5Fe(CO)_3]^+$ (b) — XXXVIII

$[CH_3HN_2C_5H_5Fe(CO)_3]^+X^-$ — XXIX

[$H_2N_2C_5H_5Fe(CO)_3$]O_2CCF_3 (Table **1**, No. **70**). The coupling constants given in the Table 1, pp. 18/9, are averaged for XXVIIIa and b. Confirmation of the fluxional nature and structure of No. 70 as shown in XXVIII derives from the ^{13}C NMR spectrum (see Table 1, p. 19) indicating the presence of a time-averaged symmetry plane and the basal-apical CO-exchange process is fast at this temperature [33]. The compound crystallizes [31, 33, 34] in the triclinic space group $P\bar{1}$-C_i^1 (No. 2) with the unit cell parameters a = 10.812(7), b = 9.403(6), c = 6.690(4) Å, α = 94.72(7)°, β = 78.16(5)°, γ = 103.59(5)°; Z = 2 molecules per unit cell [31]. The measured density is 1.78 g/cm³, calculated 1.770 g/cm³. The main bond distances and angles are shown in **Fig. 7**. The heterocycle is folded about a line joining the terminal C atoms

C-2,5 of the diene moiety with a dihedral angle of 35.25° between the planes defined by the diene C atoms, the atoms C-2,5,6 and the two N atoms [31].

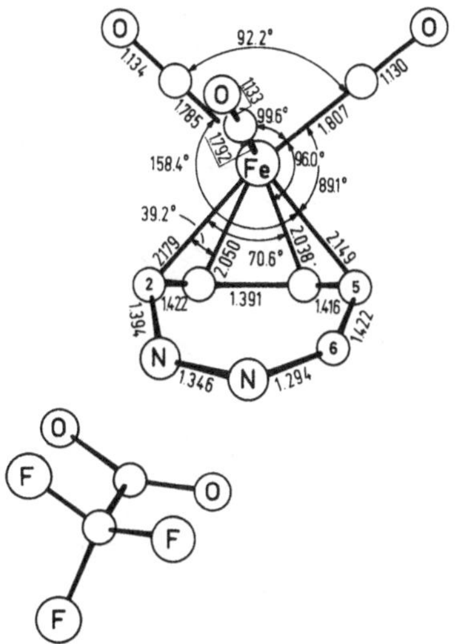

Fig. 7. Molecular structure of $[H_2N_2C_5H_5Fe(CO)_3]O_2CCF_3$ (No. 70) [31].

Hydrolysis of compound No. 70 followed by extraction with ether gives compound No. 41 in quantitative yield [33].

$[CH_3N_2HC_5H_5Fe(CO)_3]X$ (Table **1**, No. **72** with $X=OSO_2F$ or O_2CCF_3). Attempts to isolate No. 72 with $X=O_2CCF_3$ failed. Compound No. 72 with $X=OSO_2F$ is isolated as the monohydrate $[CH_3N_2HC_5H_5Fe(CO)_3]OSO_2F \cdot H_2O$. The presence of XXIX which can lead to No. 72 via an intramolecular proton transfer cannot be completely excluded. It is nevertheless clear that XXIX cannot be the thermodynamic product of protonation of No. 45 since no N-H to N-CH_3 proton coupling is observed even at low temperature. Hydrolysis of No. 72 with $X=OSO_2F$ or O_2CCF_3, followed by extraction with ether, gives No. 45 in 10 to 20% yield [33].

No. 74 ⇌ (K_1, H^+) No. 62 ⇌ (K_2) No. 75

XXX

$[6\text{-}CH_3N_2H_2C_5H_4Fe(CO)_3]O_2CCF_3$ (Table **1**, Nos. **73** and **74**). Compound No. 61 with CF_3CO_2H in liquid SO_2 solution shows four nonequivalent ring protons with an essentially identical coupling pattern. At low temperature (< -60 °C) two N-H environments are observed at $\delta=8.9$ and 14.8 ppm corresponding to the imminium and the amine site in No. 73, respectively. As the temperature is increased, broadening due to site exchange commences, but the four C-H resonances are unchanged. Finally at temperatures above 10 °C, a single

References on pp. 33/4

average signal appears at $\delta = 11.85$ ppm. N-H site exchange must then occur via a small equilibrium concentration of No. 74 which is not observable in the spectrum ($K_1 \gg K_2$, see Scheme XXX). It is shown that broadening of the more acidic and more labile imminium proton resonance of No. 73 is more appreciable than that of the amine resonance. Hydrolysis of No. 73, followed by extraction with ether, gives No. 61 in quantitative yield [33].

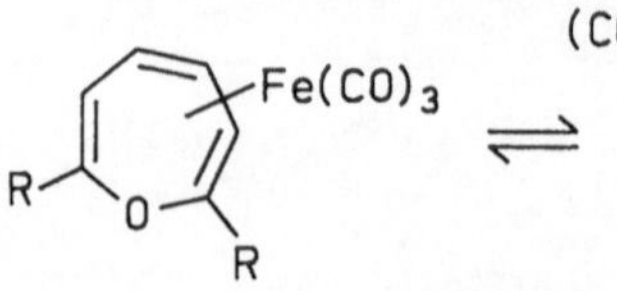

XXXI 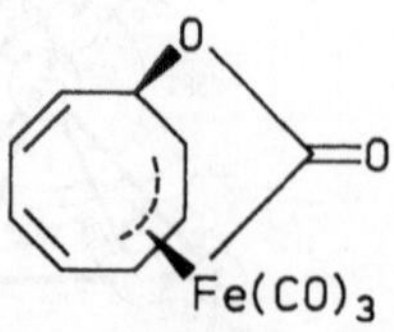 XXXII 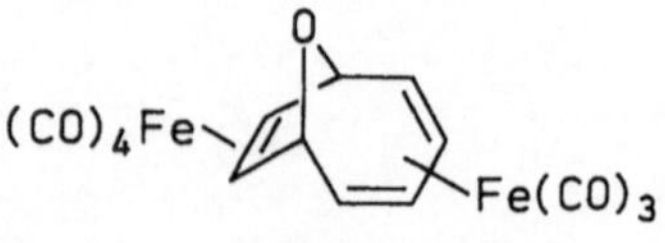 XXXIII

$OC_6H_6Fe(CO)_3$ (Table **1**, No. **75**). A degenerate valence tautomerism as shown in XXXI (R=H) is established by ^{1}H NMR spectroscopy. The limiting spectrum is observed at 2.5 °C (see Table 1, p. 20). Line broadening is observed upon warming and eventually a pair-wise coalescence of the signals occurs. Further warming of the solution sharpens the lines, and finally an averaged spectrum of type AA′XX′YY′ is obtained. These processes are reversible [27].

$(CH_3)_2C_6H_4OFe(CO)_3$ (Table **1**, No. **76**) is monomeric in C_6H_6 (by osmometry) [4]. A rearrangement as described for No. 72, and shown in XXXI, is also observed for this compound. $\Delta G^{\neq} = 15.8 \pm 0.3$ kcal/mol is estimated for this valence isomerization at the coalescence temperature for the two CH_3 resonances at 46.5 °C [27], cf. [50]. The compound is soluble in organic solvents, but decomposes slowly in halogenated hydrocarbons [4]. No. 76 does not react with $Fe_2(CO)_9$ [27].

$C_8H_8OFe(CO)_3$ (Table **1**, No. **78**) is prepared by warming XXXII in $CHCl_3$ at 40 °C for 50 min. Compound No. 78 is also obtained by short reaction of XXXIV with $ON(CH_3)_3$ in ether followed by chromatography on silica gel (Woelm, neutral) with C_6H_6. For the ^{1}H NMR spectrum see the figure in [28].

$OHCNC_6H_6Fe(CO)_3$ (Table **1**, No. **79**). Calculations of the extended Hückel type are carried out to give the HOMO coefficients +0.312, −0.217, and +0.108 for C(2), C(4), C(7) (see Table 1, p. 21). Frontier orbital calculations predict that 2,4-addition is the preferred cycloaddition mode for $(NC)_2C{=}C(CN)_2$ [49].

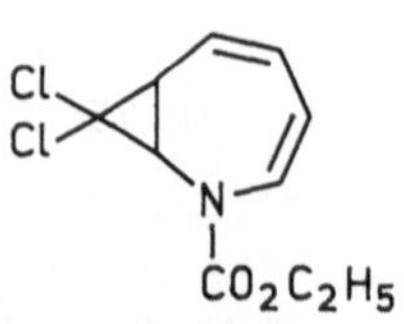

XXXIV 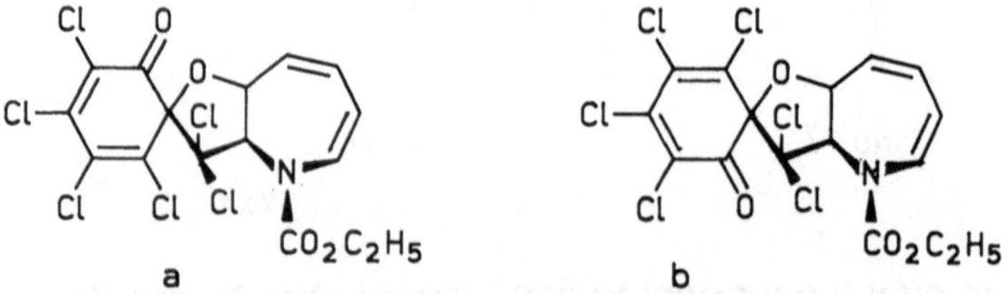 XXXV a b

$Cl_2CC_6H_6NCO_2C_2H_5Fe(CO)_3$ (Table **1**, No. **80**) is refluxed with $ON(CH_3)_3$ in C_6H_6 for 30 min to give XXXIV in 92% yield. Oxidation with 2-chloranil in CH_2Cl_2 at room temperature for 1 h gives XXXVa (42% yield) and b (28%) [51].

References:

[1] E.O. Fischer, H. Rühle (Z. Anorg. Allgem. Chem. **341** [1965] 137/45). – [2] H. Günther, R. Wenzl (Tetrahedron Letters **1967** 4155/9). – [3] L.A. Paquette (unpublished results from [18]). – [4] E.O. Fischer, C.G. Kreiter, H. Rühle, K.E. Schwarzhans (Chem. Ber. **100** [1967] 1905/9). – [5] W. Hoppe (Plenary Lect. 2nd Microsymp. Struct. Org. Solids, Prague 1968, pp. 465/88).

[6] T. Ban, K. Nagai, Y. Miyamoto, K. Harano, M. Yasuda, K. Kanematsu (J. Org. Chem. **47** [1982] 110/6). – [7] I.C. Paul, S.M. Johnson, L.A. Paquette, J.H. Barrett, R.J. Haluska (J. Am. Chem. Soc. **90** [1968] 5023/4). – [8] J. Streith, J.-M. Cassal (Angew. Chem. **80** [1968] 117). – [9] J. Streith, J.-M. Cassal (Bull. Soc. Chim. France **1969** 2175/80). – [10] J. Streith, A. Bhud, J.-M. Cassal, C. Sigwalt (Bull. Soc. Chim. France **1969** 948/52).

[11] L.A. Paquette, D.E. Kuhla, J.H. Barrett, R.J. Haluska (J. Org. Chem. **34** [1969] 2866/78). – [12] W. Hoppe, N. Brodherr, H. Englmeier, J. Gassmann, A. Gieren, S. Hechtfischer, L. Preuß, M. Röhrl, J. Schaeffer, E. Schmidt, W. Steigemann, K. Zechmeister (Pure Appl. Chem. **18** [1969] 465/88). – [13] M.A. Brown (Diss. Univ. California 1969; Diss. Abstr. Intern. B **30** [1970] 4967). – [14] G.B. Gill, N. Gourlay, A.W. Johnson, M. Mahendran (Chem. Commun. **1969** 631/3). – [15] G.R. Cliff, E.W. Collington, G. Jones (J. Chem. Soc. C **1970** 1490/3).

[16] R. Allmann (Angew. Chem. **82** [1970] 982/3). – [17] J. Lewis (unpublished results from [15]). – [18] S.M. Johnson, I.C. Paul (J. Chem. Soc. B **1970** 1783/9). – [19] M.G. Waite, G.A. Simm (J. Chem. Soc. A **1971** 1009/13). – [20] A.I. Carty, G. Kan, D.P. Madden, V. Snieckus, M. Stanton, T. Birchall (J. Organometal. Chem. **32** [1971] 241/56).

[21] A. Gieren, W. Hoppe (Acta Cryst. B **28** [1972] 2766/82). – [22] A. de Cian, P.M. L'huillier, R. Weiss (Bull. Soc. Chim. France **1973** 457/63). – [23] A.J. Carty, R.F. Hobson, H.A. Patel, V. Snieckus (J. Am. Chem. Soc. **95** [1973] 6835/7). – [24] R. Kuhrke (Staatsex.-Arbeit Univ. Münster 1973 from [42]). – [25] D.I. Woodhouse, G.A. Sim, J.G. Sime (J. Chem. Soc. Dalton Trans. **1974** 1331/5).

[26] T. Tsuchiya, V. Snieckus (Can. J. Chem. **53** [1975] 519/28). – [27] R. Aumann, H. Averbeck, C. Krüger (Chem. Ber. **108** [1975] 3336/48). – [28] R. Aumann, H. Averbeck (J. Organometal. Chem. **85** [1975] C4/C6). – [29] A.J. Carty, C.R. Jablonski, V. Snieckus (unpublished results from [31]). – [30] D.J. Harris, V. Snieckus (unpublished results from [33]).

[31] A.J. Carty, N.J. Taylor, C.R. Jablonski (Inorg. Chem. **15** [1976] 1169/74). – [32] D.J. Harris, V. Snieckus (J. Chem. Soc. Chem. Commun. **1976** 844/5). – [33] A.J. Carty, C.R. Jablonski, V. Snieckus (Inorg. Chem. **15** [1976] 601/7). – [34] A.J. Carty, C.R. Jablonski, N.J. Taylor (unpublished results from [33]). – [35] D.J. Harris (Diss. Univ. Waterloo, Can., 1976; Diss. Abstr. Intern. B **38** [1977] 689).

[36] M. Green, S.M. Heathcock, T.W. Turner, D.M.P. Mingos (J. Chem. Soc. Dalton Trans. **1977** 204/11). – [37] G.E. Herberich, E. Bauer, J. Hengesbach, U. Kölle, G. Huttner, H. Lorenz (Chem. Ber. **110** [1977] 760/72). – [38] P. Hofmann (Z. Naturforsch. **33b** [1978] 251/60). – [39] R. Aumann, J. Knecht (Chem. Ber. **111** [1978] 3927/31). – [40] F. Bellamy, J.L. Schuppiser, J. Streith (Heterocycles **11** [1978] 461/5).

[41] D. Leaver (unpublished results from [42]). – [42] W. Flitsch, E. Mukidjam (Chem. Ber. **112** [1979] 3577/88). – [43] T. Ishizu, K. Harano, M. Yasuda, K. Kanematsu (J. Org. Chem. **46** [1981] 3630/4). – [44] W. Flitsch, E.R.F. Gesing (J. Organometal. Chem. **218** [1981] 377/81). – [45] J. Streith, J.M. Cassal (Tetrahedron Letters **1968** 4541/4).

[46] F.A. Cotton (J. Am. Chem. Soc. **90** [1968] 6230/2). — [47] R. Wenzl (Diss. Univ. Köln 1969). — [48] M. Green, S. Tolson, J. Weaver, D.C. Wood, P. Woodward (Chem. Commun. **1971** 222/3). — [49] S.K. Chopra, G. Moran, P. McArdle (J. Organometal. Chem. **214** [1981] C36/C38). — [50] K.J. Karel, T.A. Albright, M. Brookhart (Organometallics **1** [1982] 419/30).

[51] T. Ishizu, K. Harano, N. Hori, M. Yasuda, K. Kanematsu (Tetrahedron **39** [1983] 1281/9).

1.4.1.4.1.2.4.2 ⁴L is an Isocyclic Ring

1.4.1.4.1.2.4.2.1 ⁴L is Cyclohepta-1,3-diene

$C_7H_{10}Fe(CO)_3$ (C_7H_{10}=cyclohepta-1,3-diene) is prepared by the reaction of cyclohepta-1,3-diene with $Fe(CO)_5$ in methylcyclohexane at 160 °C within 36 h [3, 16] in 12% yield [16], or by irradiation of cyclohepta-1,3-diene and $Fe(CO)_5$ [25]. The reaction of $Fe(CO)_5$ with cyclohepta-1,3,5-triene in a tube at 130 °C for 15 h followed by evaporation, distillation at 50 °C in high vacuum, and chromatography on neutral alumina (Woelm, activity II) gives $C_7H_{10}Fe(CO)_3$ in 6.8% yield [5]. Similar reaction of cyclohepta-1,3,5-triene (C_7H_8) with $Fe(CO)_5$ (ratio 1:1) at 135 °C gives $C_7H_8Fe(CO)_3$ and $C_7H_{10}Fe(CO)_3$ in the ratio 19:1 after 1 d and in the ratio 2:1 after 5 d [3, 6, 11]. Similar reaction at 135 to 140 °C for 7 d followed by distillation and chromatography on alumina with light petroleum ether gives $C_7H_{10}Fe(CO)_3$ and $C_7H_8Fe(CO)_3$ in the ratio 1:1 and $(C_7H_8)_2Fe_3(CO)_9$ [4]. Refluxing C_7H_{10} and $Fe(CO)_5$ in $O(C_4H_9)_2$ for 42 h gives $C_7H_{10}Fe(CO)_3$ in 88% yield [30]. Refluxing C_7H_8 and $Fe_3(CO)_{12}$ (molar ratio 3:1) in n-heptane for 20 h followed by distillation at 50 °C/high vacuum and chromatography on neutral alumina (Woelm, activity II) gives $C_7H_{10}Fe(CO)_3$ in 25.8% yield [5]. $C_7H_{10}Fe(CO)_3$ is also prepared by the reaction of $C_6H_5CH{=}CHC(R){=}OFe(CO)_3$ (R=H, CH_3, C_6H_5; Formula I) with cyclohepta-1,3-diene in toluene at 70 °C; for the kinetics of this reaction see [34]. $C_7H_{10}Fe(CO)_3$ is also obtained by the reaction of $C_7H_8Fe(CO)_3$ with H_2/Raney Ni (100 atm, 24 h) at room temperature in C_2H_5OH. The reaction mixture is filtered; H_2O is added to the filtrate which is extracted with light petroleum ether. Drying the organic layer followed by evaporation gives $C_7H_{10}Fe(CO)_3$ in 20% yield [4]. $C_7H_{10}Fe(CO)_3$ is also formed in the reduction of $[C_7H_9Fe(CO)_3]BF_4$ with excess $NaBH_4/H_2O$ after extraction with ether followed by chromatography (23% yield) [10]. Addition of $[C_7H_9Fe(CO)_3]BF_4$ in portions to 10% $NaBH_4/H_2O$ covered with ether at 0 °C [20, 23] for 30 min followed by ether extraction, washing of the organic layer with H_2O, drying ($MgSO_4$), chromatography on silica gel with C_6H_6, and sublimation of the resulting oil [23] gives $C_7H_{10}Fe(CO)_3$ in 30% [23] or 33% [20, 21] and II (R=H) in 70% [23] or 66% [20, 21] of a total yield of ca. 90% [20, 29] or 84% [23]. Similar reaction of $[C_7H_9Fe(CO)_3]BF_4$ with $NaBD_4$ in ice-water/ether for 1 h followed by separation of the organic layer, drying ($MgSO_4$), and evaporation at 15 °C/15 Torr gives II (R=D) and VI (R=D) in the ratio 2:1 (81%) [29]. Reaction of $[C_7H_9Fe(CO)_3]BF_4$ with $NaBH_4$/ice-water gives II (R=H) and $C_7H_{10}Fe(CO)_3$ in the ratio 2:1 [22]. Similar reaction with $NaBH_4$ in CH_2Cl_2/C_2H_5OH (1:1) followed by addition of H_2O, extraction with ether, drying ($MgSO_4$), and chromatography with petroleum ether (b.p. 30 to 40 °C) gives $C_7H_{10}Fe(CO)_3$ in 55% and II (R=H) in 45% yield [23]. Analogous reaction of $[C_7H_9Fe(CO)_3]Br$ with $NaBH_4$ in C_2H_5OH followed by addition of H_2O, extraction with light petroleum ether, drying, and chromatography also gives $C_7H_{10}Fe(CO)_3$ [4]. Reaction of $[C_7H_9Fe(CO)_3]BF_4$ with KCN in acetone/H_2O (1:1) at room temperature followed by work-up as before, but elution with 5% $CH_3CO_2C_2H_5/C_6H_6$ and distillation of the resulting oil, gives a mixture of $C_7H_{10}Fe(CO)_3$ (22%) and II (R=CN, 78%) in a total yield of 53% [23]. Refluxing $[C_7H_9Fe(CO)_3]BF_4$ with $C_2H_5N(C_6H_{11}\text{-cyclo})_2$ in i-C_3H_7OH for 3 h gives mainly $C_7H_8Fe(CO)_3$ and traces of $C_7H_{10}Fe(CO)_3$ [9]. Refluxing $[C_7H_9Fe(CO)_3]BF_4$ in CH_3CN gives III, IV, and traces of $C_7H_{10}Fe(CO)_3$ [33].

References on pp. 36/7

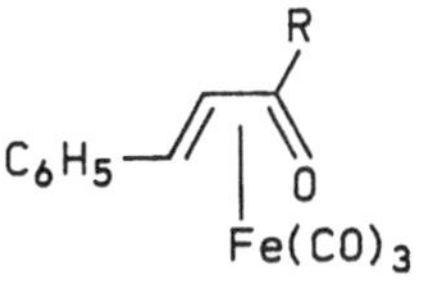

I 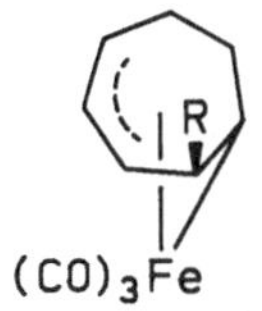 II 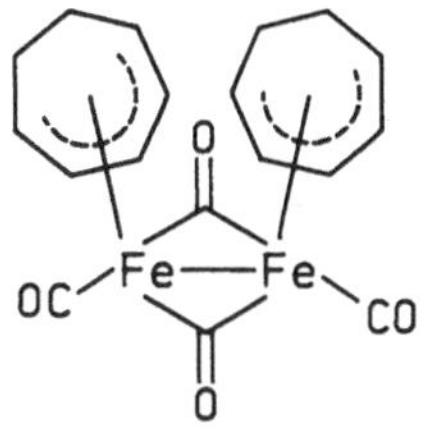 III 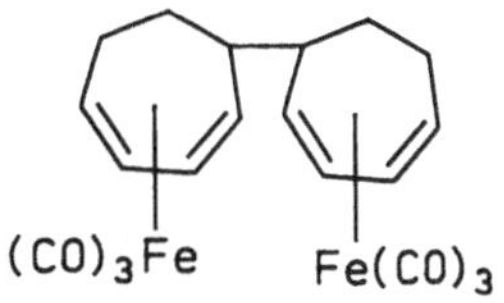 IV

II (R=H) isomerizes upon heating to $C_7H_{10}Fe(CO)_3$ with a half life time of ca. 15 h at 120 °C in C_6H_6 [21]. Refluxing $[C_7H_9(C_6H_8)FeCO]BF_4$ in tetrahydrofuran gives $C_6H_8Fe(CO)_3$ and $C_7H_{10}Fe(CO)_3$ in the ratio of ca. 1:1 [27]. $C_7H_{10}Fe(CO)_3$ can be purified by gas chromatography. No apparent thermal decomposition at temperatures up to 150 °C is observed. Teflon columns employing E-301 methyl silicone gum as stationary phase give best results [8].

$C_7H_{10}Fe(CO)_3$ is an air-sensitive [3, 5], yellow liquid [3] or oil [10] which boils at 50 to 60 °C/high vacuum [5], 60 °C/0.5 Torr [3], or at 65 to 68 °C/0.3 Torr [16], m.p. ca. 5 °C [3], or m.p. 23 °C [4]. The ^1H NMR spectrum [3, 4, 15 to 17, 19, 23, 32, 35] in benzene-d_6 shows chemical shifts (compare V) at δ=1.45 (m, H-6), 1.95 (m, H-5,7), 3.01 (m, H-1,4), 5.24 (m, H-2,3) ppm [23] or at δ=2.66 (H-5,7; ^4J(H-2,7)+^5J(H-3,7)=0.65±0.05, ^5J(H-1,4)=0.44±0.07), 4.75 (H-2,3; ^3J(H-1,2)=7.61±0.08, ^4J(H-1,3)=1.13±0.09, ^3J(H-2,3)=4.83 ±0.08) ppm [16, 17], in CCl_4 at δ=1.45 (H-6), 1.95 (H-5,7), 3.01 (unresolved t, H-1,4; ^3J(H-1,2)=7.6, ^4J(H-1,3)=1.2), 5.24 (H-2,3; ^3J(H-2,3)=4.75) ppm [3, 4] and in $CDCl_3$ at δ=1.25 (m, H-6), 1.83 (m, H-5,7), 2.94 (m, H-1,4), 5.25 (m, H-2,3) ppm [32, 35]. Similar coupling constants with ^3J(H-1,2)=7.69, ^3J(H-2,3)=4.76, ^4J(H-1,3)=1.20 and ^5J(H-1,4)=0.16 Hz are given in [19]. The olefinic H in $C_7H_{10}Fe(CO)_3$ formally represent a four spin system of the type AA'XX' which is disturbed by long range coupling through four bonds [16]. The vicinal coupling constants decrease when compared with the appropriate values in the free organic ligand. The change of the value ^3J(H-1,2) can be explained with the correlation $^3J = -35.10\ R_{\mu,\nu} + 56.65$ by a prolongation of the C-C distances $R_{\mu,\nu}$ in the complex, but this correlation fails for ^3J(H-2,3). The dependence of the ^3J values on the size of the ring, well-known from cyclic olefins, is confirmed [17].

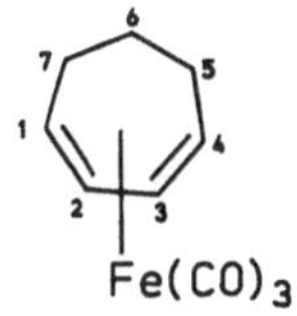

V 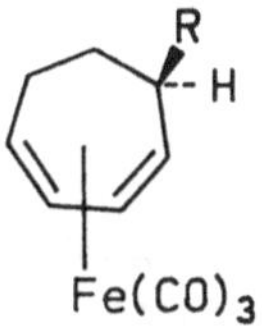 VI 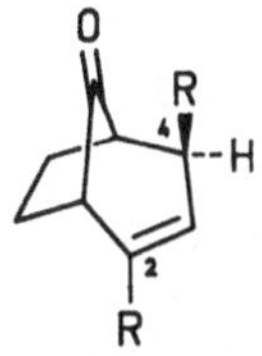 VII 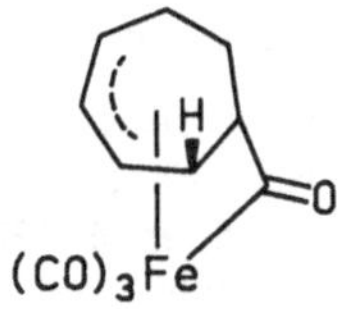 VIII

The activation parameters for the CO-exchange process in $C_7H_{10}Fe(CO)_3$ are determined by temperature-dependent ^{13}C NMR spectroscopy. The coalescence temperature is 219 K, the energy of activation is E_a=42.9 kJ/mol, and log A is 12.9; $\Delta H^{\neq}$=41.0±2 kJ/mol, $\Delta S^{\neq}$= −4±5 $J \cdot mol^{-1} \cdot deg^{-1}$ and $\Delta G^{\neq}_{298}$=42.1±3 kJ/mol [7]. The ^{57}Fe NMR spectrum in C_6H_6 (0.52 M) shows a chemical shift relative to $Fe(CO)_5$ (δ=0) at δ=+86.4±1 ppm [31]. The IR spectrum of the pure liquid [3, 4] shows bands at 790, 803, 843, 860, 868, 933, 958, 1060, 1090, 1163, 1194, 1225, 1262, 1338, 1348, 1372, 1404, 1442, 1447, 1458, 1977, 2046, 2840, 2875, 2918, 2975, and 3026 cm^{-1} [4] and in cyclohexane at 1977, 1981, and 2044 ν(CO) cm^{-1} [23]. The He (I) photoelectron spectrum shows two bands at 7.78 and 8.46 eV

References on pp. 36/7

due to the ionization of electrons largely localized on the Fe atom, one band at 9.12 eV due to an orbital which correlates with the higher filled π-level of the cis-butadiene part ($1a_2$), one band at 10.86 eV which arises through ionization from an MO correlating with the lower filled π-level ($1b_1$) of the free ligand, and at 11.71 eV due to loss of electrons from σ-levels of the C_7H_{10} ligand and from CO ligands [28].

$C_7H_{10}Fe(CO)_3$ is soluble in organic solvents [5], and shows in the mass spectrum $[M-nCO]^+$ (n=0 to 3) [4, 23]. The removal of C_7H_{10} with $ON(CH_3)_3$ is first order with respect to $C_7H_{10}Fe(CO)_3$ and second order with respect to $ON(CH_3)_3$. The activation parameter for the release of the C_7H_{10} ligand $\Delta E^* = 19.0$ kJ/mol correlates with the resonance stabilization energy of the complexed π system [36]. Reaction with $HOSO_2F$ in liquid SO_2 at -80 °C gives $[C_7H_{11}Fe(CO)_3H]^{2+}$, diprotonated at olefin and Fe [13, 14]. Reaction with $[C(C_6H_5)_3]BF_4$ [1, 3, 6, 12, 23] or of a mixture of $C_7H_{10}Fe(CO)_3$ and $C_7H_8Fe(CO)_3$ [3] or of $C_7H_{10}Fe(CO)_3$ and II (R=H) [23] in CH_2Cl_2 gives $HC(C_6H_5)_3$ (93%) [3] and $[C_7H_9Fe(CO)_3]BF_4$ (92% [3]) [1, 3, 6, 12, 23]. Similar reaction of a mixture of $C_7H_8Fe(CO)_3$ and $C_7H_{10}Fe(CO)_3$ with $[C(C_6H_5)_3]ClO_4$ in CH_2Cl_2 gives $[C_7H_9Fe(CO)_3]ClO_4$ [10]. $C_7H_{10}Fe(CO)_3$ does not give CO insertion products with $AlCl_3$ in C_6H_6 at room temperature within 16 h [26] and does not react with $P(C_6H_5)_3$ at reflux in C_6H_6 [9, 18] within 15 h [18] or in n-heptane at 60 °C [24]. But refluxing with PR_3 in $O(C_4H_9)_2$ for 24 h gives $C_7H_{10}Fe(CO)_2PR_3$ (R=OC_6H_5, C_6H_5) [12, 30]. Irradiation with $P(C_6H_5)_3$ in C_6H_6 for 18 h gives $C_7H_{10}Fe(CO)_2P(C_6H_5)_3$ and $C_7H_{10}Fe(P(C_6H_5)_3)_2CO$ [9]. Similar irradiation with $P(OCH_2)_3CC_2H_5$ in C_6H_6 at room temperature for 4 to 15 h gives $C_7H_{10}Fe(CO)_2P(OCH_2)_3CC_2H_5$ and $C_7H_{10}Fe(P(OCH_2)_3CC_2H_5)_2CO$ [25]. Reaction of a 2:1 mixture of II (R=H) and $C_7H_{10}Fe(CO)_3$ at 80 °C with CO (80 atm) for 4 h followed by chromatography with petroleum ether gives $C_7H_{10}Fe(CO)_3$, and elution with ether gives VII (R=H). Analogous reaction of a 2:1 mixture of II (R=D) and VI (R=D) with CO gives VII with equal amounts of deuterium in R-2 and R-4 exo position. Similar reaction of a mixture of II and $C_7H_{10}Fe(CO)_3$ in a 30% pentane solution with CO (80 atm) at 25 °C gives VIII [29]. Reaction with $NaN(Si(CH_3)_3)_2$ in C_6H_6 at 110 °C gives $Na[C_7H_{10}Fe(CO)_2CN]$ and $O(Si(CH_3)_3)_2$ [32, 35].

References:

[1] R. Burton, L. Pratt, G. Wilkinson (unpublished results from [2]). – [2] G. Winkhaus, G. Wilkinson (Proc. Chem. Soc. **1960** 311/2). – [3] H.J. Dauben, D.J. Bertelli (J. Am. Chem. Soc. **83** [1961] 497/8). – [4] R. Burton, L. Pratt, G. Wilkinson (J. Chem. Soc. **1961** 594/602). – [5] W. Fröhlich (Diss. Univ. München 1961).

[6] D.J. Bertelli (Diss. Univ. Washington 1961; Diss. Abstr. **22** [1962] 3850). – [7] P. Bischofsberger, H.-J. Hansen (Helv. Chim. Acta **65** [1982] 721/5). – [8] E.J. Forbes, M.K. Sultan, P.C. Uden (Anal. Letters **5** [1972] 927/30). – [9] F.M. Chaudhari, P.L. Pauson (J. Organometal. Chem. **5** [1966] 73/8). – [10] M.A. Hashmi, J.D. Munro, P.L. Pauson, J.M. Williamson (J. Chem. Soc. A **1967** 240/3).

[11] D.A. White (Organometal. Chem. Rev. A **3** [1968] 497/554, 516/7). – [12] A.J. Pearson (Pure Appl. Chem. **55** [1983] 1767/79). – [13] D.A.T. Young, J.R. Holmes, H.D. Kaesz (J. Am. Chem. Soc. **91** [1969] 6968/77). – [14] D.A. Young (Diss. Univ. California 1969; Diss. Abstr. Intern. B **30** [1970] 3540/1). – [15] H. Günther, R. Wenzl (Ann. Meeting German Chem. Soc., Hamburg 1969, p. 167).

[16] R. Wenzl (Diss. Univ. Köln 1969, pp. 1/136). – [17] H. Günther, R. Wenzl (Angew. Chem. **81** [1969] 919/20). – [18] J.A.S. Howell (unpublished results from [24]). – [19] P. Crews (J. Am. Chem. Soc. **95** [1973] 636/8). – [20] R. Aumann (unpublished results from [21]).

[21] R. Aumann (J. Organometal. Chem. **47** [1973] C29/C32). – [22] M. Sternberg (Dipl.-Arbeit Univ. Münster 1974 from [29]). – [23] R. Edwards, J.A.S. Howell, B.F.G. Johnson, J. Lewis (J. Chem. Soc. Dalton Trans. **1974** 2105/12). – [24] B.F.G. Johnson, J. Lewis, M.V. Twigg (J. Chem. Soc. Dalton Trans. **1974** 2546/9). – [25] T.H. Whitesides, R.A. Budnik (Inorg. Chem. **14** [1975] 664/76).

[26] B.F.G. Johnson, J. Lewis, D.J. Thompson, B. Heil (J. Chem. Soc. Dalton Trans. **1975** 567/71). – [27] B.F.G. Johnson, J. Lewis, I.E. Ryder, M.V. Twigg (J. Chem. Soc. Dalton Trans. **1976** 421/5). – [28] J.C. Green, P. Powell, J. van Tilborg (J. Chem. Soc. Dalton Trans. **1976** 1974/6). – [29] R. Aumann, J. Knecht (Chem. Ber. **109** [1976] 174/9). – [30] A.J. Pearson, S.L. Kole, B. Chen (J. Am. Chem. Soc. **105** [1983] 4483/4).

[31] T. Jenny, W. von Philipsborn, J. Kronenbitter, A. Schwenk (J. Organometal. Chem. **205** [1981] 211/22). – [32] M. Moll, H. Behrens, W. Popp (Z. Anorg. Allgem. Chem. **458** [1979] 202/10). – [33] B.R. Reddy, J.S. McKennis (J. Organometal. Chem. **182** [1979] C61/C64). – [34] P.M. Burkinshaw, D.T. Dixon, J.A.S. Howell (J. Chem. Soc. Dalton Trans. **1980** 999/1004). – [35] H. Behrens, M. Moll, H.-J. Seibold, E. Sepp, P. Würstl (J. Organometal. Chem. **192** [1980] 389/98).

[36] K.H. Pannell, A.J. Mayr, W.C. Herndon, D. Lin, N. Chiocca (10th Intern. Conf. Organometal. Chem., Toronto 1981, Abstr., p. 211).

1.4.1.4.1.2.4.2.2 ^{4}L is a Substituted Cyclohepta-1,3-diene

The compounds described in this section are of the type $^4LFe(CO)_3$ where ^{4}L is a substituted cyclohepta-1,3-diene. Table 2 deals first with the monosubstituted compounds (Nos. 1 to 50, continued at No. 155). The disubstituted compounds (Nos. 51 to 76, continued at Nos. 156 to 159) are arranged as follows. 5,5-substituted complexes (Nos. 51 to 59) are followed by 6,6-substituted complexes (Nos. 60, 61) and by 5,6-substituted complexes (Nos. 62 to 69, 156 to 158). Compounds No. 70 to 76 and 159 are 5,7-substituted. The third part deals with trisubstituted complexes (Nos. 77 to 116, continued at Nos. 160 to 163) arranged as 5,6,7- (Nos. 77 to 82, 160 to 163), 5,5,6- (No. 83), 5,6,6- (Nos. 84 to 92), and 5,5,7- (Nos. 93 to 116) substituted complexes. The fourth part deals with tetrasubstituted complexes (No. 117 to 141, continued at Nos. 164 to 168) arranged as 4,5,6,6- (No. 117), 4,5,5,7- (Nos. 118 to 120), 5,5,6,7- (Nos. 121 to 136, 164, 167, 168), 5,6,6,7- (Nos. 137, 165, 166), 5,5,6,6- (No. 138), and 5,5,7,7-complexes (Nos. 139 to 141). The last parts describe pentasubstituted (Nos. 142 to 148, 169 to 172), hexasubstituted (Nos. 149 to 151), and four hepta-, octa-, nona-, and decasubstituted complexes (Nos. 152 to 154, 173).

Most of these complexes are prepared by the following methods:

Method I: $Fe(CO)_5$ reacts with the organic ligand ^{4}L:

a. $Fe(CO)_5$ and bicyclo[5.1.0]octa-2,4-diene are irradiated in C_6H_6 for 3 h. Chromatography on basic alumina (grade 1.5) with hexane gives compound No. 63 [44].

b. $Fe(CO)_5$ and 7H-benzocycloheptene or 5-methyl-7H-benzocycloheptene react at 120 °C for 72 h to give Nos. 138 or 139. Reaction of 9H-benzocycloheptene at 120 °C for 36 h gives No. 138, too. Hexane is added to the reaction mixture followed by filtration. The compounds are purified by sublimation [14].

Method II: $Fe_2(CO)_9$ and the appropriate organic substrate react thermally in an inert organic solvent. The products are purified by chromatography.

References on pp. 99/102

a. $Fe_2(CO)_9$ and 5-anilinocyclohepta-1,3-diene (^4L) are stirred in hexane at 50 °C for 2 h. The residue, after filtration and evaporation of the solvent, is chromatographed on neutral alumina. Petroleum ether elutes No. 16b. A mixture of the isomers No. 33 and 16a elutes with 15% CH_2Cl_2/petroleum ether [78]. Refluxing excess $Fe_2(CO)_9$ with bicyclo[4.2.1]nonatriene for 10 h gives No. 74. The latter compound No. 74 is also obtained by refluxing excess $Fe_2(CO)_9$ with tricyclo[6.1.0.0^{4,9}]nona-2,6-diene (see Formula I) in hexane for 10 h. Chromatography gives II, III, and a small amount of No. 74 [45].

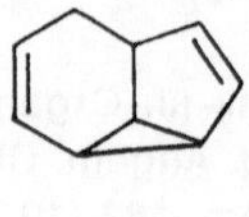

(CO)$_3$Fe—Fe(CO)$_3$ Fe(CO)$_3$ N–C_6H_5 O

I 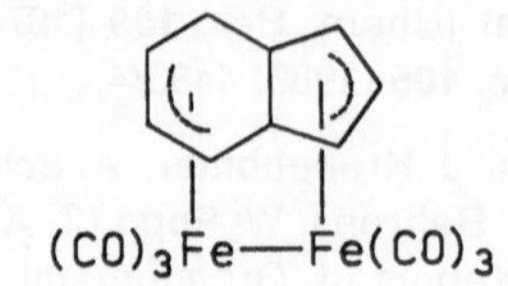 II 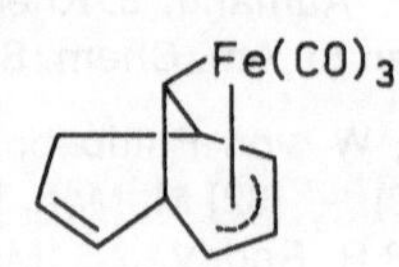 III 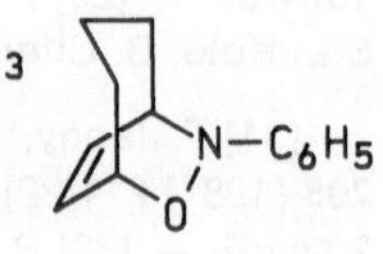 IV

b. $Fe_2(CO)_9$ and cis-6-anilinocyclohepta-2,4-dienol (^4L) react in C_6H_6 at 40 °C for 10 h. Chromatographic work-up gives No. 68. Similar reaction of $Fe_2(CO)_9$ with IV in C_6H_6 at 50 °C for 1 h, followed by evaporation and chromatography of the residue on basic alumina with CH_2Cl_2/petroleum ether (1:9) elutes No. 16b. Further washing with 15% CH_2Cl_2/petroleum ether results in No. 33 and further elution with CH_2Cl_2 gives V (30%). Analogous reaction with VI followed by filtration and partial removal of the solvent separates VII in 31% yield. The filtrate from VII is reduced to a small volume and chromatographed on basic alumina. CH_2Cl_2/petroleum ether (1:4) elutes an oil which is a mixture. Preparative thin layer chromatography of this oil on silica with C_6H_6 as eluent gives No. 73 as the most polar compound followed by an additional component which cannot be purified or identified. Further elution of the basic alumina column with CH_2Cl_2/petroleum ether (1:1) gives an oil which on re-chromatography (basic alumina, 30% CH_2Cl_2/petroleum ether) gives No. 97b. Further development with CH_2Cl_2/petroleum ether (1:1) gives No. 70 [78]. Similarly, $Fe_2(CO)_9$ reacts with bicyclo[4.2.1]nona-2,4-dien-7-one in C_6H_6 at 60 °C for 2 h. Filtration and concentration followed by chromatography on silica gel with C_6H_6 give a yellow main fraction containing No. 75a and b in the ratio 2:1. Re-chromatography on silica gel (Merck Type 60) with C_6H_6 separates the mixture of isomers to give No. 75a and b in pure form [77]. Refluxing $Fe_2(CO)_9$ and VIII or IX in C_6H_6 for 24 h, followed by filtration, evaporation of the filtrate, and chromatography on acidic silica gel with pentane gives No. 150. Analogous reaction with X or XI gives No. 149 [83].

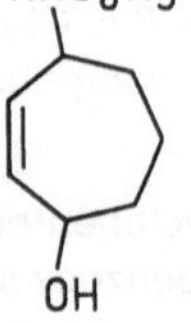

HNC$_6$H$_5$ OH C$_6$H$_5$–N O N–C$_6$H$_5$ O O CH$_3$ CH$_3$ CH$_3$ CH$_3$

V 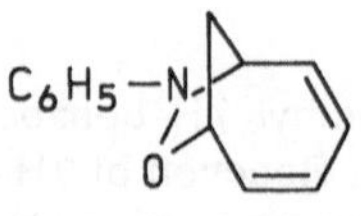 VI 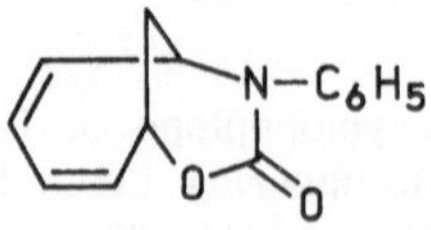 VII 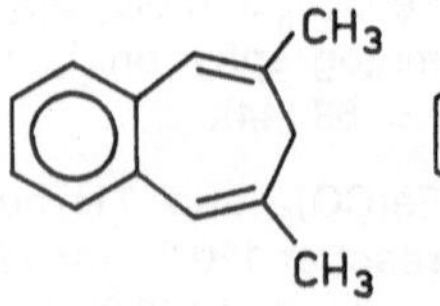 VIII 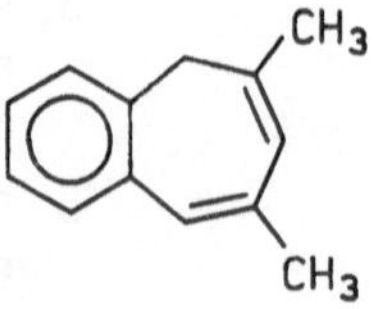 IX

c. $Fe_2(CO)_9$ and bicyclo[5.1.0]octa-2,4-diene (^4L) react in ether at 25 °C for 48 h, followed by filtration and evaporation of the filtrate. Chromatography of the residue on silica gel with petroleum ether gives No. 63 before

$Fe_3(CO)_{12}$, and elution with ether gives No. 75a [77], cf. [66, 67]. Analogous reaction of XII with R=H (^4L) followed by work-up as before but dissolving the residue in CH_2Cl_2, filtration (from $Fe_3(CO)_{12}$), evaporation, and chromatography of this residue on silica gel with petroleum ether/ether (5:1) gives No. 123a and b [77]. Refluxing $Fe_2(CO)_9$ with 8,8-dichlorobicyclo[5.1.0]octa-2,4-diene (^4L) in ether for 3 h gives No. 65, identified by thin layer chromatography of the crude mixture [84].

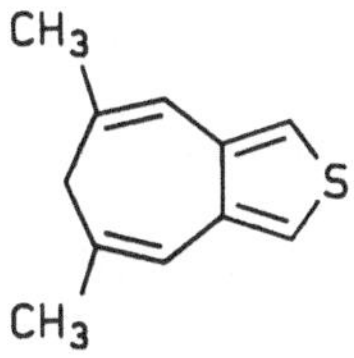

X 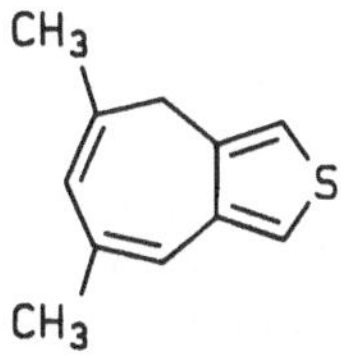 XI 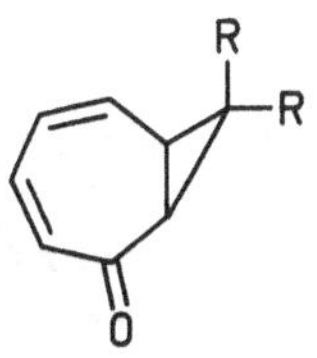 XII 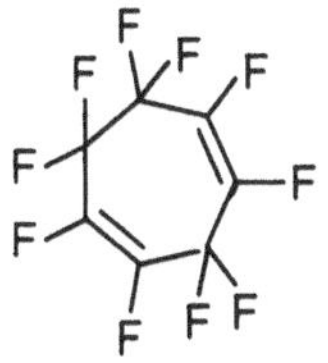 XIII

Method III: $Fe_3(CO)_{12}$ and the organic ligand ^4L are heated in an inert solvent (or without a solvent).

a. $Fe_3(CO)_{12}$ and the appropriate fluorocycloheptatriene ^4L are heated in a sealed tube at 130 °C for 15 h to give Nos. 152 to 154. Attempted reaction of XIII with $Fe_3(CO)_{12}$ gives only trace amounts of No. 154 [51].

b. $Fe_3(CO)_{12}$ and excess 8,8-dibromobicyclo[5.1.0]octa-2,4-diene are refluxed in n-heptane. The reaction mixture is filtered, concentrated, and chromatographed on silica gel with pentane to give No. 66 [32], cf. [23, 24]. Refluxing $Fe_3(CO)_{12}$ and β-tropolone (^4L) in C_6H_6/C_2H_5OH (20:1) for 45 min followed by work-up as before but elution with $CH_3CO_2C_2H_5$ gives No. 139 [15].

Method IV: $^3LFe(CO)_3$ (^3L=benzylideneacetone) and 8,8-dichlorobicyclo[5.1.0]octa-2,4-diene (^4L) are refluxed for 3.5 h. Chromatography on silica gel with light petroleum ether (b.p. 40 to 60 °C)/C_6H_6 (1:4) gives an oil containing No. 65 and ^4L. Distillation at 70 °C/0.05 Torr gives ^4L. The residual oil is identified as No. 65 [84].

Method V: $^4LFe(CO)_3$ (^4L is cycloheptatriene or its derivative) reacts with an organic substrate in an inert organic solvent. The products are purified by chromatography.

a. $C_7H_8Fe(CO)_3$ (C_7H_8=cycloheptatriene) is reduced with CH_2I_2/Cu/Zn in boiling ether in the presence of one crystal of I_2 within 18 h. Chromatography on a column consisting of 75% Woelm and 25% Alcoa alumina with hexane as eluent gives No. 63 [58, 86, 91].

b. $C_7H_8Fe(CO)_3$ (C_7H_8=cycloheptatriene) is added to a suspension of KOC_4H_9-t in heptane, and the mixture is cooled in ice. A mixture of $CHCl_3$/hexane (6:10) is added dropwise with stirring over 0.5 h. After stirring at room temperature for 1 h, dilute aqueous HCl is added, the organic layer is separated, washed with H_2O, and dried ($MgSO_4$). Distillation gives No. 65. Similar addition of $C_7H_8Fe(CO)_3$ in hexane to a solution of KOC_4H_9-t in HOC_4H_9-t gives a dark red solution. Addition of $CHCl_3$/hexane to this mixture at 0 °C gives No. 65 (38%) identified by thin layer chromatography [84]. Compound No. 65 is also obtained by the reaction of a mixture of 50% aqueous NaOH in C_6H_6, $[C_6H_5CH_2N(C_2H_5)_2]Cl$, and $C_7H_8Fe(CO)_3$ with $CHCl_3$. After 30 min, the

References on pp. 99/102

reaction mixture is poured into H_2O and extracted with CH_2Cl_2. The extract is dried (Na_2SO_4) and evaporated. Chromatography on silica gel with hexane gives No. 65 in 56% yield [129].

c. $C_7H_8Fe(CO)_3$ (C_7H_8=cycloheptatriene) reacts with $O{=}C{=}C(C_6H_5)_2$ in C_6H_6 at room temperature. Chromatography gives No. 67 [105], cf. [126]. Similar reaction with XIV in C_6H_6 for 24 h, evaporation, and chromatography of the residue on silica gel with n-hexane gives No. 68. Reaction with XV in C_6H_6 in a sealed tube at 80 °C for 30 h followed by work-up as before but elution with n-hexane/$CH_3CO_2C_2H_5$ (3:1) gives No. 69. Analogous reaction with XVI in CH_2Cl_2 at room temperature for 15 min followed by work-up as before but elution with $CHCl_3$ gives No. 84 [109]. Irradiation of $C_7H_8Fe(CO)_3$ and $CH_3O_2CC{\equiv}CCO_2CH_3$ (1:1) in tetrahydrofuran at 0 °C gives No. 76 [53, 55, 113]. Compound LXXXII (p. 98) reacts with $(C_6H_5)_2C{=}C{=}O$ in C_6H_6 at room temperature for one week. The solvent is removed, and the residue is treated with hexane and filtered. Slow evaporation of the hexane gives No. 171 [126].

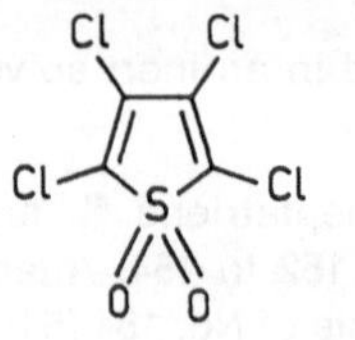

XIV 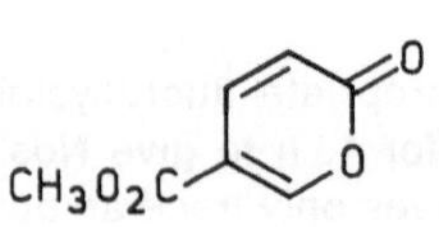 XV 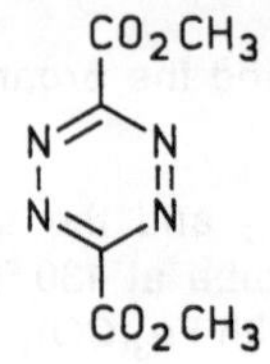 XVI 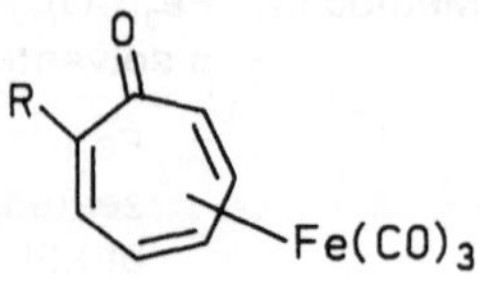 XVII

d. XVII (R=H) reacts with XIV in CH_2Cl_2 at room temperature for 2 d. After removal of the solvent, the residue is chromatographed on silica gel with n-hexane/$CH_3CO_2C_2H_5$ (20:1) to give No. 136. Similar reaction with XVI in CH_2Cl_2 overnight followed by filtration from the crystals and chromatography of the mother liquor as above, but with n-hexane/$CH_3CO_2C_2H_5$ (2:1) and combining both, gives No. 144 [109]. Similar reaction with the dimer of 2,5-dimethyl-3,4-diphenylcyclopenta-2,4-dienone in C_6H_6 at 95 °C for 50 h gives a mixture of No. 135 and XVIII (14%) [72]. XVII (R=H) reacts with benzonitrile oxide within 3 h or with 2,4,6-trimethylbenzonitrile oxide within 50 h at room temperature to give Nos. 126 and 127 [72]. Reaction with $C_6H_5CNNC_6H_5$ (prepared in situ from $C_6H_5CCl{=}NNHC_6H_5$ with 20% excess $N(C_2H_5)_3$) in C_6H_6 in the dark at room temperature for 5 d followed by chromatography gives a mixture of complexes Nos. 131 and 132 in 90% yield, No. 132 being the major product [71]. Reaction of XVII (R=H) with excess CH_2N_2 in ether in a refrigerator for 60 h followed by filtration through Celite, concentration of the filtrate, and chromatography on silica gel with $CH_3CO_2C_2H_5$/CH_2Cl_2 (1:9) gives No. 128 and two other nonisolated products [69]. Similar reaction with R_2CN_2 (R=H or CH_3) in ether at 20 °C gives Nos. 128 and 130, respectively. Analogous reaction with CH_3CHN_2 gives No. 129a and b (ratio 1:2.7) in

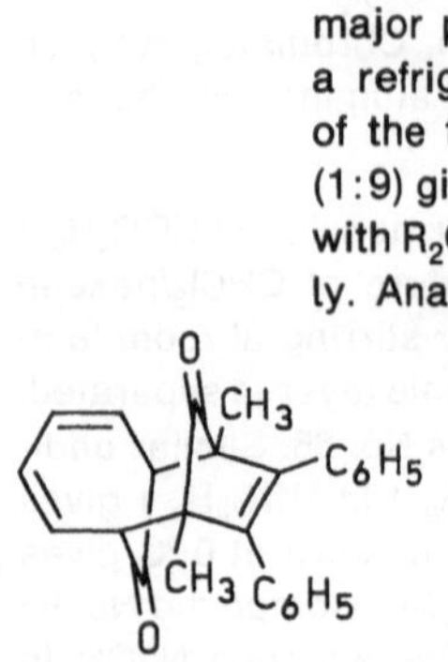

XVIII 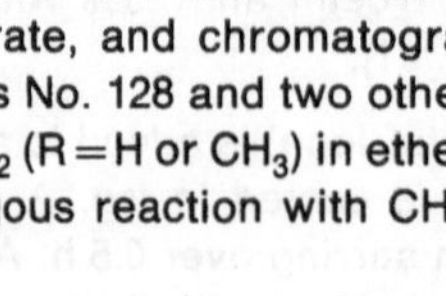 XIX 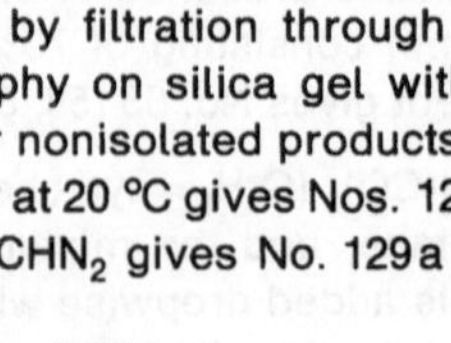 XX 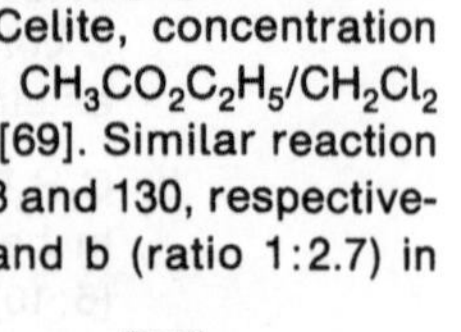 XXI

a total yield of 75% which are difficult to separate by chromatography on silica gel [62]. When XVII (R=H or Cl) is treated with o-quinodimethane (see Formula LXXXIII, p. 98) at room temperature, a smooth reaction occurs to give Nos. 164 or 170, exclusively [127].

e. XVII (R=Cl) is treated with $HOSO_2F$ in CH_2Cl_2 at −78 °C for 15 min. The reaction mixture is poured quickly into CH_3OH/Na_2CO_3 suspension at −78 °C. After 10 min, the resulting mixture is diluted with H_2O and after evolution of CO_2 ceased, it is extracted with CH_2Cl_2. The combined extracts are washed with H_2O followed by saturated aqueous NaCl, dried ($MgSO_4$), and evaporated. The resulting residue is chromatographed on silica to give Nos. 121 and 118. Quenching the $HOSO_2F$ solution of XVII (R=CH_3) in CH_3OH at −78 °C gives an inseparable 1:1 mixture of the two isomers No. 122 and 119. Similar reaction of XVII (R=C_6H_5) gives No. 120 [63]. XVII (R=$COCH_3$) reacts readily even below 0 °C with an excess of $(CH_3)_2CN_2$ in ether. Cooling the solution to −78 °C yields No. 143 [82]. XIX reacts with XVI in CH_2Cl_2 at room temperature overnight. Crystals which separated out are filtered to give No. 145 [109]. Irradiation of XX and $CH_3O_2CC{\equiv}CCO_2CH_3$ in tetrahydrofuran at 0 °C gives No. 137 and XXI (5%) [53, 55].

f. XXII (R=CO_2CH_3, $CO_2C_2H_5$, CN, or C_6H_5) reacts with N-phenyltriazolinedione in CH_2Cl_2 at room temperature. The solution is treated with activated charcoal, filtered, and concentrated to give the compounds No. 85 to 88. Similar reaction with $(NC)_2C{=}C(CN)_2$ in CH_2Cl_2 at room temperature for 1 h, removal of the solvent and recrystallization gives the compounds No. 89 to 92 [87].

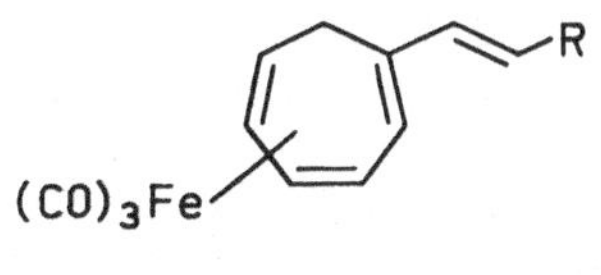

XXII

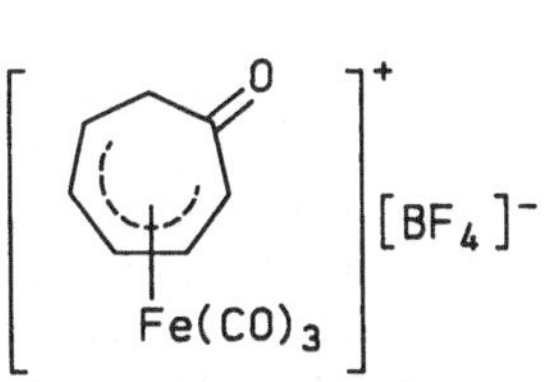

XXIII

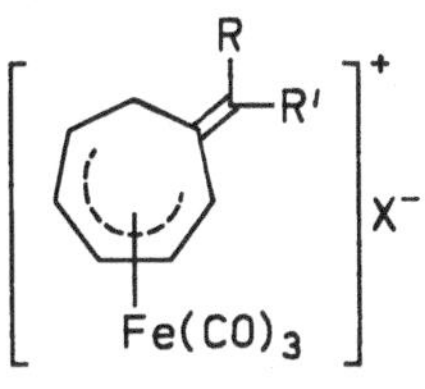

XXIV

Method VI: $[C_7H_9Fe(CO)_3]BF_4$ reacts with various nucleophiles.

a. $[C_7H_9Fe(CO)_3]BF_4$ is treated with $[NH_4]NCS$ in CH_2Cl_2/H_2O for 20 min. The CH_2Cl_2 layer is separated, washed with H_2O, dried ($MgSO_4$), filtered, and evaporated. The residue is dissolved in deoxygenated n-pentane and crystallized at −78 °C to give No. 11. Similar reaction with $SeCN^-$ gives No. 12 [60]. Similar reaction with KCN in acetone at room temperature for 12 h, followed by chromatography on neutral alumina with ligroin (b.p. 40 to 60 °C), gives an unstable oil, and elution with C_6H_6 gives No. 31 [13]. Reaction with aqueous NaN_3 [79] or NaN_3 in CH_2Cl_2 for 30 min [68] gives No. 10 [68, 79]. When $[C_7H_9Fe(CO)_3]BF_4$ is dissolved in H_2O, no apparent reaction occurs, but addition of $NaHCO_3$ causes a vigorous evolution of CO_2 and an immediate separation of No. 1 [79, 112]. Reaction of $[C_7H_9Fe(CO)_3]^+$ with aqueous KOH in CH_3NO_2 gives $C_7H_9Fe(CO)_2CO_2H$ together with No. 1 as final product [125]. Compound No. 1 is also prepared by the reaction of $[C_7H_9Fe(CO)_3]BF_4$ with potassium azodicarboxylate in H_2O/C_6H_6 (1:1) at room temperature for 3 h followed by evaporation [18].

References on pp. 99/102

b. Refluxing $[C_7H_9Fe(CO)_3]BF_4$ in CH_3OH for 23 h gives No. 2b in good yield [57]. Reaction with $NaOCH_3/H_2O$ at 5 °C [79] or with $NaOC_2H_5/C_2H_5OH$ [13] or with $NaOC_6H_5/H_2O$ [79] at room temperature for 1 h [13, 79] gives Nos. 2a, 3, and 5. For isolation of No. 3, the reaction mixture is evaporated, and the residue is extracted with ligroin; the oily residue is sublimed [13], cf. [25, 68]. The reaction mixture containing No. 5 is poured into H_2O and extracted with ether. The extracts are washed with dilute NaOH solution, dried ($MgSO_4$), and condensed. The residue is chromatographed over deactivated alumina with petroleum ether to give No. 5 [79]. Generally, the reaction of $[C_7H_9Fe(CO)_3]BF_4$ with a slight excess of NaOR/HOR (R = CH_3, C_2H_5, C_3H_7-i) gives an initial red color and rapid formation of the corresponding $C_7H_9Fe(CO)_2$-CO_2R complexes. However, these complexes are not thermodynamically stable, and when the temperature is raised, they rearrange spontaneously both in the solid state and in solution to give the corresponding 5-exo-$ROC_7H_9Fe(CO)_3$ compounds (Nos. 2a, 3, 4) [61, 90], cf. [25, 57]. Similar reaction with $NaSC_6H_5/H_2O$ at 5 °C gives No. 7 [79], and reaction with $NaSC_6F_5$ in H_2O/C_6H_6 at room temperature for 24 h separates a yellow organic phase, which is washed, dried, and evaporated to give No. 8 [19]. Similar reaction with acetylacetonate or dimedone in CH_3CN gives Nos. 27 and 30 [30], and reaction with $NaHC(CO_2C_2H_5)_2$ in C_2H_5OH followed by chromatography on alumina gives No. 28 [13, also 79]. Analogous reaction with $NaHC(CO_2CH_3)_2$ gives No. 155 [124].

c. A slight excess of $[C_7H_9Fe(CO)_3]BF_4$ reacts with the appropriate primary and secondary aliphatic and aromatic amines (RNH_2 or R_2NH) or pyridine in CH_2Cl_2 at room temperature for 20 min in the case of RNH_2 with R = n-C_3H_7, n-C_4H_9, t-C_4H_9, and R_2NH with R = C_2H_5, R_2 = C_4H_8, or $CH_3(C_6H_5)NH$ or for 30 min in the case of $C_5H_{10}NH$ or pyridine. After the removal of solvent and excess amine, extraction with dry pentane, and evaporation of the solvent, compounds No. 13 to 15, 22 to 24, and 35 to 41 are obtained. In the case of aniline and N-methylaniline much larger ratios (e.g., 1:7) are required to obtain Nos. 16a and 23. In the case of N-methylaniline chromatography of the reaction mixture gives both No. 23 and $C_7H_8Fe(CO)_3$ [61]. The amine t-$C_4H_9NH_2$ reacts for 2 h. The reaction mixture is poured into H_2O and the organic layer is separated and dried ($MgSO_4$). After evaporation, distillation of the residue gives No. 15 [79]. Compound No. 21 is prepared by the reaction of $[C_7H_9Fe(CO)_3]BF_4$ with 25% aqueous $HN(CH_3)_2$ within a few minutes. The reaction mixture is extracted with ligroin, dried (Na_2SO_4), and evaporated. Chromatography on alumina with ligroin/C_6H_6 (4:1) gives No. 21 [8, 79]. Similar reaction with $H_2NC_6H_4R$ (R = OCH_3-2, OCH_3-4, CH_3-2, CH_3-4) [93, 98, 99] or with pyridine or its methyl derivatives [93, 96] in CH_3CN [93, 96, 98, 99] for 10 min [96] gives Nos. 17 to 20 and 42 to 44. In the case of $H_2NC_6H_4CH_3$-2 a tenfold excess of the amine [98] at 10.0 °C [93] is employed, and $H_2NC_6H_4CH_3$-4 reacts at 0.0 °C [93, 98] or 20 °C [99]. For No. 44 a twofold excess of 2,4,6-trimethylpyridine is employed, and both Nos. 43 and 44 can also be prepared in CD_3NO_2 [96]. The reaction mixture for compounds No. 42 to 44 is evaporated yielding an oily solid. Addition of light petroleum ether (b.p. 60 to 80 °C) gives the crystalline product [96], cf. [93]. For the kinetics of these reactions see [93, 96, 98, 99].

d. $[C_7H_9Fe(CO)_3]BF_4$ reacts with PR_3 (ratio 1:1) in CH_2Cl_2 at room temperature for 30 min. The solvent volume is reduced followed by dropwise addition to pentane solution to give the 5-exo isomers No. 45 to 50 [33, 61], cf. [68, 128]. No. 50 is isolated as the CH_2Cl_2 adduct [85]. The 5-endo isomers No. 45 to 50 are obtained by covering $[C_7H_9Fe(CO)_3]BF_4$ (1.41 mmol) with PR_3 (1.69 mmol) and CH_3CN. After 30 min the solvent is removed in vacuum, and the residue is dissolved in CH_3CN and filtered. Dropwise addition of ether gives the desired 5-endo isomers No. 45 to 50 [61]. For the kinetics of the reaction of $[C_7H_9Fe(CO)_3]BF_4$ with $P(C_6H_5)_3$ in CH_3NO_2 at 20 °C yielding No. 50 see [101], cf. [6, 104].

e. $[C_7H_9Fe(CO)_3]BF_4$ reacts with 10% aqueous $KHCO_3$/ether. The ethereal layer is concentrated, dissolved in pyridine, and oxidized with CrO_3/pyridine at room temperature for 15 h, followed by dilution with $CH_3CO_2C_2H_5$ and filtration through alumina, washing with 2N HCl, then H_2O, and drying ($MgSO_4$). Sublimation gives No. 55 [37].

Method VII: $[^5LFe(CO)_3]X$ (^5L = cycloheptadienyl, X = BF_4, PF_6) reacts with nucleophiles.

a. $[1\text{-}CH_3O_2CCH_2C_7H_8Fe(CO)_3]BF_4$ is warmed in H_2O. The reaction mixture is extracted with ether, the organic layer is dried ($MgSO_4$), evaporated, and recrystallized to give No. 57 [37]. The reaction of $[5\text{-}CH_3COC_7H_8Fe(CO)_3]PF_6$ with $NaHCO_3$ (pH = 4) followed by work-up as before gives No. 71. Similar reaction of $[5\text{-}C_6H_5COC_7H_8Fe(CO)_3]BF_4$ with $NaOCH_3/CH_3OH$ followed by work-up as before but chromatography on silica gel with 10% $CH_3CO_2C_2H_5$/toluene gives No. 72 [42].

b. XXIII reacts with aqueous NaCN covered with an ether layer at 0 °C for 30 min. The reaction mixture is extracted with ether; the organic layer is washed with H_2O, dried ($MgSO_4$), and evaporated. Chromatography on silica gel with CH_2Cl_2 gives No. 98. Further elution with CH_2Cl_2 gives XXXIV (p. 85) [70]. Similar reaction with aqueous NaN_3/ether [27, 70] at room temperature for 20 min followed by work-up as before but without chromatography gives No. 95 [70]. Analogous reaction with excess $t\text{-}C_4H_9NH_2$ or $C_6H_5NH_2$ gives Nos. 96 and 97a [70]. Reaction of XXIII with Na_2CO_3/CH_3OH at 0 °C gives No. 94 [27].

c. XXIV (R = R′ = H, X = PF_6) reacts with $HOCH_3$ for 20 min. The reaction mixture is poured into H_2O, extracted with ether, and the organic layer is dried ($MgSO_4$). Sublimation gives No. 107 [42], cf. [20]. Similar reaction of XXIV (R = H; R′ = CH_3, C_6H_5, $CH_2C_6H_5$; X = PF_6) gives Nos. 108, 110 [42], and 109 [29]. XXIV (R = R′ = H) reacts with aqueous NaCN in ether at room temperature for 20 min. Similar work-up as before followed by chromatography on silica gel with CH_2Cl_2 gives XXV, No. 116, and minor amounts of a dimeric complex [70]. Similar reaction of XXIV (R = R′ = H, X = BF_4 or PF_6) with $NaOCH_3$ gives No. 107 [20]. Treatment of the appropriate salt XXIV in acetone with a large excess of $NaHCO_3$ gives Nos. 99 to 106. H_2O is added until the $NaHCO_3$ begins to dissolve. Extraction with ether, washing with H_2O, drying ($MgSO_4$) of the organic layer, evaporation, and chromatography on alumina (activity V) with ether gives the pure products No. 99 to 106 [49], cf. [29]. Compound No. 103 is obtained as a mixture of two isomers No. 103a and b (80 to 90% yield) [49]. Reaction of XXIV (R = H, R′ = $C_6H_4CH_3$-4; X = PF_6) with $NaOCH_3/HOCH_3$ followed by work-up as usual (ether extraction, drying ($MgSO_4$)) gives No. 111 [38]. Similar reaction of XXIV (R = CH_3, R′ = CH_3, C_2H_5, $C_6H_4CH_3$-4 or R = R′ = C_6H_5; X = PF_6) with $NaOCH_3/CH_3OH$ gives Nos. 112 [29], 114 [38], 115 [38], or 113 [29]. Treatment of XXIV (R = H, R′ =

References on pp. 99/102

$C_6H_4CH_3$-4) with $NaOCH_3/CH_3OH$ in ether at −50 °C for 20 min followed by the usual work-up as before gives a mixture of No. 146a and b [38].

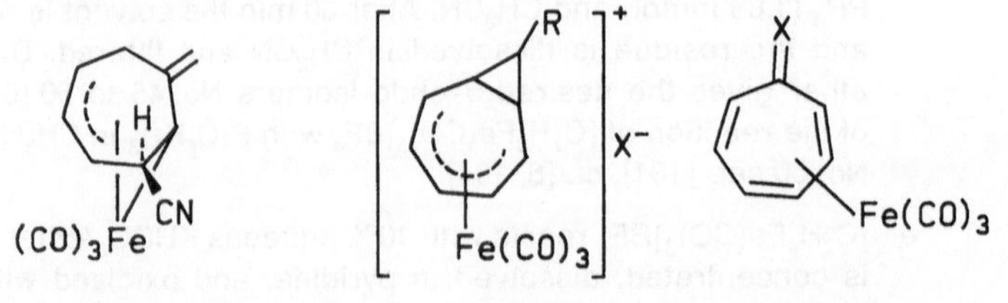

XXV XXVI XXVII

d. XXVI (R = $C_6H_4NO_2$-4, X = PF_6) reacts with excess $NaBH_4$ in tetrahydrofuran at −78 °C for 1 h. The reaction mixture is evaporated, and the residue is dissolved in CH_2Cl_2. Chromatography on silica gel with hexane gives a yellow band which elutes with hexane/ether (2:1). Concentration of the eluate and cooling to −78 °C gives No. 64 [4]. Treatment of XXVI (R = H, X = BF_4) with NaOH in acetone/H_2O [7] or acetone/ether/H_2O (30:30:13) at −50 to −10 °C for 2 to 2.5 h [12] followed by the usual work-up (extraction with ether, washing with H_2O, drying ($MgSO_4$), evaporation) gives No. 77 [12]. Similar reaction of XXVI (R = D, X = PF_6) with pyridine (molar ratio 1:1) in acetone-d_6 at −15 °C gives No. 82 [46]. Treatment of XXVI (R = $COCH_3$, X = PF_6) with $NaOCH_3/CH_3OH$ at room temperature for 10 min followed by pouring the reaction mixture into H_2O and work-up as before gives a mixture of compounds. Chromatography with toluene gives a very small amount of $C_8H_8Fe(CO)_3$, and elution with 10% $CH_3CO_2C_2H_5$/toluene gives No. 78 [31].

Frontier orbital calculations predict that 1,3-addition is the preferred cycloaddition mode for $(NC)_2C$=$C(CN)_2$ with a range of cyclic triene complexes such as $C_7H_8Fe(CO)_3$, XVII (R = H), or XXVII (X = CH_2 or NH) and explains why no 1,6-addition products are actually known (compare Method V) [108]. Throughout this section and volume, protons or substituents which are on the same or opposite side of the ring as the $Fe(CO)_3$ group are referred to as being endo or exo, respectively, as described throughout in most of the original papers. For the alternatives syn and anti or Fe-endo or Fe-exo see [69].

Table 2
Compounds of the Type $^4LFe(CO)_3$ where ^4L is a Substituted Cyclohepta-1,3-diene.
Further information on numbers preceded by an asterisk is given at the end of the table, pp. 82/99.
For abbreviations and dimensions see p. VIII.

No.	$^4LFe(CO)_3$	method of preparation (yield in %), properties and remarks	Ref.
^{4}L is a monosubstituted cyclohepta-1,3-diene (continued with No. 155, p. 78):			
*1	OH $Fe(CO)_3$	VI a (26 to 86) m.p. 55°, subl. p. 60°/0.1 Torr, yellow crystals (from pentane) ^{1}H NMR (CCl_4 or CS_2): 1.70 (H-6), 2.18 (H-7), 2.45 (OH), 2.90 (H-1,4), 4.00 (H-5), 5.28 (H-2,3) IR: 795, 826, 842, 858, 895, 942, 995, 1042, 1068, 1160, 1180, 1195, 1225, 1287, 1325, 1333, 1360, 1435, 1445; ca. 1950, 2015 (CO); 2820, 2865, 2900, 2940 (CH); 3200 (OH) mass spectrum: $[M-nCO]^+$, $[M-nCO-H_2]^+$ (n=0 to 3), $[M-CO-OH]^+$, $[M-CO-H_2O]^+$, $[M-CO-H_2O-C_2H_2]^+$ (n=1 to 3), $[M-nCO-OH-H_2-C_2H_2]^+$, $[M-nCO-H_2O-H_2-C_2H_2]^+$ (n=1,2), $[M-2CO-OH-C_2H_2-2H_2]^+$, $[(C_5H_5)_2Fe]^+$, $[C_nH_6Fe]^+$ (n=5,6), $[C_7H_nOH]^+$ (n=4 to 7, 9), $[C_7H_n]^+$ (n=7 to 9), $[C_6H_n]^+$ (n=5 to 7), $[C_5H_n]^+$ (n=3,5,6), $[C_4H_n]^+$ (n=2 to 5), $[C_3H_3]^+$, $[Fe]^+$	[18, 79, 112, 125]
*2a	OCH_3 $Fe(CO)_3$	VI b (62) b.p. 91°/0.7 Torr, oil ^{1}H NMR (CCl_4 or CS_2): 1.30 (H-6), 2.10 (H-7), 2.92 (H-1,4), 3.25 (CH_3), 3.55 (H-5), 5.31 (H-2,3) IR (hexane): 1980, 2050 (CO)	[25, 57, 61, 79, 90]
b	OCH_3 $Fe(CO)_3$	VI b (good yield)	[57]
*3	OC_2H_5 $Fe(CO)_3$	VI b (61) dec. p. 45°, m.p. 62 to 65° (from ligroin), subl. p. 50°/0.005 Torr, orange crystals ^{1}H NMR ($CHCl_3$): 3.60 (H-5; J(H-5,6endo)=4.4, J(H-5,6exo)=11.5, J(H-4,5)=0), 5.35 (H-2,3)	[13, 25, 61, 68, 90]

References on pp. 99/102

Table 2 [continued]

No.	$^4LFe(CO)_3$	method of preparation (yield in %), properties and remarks	Ref.
		^1H NMR (CS_2): 1.0 to 2.4 (m, H-5,6), 1.1 (t, CH_3), 3.3 (q, OCH_2), 2.6 to 3.8 (m, H-1,4,7), 5.3 (m, H-2,3) IR (KCl): 1961, 2062 (CO) IR (hexane): 1980, 2045 (CO) mass spectrum: $[M-CO]^+$, $[M-CO-H_2]^+$	
*4	$OCH(CH_3)_2$ / $Fe(CO)_3$	VIb oil IR (hexane): 1980, 2048 (CO)	[25, 61]
*5	OC_6H_5 / $Fe(CO)_3$	VIb (54) m.p. 76° (from pentane), yellow crystals ^1H NMR (CCl_4 or CS_2): 1.65 (H-6), 2.20 (H-7), 3.00 (H-1,4), 4.55 (H-5), 5.35 (H-2,3), 7.00 (C_6H_5)	[79]
6	SCN / $Fe(CO)_3$	see No. 11, p. 83 red ^1H NMR ($CDCl_3$): 1.62 (H-6endo, 7endo), 1.98 (H-6exo, 7exo), 3.03 (H-1,4), 4.06 (H-5; J(H-5,6endo) = 4.39, J(H-5,6exo) = 11.30), 5.38 (H-2,3) IR: 1980.0, 2045.6 (CO); 2109.0 (SCN) mass spectrum: similar to that of No. 11	[60]
*7	SC_6H_5 / $Fe(CO)_3$	VIb (57) m.p. 42° ^1H NMR (CCl_4 or CS_2): 1.60 (H-6), 2.05 (H-7), 3.00 (H-1,4), 3.50 (H-5), 5.25 (H-2,3), 7.28 (C_6H_5)	[79]
8	SC_6F_5 / $Fe(CO)_3$	VIb m.p. ca. 5° (from petroleum ether at 0°), unstable yellow crystalline substance IR (KBr): 1972, 2044 (CO) mass spectrum: $[M]^+$; base peak is $[M-3CO]^+$	[19]
9	SeCN / $Fe(CO)_3$	see No. 12, p. 83 dark brown	[60]

Table 2 [continued]

No.	$^4LFe(CO)_3$	method of preparation (yield in %), properties and remarks	Ref.
*10	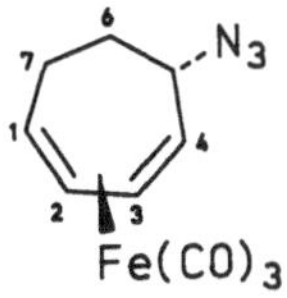 N_3; $Fe(CO)_3$	VIa yellow oil ^1H NMR ($CDCl_3$): 0.7 to 2.2 (m, H-6,7), 2.9 (m, H-1,4), 3.6 (m, H-5), 5.3 (m, H-2,3) IR ($CHCl_3$): 1975, 2049 (CO); 2085 ($C-N_3$) the compound possibly shows fluxional behavior	[68]
*11	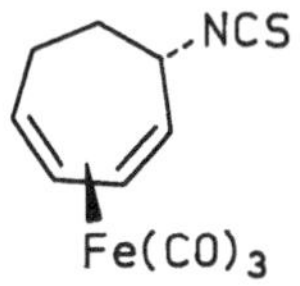 NCS; $Fe(CO)_3$	VIa yellow ^1H NMR ($CDCl_3$): 1.69 (H-6endo, 7endo), 2.04 (H-6exo, 7exo), 3.04 (H-1,4), 4.06 (H-5; J(H-5,6endo) = 4.39, J(H-5,6exo) = 11.30), 5.45 (H-2,3) IR: 1981.0, 2044.0 (CO); 2130.0 (NCS) mass spectrum: similar to that of No. 6	[60]
*12	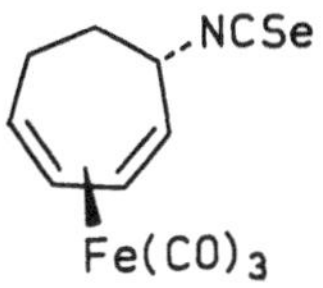 NCSe; $Fe(CO)_3$	VIa yellow	[60]
13	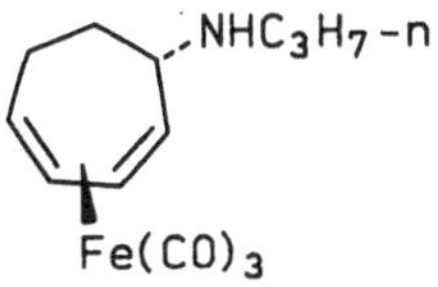 NHC_3H_7-n; $Fe(CO)_3$	VIc yellow oil IR (hexane): 1977, 2045 (CO) IR (CCl_4): 3456 (NH)	[61]
14	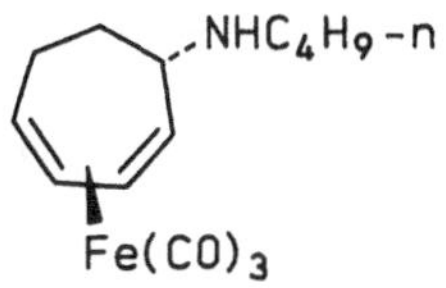 NHC_4H_9-n; $Fe(CO)_3$	VIc yellow oil IR (hexane): 1977, 2043 (CO) IR (CCl_4): 3458 (NH)	[61]
*15	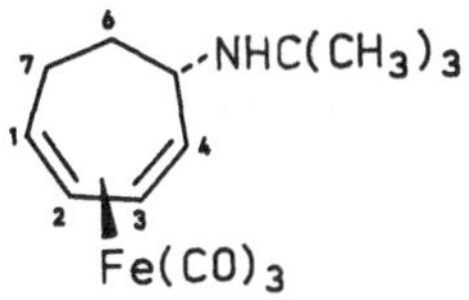 $NHC(CH_3)_3$; $Fe(CO)_3$	VIc (54) b.p. 103°/0.5 Torr, yellow liquid or oil ^1H NMR (CCl_4 or CS_2): 0.70 and 1.15 (C_4H_9), 1.33 (H-6), 2.18 (H-7), 3.00 (H-1,4,5), 5.30 (H-2,3) IR (hexane): 1973, 2043 (CO) IR (pure liquid): 3446 (NH)	[61, 79]
*16a	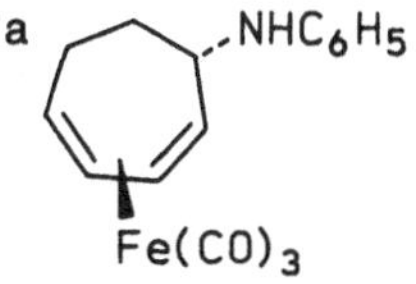 NHC_6H_5; $Fe(CO)_3$	IIa (18), VIc m.p. 88 to 89° (from hexane), m.p. 96° ^1H NMR: 3.58 (br, NH, exchanges with D_2O), 3.77 (H-5; J(H-5,6endo) = 3.7, J(H-5,6exo) = 10.7), 6.92 (C_6H_5)	[61, 78]

References on pp. 99/102

Table 2 [continued]

No.	$^{4}LFe(CO)_3$	method of preparation (yield in %), properties and remarks	Ref.
		IR (hexane): 1981, 2045 (CO) or 1970, 2045 (CO) IR (KBr): 3394 (NH) IR ($CHCl_3$): 3390 (NH) mass spectrum: $[M-nCO]^+$ (n=0 to 3)	
b	NHC_6H_5, $Fe(CO)_3$	IIa (49.5), IIb (22) m.p. 118 to 119° (from C_6H_6/petroleum ether), yellow solid IR (hexane): 1980, 1985, 2040 (CO) IR ($CDCl_3$): 3420 (NH) mass spectrum: $[M-nCO]^+$ (n=0 to 3), $[M-3CO-C_6H_5NH_2]^+$	[78]
17	$NHC_6H_4OCH_3$-2, $Fe(CO)_3$	VIc	[98]
18	$NHC_6H_4OCH_3$-4, $Fe(CO)_3$	VIc ^{1}H NMR (acetone-d_6): 1.05 to 1.8 (H-6exo, 7exo), 2.0 to 2.5 (H-6endo, 7endo, masked by acetone), 3.1 (H-1,4), 3.68 (s, OCH_3), 4.1 (m, H-5), 5.5 (H-2,3), 6.7 to 7.5 (C_6H_4, overlaps with NH)	[98]
19	$NHC_6H_4CH_3$-2, $Fe(CO)_3$	VIc ^{1}H NMR (acetone-d_6): 1.7 (H-6exo, 7exo), 2.02 (s, CH_3), 2.2 to 2.6 (H-6endo, 7endo, masked by acetone), 3.05 (H-1,4), 3.7 (m, H-5), 5.3 (m, H-2,3), 6.6 to 7.0 (C_6H_4, overlaps with NH)	[93, 98]
20	$NHC_6H_4CH_3$-4, $Fe(CO)_3$	VIc (50) pale yellow solid ^{1}H NMR (acetone-d_6): 1.69 (H-6exo, 7exo), 2.13 (s, CH_3), 2.67 (br, m, H-6endo, 7endo), 3.10 (H-1,4), 3.70 (m, H-5), 4.60 (br, NH), 5.46 (H-2,3), 6.50 (d, 2H in C_6H_4), 6.93 (d, 2H in C_6H_4) IR (acetone): 1970, 2045 (CO) IR (Nujol): 3400 (NH) mass spectrum: $[M]^+$	[93, 98, 99]
*21	$N(CH_3)_2$, $Fe(CO)_3$	VIc (87 to 92) b.p. 75°/0.01 Torr, b.p. 75°/0.1 Torr ^{1}H NMR (CCl_4 or CS_2): 1.40 (H-6), 2.07 (H-7), 2.19 (CH_3), 2.85 (H-1,4), 3.00 (H-5), 5.30 (H-2,3) IR: 1961, 2041 (CO)	[8, 79]

Table 2 [continued]

No.	$^4LFe(CO)_3$	method of preparation (yield in %), properties and remarks	Ref.
22	cycloheptadiene-Fe(CO)$_3$ with $N(C_2H_5)_2$	VIc dec. p. 34° IR (hexane): 1977, 2046 (CO)	[61]
23	cycloheptadiene-Fe(CO)$_3$ with pyrrolidin-1-yl	VIc oil IR (hexane): 1980, 2047 (CO)	[61]
24	cycloheptadiene-Fe(CO)$_3$ with 5-$N(CH_3)C_6H_5$	VIc dec. p. 92° ^1H NMR: 3.10 (H-5; J(H-5,6endo) = 4.8, J(H-5,6exo) = 15.7) IR (hexane): 1982, 2047 (CO)	[61]
*25	cycloheptadiene-Fe(CO)$_3$ with N(O·)C_6H_5	–	[73]
*26a	cycloheptadiene-Fe(CO)$_3$ with 5-CH_3	b.p. 25°/0.1 Torr, pale yellow liquid ^1H NMR (CCl_4 or CS_2): 0.97 (CH_3), 1.32 (H-6), 1.95 (H-5), 2.00 (H-7), 2.90 (H-1,4), 5.25 (H-2,3) ^1H NMR ($CDCl_3$): 0.95 (d, CH_3; J = 6), 1.28 (m, 1H), 1.43 (m, 1H), 1.97 (br, m, 3H); 2.86 (m), 3.07 (m) (H-1,4); 5.28 (m, H-2,3) IR ($CHCl_3$): 1964, 2039	[79, 123]
b	cycloheptadiene-Fe(CO)$_3$ with 5-CH_3	b.p. 25°/0.1 Torr, yellow oil ^1H NMR ($CDCl_3$): 0.48 (d, CH_3; J(CH_3, H-5) = 6.2), 0.8 to 1.8 (m, H-5,6,7), 2.64 (m, H-1,4), 4.65 (m, H-2,3) IR (cyclohexane): 1977, 2044 mass spectrum: $[M]^+$	[54]
27	cycloheptadiene-Fe(CO)$_3$ with $C(COCH_3)_2H$	VIb not isolated	[30]

References on pp. 99/102

Table 2 [continued]

No.	${}^4LFe(CO)_3$	method of preparation (yield in %), properties and remarks	Ref.
*28	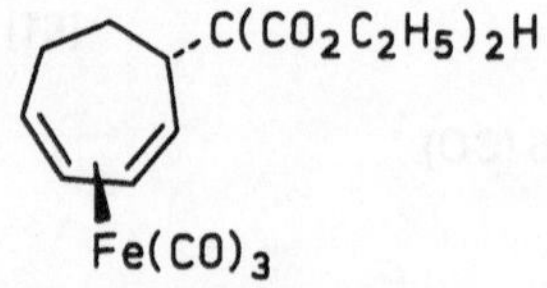 $C(CO_2C_2H_5)_2H$ substituted cycloheptadiene–$Fe(CO)_3$	VIb (62) b.p. 70°/0.005 Torr, oil IR (liquid film): 1724, 1754 (C=O); 1996, 2000, 2053 (CO)	[13, 79]
*29	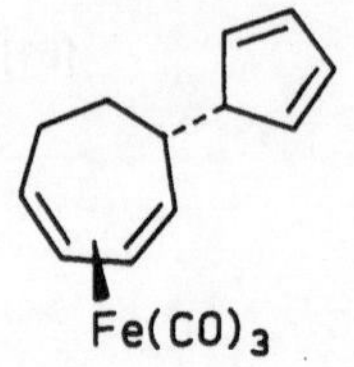 cyclopentadienyl-substituted cycloheptadiene–$Fe(CO)_3$	–	[52]
30	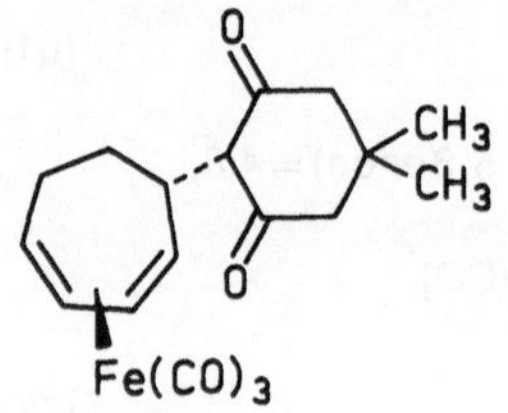 CH_3, CH_3, O, O (dimedone-substituted) cycloheptadiene–$Fe(CO)_3$	VIb not isolated	[30]
*31	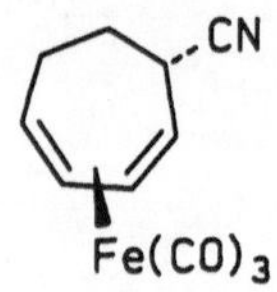 CN substituted cycloheptadiene–$Fe(CO)_3$	VIa (48) viscous oil (can not be purified) IR (liquid film): 2047, 2083, 2198 (CO); 2237 (CN)	[13]
*32	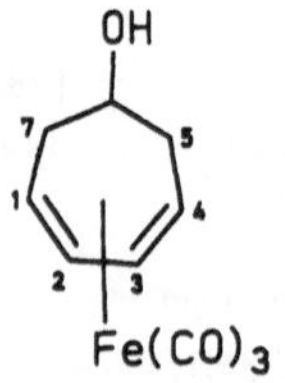 OH; positions 1, 2, 3, 4, 5, 7; $Fe(CO)_3$	m.p. 102°, m.p. 108 to 109°, pale yellow needles ${}^1H$ NMR ($CDCl_3$): 2.13 (m, 5H), 2.77 (m, 2H), 3.67 (br, 1H), 5.43 (m, 2H) ${}^1H$ NMR (CCl_4 or CS_2): 1.40 (OH), 2.20 (H-5,7), 2.80 (m, H-1,4), 3.60 (H-6), 5.50 (m, H-2,3)	[79, 103]
*33	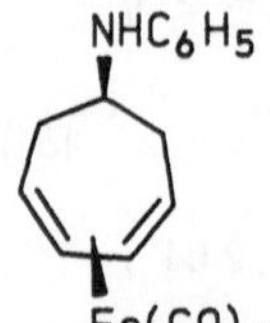 NHC_6H_5; $Fe(CO)_3$	IIa (13), IIb (16) m.p. 153 to 154° (from pentane), yellow needles IR (hexane): 1981, 2043 (CO) IR ($CHCl_3$): 3420 (NH) mass spectrum: $[M-nCO]^+$ (n=0 to 3)	[78]
*34	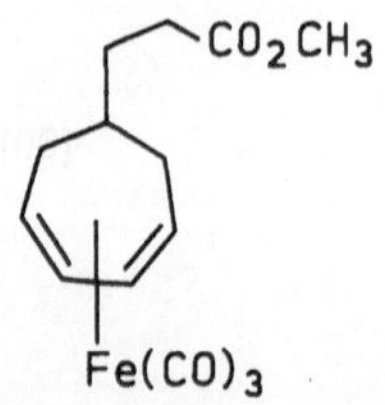 CO_2CH_3; $Fe(CO)_3$	–	[102]

Table 2 [continued]

No.	$^4LFe(CO)_3$	method of preparation (yield in %), properties and remarks	Ref.
35	$[(C_7H_9)\text{-}5\text{-}NH_2C_3H_7\text{-}n\ Fe(CO)_3]^+[BF_4]^-$	VIc (94) m.p. 123° (from pentane/CH_2Cl_2), white microcrystalline product ^1H NMR: 1.00 (t, CH_3), 1.18 to 2.65 (9H, NH_2, CH_2, CH_2-ring), 2.74 (d, 1H, CH_2-ring), 3.02 (m, H-1,4), 3.59 (m, H-5; J(H-5,6endo) = 3.3, J(H-5,6exo) = 11.2), 5.45 (m, H-2,3) IR (CH_2Cl_2): 1985, 2058 (CO) IR (KBr): 1081 (BF_4), 2963 (NH)	[61]
36	$[(C_7H_9)\text{-}NH_2C_4H_9\text{-}n\ Fe(CO)_3]^+[BF_4]^-$	VIc m.p. 114° (from pentane/CH_2Cl_2) IR (CH_2Cl_2): 1984, 2057 (CO) IR (Nujol): 1055 (BF_4)	[61]
37	$[(C_7H_9)\text{-}NH_2C(CH_3)_3\ Fe(CO)_3]^+[BF_4]^-$	VIc IR (CH_2Cl_2): 1977, 2054 (CO)	[61]
38	$[(C_7H_9)\text{-}5\text{-}NH(C_2H_5)_2\ Fe(CO)_3]^+[BF_4]^-$	VIc m.p. 89° (from pentane/CH_2Cl_2) ^1H NMR: 3.87 (m, H-5; J(H-5,6endo) = 3.4, J(H-5,6exo) = 12.0) IR (CH_2Cl_2): 1983, 2054 (CO) IR (Nujol): 1051 (BF_4) IR (KBr): 2879 (NH)	[61]
39	$[(C_7H_9)\text{-}HN(C_4H_8)\ Fe(CO)_3]^+[BF_4]^-$ (pyrrolidinium)	VIc m.p. 111° (from pentane/CH_2Cl_2) IR (CH_2Cl_2): 1988, 2058 (CO) IR (KBr): 1082 (BF_4), 3186 (NH)	[61]
40	$[(C_7H_9)\text{-}5\text{-}HN(C_5H_{10})\ Fe(CO)_3]^+[BF_4]^-$ (piperidinium)	VIc m.p. 128° (from pentane/CH_2Cl_2) ^1H NMR: 3.68 (m, H-5; J(H-5,6endo) = 3.8, J(H-5,6exo) = 12.6) IR (CH_2Cl_2): 1989, 2063 (CO) IR (KBr): 1082 (BF_4), 2942 (NH)	[61]

References on pp. 99/102

Table 2 [continued]

No.	$^4LFe(CO)_3$	method of preparation (yield in %), properties and remarks	Ref.
41	$[C_7H_{10}(NC_5H_5)Fe(CO)_3]^+[BF_4]^-$	VIc (60) m.p. 124° (from pentane/CH_2Cl_2), m.p. 124 to 125° (dec.), cream-yellow product ^{1}H NMR (CD_3CN): 1.5 to 2.6 (H-6,7), 2.85 (m, H-4), 3.28 (m, H-1), 5.05 (m, H-5), 5.68 (2m, H-2,3), 7.99 (m, 2H in C_5H_5N), 8.45 (1H in C_5H_5N), 8.82 (m, 2H in C_5H_5N) ^{1}H NMR: 5.18 (m, H-5; J(H-5,6endo) = 4.2, J(H-5,6exo) = 15.6) IR (CH_2Cl_2): 1965, 2063 (CO) IR (CH_3CN): 1980, 2050 (CO) IR (Nujol): ca. 1060 (BF_4) IR (KBr): 1083 (BF_4) mass spectrum: $[M-nCO-C_5H_5N]^+$ (n = 0, 1)	[61, 93, 96]
42	$[C_7H_{10}(NC_5H_4CH_3\text{-}2)Fe(CO)_3]^+[BF_4]^-$	VIc (70) m.p. 120° (dec.), off-white powder ^{1}H NMR (CD_3CN): 1.6 to 3.0 (H-6,7), 2.47 (s, CH_3), 3.26 (2m, H-1,4), 5.20 (m, H-5), 5.70 (2m, H-2,3), 7.75 to 8.90 (NC_5H_4) IR (CH_3CN): 1975, 2045 (CO) IR (Nujol): 1060 (BF_4)	[96]
43	$[C_7H_{10}(NC_5H_3(CH_3)_2\text{-}2,6)Fe(CO)_3]^+[BF_4]^-$	VIc not isolated ^{1}H NMR (CD_3NO_2): 1.8 to 3.2 (H-6,7), 2.57 (s, CH_3), ca. 3.3 (2m, H-1,4), 5.0 to 6.0 (H-2,3,5, overlapping resonances), 7.40 (d, 2H in NC_5H_3), 8.00 (t, 1H in NC_5H_3) IR (CH_3CN): ca. 1980, ca. 2055	[96]
44	$[C_7H_{10}(NC_5H_2(CH_3)_3\text{-}2,4,6)Fe(CO)_3]^+[BF_4]^-$	VIc not isolated ^{1}H NMR (CD_3NO_2): 1.7 to 3.2 (H-6,7), 2.39 (s, CH_3), 2.52 (s, CH_3), 2.9 to 3.6 (2m, H-1,4), 5.0 to 5.5 (m, H-5), 5.3 to 6.0 (2m, H-2,3), 7.17 (s, NC_5H_2) IR (CH_3CN): ca. 1980, ca. 2055	[96]
45a	$[C_7H_{10}(P(C_2H_5)_3)Fe(CO)_3]^+[BF_4]^-$	VId (93.5) m.p. 123 to 125° (from CH_2Cl_2/pentane), pale yellow crystals ^{1}H NMR: 1.30 (m, CH_3; J(^{31}P, CH_3) = 18), 1.66 (H-6,7), 2.27 (PCH_2; J(^{31}P, CH_2) = 13), ca. 2.50 (d, H-4), ca. 2.73 (d, H-5; J(H-5,6endo) = ca. 5, J(H-5,6exo) = ca. 11), 3.17 (t, H-1), 5.46 (m, H-2,3)	[61, 68]

References on pp. 99/102

Table 2 [continued]

No.	$^4LFe(CO)_3$	method of preparation (yield in %), properties and remarks	Ref.
		^{13}C NMR: 6.31 (d, CH_3; J(^{31}P, CH_3) = 4.9), 10.63 (d, PCH_2; J(^{31}P, CH_2) = 46.4), 23.20 (s, C-7), 27.71 (d, C-6; J(^{31}P, C-6) = 12.2), 32.23 (d, C-5; J(^{31}P, C-5) = 39.1), 46.69 (d, C-4; J(^{31}P, C-4) = 5.0), 58.63 (s, C-1); 86.69 and 91.35 (s, C-2,3), 210.27 (s, CO) IR (CH_2Cl_2): 1989, 2059 (CO) IR (KBr): 1085 (BF_4)	
b	$[P(C_2H_5)_3 \ldots Fe(CO)_3]^+ [BF_4]^-$	VId (67) m.p. 72 to 74°, brown product ^1H NMR: 1.28 (m, CH_3), 2.26 (m, H-6,7, PCH_2), 2.73 (m, H-4,5), 3.17 (m, H-1), 5.46 (m, H-2,3) ^{13}C NMR: 6.31 (d, CH_3; J(^{31}P, CH_3) = 4.9), 10.73 (d, PCH_2; J(^{31}P, CH_2) = 46.6), 23.21 (s, C-7), 27.81 (d, C-6; J(^{31}P, C-6) = 12.2), 32.28 (d, C-5; J(^{31}P, C-5) = 36.6), 46.74 (d, C-4; J(^{31}P, C-4) = 7.3), 58.73 (s, C-1); 86.78 and 91.35 (s, C-2,3), 210.07 (s, CO) IR (CH_2Cl_2): 1987, 2051 (CO) IR (KBr): 1084 (BF_4)	[61]
46a	$[P(C_3H_7\text{-}n)_3 \ldots Fe(CO)_3]^+ [BF_4]^-$	VId m.p. 154° (from CH_2Cl_2/pentane) IR (CH_2Cl_2): 1988, 2058 (CO)	[61]
b	$[P(C_3H_7\text{-}n)_3 \ldots Fe(CO)_3]^+ [BF_4]^-$	VId m.p. 87° IR (CH_2Cl_2): 1984, 2053 (CO)	[61]
47a	$[P(C_4H_9\text{-}n)_3 \ldots Fe(CO)_3]^+ [BF_4]^-$	VId dec. p. 114° (from CH_2Cl_2/pentane), or yellow oil ^1H NMR ($CDCl_3$): 1.0 (10H), 1.5 (19H), 2.2 (2H), 2.8 (2H), 3.2 (1H), 5.4 (2H) IR (CH_2Cl_2): 1984, 2053 (CO) IR (CH_2Cl_2): 1990, 2059 (CO) IR (KBr): 1085 (BF_4)	[61, 68]

References on pp. 99/102

Table 2 [continued]

No.	$^4LFe(CO)_3$	method of preparation (yield in %), properties and remarks	Ref.
b	$[(C_7H_{10}P(C_4H_9\text{-}n)_3)Fe(CO)_3]^+[BF_4]^-$	VId dec. p. 91° IR (CH_2Cl_2): 1986, 2055 (CO) IR (KBr): 1085 (BF_4)	[61]
*48a	$[(C_7H_{10}P(CH_3)_2C_6H_5)Fe(CO)_3]^+[BF_4]^-$	VId (95) m.p. 117° (from CH_2Cl_2/pentane), m.p. 134 to 136°, pale yellow solid ^1H NMR (acetone-d_6): 1.80 to 2.60 (m, 4H), 2.38 (d, 6H; J(^{31}P, H) = 14), 2.80 to 3.45 (m, 3H), 5.54 (m, 2H), 7.91 (m, 5H) IR (CH_2Cl_2): 1985, 2055 (CO) IR (CH_2Cl_2): 1987, 2056 (CO) IR (KBr): 1086 (BF_4)	[33, 61]
b	$[(C_7H_{10}P(CH_3)_2C_6H_5)Fe(CO)_3]^+[BF_4]^-$	VId m.p. 45° IR (CH_2Cl_2): 1988, 2056 (CO) IR (KBr): 1084 (BF_4)	[61]
49	$[(C_7H_{10}P(C_6H_5)_2C_2H_5)Fe(CO)_3]^+[BF_4]^-$	VId dec. p. 160° IR (CH_2Cl_2): 1986, 2058 (CO)	[61]
*50	$[(C_7H_{10}P(C_6H_5)_3)Fe(CO)_3]^+[BF_4]^-\cdot CH_2Cl_2$	VId off-white crystals ^{13}C NMR (CD_3NO_2): 24.7 (C-6; J(P, C) = 1.5), 28.5 (C-7; J(P, C) = 14.0), 36.0 (C-5; J(P, C) = 37.0), 49.6 (C-4; J(P, C) = 6.5), 60.2 (C-1; J(P, C) = 1.0), 87.6 (C-3; J(P, C) = 2.0), 92.2 (C-2) IR (CH_2Cl_2): 1985, 2058 (CO) IR (CH_2Cl_2): 1986, 2060 (CO)	[6, 61, 85, 101, 128]

^{4}L is a disubstituted cyclohepta-1,3-diene (continued with No. 156, p. 79):

No.	$^4LFe(CO)_3$	method of preparation (yield in %), properties and remarks	Ref.
*51	$(C_7H_9(OH)(CH_3))Fe(CO)_3$ (6-OH, 6-CH_3)	^1H NMR ($CDCl_3$): 1.26 (H-6), 1.31 (CH_3), 1.67 (OH; J = 57.5), 2.00 (H-7), 2.87 (H-4), 3.13 (H-1), 5.29 (H-2,3)	[28]

Table 2 [continued]

No.	${}^4LFe(CO)_3$	method of preparation (yield in %), properties and remarks	Ref.
*52	cycloheptadiene ring (positions 1–7), C-6 bearing OH and $CH_2CO_2CH_3$, $Fe(CO)_3$	see No. 55, p. 86 m.p. 73.5 to 75°, pale yellow crystals ${}^1H$ NMR ($CDCl_3$): 1.0 to 2.5 (H-6,7), 2.5 (CH_2), 3.0 (H-1,4), 3.5 (OH), 3.7 (CH_3), 5.3 (H-2,3) IR (Nujol): 1710 (C=O); 3480 (OH) mass spectrum: $[M]^+$, $[M-CO]^+$	[16, 37]
*53	OH, C_6H_5, $Fe(CO)_3$	see No. 55, pp. 86/7	[76]
*54	OH, $C_6H_4CH_3$-4, $Fe(CO)_3$	see No. 55, pp. 86/7	[76]
*55	cycloheptadienone ring (positions 1–7), =O, $Fe(CO)_3$	VIe (14), see also No. 60, p. 87 m.p. 103 to 106°, m.p. 105 to 106°, m.p. 108 to 109°, m.p. 108 to 110° (from C_6H_6/petroleum ether), subl. p. 50°/0.2 Torr, subl. p. 60°/0.06 Torr, subl. p. 90°/high vacuum, yellow needles or solid ${}^1H$ NMR ($CDCl_3$): 2.03 (H-6; J(H-6exo, 6endo) = 13.0), 2.62 (q, H-7), 3.06 (H-4), 3.31 (quint (pair of overlapped t), H-1), 5.55 (q (pair of d), H-2), 5.90 (t (pair of overlapped d), H-3) ${}^{13}C$ NMR (acetone-d_6/CS_2 at 280 K): 58.0, 62.4 (C-1,4); 91.8, 92.2 (C-2,3); 136.0, 139.3 (C-6,7); 208.1 (C=O); 210.5 (CO) IR (CCl_4): 1661 (C=O); 1988, 2000, 2058 (CO); 2849, 2898, 2915, 2967, 3021 (CH) IR ($CHCl_3$): 1653 (C=O); 1990, 2000, 2060 (CO) mass spectrum: $[M]^+$	[3, 21, 28, 37, 39, 70, 107, 111]
56	$=NNHC_6H_3(NO_2)_2$-2,4, $Fe(CO)_3$	see No. 55, p. 86 m.p. 189 to 191°, m.p. 190 to 192° (dec.), blood-red needles (from n-C_4H_9OH) IR (KBr): 1976, 1992, 2058 (CO) UV (CH_2Cl_2): $\lambda_{max}(\varepsilon)$ = 25100 (23.7 × 10^3) cm^{-1}	[3, 37, 39]

References on pp. 99/102

Table 2 [continued]

No.	$^4LFe(CO)_3$	method of preparation (yield in %), properties and remarks	Ref.
*57	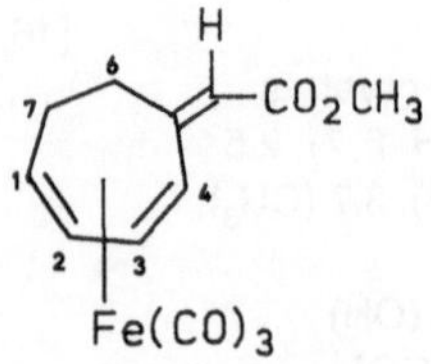 [structure: 7 ring, =CH–CO_2CH_3, $Fe(CO)_3$]	VII a m.p. 62 to 64° (from pentane at −78°), crystals ^1H NMR ($CDCl_3$): 1.7 to 2.5 (H-6,7), 3.2 (H-1), 3.7 (CH_3), 4.7 (H-4), 5.5 (H-2,3, =CH-) IR (Nujol): 1608 (C=C); 1705 (C=O) mass spectrum: $[M]^+$	[37]
58	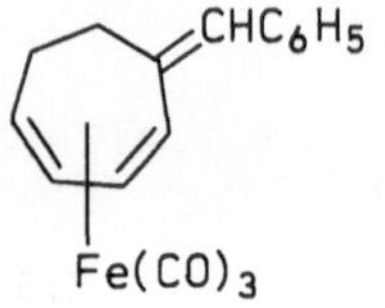 [structure: =CHC_6H_5, $Fe(CO)_3$]	see No. 48, p. 84 yellow oil (inseparable 1:2 mixture of isomers) ^1H NMR ($CDCl_3$): 1.85 to 2.55 (m, 4H), 3.28 (m, 1H), 3.98 (m, 1H), 5.44 (m, 2H), 6.20 (s, 2,3H), 6.46 (s, 1,3H), 7.32 (m, 5H) IR ($CHCl_3$): 1958, 1967, 2042 mass spectrum: $[M]^+$	[33, 48]
59	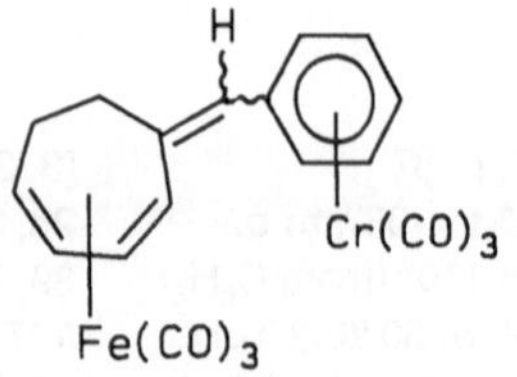 [structure: =CH–$C_6H_5Cr(CO)_3$, $Fe(CO)_3$]	see No. 48, p. 85 one isomer: m.p. 138 to 140°, orange crystals ^1H NMR ($CDCl_3$): 1.50 to 2.50 (m, 4H), 3.33 (m, 1H), 3.82 (m, 1H), 5.43 (m, 7H), 5.75 (s, 1H) IR ($CHCl_3$): 1890, 1966, 1980, 2045 mass spectrum: $[M]^+$ second isomer: m.p. 158 to 160°, orange crystals ^1H NMR ($CDCl_3$): 1.40 to 2.55 (m, 4H), 3.25 (m, 1H), 3.75 (m, 1H), 5.30 (m, 7H), 5.99 (s, 1H) IR ($CHCl_3$): 1890, 1966, 1980, 2045 mass spectrum: $[M]^+$	[33]
*60	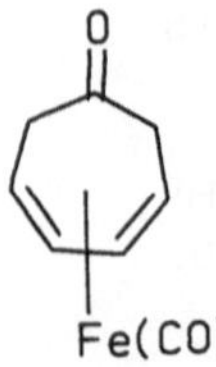 [structure: =O, $Fe(CO)_3$]	decomposes at ca. 55°, subl. p. 30°/high vacuum (with partly dec.), pale yellow crystals (from petroleum ether) IR (CCl_4): 1712 (C=O); 1949, 1984, 2053 (CO); 2809, 2898, 2985, 3030 (CH)	[3]
61	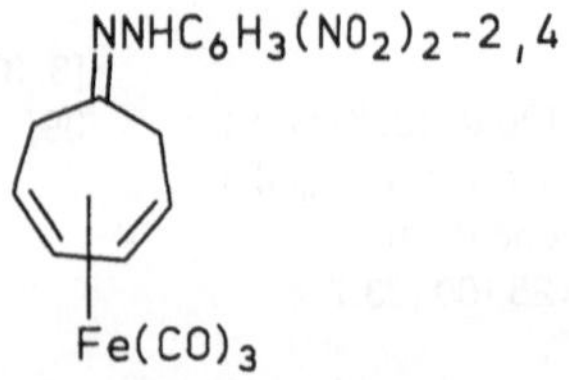 [structure: =$NNHC_6H_3(NO_2)_2$-2,4, $Fe(CO)_3$]	see No. 60, p. 87 dec. p. 190 to 198°, air-stable orange-yellow crystals IR (KBr): 1976, 2049 (CO)	[3]

Table 2 [continued]

No.	$^{4}LFe(CO)_3$	method of preparation (yield in %), properties and remarks	Ref.
*62	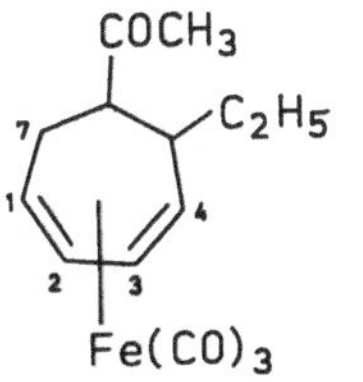	liquid ^{1}H NMR (CS_2): 0.60 (t, CH_3; J = 8), 1.90 (s, $COCH_3$), 2.00 to 3.00 (m, H-5, 6, 7, CH_2), 5.36 (m, H-2, 3) IR (liquid film): 1720 (C=O); 1920, 2050 (CO)	[38]
*63	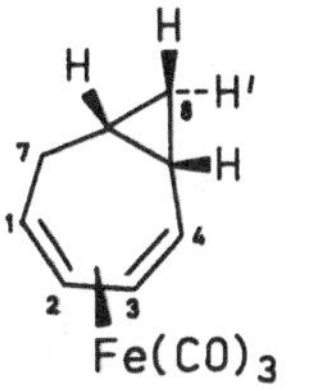	I a (86), II c (64), V a (23) m.p. 36.5° (from light petroleum ether (b.p. 30 to 40°) at −78°), sublimes at room temperature, deep yellow crystals, yellow solid, or yellow oil ^{1}H NMR (C_6H_6): −0.46 (m, H-8), 0.46 (m, H-6, H-8'), 1.04 (m, H-5), 2.06 (m, H-7), 2.76 (m, H-1), 3.42 (t, H-4; J = 5, J = ca. 2), 4.57 (m, H-2, 3) ^{13}C NMR (benzene-d_6): 14.0, 14.6 (C-5, 6); 18.0 (C-8); 24.0 (C-7); 59.5, 64.2 (C-1, 4); 85.5, 88.1 (C-2, 3), 212.1 (CO) IR (CS_2 or CCl_4/Nujol): 673, 710, 722, 758, 802, 828, 850, 872, 896, 928, (937, sh), 966, 993, 1020, 1035, 1120, 1162, 1198, 1226, 1242, 1320, (1340), 1362, (1375), 1416, 1435, 1460; 1920, 1965, 2025 (CO); 2820, 2855, 2900, 2935, (3040, sh) IR (hexane): 1977, 1988, 1996, 2048 (CO) mass spectrum: $[M-nCO]^+$ (n = 0 to 3), $[M-2CO-Fe]^+$, $[C_5H_5Fe]^+$, $[C_7H_7]^+$	[1, 2, 44, 58, 66, 67, 77, 86, 91]
64	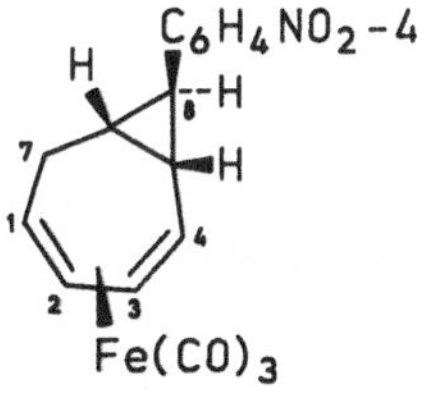	VII d (61) m.p. 168 to 170° (dec.), bright yellow solid, soluble in hexane and polar solvents to give air-stable yellow solutions ^{1}H NMR ($CDCl_3$): 1.28 (t, H-8; J(H-6, 8) = 5), 1.40 (m, H-6; J(H-5, 6) = 8), 1.83 (m, H-5; J(H-4, 5) = 6, J(H-5, 8) = 5), 2.40 (t, H-7; J(H-6, 7) = 4), 3.17 (m, H-1; J(H-1, 2) = 6, J(H-1, 7) = 4), 3.65 (t, H-4; J(H-3, 4) = 6), 5.09 (t, H-3; J(H-2, 3) = 6), 5.28 (t, H-2), 6.99 (d, H-3', 5' in C_6H_4; J(H-2', 3') = 9), 8.06 (d, H-2', 6' in C_6H_4) ^{13}C NMR ($CDCl_3$ at −50°): 23.68 (C-7); 28.10, 29.22, 36.13 (C-5, 6, 8); 60.59, 61.10 (C-1, 4); 85.97, 88.59 (C-2, 3); 123.63, 125.01 (C-2', 3', 5', 6' in C_6H_4); 145.50, 151.83 (C-1', 4' in C_6H_4); 211.02 (CO)	[4]

References on pp. 99/102

Table 2 [continued]

No.	$^4LFe(CO)_3$	method of preparation (yield in %), properties and remarks	Ref.
		IR (hexane): 1985, 1991, 2055 (CO) IR (CH_2Cl_2): 1975, 2049 (CO) mass spectrum: $[M]^+$	
*65	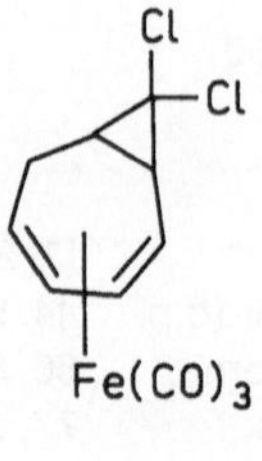	IIc, IV (95), Vb (38 or 56) m.p. 62 to 64° (from CH_3OH), m.p. 66° (from CH_3OH), b.p. 90 to 100°/0.2 Torr, orange oil, solidifying to yellow crystals ^1H NMR ($CDCl_3$): 1.5 to 2.0 (m, 2H), 2.0 to 2.5 (m, 2H), 2.8 to 3.1 (m, 1H), 3.22 (dt, 1H; J=8, J=1), 5.1 to 5.4 (m, 2H) IR (heptane): 1943, 1979, 1983, 2040, 2050 mass spectrum: $[M]^+$	[84, 129]
*66	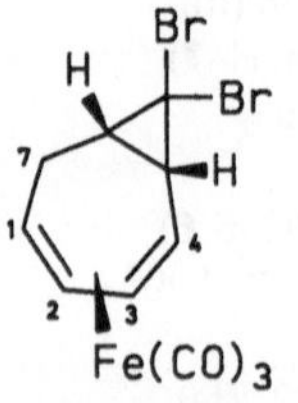	IIIb (15) m.p. 101 to 103° (dec.), bright yellow needles and plates (from pentane) ^1H NMR (CS_2): 1.6 to 2.6 (m, H-5,6,7), 2.95 (br, m, H-1 or H-4), 3.25 (m, H-1 or H-4), 5.25 (m, H-2,3) IR (CS_2): 1945, 1985, 2040 (CO) mass spectrum: $[M]^+$, $[C_8H_8]^+$	[23, 24, 32, 59]
*67	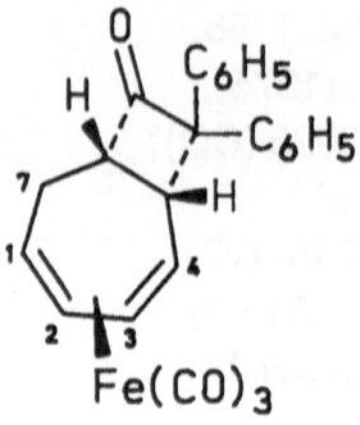	Vc (25) m.p. 136 to 138° (from CH_2Cl_2/hexane) ^1H NMR ($CDCl_3$): 1.99 (ddd, H-7; J=2.5, J=9, J=18), 2.48 (dd, H-7'; J=5.5, J=18), 2.72 (dd, H-4; J=4.5, J=8), 2.99 (m, H-1), 3.35 (dd, H-6; J=9, J=10), 3.83 (dd, H-5; J=4.5, J=10), 4.70 (dd, H-3; J=5, J=8), 5.10 (dd, H-2; J=5, J=8), 7.3 (m, $2C_6H_5$) IR: 1775 (C=O); 1980, 2060 (CO)	[105, 126]
68	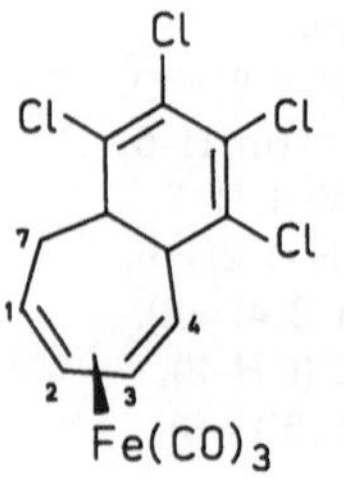	Vc (85) m.p. 131 to 132° (from ether), pale yellow crystals ^1H NMR (benzene-d_6): 1.44 (dd, H-7), 1.80 to 2.20 (m, H-1,4,6,7), 2.53 (dd, H-5), 4.26 (dd, H-2 or H-3), 4.52 (dd, H-2 or H-3) ^1H NMR ($CDCl_3$): 2.15 (dd, H-7; J(H-7,7)=17.0), 2.44 to 2.60 (m, H-4,6; J(H-4,5)=3.0), 2.64 to 3.04 (m, H-1,5,7; J(H-5,6)=7.0, J(H-6,7)=4.0), 5.23 to 5.48 (m, H-2,3) IR (Nujol): 1940, 1970, 2080 (CO) mass spectrum: $[M+n]^+$ (n=0, 2, 4)	[109]

References on pp. 99/102

Table 2 [continued]

No.	$^4LFe(CO)_3$	method of preparation (yield in %), properties and remarks	Ref.
*69	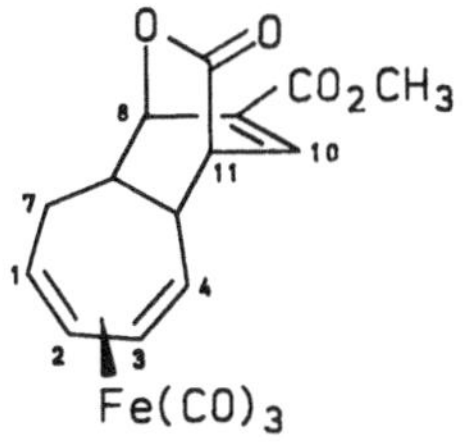	Vc (86) m.p. 90 to 92° (from CH_3OH), pale yellow crystals ^1H NMR (benzene-d_6): 1.41 (dd, H-7), 1.70 to 2.04 (m, H-4,6,7), 2.14 (m, H-1), 2.38 (m, H-5), 3.13 (dd, H-11), 3.34 (s, CH_3), 4.08 (m, H-3), 4.24 (m, H-2), 4.86 (dd, H-8), 6.76 (dd, H-10) ^1H NMR ($CDCl_3$): 1.86 (dd, H-7; J(H-7,7)=19.0), 2.32 to 2.92 (m, H-1,4,6,7; J(H-6,7)=3.5, J(H-6,8)=3.0), 3.02 (m, H-5; J(H-5,11)=3.0), 3.68 (dd, H-11; J(H-10,11)=7.0), 3.84 (s, CH_3), 4.98 to 5.20 (m, H-2,3,8; J(H-8,10)=2.0), 7.25 (dd, H-10) IR (Nujol): 1718 (ester), 1760 (lactone); 1960, 2070 (CO) mass spectrum: $[M-nCO]^+$ (n=1 to 3)	[109]
*70	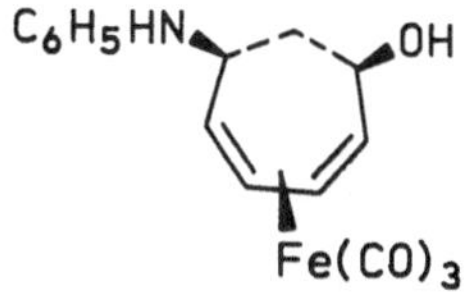	IIb (4) m.p. 90 to 92° (from $CH_3CO_2C_2H_5$/hexane), yellow crystals ^1H NMR ($CDCl_3$): 1.60 (m, 2H), 3.08 (m, 1H), 3.28 (m, 1H), 3.85 (m, 1H), 4.13 (m, 1H), 4.74 (m, 2H), 5.35 (m, 2H), 6.65 (m, 3H), 7.14 (q, 2H) IR (hexane): 1985, 1995, 2026 (CO) IR ($CHCl_3$): 3420 (NH), 3590 (OH) mass spectrum: $[M-nCO]^+$ (n=0 to 3)	[78]
*71	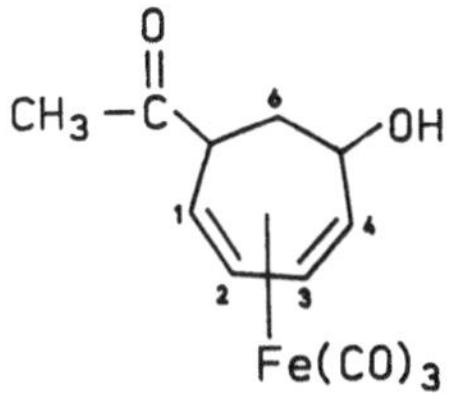	VIIa (40) m.p. 108 to 110° (from light petroleum ether), pale yellow crystals ^1H NMR ($CDCl_3$): 1.02 (q, H-6; J=12), 1.76 (d, H-6'; J=12), 2.12 (s, CH_3), 2.58 (s, OH), 2.90 (m, H-1,4), 3.05 (m, H-7), 4.10 (m, H-5), 5.38 (m, H-2,3) IR ($CHCl_3$): 1708 (C=O); 1993, 2058 (CO); 3403, 3593 (OH)	[42]
*72	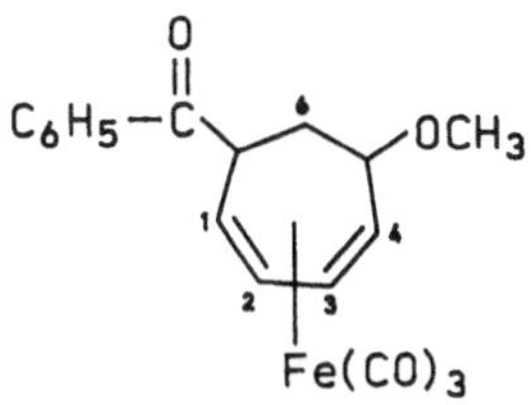	VIIa (90) almost colorless oil ^1H NMR (CS_2): 1.56 (q, H-6; J=13), 1.69 (d, H-6; J=13), 2.93 (m, H-1,4), 3.20 (s, CH_3), 3.70 (m, H-5), 5.38 (m, H-2,3), 7.64 (m, C_6H_5) IR (liquid film): 1686 (C=O) IR (CCl_4): 1978, 2050 (CO)	[42]

References on pp. 99/102

Table 2 [continued]

No.	$^4LFe(CO)_3$	method of preparation (yield in %), properties and remarks	Ref.
73	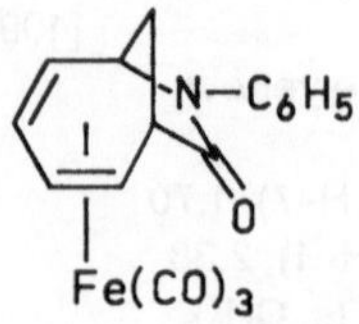	IIb (0.5) m.p. 155° (dec.), yellow needles ^1H NMR ($CDCl_3$): 1.76 (m, 2H), 3.0 (dd, 1H; J=4, J=2.5), 3.25 (t, 1H), 3.50 (t, 1H), 4.15 (dt, 1H; J=6, J=2, J=1), 5.48 (ddd, 1H; J=5, J=4, J=1), 5.58 (ddd, 1H; J=5.5, J=3.5, J=2), 7.36 (m, C_6H_5) IR (hexane): 1970, 2000, 2050 (CO) IR ($CHCl_3$): 1695 (C=O) mass spectrum: $[M-nCO]^+$ (n=0 to 3)	[78]
74	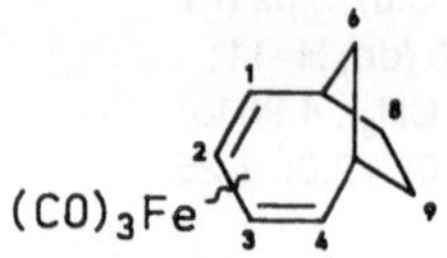	IIa yellow oil ^1H NMR (C_6H_6): 0.8 (m) and 1.3 (m) (H-6,8,9); 2.10 (m, H-5,7), 3.44 (m, H-1,4), 4.52 ("dd", H-2,3) mass spectrum: $[M]^+$	[45]
*75a	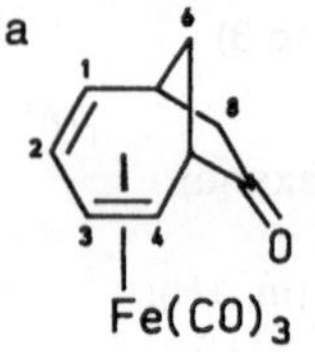	IIb (53), IIc (10) m.p. 108° (from hexane/toluene (1:1)), yellow crystals ^1H NMR (CS_2): 1.25 (ddd, H-6; J=12.5, J=4, J=4), 1.7 (dd, H-6'; J=−12.5, J=3), 2.06 (dd, H-8; J=−18, J=6), 2.5 (m, H-8'; J=18), 2.6 (m, H-5,7), 2.70 (dd, H-4; J=8.5, J=9), 3.00 (dd, H-1; J=8.5, J=9), 5.3 (m, H-2,3) ^{13}C NMR (benzene-d_6): 34.2 (C-7), 37.0 (C-6), 45.7 (C-5), 47.4 (C-8); 53.0, 64.1 (C-1,4); 86.2, 88.8 (C-2,3); 210.1 (C-9), 210.7 (CO) IR (KBr): 1735 (C=O) IR (hexane): 1983, 1996, 2061 (CO) mass spectrum: $[M-nCO]^+$ (n=0 to 3) and other fragments given	[66, 67, 77]
b	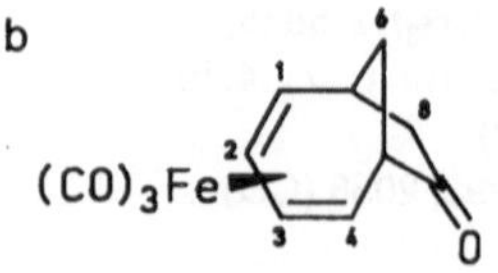	IIb (23) ^1H NMR (CS_2): 1.37 (dd, H-6'; J=−18), 1.6 (m, H-6,8), 1.82 (d, H-8', J=−18), 2.33 (m, H-5), 2.67 (m, H-7), 3.38 (dd, H-4; J=8, J=10), 3.61 (dd, H-1; J=8, J=9), 5.2 (m, H-2,3) ^{13}C NMR (benzene-d_6): 31.8 (C-7), 42.6 (C-6), 44.3 (C-5), 44.5 (C-8); 67.2, 72.7, 88.5, 90.2 (C-1 to 4); 211.1 (C-9), 214.5 (CO) IR (KBr): 1735 IR (hexane): 1989, 1997, 2058 (CO) mass spectrum: $[M-nCO]^+$ (n=0 to 3) and other fragments given	[77]

References on pp. 99/102

Table 2 [continued]

No.	$^4LFe(CO)_3$	method of preparation (yield in %), properties and remarks	Ref.
*76	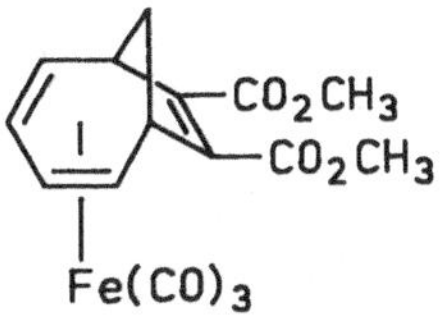	Vc (10) m.p. 109° (from ether/pentane), yellow crystals ^1H NMR (CS_2): 1.50 (m, 2H), 3.05 (t, 2H), 3.36 (t, 2H), 3.70 (s, 6H), 5.54 (m, 2H)	[53, 55, 113]

^4L is a trisubstituted cyclohepta-1,3-diene (continued with No. 160, p. 79):

No.	$^4LFe(CO)_3$	method of preparation (yield in %), properties and remarks	Ref.
*77	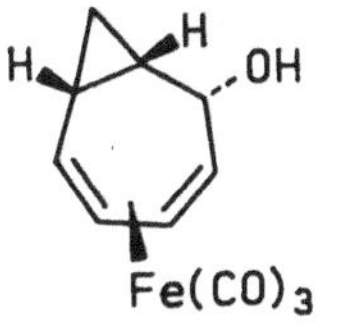	VIId (96), see also No. 63, p. 88 m.p. 84 to 87° (dec.), m.p. 90 to 91.5° (from petroleum ether), yellow needles or solid	[7, 12, 40]
*78	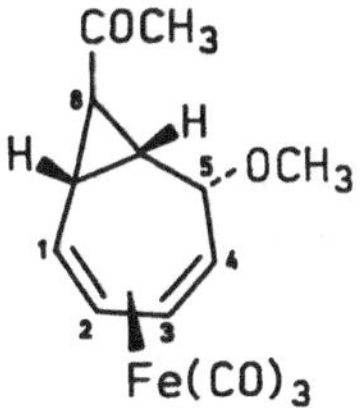	VIId b.p. 120°/0.05 Torr, yellow oil ^1H NMR (CS_2 at 35°): 2.55 (m, H-6,7), 2.7 (m, H-1 or H-4), 2.8 (s, H-8), 3.1 (m, H-1 or H-4), 3.2 (m, H-2,3), 3.3 (s, OCH_3), 3.8 (m, H-5) IR (CCl_4): 1070, 1090 (C-O); 1703 (C=O); 1963, 1978, 2043 (CO) mass spectrum: $[M-nCO]^+$ (n=0 to 3)	[31]
*79a	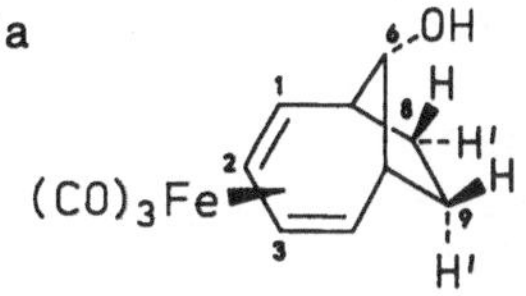	m.p. 135° (from pentane), colorless needles ^1H NMR ($CDCl_3$): 1.10 (d, H-8',9'; J=10), 1.50 (s, OH), 1.80 (dd, H-8,9; J=4, J=10), 2.28 (m, H-5,7), 3.20 (s, H-6), 3.56 (m, H-1,4), 5.10 (dd, H-2,3; J=6, J=4) IR (hexane): 1980, 2049 (CO) mass spectrum: $[M-nCO]^+$ (n=0 to 3)	[56]
b	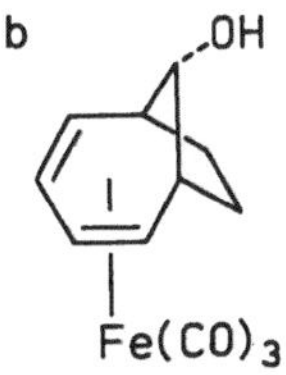	see No. 165, p. 98	[118]
c	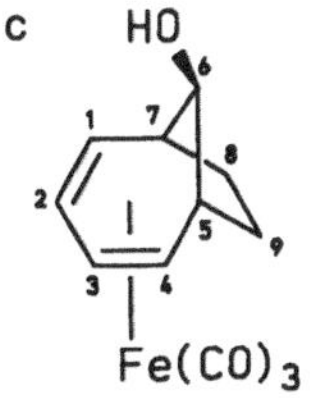	see Nos. 162, p. 98, 165, p. 98 pale yellow crystals (from n-hexane) ^{13}C NMR (benzene-d_6): 30.9 (C-8,9), 43.0 (C-5,7), 60.6 (C-1,4), 78.6 (C-6), 89.2 (C-2,3) IR: 1984(br), 2051 (CO)	[118, 120]

References on pp. 99/102

Table 2 [continued]

No.	$^4LFe(CO)_3$	method of preparation (yield in %), properties and remarks	Ref.
80	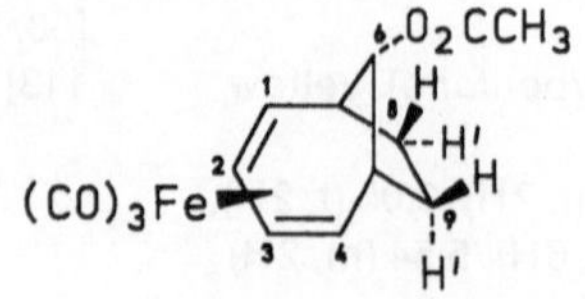	see No. 79, p. 90 m.p. 120 to 122° (from pentane), colorless needles ^{1}H NMR ($CDCl_3$): 1.14 (d, H-8′,9′; J = 10), 1.66 (dd, H-8,9; J = 4, J = 10), 1.92 (s, CH_3), 2.40 (m, H-5,7), 3.56 (m, H-1,4), 4.20 (s, H-6), 5.08 (dd, H-2,3; J = 6, J = 4) IR (hexane): 1744 (C=O); 1980, 1988, 2052(CO) mass spectrum: $[M-nCO]^+$ (n = 0 to 3)	[56]
*81	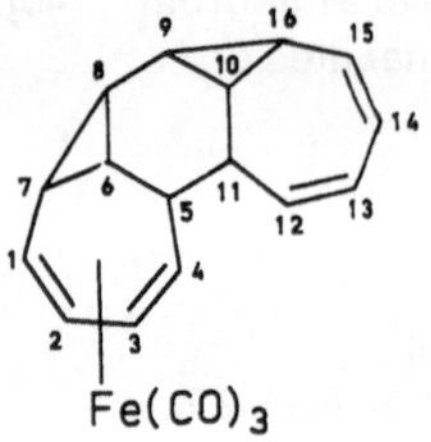	m.p. 140 to 144° (from hexane at −78 or −20°), yellow crystals or rhombs ^{1}H NMR ($CDCl_3$): 0.33 (ddd, H-8; J(H-6,8) = 8), 0.86 (ddd, H-9; J(H-9,10) = 7, J(H-9,16) = 6), 0.88 (ddd, H-6; J(H-5,6) = 8), 1.19 (ddd, H-10; J(H-10,11) = 7), 1.62 (ddd, H-7; J(H-6,7) = 8, J(H-7,8) = 7), 1.93 (m, H-16; J(H-10,16) = 7), 2.48 (ddd, H-5; J(H-4,5) = 6), 2.90 (ddd, H-11; J(H-11,12) = 3), 3.07 (ddd, H-4; J(H-3,4) = 8, J(H-2,4) = 2), 3.28 (ddd, H-1; J(H-1,2) = 8, J(H-1,7) = 8, J(H-1,3) = 2), 5.19 (m, H-2,3), 5.55 (m, H-12,13), 5.63 (m, H-14; J(H-14,15) = 9), 5.88 (ddd, H-15; J(H-15,16) = 6) ^{13}C NMR ($CDCl_3$): 14.99, 15.35, 17.11, 17.41 (C-6,8,9,10); 22.21, 26.21, 33.01, 41.62 (C-5,7,11,16); 60.19, 67.65 (C-1,4); 85.67, 89.13 (C-2,3); 122.50, 124.01, 131.11, 135.72 (C-12,13,14,15); 211.45(CO) IR (hexane): 1971, 1981, 2045 (CO) IR (CH_2Cl_2): 1967, 2037 (CO)	[95]
*82	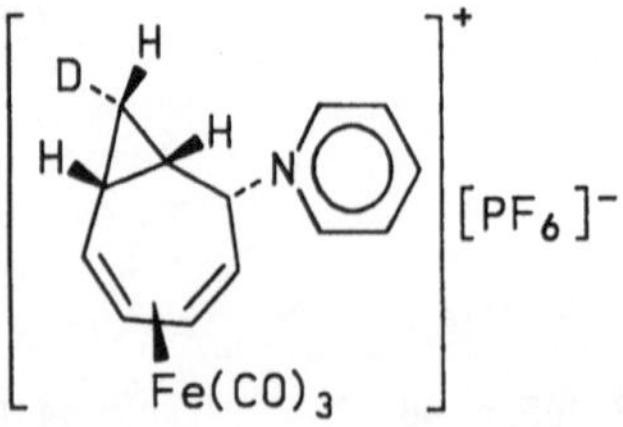 $[PF_6]^-$	VIId pale yellow crystals (from ether/acetone)	[46]
83	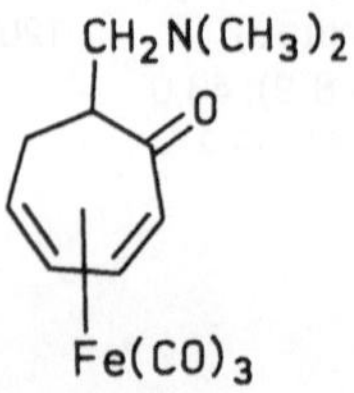 $CH_2N(CH_3)_2$	see No. 55, p. 86 m.p. 109 to 112°, subl. p. 60°/0.5 Torr, yellow crystals mass spectrum: $[M-CO]^+$	[16, 37]

Table 2 [continued]

No.	$^4LFe(CO)_3$	method of preparation (yield in %), properties and remarks	Ref.
*84	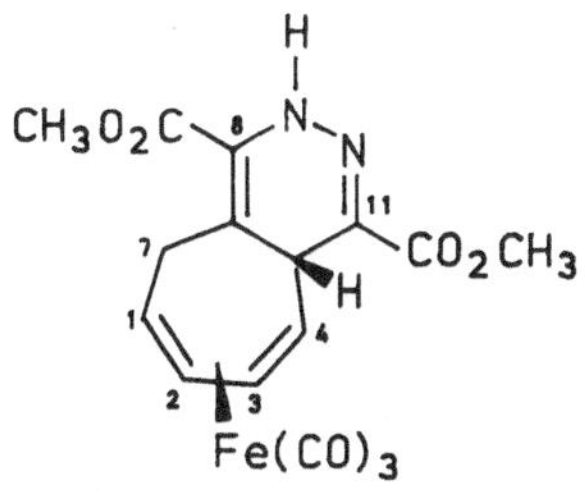	Vc (73) m.p. 165 to 167° (dec.), pale yellow crystals (from ether) ^{1}H NMR (benzene-d_6): 1.95 (d, H-7), 2.60 (t, H-1), 2.73 (d, H-4), 3.11 (s, CH_3), 3.51 (s, CH_3), 3.86 (dd, H-7), 4.04 (s, H-5), 4.48 to 4.76 (m, H-2,3), 7.88 (s, NH, exchanges with D_2O) ^{1}H NMR ($CDCl_3$): 2.27 (d, H-7; J(H-7,7) = 16.0), 2.78 (d, H-4; J(H-3,4) = 7.0), 3.16 (t, H-1; J(H-1,2) = 6.0), 3.80 to 4.20 (m, 2CH_3, H-7; J(H-1,7) = 6.0), 4.04 (s, H-5), 5.10 to 5.46 (m, H-2,3), 8.01 (s, NH, exchanges with D_2O) ^{13}C NMR ($CDCl_3$): 31.93 (t), 37.68 (d), 52.32 (q) (C-5, 7, CH_3); 57.07 (d), 61.82 (d), 87.83 (d), 88.48 (d), 121.35 (s), 122.87 (s), 129.49 (s) (C-1, 2, 3, 4, 6, 8, 11); 161.48 (s), 164.47 (s) (C=O); 209.88 (s, CO) IR (Nujol): 1720 (C=O); 1950, 1980, 2060 (CO); 3270 (NH) mass spectrum: $[M-nCO]^+$ (n = 1 to 3)	[109]
85	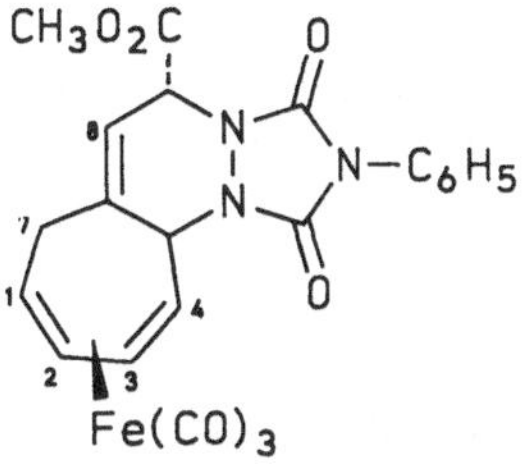	Vf (20) m.p. 181° (dec., from CH_2Cl_2/hexane), stable crystals ^{1}H NMR ($CDCl_3$): 2.81 (m, H-7), 3.01 (m, H-1), 3.83 (br,d, H-4, CH_3; J(H-3,4) = 7), 4.84 (br,s, H-5), 4.87 (m, H-9), 5.36 (m, H-2,3), 5.57 (br,d, H-8; J(H-8,9) = 6), 7.45 (m, C_6H_5) IR (Nujol): 1700, 1760, 1960, 1980, 2040	[87]
86	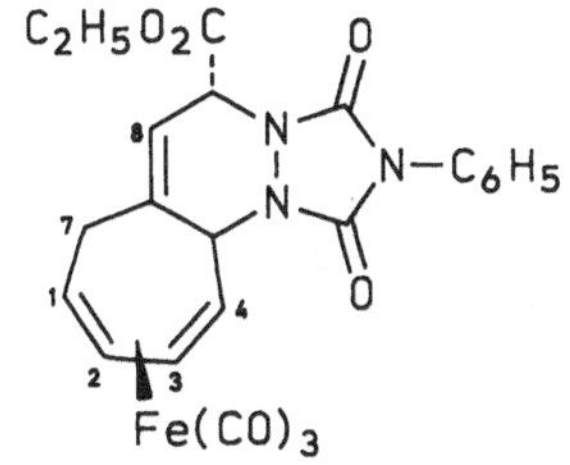	Vf (16) m.p. 145° (dec., from CH_2Cl_2/hexane), stable crystals ^{1}H NMR ($CDCl_3$): 1.30 (t, CH_3; J = 7), 2.80 (m, H-7), 3.03 (m, H-1), 3.80 (br,d, H-4; J(H-3,4) = 7), 4.27 (q, CH_2), 4.80 (br,s, H-5), 4.87 (m, H-9), 5.31 (m, H-2,3), 5.55 (br,d, H-8; J(H-8,9) = 6), 7.42 (m, C_6H_5) IR (Nujol): 1710, 1965, 1980, 2040 mass spectrum: $[M]^+$	[87]

References on pp. 99/102

Table 2 [continued]

No.	${}^4LFe(CO)_3$	method of preparation (yield in %), properties and remarks	Ref.
87	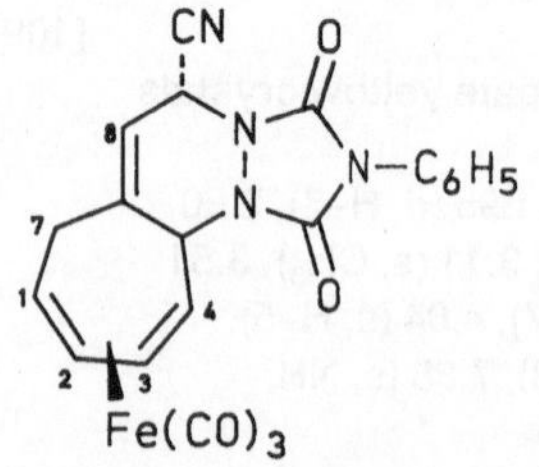	Vf (27) m.p. 152° (dec., from CH_2Cl_2/hexane), stable crystals ^1H NMR ($CDCl_3$): 2.82 (m, H-7), 3.08 (m, H-1), 3.55 (br, d, H-4; J(H-3,4) = 7), 4.80 (br, s, H-5), 5.00 (m, H-9), 5.42 (m, H-2,3,8; J(H-8,9) = 6), 7.50 (m, C_6H_5) IR (Nujol): 1715, 1760, 1950, 1990, 2040	[87]
88	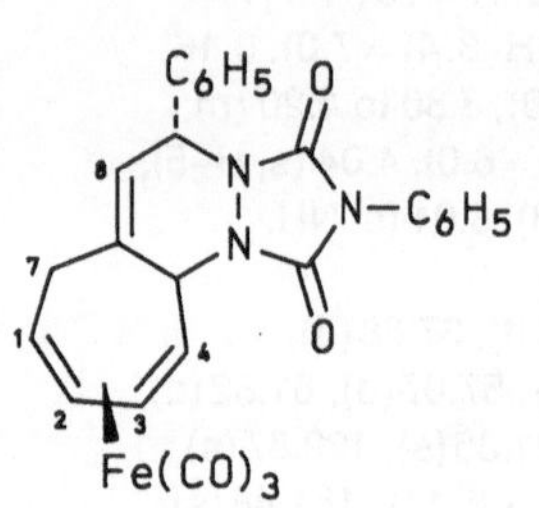	Vf (26) m.p. 205° (dec., from CH_2Cl_2/hexane), stable crystals ^1H NMR ($CDCl_3$): 2.82 (m, H-7), 3.04 (m, H-1), 3.55 (br, d, H-4; J(H-3,4) = 7), 4.90 (br, s, H-5), 5.22 (m, H-9), 5.37 (m, H-2,3), 5.52 (br, d, H-8; J(H-8,9) = 6), 7.40 (m, 2C_6H_5) IR (Nujol): 1704, 1760, 1953, 1980, 2040	[87]
89	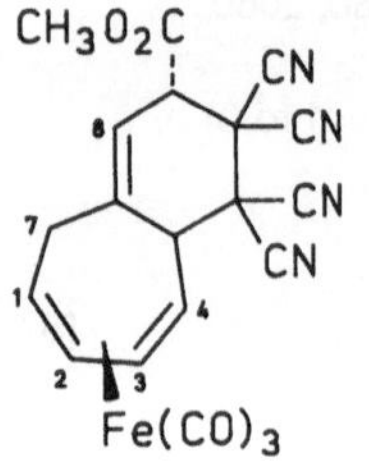	Vf (58) m.p. 146° (dec., from CH_2Cl_2/hexane), stable crystals ^1H NMR ($CDCl_3$): 2.88 (m, H-7), 3.08 (m, H-1,4), 3.50 (br, s, H-5), 3.93 (br, s, H-9, CH_3), 5.45 (m, H-2,3), 5.66 (br, s, H-8) IR (Nujol): 1965, 1995, 2032 mass spectrum: $[M-nCO]^+$ (n = 0 to 3)	[87]
90	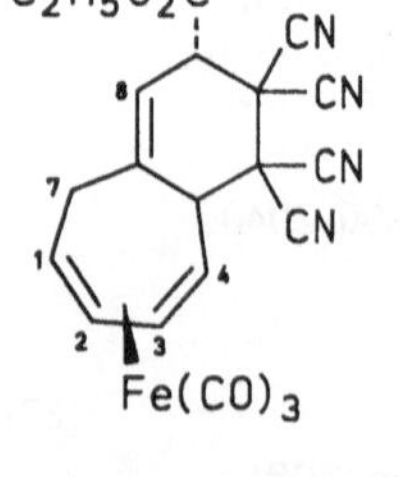	Vf (49) m.p. 150° (dec., from CH_2Cl_2/hexane), stable crystals ^1H NMR ($CDCl_3$): 1.42 (t, CH_3; J = 7), 2.89 (m, H-7), 3.10 (m, H-1,4), 3.51 (br, s, H-5), 3.94 (br, s, H-9), 4.42 (q, CH_2), 5.45 (m, H-2,3), 5.67 (br, s, H-8)	[87]
91	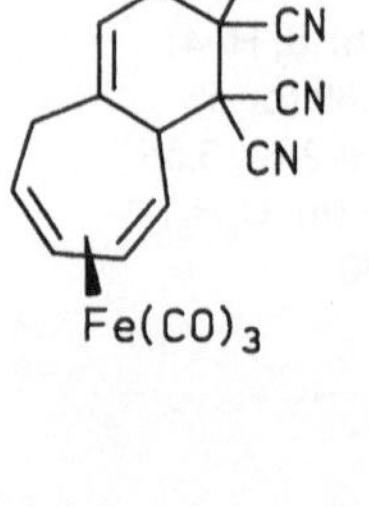	Vf (44) m.p. 138° (dec., from CH_2Cl_2/hexane), unstable, decomposes rapidly in acetone ^1H NMR (acetone-d_6): 3.0 to 3.4 (m, 4H), 3.85 (br, s, 1H), 4.4 (br, s, 1H), 5.5 to 6.0 (m, 3H)	[87]

Table 2 [continued]

No.	$^4LFe(CO)_3$	method of preparation (yield in %), properties and remarks	Ref.
92	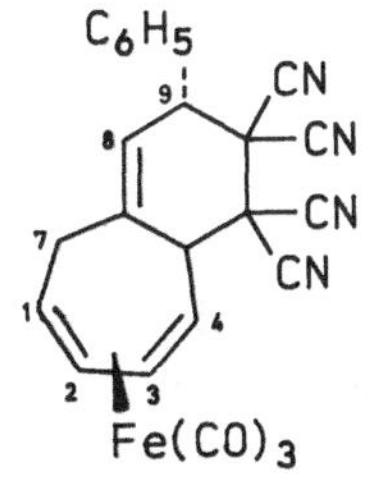	Vf (70) m.p. 167° (dec., from CH_2Cl_2/hexane), stable crystals ^1H NMR ($CDCl_3$): 2.92 (m, H-7), 3.17 (m, H-1,4), 3.55 (br, s, H-5), 4.19 (br, s, H-9), 5.47 (br, s, H-8), 5.53 (m, H-2,3), 7.46 (m, C_6H_5) IR (Nujol): 1970, 1992, 2025 mass spectrum: $[M-C_6N_4-nCO]^+$ (n=1 to 3)	[87]
*93	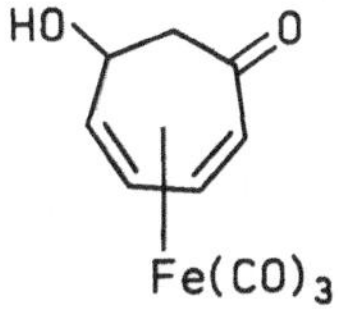	m.p. 141 to 143° (dec.), m.p. 165 to 167°, subl. p. 120°/0.7 Torr, pale yellow needles (from C_6H_6) ^1H NMR ($CDCl_3$): 2.18 (m, 1H), 2.42 (m, 1H), 3.32 (m, 2H), 4.26 (m, 1H), 4.78 (s, OH), 5.59 (t, 1H), 5.84 (t, 1H) IR (Nujol): 1613 (C=O); 1984, 2000, 2066 (CO); 3367 (OH) mass spectrum: $[M]^+$ and fragments given	[3, 111]
*94	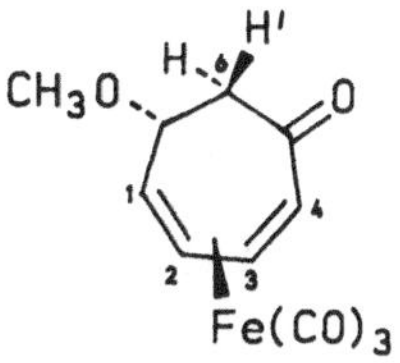	VIIb m.p. 78 to 80° (from hexane) ^1H NMR ($CDCl_3$): 1.9 (t, H-6), 2.4 (m, H-6′), 3.18 (H-4), 3.28 (H-1), 3.35 (s, CH_3), 3.91 (ddd, H-7), 5.6 (H-2), 5.87 (H-3) IR (hexane): 2002, 2010, 2073 (CO)	[27, 63, 70]
95	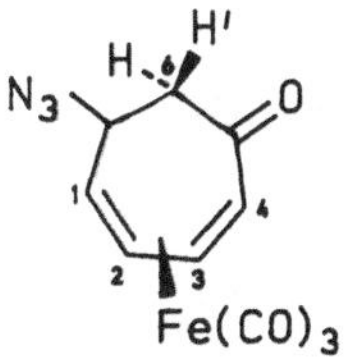	VIIb (80) m.p. 61 to 62° (from hexane), yellow crystals ^1H NMR ($CDCl_3$): 2.0 (t, H-6; J(H-6,6′) = 12.0), 2.36 (m (12 lines), H-6′; J(H-7,6′) = 5.5), 3.24 (H-1,4; J(H-1,7) = 2.0, J(H-4,6′) = 2.0), 4.1 (ddd, H-7; J(H-7,6) = 12.0), 5.42 (H-2; J(H-1,2) = 6.0), 5.66 (H-3) IR (hexane): 1680 (C=O); 1996, 2005, 2060 (CO) IR ($CHCl_3$): 1660 (C=O); 2090 (N_3)	[27, 70]
96	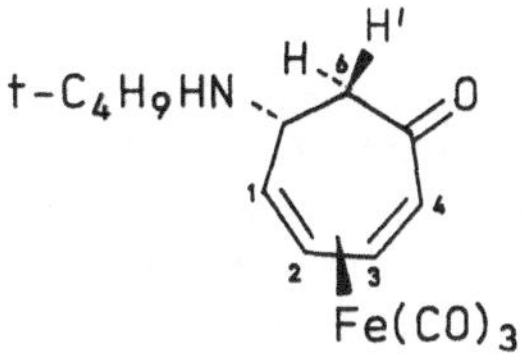	VIIb m.p. 142 to 144°, orange crystals ^1H NMR ($CDCl_3$): 1.18 (s, C_4H_9), 1.83 (t, H-6), 2.24 (m, H-6′), 3.1 (H-1), 3.1 to 3.78 (H-4), 3.78 (H-7), 5.47 (H-2), 5.8 (H-3) IR (hexane): 1775 (C=O); 1988, 1998, 2060 (CO)	[70]

References on pp. 99/102

Table 2 [continued]

No.	$^4LFe(CO)_3$	method of preparation (yield in %), properties and remarks	Ref.
		IR ($CHCl_3$): 1650 (C=O); 3580 (NH) mass spectrum: $[M-nCO]^+$ (n=2, 3), $[M-C_4H_9NH]^+$, $[M-3CO-Fe]^+$	
97a	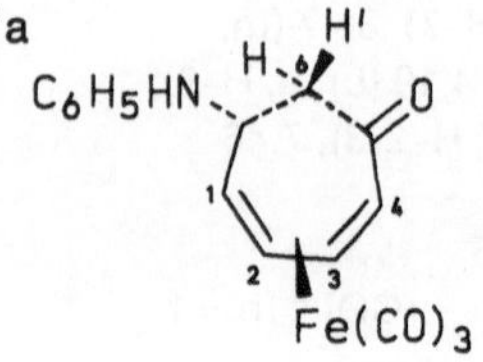 H′ H 6 C_6H_5HN O 1 4 2 3 $Fe(CO)_3$	VIIb m.p. 138 to 140° (from hexane) ^1H NMR ($CDCl_3$): 1.9 (t, H-6), 2.4 (m, H-6′), 3.23 (H-1,4), 4.27 (H-7), 5.41 (H-2), 5.78 (H-3) (see also figure in [78]) IR (hexane): 1998, 2010, 2070 IR (hexane): 2000, 2008, 2070 (CO) IR ($CHCl_3$): 1668 (C=O); 3440 (NH) IR ($CHCl_3$): 1650 (C=O); 3400 (NH) mass spectrum: $[M-nCO]^+$ (n=0 to 3), $[M-CO-H_2NC_6H_5]^+$, $[M-HNC_6H_5]^+$	[70, 78]
b	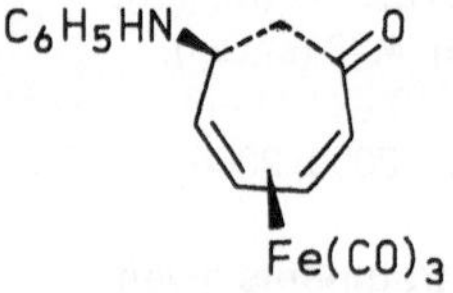 C_6H_5HN O $Fe(CO)_3$	IIb (2) m.p. 151 to 152° (from hexane), yellow needles ^1H NMR ($CDCl_3$): see figure in [78] IR (hexane): 2000, 2060 (CO) IR ($CHCl_3$): 1655 (C=O); 3418 (NH) mass spectrum: $[M-nCO]^+$ (n=0 to 3)	[78]
*98	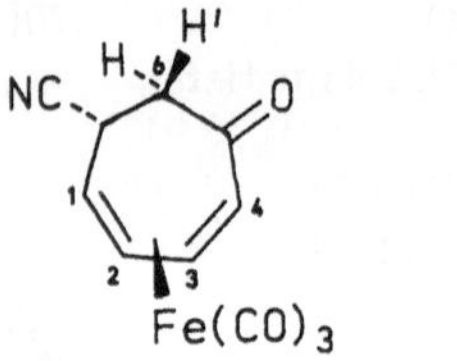 H′ H 6 NC O 1 4 2 3 $Fe(CO)_3$	VIIb m.p. 155 to 157° (from C_2H_5OH) ^1H NMR ($CDCl_3$): 2.4 (H-6; J(H-6,7)=10.5), 3.27 (d, H-1,4; J(H-1,7)=2.0), 3.68 (H-7; J(H-6′,7)=7.5), 5.8 (m, H-2; J(H-2,3)=5.5, J(H-1,2)=7.5), 6.03 (t, H-3; J(H-1,3)=1.1) IR ($CHCl_3$): 1665 (C=O); 2000, 2080 (CO); 2230 (CN) mass spectrum: $[M-nCO]^+$ (n=1 to 3), $[M-3CO-HCN]^+$, $[M-4CO-HCN]^+$	[70]
*99	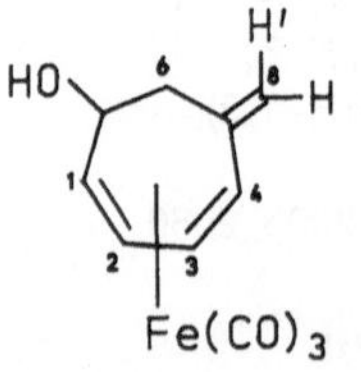 H′ HO 6 8 H 1 4 2 3 $Fe(CO)_3$	VIIc (80 to 90) ^1H NMR ($Eu(dpm)_3$-induced shifts (dpm=2,2,6,6-tetramethyl-3.5-heptanedionate) relative to H-7 shift taken as 0): 2.65 (H-6), 3.25 (H-6′; J(H-6,6′)=54), 3.3 (H-1), 7.05 (H-2), 8.15 (H-3), 8.4 (H-4), 8.7 (H-8′), 8.95 (H-8)	[49]
*100	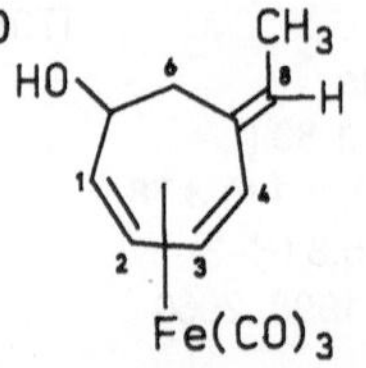 CH_3 HO 6 8 H 1 4 2 3 $Fe(CO)_3$	VIIc (80 to 90) ^1H NMR ($Eu(dpm)_3$-induced shifts (dpm=2,2,6,6-tetramethyl-3,5-heptanedionate) relative to H-7 shift taken as 0): 0 (H-7), 2.4 (H-6), 2.8 (H-6′; J(H-6,6′)=118), 3.0 (H-1), 7.0 (H-2), 8.15 (H-3), 8.3 (H-4), 8.8 (H-8), 9.15 (CH_3)	[49]

References on pp. 99/102

Table 2 [continued]

No. $^{4}LFe(CO)_3$	method of preparation (yield in %), properties and remarks	Ref.
*101 HO, $CHCH_2C_6H_5$, $Fe(CO)_3$	VIIc	[29]
*102 HO, CHC_6H_5, $Fe(CO)_3$	VIIc	[29]
*103a HO, $C_6H_4CH_3$-4, H, $Fe(CO)_3$	VIIc ^{1}H NMR (Eu $(dpm)_3$-induced shifts (dpm = 2,2,6,6-tetramethyl-3,5-heptanedionate) relative to H-7 shift taken as 0): 2.2 (H-6), 2.6 (H-6'; J(H-6,6') = 140), 8.15 (H-4), 8.7 (H-8), 10 (CH_3)	[49]
b HO, H, $C_6H_4CH_3$-4, $Fe(CO)_3$	VIIc ^{1}H NMR (Eu $(dpm)_3$-induced shifts (dpm = 2,2,6,6-tetramethyl-3,5-heptanedionate) relative to H-7 shift taken as 0): 2.4 (H-6,6'; J(H-6,6') = 40), 8.2 (H-4), 8.6 (H-8), 9.6 (CH_3)	[49]
*104 HO, CH_3', CH_3, $Fe(CO)_3$	VIIc (80 to 90) ^{1}H NMR (Eu $(dpm)_3$-induced shifts (dpm = 2,2,6,6-tetramethyl-3,5-heptanedionate) relative to H-7 shift taken as 0): 2.55 (H-6), 2.8 (H-6'; J(H-6,6') = 115), 3.05 (H-1), 7.0 (H-2), 8.05 (H-3), 8.35 (H-4), 9.0 (CH_3-8'), 9.1 (CH_3-8)	[29, 49]
*105 HO, $C(C_6H_5)_2$, $Fe(CO)_3$	VIIc	[29]

References on pp. 99/102

Table 2 [continued]

No.	$^4LFe(CO)_3$	method of preparation (yield in %), properties and remarks	Ref.
*106	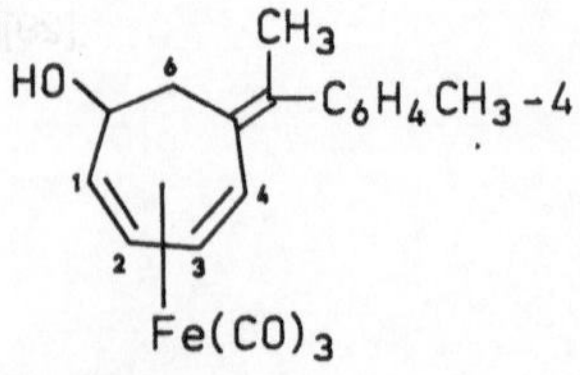 (structure: HO, CH_3, $C_6H_4CH_3$-4, positions 1, 2, 3, 4, 6; $Fe(CO)_3$)	VIIc (80 to 90) ^1H NMR (Eu $(dpm)_3$-induced shifts (dpm = 2,2,6,6-tetramethyl-3,5-heptanedionate) relative to H-7 shift taken as 0): 2.35 (H-6), 2.7 (H-6′; J (H-6,6′) = 118), 2.95 (H-1), 7.0 (H-2), 8.1 (H-3), 8.3 (H-4), 9.0 (CH_3), 9.7 (C_6CH_3)	[49]
107	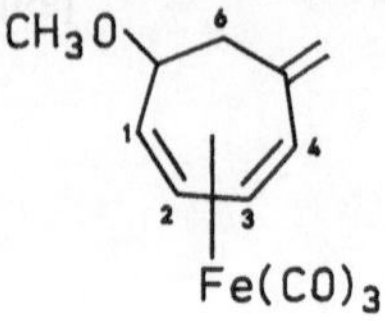 (structure: CH_3O, =CH_2, positions 1, 2, 3, 4, 6; $Fe(CO)_3$)	VIIc (95) subl. p. 90°/0.05 Torr, yellow solid ^1H NMR (CS_2): 1.56 (t, H-6; J = 11), 2.16 (ddd, H-6; J = 2, J = 4, J = 5, J = 11), 3.02 (dd, H-1; J = 2, J = 7), 3.22 (s, OCH_3), 3.50 (ddd, H-7; J = 2, J = 4, J = 5, J = 11), 3.63 (d, H-4; J = 7), 4.72 (d, CH_2; J = 0), 5.31 (m, H-2, 3) IR (Nujol): 1079 (C-O); 1610 (C=C) IR (CCl_4): 1978, 1983, 2053 (CO)	[20, 42]
*108	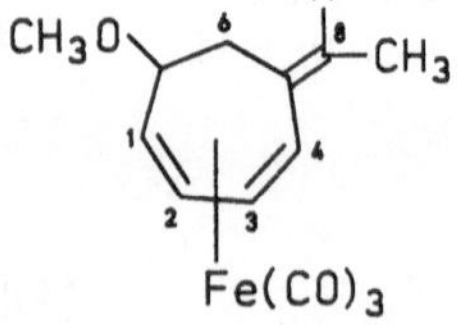 (structure: CH_3O, =C(H)CH_3, positions 1, 2, 3, 4, 6, 8; $Fe(CO)_3$)	VIIc b.p. 80°/0.05 Torr, yellow oil ^1H NMR (CS_2): 1.0 to 2.7 (m, H-6), 1.52 (dd, CH_3-8; J = 6.5), 2.96 (d, H-1; J = 4), 3.22 (2s, OCH_3), 3.35 (m, H-7), 3.66 (dd, H-4; J = 6), 5.25 (m, H-2, 3, 8) IR (Nujol): 1085 (C-O); 1632 (C=C) IR (CCl_4): 1983, 1993, 2053 (CO)	[38, 42]
*109	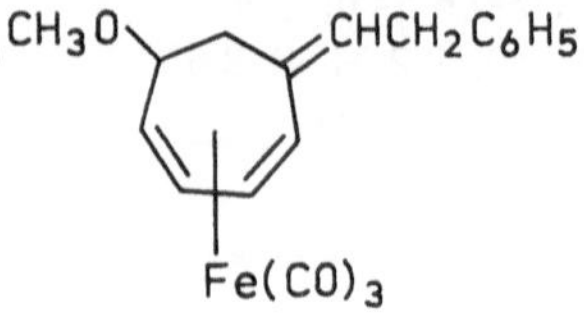 (structure: CH_3O, =$CHCH_2C_6H_5$; $Fe(CO)_3$)	VIIc	[29]
*110	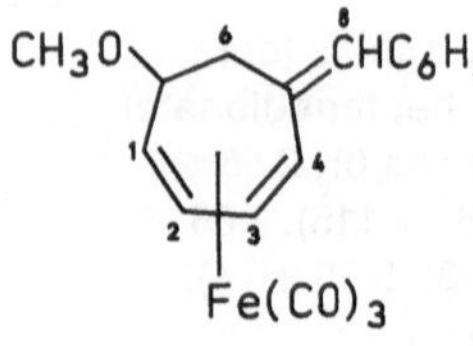 (structure: CH_3O, =CHC_6H_5, positions 1, 2, 3, 4, 6, 8; $Fe(CO)_3$)	VIIc b.p. 120°/0.05 Torr, yellow oil ^1H NMR (CS_2): 1.32 and 1.74 (2t, 0.6H + 0.4H, H-6; J = 11), 2.12 and 2.74 (2d, 0.4H + 0.6H, H-6; J = 11), 3.12 (m, H-1), 3.22 and 3.26 (2s, 1.8H + 1.2H, OCH_3, non-coupled splitting, indicative of 2 geometrical isomers (ratio 1:1)), 3.58 (2dd, H-7; J = 2, J = 11), 3.84 and 3.94 (2d, 0.6H + 0.4H, H-4; J = 7, J = 7), 5.32 (m, H-2, 3), 6.05 and 6.35 (2s, 0.4H + 0.6H, H-8), 7.15 (m, C_6H_5) IR (liquid film): 1080 (C-O); 1600 (C=C) IR ($CHCl_3$): 1980, 1988, 2049 (CO)	[29, 38, 42]

References on pp. 99/102

Table 2 [continued]

No. ⁴LFe(CO)₃	method of preparation (yield in %), properties and remarks	Ref.
*111 CH_3O–C₇ ring (positions 1–4, 6) =$CH\overset{8}{C}_6H_4CH_3$-4; $Fe(CO)_3$	VIIc (90) yellow oil ^{1}H NMR (CS_2): 1.30 and 1.70 (2t, 0.5H + 0.5H, H-6; J = 12), 2.19 and 2.23 (2s, 1.5H + 1.5H, OCH_3, non-coupled splitting, indicative of 2 geometrical isomers (ratio 1:1)), 2.22 and 2.30 (2s, CH_3), 2.78 and 2.2 (d, 0.5H + 0.5H, H-6; J = 12), ca. 3.15 (H-1), 3.59 (m, H-7), 3.88 (m, H-4), 5.30 (m, H-2,3), 7.02 (m, C_6H_4) IR (liquid film): 1980, 2049 (CO)	[38]
*112 CH_3O–C₇ ring (positions 1–4, 6) =$C(CH_3)CH_3$; $Fe(CO)_3$	VIIc	[29, 38]
*113 CH_3O–C₇ ring =$C(C_6H_5)C_6H_5$; $Fe(CO)_3$	VIIc	[29]
*114 CH_3O–C₇ ring =$C(CH_3)C_2H_5$; $Fe(CO)_3$	VIIc	[38]
*115 CH_3O–C₇ ring (positions 1–4, 6) =$\overset{8}{C}(CH_3)C_6H_4CH_3$-4; $Fe(CO)_3$	VIIc (90) yellow oil ^{1}H NMR (CS_2): 1.38 (t, H-6; J = 10), 1.84 (s, CH_3-8), 2.24 (s, C_6CH_3), 2.60 (d, H-6; J = 12), 3.04 (m, H-1,7), 3.24 (s, OCH_3), 3.78 (d, H-4; J = 8), 4.92 (t, H-3), 5.29 (t, H-2; J = 7), 7.02 (m, C_6H_4) IR (liquid film): 1980, 2048 (CO)	[38]
116 NC–C₇ ring =CH_2; $Fe(CO)_3$	VIIc	[70]

References on pp. 99/102

Table 2 [continued]

No.	$^4LFe(CO)_3$	method of preparation (yield in %), properties and remarks	Ref.
	^{4}L is a tetrasubstituted cyclohepta-1,3-diene (continued with No. 164, p. 80):		
*117	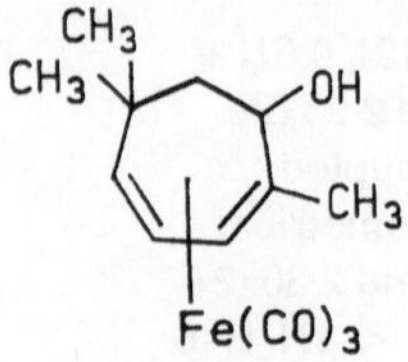	see No. 142, p. 96 yellow oil ^{1}H NMR ($CDCl_3$): 0.98 (s, CH_3), 1.30 (s, CH_3 and 2H obscured), 1.65 (s, CH_3), 2.68 (d, 1H; J=7), 3.85 (br,s, 1H), 4.90 (dd, 1H; J=6, J=7), 5.12 (d, 1H; J=6)	[126]
*118	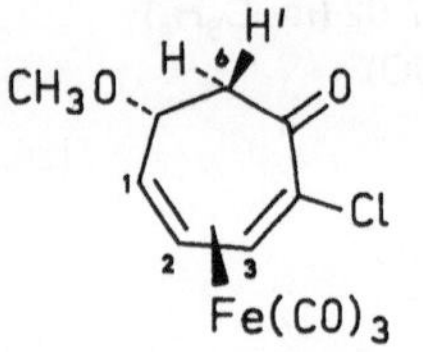	Ve m.p. 123 to 125° ^{1}H NMR ($CDCl_3$): 1.78 (H-6), 2.57 (H-6′), 3.15 (H-1), 3.28 (CH_3), 3.84 (H-7), 5.3 (H-2), 6.1 (H-3)	[63]
*119	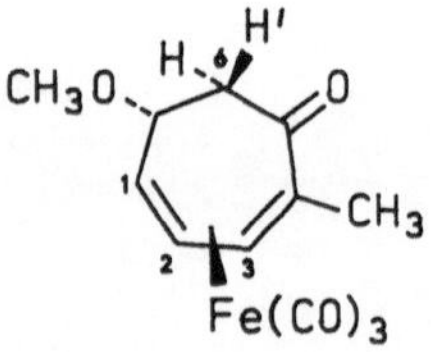	Ve oil ^{1}H NMR ($CDCl_3$): 1.66 (s, CH_3-4), 1.83 (t, H-6; J(H-6,6′)=11), 2.5 (dddd, H-6′), 3.23 (m, H-1; J(H-1,2)=8.0), 3.35 (s, OCH_3), 3.91 (dddd, H-7), 5.4 (dd, H-2), 5.82 (d, H-3; J(H-2,3)=5.5)	[63]
*120	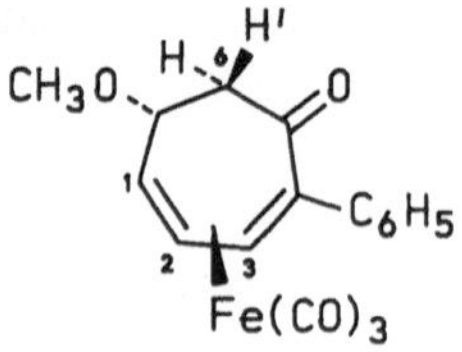	Ve m.p. 160 to 163° ^{1}H NMR ($CDCl_3$): 1.97 (H-6), 2.52 (H-6′), 3.22 (H-1), 3.35 (OCH_3), 4.0 (H-7), 5.48 (H-2), 6.33 (H-3), 7.26 (C_6H_5)	[63]
121	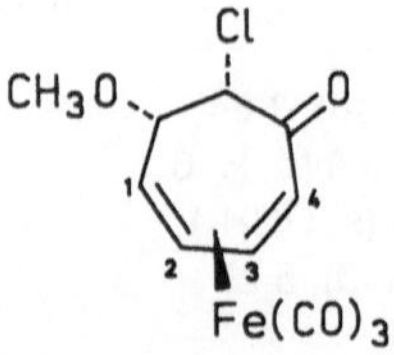	Ve m.p. 109 to 111° ^{1}H NMR ($CDCl_3$): 3.08 (m, H-1,4), 3.38 (OCH_3), 4.13 (dd, H-6), 4.32 (quint, H-7; J(6,7)=4.5), 5.67 (H-2), 5.88 (t, H-3)	[63]
122	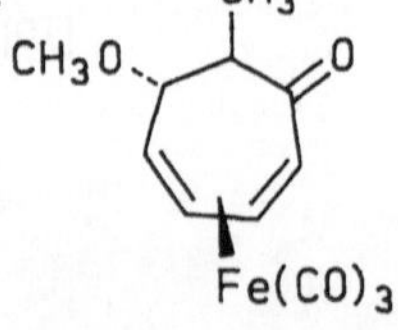	Ve	[63]

Table 2 [continued]

No. $^4LFe(CO)_3$	method of preparation (yield in %), properties and remarks	Ref.
*123a 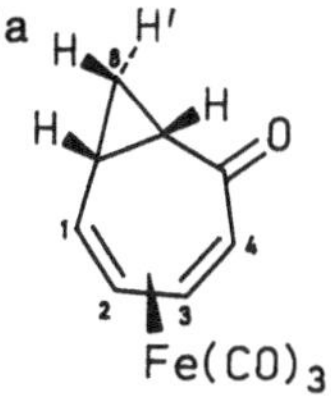	IIc (28), see also Nos. 77 and 128, pp. 90 and 94 m.p. 131°, m.p. 131 to 132.5° (from C_6H_6/petroleum ether), m.p. 132°, m.p. 132 to 133° (from C_6H_6/light petroleum ether), m.p. 132.5 to 133.5° (from C_6H_6/hexane), yellow needles or honey-yellow crystals ^{1}H NMR ($CDCl_3$): 1.00, 1.25 (m, H-8,8'); 1.55 (m, H-6), 2.01 (m, H-7), 3.12 (d, H-4; J=7.5), 3.61 (dd, H-1; J=7.5, J=7.5), 5.46 (m, H-2 or H-3), 5.80 (m, H-2 or H-3) ^{1}H NMR ($CDCl_3$): 0.96 (H-8'), 1.27 (H-8), 1.53 (H-7; J(H-1,7)=7.5), 1.93 (H-6), 3.09 (H-4; J(H-3,4)=7.5), 3.58 (H-1; J(H-1,2)=7.5), 5.48 (H-2), 5.79 (H-3) ^{13}C NMR ($CDCl_3$): 29.4 (C-8); 26.7, 27.8 (C-6,7); 62.2, 63.6 (C-1,4); 87.3, 92.2 (C-2,3); 206.1 (C-5); 208.7 (CO) IR (hexane): 1986, 2002, 2059 IR (cyclohexane): 1652, 1986, 2001, 2060	[7, 12, 62, 69, 77]
b 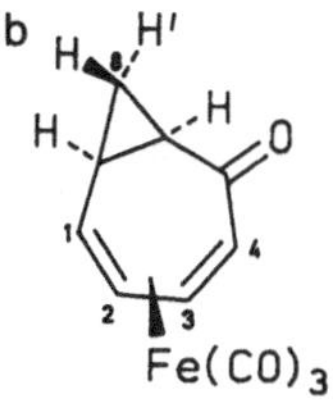	IIc (9) m.p. 77°, pale yellow crystals ^{1}H NMR ($CDCl_3$): 0.66 (H-8), 1.00 (H-8'), 1.27 (H-6), 3.16 (H-4; J(H-3,4)=7), 3.47 (H-1; J(H-1,7)=5), 5.46 (H-3; J(H-2,3)=5.5), 5.66 (H-2; J(H-1,2)=7) ^{13}C NMR ($CDCl_3$): 15.1 (C-8); 12.0, 21.9 (C-6,7); 57.9, 60.8 (C-1,4); 90.1, 91.2 (C-2,3); 205.5 (C-5); 208.7 (CO) IR (hexane): 1990, 1997, 2063.5 mass spectrum: $[M-nCO]^+$ (n=0 to 3) and further fragments given	[77]
*124a 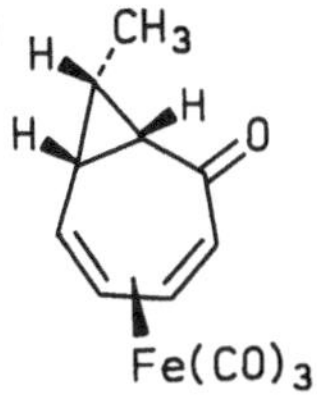	see No. 129, p. 94	[62]

Table 2 [continued]

No.	$^4LFe(CO)_3$	method of preparation (yield in %), properties and remarks	Ref.
b	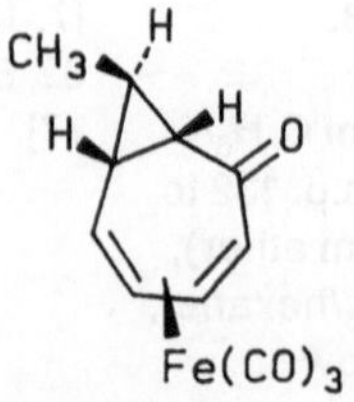	see No. 129, p. 94 m.p. 86°, orange crystals IR ($CHCl_3$): 1620 (C=O); 1960 to 2000, 2060 (CO)	[62]
*125	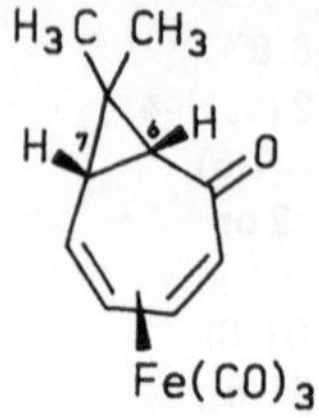	see No. 130, p. 94 m.p. 76°, orange crystals ^1H NMR ($CDCl_3$): 1.08 (s, CH_3), 1.13 (s, CH_3), 1.47 (d, H-6 or H-7; J=9.5), 1.90 (m, H-6 or H-7) IR ($CHCl_3$): 1617 (C=O); 1980, 2050 (CO)	[62]
*126	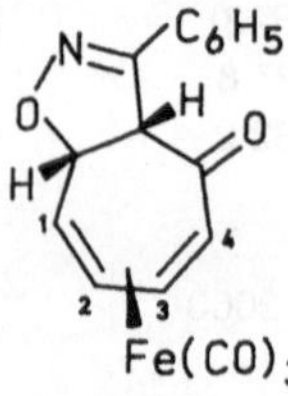	Vd (86) m.p. 156 to 157° (dec.), yellow crystals ^1H NMR ($CDCl_3$): 3.10 (m, H-1,4), 4.19 (dd, H-6; J(H-4,6)=1.3), 5.50 (dd, H-7; J(H-6,7)=9.6, J(H-1,7)=3.9), 5.58 (m, H-2,3) IR (Nujol): 1640 (C=O); 1990, 2000, 2070 (CO)	[72]
*127	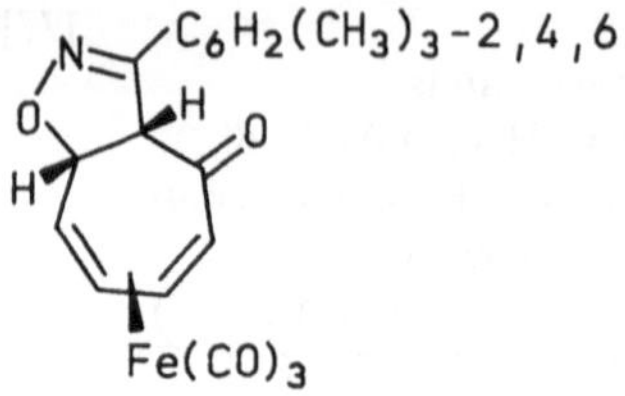	Vd (97) m.p. 153 to 155° (dec.), yellow crystals the spectra are similar to those of No. 126	[72]
*128	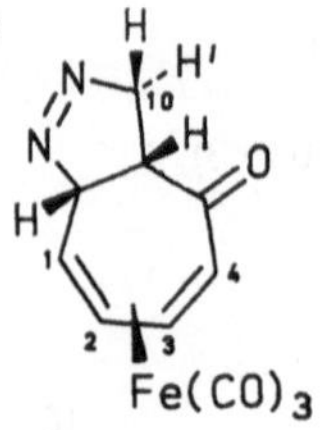	Vd (60 to 83) dec. p. 120°, m.p. 130 to 137° (dec. with gas evolution), yellow plates (from C_6H_6/light petroleum ether) ^1H NMR ($CDCl_3$): 2.57 (ddd, H-6; all J=ca. 10), 3.16 (dt, H-4; J=1.0, J=7.5), 3.42 (dd, H-1; J=4.0, J=7.5), 4.0 (ddd, H-10; J=2.0, J=10.5, J=18.0), 4.75 (dd, H-10'; J=10.5, J=18.0), 5.40 to 5.90 (m, H-2,3,7) IR (cyclohexane): 1650 (C=O); 1998, 2008, 2067 (CO) IR ($CHCl_3$): ca. 1550 (N=N); 1990 to 2020, 2070 (CO)	[62, 69]

Table 2 [continued]

No.	$^4LFe(CO)_3$	method of preparation (yield in %), properties and remarks	Ref.
*129a		Vd IR ($CHCl_3$): ca. 1550 (N=N); 1990 to 2020, 2070 (CO)	[62]
b		Vd m.p. 103° IR ($CHCl_3$): ca. 1550 (N=N); 1990 to 2020, 2070 (CO)	[62]
*130		Vd (90) m.p. 115° (dec.) ^{1}H NMR ($CDCl_3$): 1.17 (s, CH_3), 1.42 (s, CH_3), 2.53 (d, 1H; J=11), 5.48 (dd, 1H; J=4.5, J=11) IR ($CHCl_3$): ca. 1550 (N=N); 1990 to 2020, 2070 (CO)	[62]
*131		Vd m.p. 155 to 157° (from CH_3OH), orange-yellow prisms IR: 1641 (C=O); 1990, 2000, 2070 (CO)	[71]
*132		Vd m.p. 165 to 166° (from CH_3OH), yellow needles ^{1}H NMR ($CDCl_3$): 3.34 (m, 2H), 6.22 (m, 2H) IR: 1633 (C=O); 1990, 2000, 2070 (CO)	[71]
*133		see No. 55, p. 86 m.p. 81°, yellow crystals ^{1}H NMR (benzene-d_6): 0.93 (d, H-9; J=8), 1.20 (d, H-9′; J=8), 2.33 (br,s, H-8 or H-10), 2.40 to 2.48 (H-7), 2.44 (not resolved, 1H, H-1,2,3, or 4), 2.63 (dd, H-6; J=10, J=3.5), 2.96 (br,s, H-8 or	[74]

References on pp. 99/102

Table 2 [continued]

No.	$^4LFe(CO)_3$	method of preparation (yield in %), properties and remarks	Ref.
		H-10), 3.14 (d, 1H; J=8), 4.25 (t, 1H; J=6), 4.64 (dd, 1H, H-1,2,3, or 4; J=6, J=8); 5.65 (dd), 5.82 (dd) (H-11, 12) IR (CCl_4): 1635 (C=O); 1995, 2060 (CO)	
134	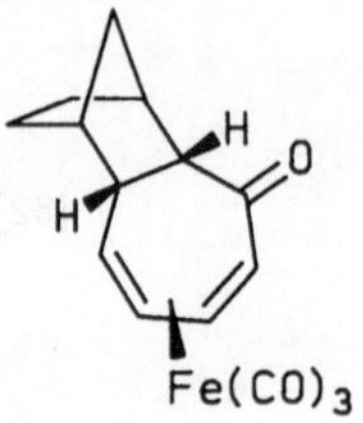	see No. 133, p. 95	[74]
*135	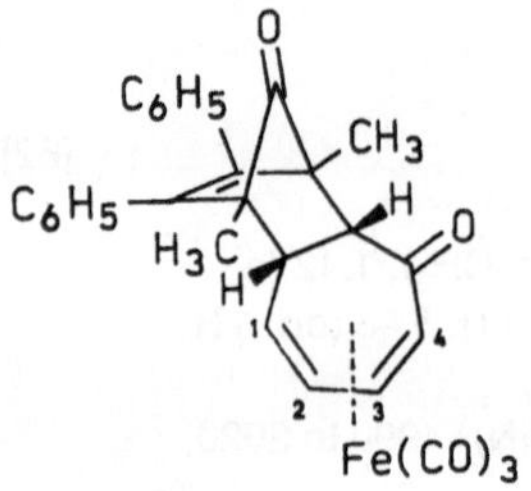	Vd (17) m.p. 180 to 181° (dec.) ^{1}H NMR ($CDCl_3$): 1.15 (s, CH_3), 1.55 (s, CH_3), 2.61 to 3.31 (m, H-1,6,7), 3.55 (d, H-4; J(H-3,4)=7.3), 5.24 (m, H-2), 5.65 (m, H-3) IR (Nujol): 1615 (conjugated C=O); 1775 (bridged C=O); 1995, 2010, 2060 (CO)	[72]
136	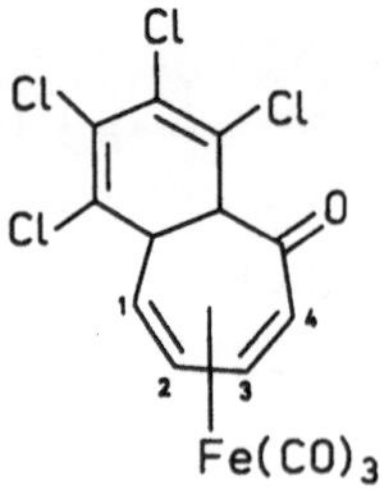	Vd (92) m.p. 168 to 169.5° (from ether), yellow crystals ^{1}H NMR ($CDCl_3$): 2.95 (d, H-1; J(H-1,2)=8.0), 3.20 (d, H-4; J(H-3,4)=6.5), 3.57 (s, H-6,7), 5.56 (dd, H-2; J(H-2,3)=5.0), 5.86 (dd, H-3) IR (Nujol): 1640 (C=O); 1990, 2080 (CO) mass spectrum: $[M+n]^+$ (n=0, 2, 4)	[109]
137	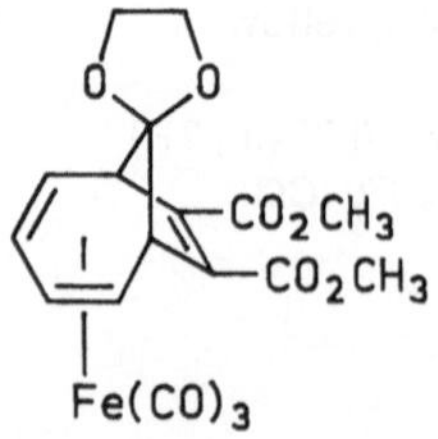	Ve (20) ^{1}H NMR (acetone-d_6): 3.10 (d, 2H), 3.30 (m, 2H), 3.80 (s, 2CH_3), 3.82 (m, 4H), 5.84 (m, 2H)	[53, 55]
*138	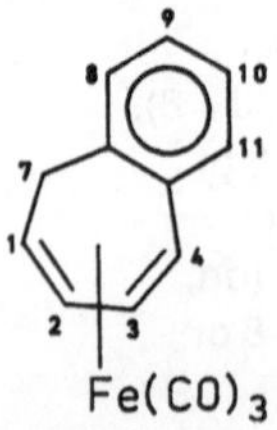	Ib (66) m.p. 82.0 to 82.5°, subl. p. 100°/0.025 Torr, yellow crystals ^{1}H NMR (CCl_4/CH_3CN (1:1)): 3.10 (t, H-7), 3.42 (m, H-1), 3.64 (quint, H-4), 5.37 (sept, H-2,3), 7.00 (m, H-8 to 11)	[14]

Table 2 [continued]

No. $^4LFe(CO)_3$	method of preparation (yield in %), properties and remarks	Ref.
	IR (CCl_4): 562, 617, 696, 728, 882, 932, 988, 1031, 1098, 1150, 1162, 1183, 1250, 1450, 1468, 1501; 1974, 1988, 2055 (CO); 2915, 2960, 3035 UV (hexane): $\lambda_{max}(\varepsilon)$ = 228 (33500), 283 (sh, 13000) nm	
*139 [structure: cyclohepta-dienedione, Fe(CO)$_3$]	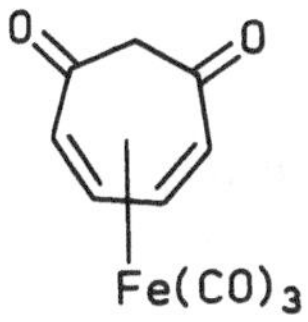 IIIb fine pale yellow needles (from C_6H_6) mass spectrum: $[M-nCO]^+$ (n = 0 to 3) and further fragments given	[11, 15]
140 2,4-$(O_2N)_2C_6H_3$NHN=[ring]=NNH$C_6H_3(NO_2)_2$-2,4, Fe(CO)$_3$	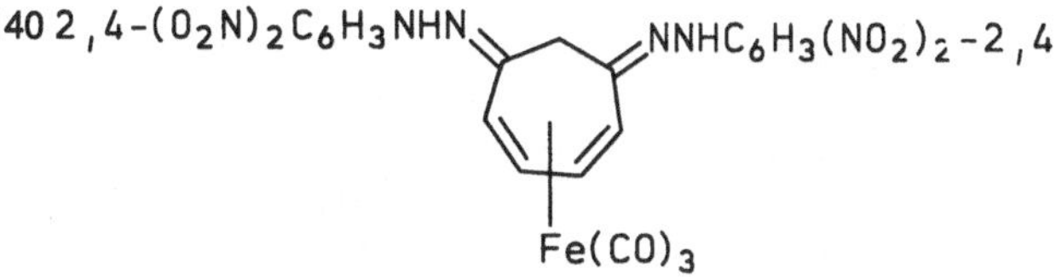 see No. 139, p. 95	[15]
*141 [structure: CH_3, CH_3, 6, O, Fe(CO)$_3$]	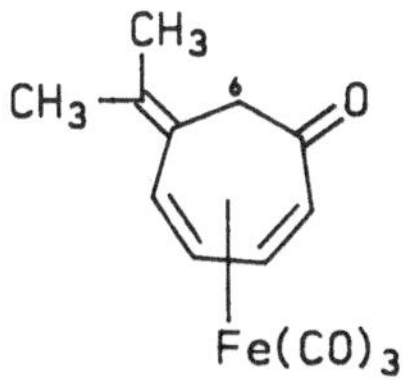 see Nos. 143, p. 96, and 147, p. 96 m.p. 85°, yellow-orange crystals ^{1}H NMR ($CDCl_3$): 1.73 (s, CH_3), 1.84 (s, CH_3), 2.44 (H-6; J = 12), 3.06 (H-6′) IR ($CHCl_3$): 1665 (C=O)	[82]

^{4}L is a pentasubstituted cyclohepta-1,3-diene (continued with No. 169, p. 81):

No.	method of preparation (yield in %), properties and remarks	Ref.
*142 [structure: CH_3, CH_3, O, CH_3, Fe(CO)$_3$]	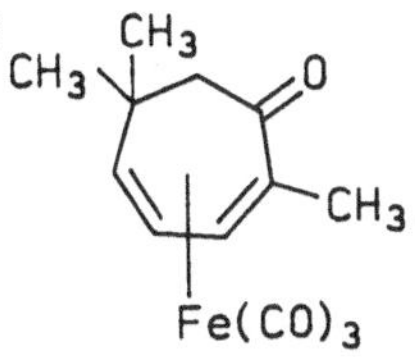 m.p. 46°, b.p. 43°/0.3 Torr, orange oil which solidifies on standing ^{1}H NMR (CCl_4): 1.13 (s, CH_3), 1.20 (s, CH_3), 1.55 (s, CH_3), 1.80 (br, s, 2H), 2.80 (d, 1H; J = 8), 5.13 (dd, 1H; J = 5, J = 8), 5.77 (d, 1H; J = 5) IR (CCl_4): 1650, 1960, 2040	[126]
*143 [structure: CH_3, CH_3, N=N, C(=O)–CH_3, 7, O, Fe(CO)$_3$]	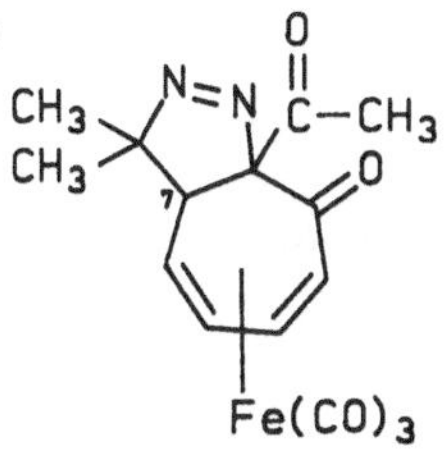 Ve (85) m.p. ca. 85° (dec.), yellow crystals (from ether) ^{1}H NMR ($CDCl_3$ at −30°): 0.93 (s, CH_3), 1.82 (s, CH_3), 2.21 (s, CH_3), 3.60 (d, H-7; J = 5)	[82]

References on pp. 99/102

Table 2 [continued]

No.	${}^4LFe(CO)_3$	method of preparation (yield in %), properties and remarks	Ref.
144	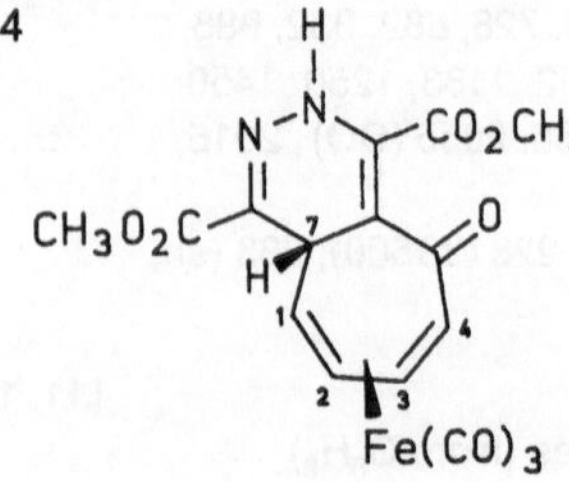	Vd (89) m.p. 218 to 219° (from CH_3OH), pale yellow crystals ${}^1H$ NMR ($CDCl_3$): 3.12 to 3.32 (m, H-1, 4; J=3.0, J=2.5), 3.78 (s, CH_3), 3.86 (s, CH_3), 4.74 (s, H-7), 5.50 (t, H-2 or H-3), 6.00 (t, H-2 or H-3), 7.85 (s, NH, exchanges with D_2O) IR (Nujol): 1635 (tropone CO); 1720, 1740 (ester CO); 1990, 2110 (CO); 3240 (NH) mass spectrum: $[M-nCO]^+$ (n=0 to 3)	[109]
*145	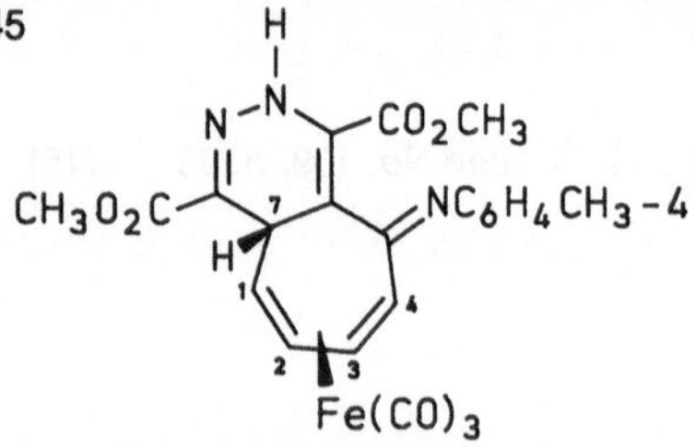	Ve (74) m.p. 190 to 193° (dec.), yellow crystals (from $CHCl_3$) ${}^1H$ NMR ($OS(CD_3)_2$): 2.19 (s, CH_3), 2.28 (s, CH_3), 3.20 (m, H-1), 3.60 to 3.96 (m, H-4, 2CH_3); 4.37, 4.50 (s, H-7); 5.84 (m, H-2), 6.08 (m, H-3); 6.46, 6.66 (d, 2H in C_6H_4); 7.00, 7.18 (d, 2H in C_6H_4; J=8.0); 10.22, 10.39 (s, NH, exchanges with D_2O) IR (Nujol): 1715, 1730 (C=O); 1990, 2020, 2100 (CO); 3220 (NH) mass spectrum: $[M-nCO]^+$ (n=2, 3)	[109]
*146a	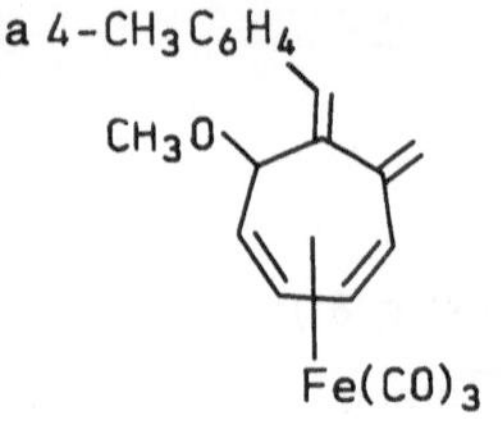	see isomer b	[38]
b	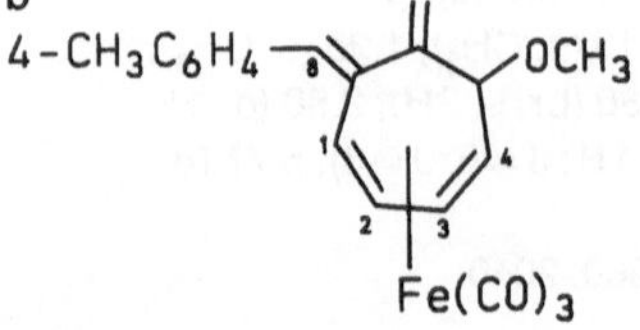	VIIc (90) yellow oil ${}^1H$ NMR (CS_2): 2.29 (s, CH_3), 2.93 (t, H-4; J=7), 3.11 (s, OCH_3), 3.83 (s, OCH_3), 4.22 (d, H-1; J=8), 5.47 (m, H-2,3,8,9), 6.97 (m, C_6H_4) IR (liquid film): 1955, 2070 (CO)	[38]
*147	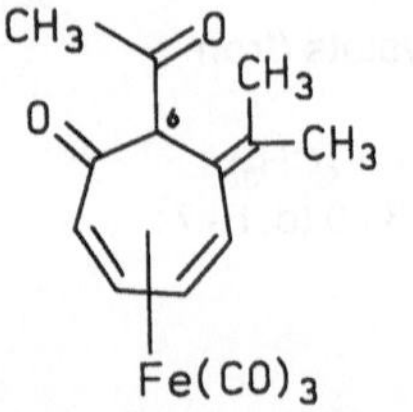	m.p. 126°, orange crystals ${}^1H$ NMR ($CDCl_3$): 1.76 (s, CH_3), 1.91 (s, CH_3), 2.04 (s, CH_3) ${}^1H$ NMR (benzene-d_6): 4.10 (s, H-6) IR ($CHCl_3$): 1655, 1720 (C=O)	[82]

References on pp. 99/102

Table 2 [continued]

No.	$^4LFe(CO)_3$	method of preparation (yield in %), properties and remarks	Ref.
148	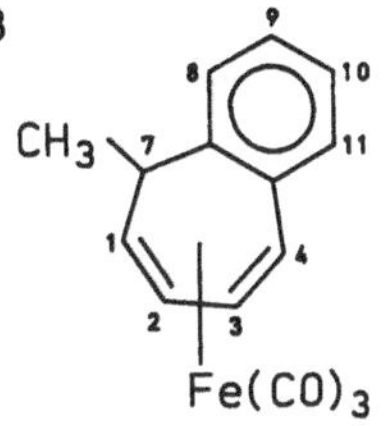	I b (95) m.p. 89.0 to 90.0° (from hexane at 0°), sublimes, yellow crystals ^{1}H NMR ($CCl_4/CH_3CN/Si(CH_3)_4$ (12:7:1)): 1.23 (d, CH_3; J=6.7), ca. 3.1 (m, H-7), 3.27 (sept, H-1), 3.68 (quint, H-4), 5.42 (quint, H-2,3), 7.00 (m, H-8 to 11) IR (CCl_4): 560, 617, 695, 730, 885, 930, 984, 1031, 1100, 1148, 1160, 1185, 1248, 1450, 1465, 1495; 1972, 1980, 2049 (CO); 2850, 2910, 2950, 3020, 3700 UV (hexane): $\lambda_{max}(\varepsilon)$ = 230 (33000), 285 (sh, 12500) nm	[14]

^{4}L is a hexasubstituted cyclohepta-1,3-diene:

No.	$^4LFe(CO)_3$	method of preparation (yield in %), properties and remarks	Ref.
*149	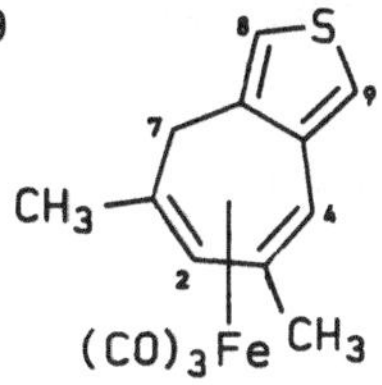	II b (18) m.p. 72° (from hexane), yellow solid ^{1}H NMR ($CDCl_3$): 1.75 (s, CH_3-1), 2.17 (s, CH_3-3), 3.10 (q, H-7), 3.76 (m, H-4), 5.26 (m, H-2), 6.77 (m, H-8), 6.92 (d, H-9) mass spectrum: $[M-nCO]^+$ (n=0 to 3) and further fragments given	[83, 92]
150	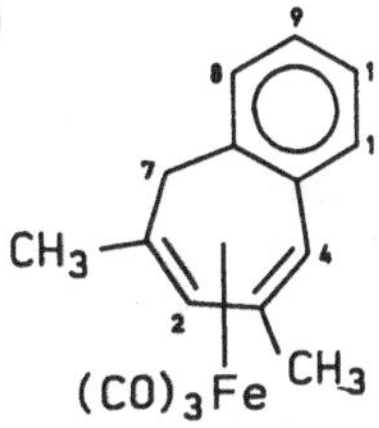	II b (23) m.p. 78° (from pentane), yellow solid ^{1}H NMR ($CDCl_3$): 1.80 (s, CH_3-1), 2.20 (s, CH_3-3), 3.65 (m, H-2,4), 5.27 (q, H-7), 7.08 (m, H-8 to 11)	[83]
151	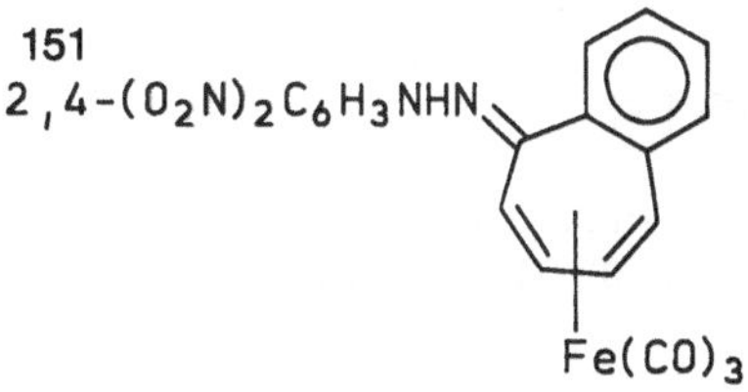 $2,4-(O_2N)_2C_6H_3NHN$=	UV (CH_2Cl_2): $\lambda_{max}(\varepsilon)$ = 24300 (26.6 × 10^3) cm^{-1}	[39]

^{4}L is an octasubstituted cyclohepta-1,3-diene:

No.	$^4LFe(CO)_3$	method of preparation (yield in %), properties and remarks	Ref.
*152	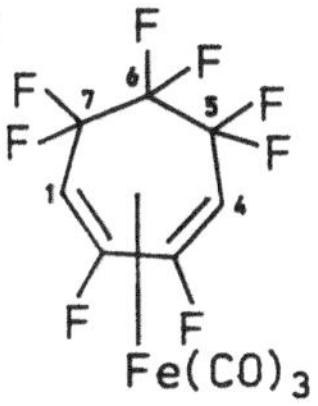	III a m.p. 86 to 87.5° (from light petroleum ether (b.p. 40 to 60°)), subl. p. 45°/17 Torr, yellow crystals ^{1}H NMR (acetone-d_6): 3.95 (H-1,4) ^{19}F NMR (acetone-d_6): 93.0 (F-5,7), 113.9 (F-6; J=276), 140.1 (F-6), 148.1 (F-2,3)	[51]

References on pp. 99/102

Table 2 [continued]

No.	$^4LFe(CO)_3$	method of preparation (yield in %), properties and remarks	Ref.
		IR: as expected mass spectrum: $[M-nCO]^+$ (n=0 to 3), $[Fe(CO)_n]^+$ (n=1 to 3), $[C_7F_6H_2]^+$, $[C_7F_5H_2]^+$, $[C_6F_4H_2]^+$	
^{4}L is a nonasubstituted cyclohepta-1,3-diene:			
*153	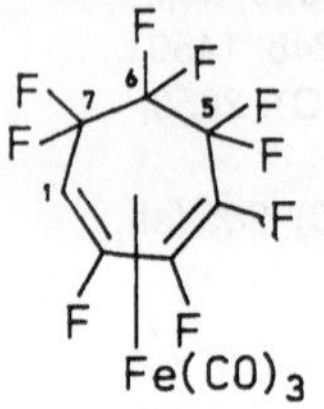	IIIa m.p. 81 to 82° (from light petroleum ether (b.p. 40 to 60°)), subl. p. 55°/17 Torr, yellow crystals ^{1}H NMR (acetone-d_6): 3.51 (H-1) ^{19}F NMR (acetone-d_6): 93.5 (F-7), 104.7 (F-5; J=259), 114.5 (F-6; J=276), 115.9 (F-5; J=259), 141.5 (F-6; J=276), 148.6 (F-2), 168.9 (F-3), 177.2 (F-4) IR: as expected mass spectrum: $[M-nCO]^+$ (n=0 to 3), $[Fe(CO)_n]^+$ (n=1 to 3), $[C_7F_7H]^+$, $[C_7F_6H]^+$, $[C_6F_5H]^+$	[22, 51]
^{4}L is a decasubstituted cyclohepta-1,3-diene:			
*154	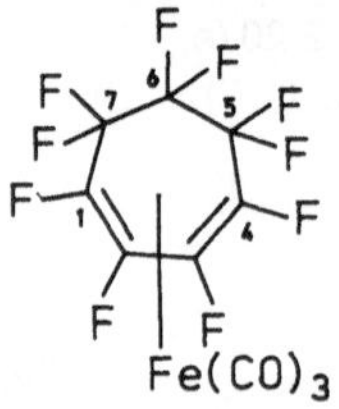	IIIa m.p. 90 to 91° (from light petroleum ether (b.p. 40 to 60°)), subl. p. 50°/17 Torr, yellow crystals ^{19}F NMR (acetone-d_6): 104.9 (F-5,7; J=263), 114.4 (F-6; J=281), 116.4 (F-5,7; J=263), 144.4 (F-6; J=281), 171.7 (F-2,3), 179.6 (F-1,4) IR: as expected mass spectrum: $[M-nCO]^+$ (n=0 to 3), $[Fe(CO)_n]^+$ (n=1 to 3), $[C_7F_8]^+$, $[C_7F_7]^+$, $[C_6F_6]^+$	[50, 51]
supplements:			
^{4}L is a monosubstituted cyclohepta-1,3-diene:			
155	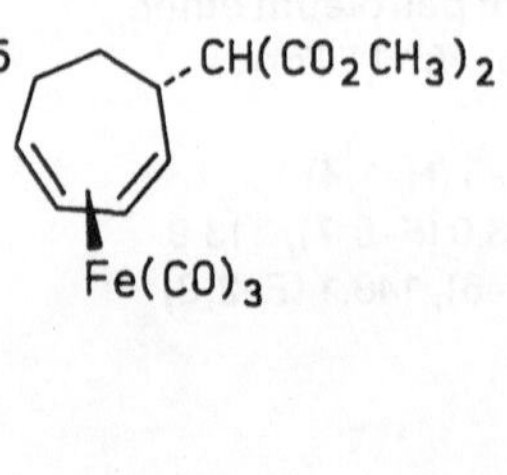 $CH(CO_2CH_3)_2$ $Fe(CO)_3$	VIb (45 to 50)	[124]

References on pp. 99/102

Table 2 [continued]

No.	$^4LFe(CO)_3$	method of preparation (yield in %), properties and remarks	Ref.
^4L is a disubstituted cyclohepta-1,3-diene:			
*156	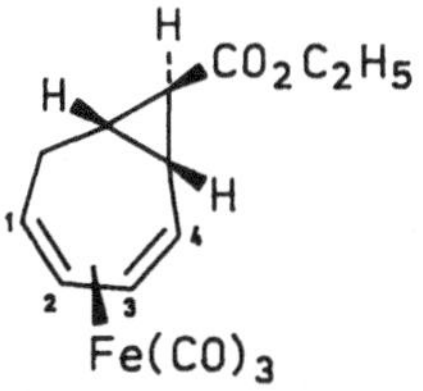 H, H, $CO_2C_2H_5$, H, 1, 2, 3, 4, $Fe(CO)_3$	m.p. 35°, b.p. 175°/0.1 Torr, orange oil ^1H NMR ($CDCl_3$): 1.23 (t, CH_3; J=7); 1.2, 1.54 (m), 1.92 (dd) (H-5,6,8; J=4, J=6), 2.30 (br, t, H-7; J=4), 3.09 (m, H-1), 3.49 (dd, H-2; J=6, J=8), 4.20 (CH_2O; J=7), 5.06 (m, H-2,3) IR (neat): 1710 (C=O); 1965 (br), 2040 (CO) mass spectrum: $[M-nCO]^+$ (n=0 to 3)	[130]
157	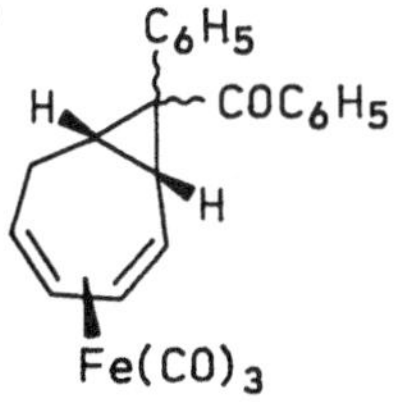 C_6H_5, H, COC_6H_5, H, $Fe(CO)_3$	not isolated, see "Further information" for No. 55 in Table 4, p. 176	[5]
158	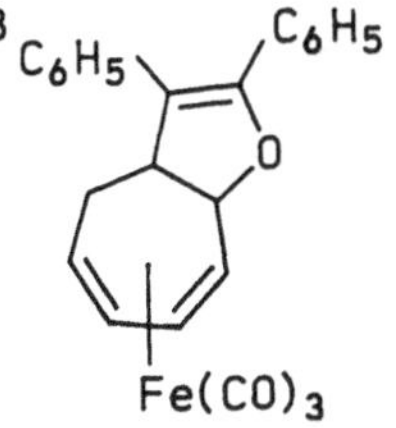 C_6H_5, C_6H_5, O, $Fe(CO)_3$	not isolated, see "Further information" for No. 55 in Table 4, p. 176	[5]
*159	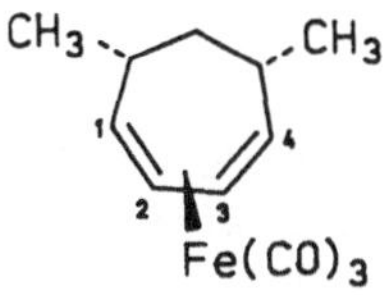 CH_3, CH_3, 1, 2, 3, 4, $Fe(CO)_3$	^1H NMR ($CDCl_3$): 0.86 (d, 2CH_3; J=7), 1.19 (m, 2H), 1.98 (m, 2H), 2.79 (m, H-1,4), 5.18 (m, H-2,3) IR ($CHCl_3$): 1970, 2040	[123, 124]
^4L is a trisubstituted cyclohepta-1,3-diene:			
*160	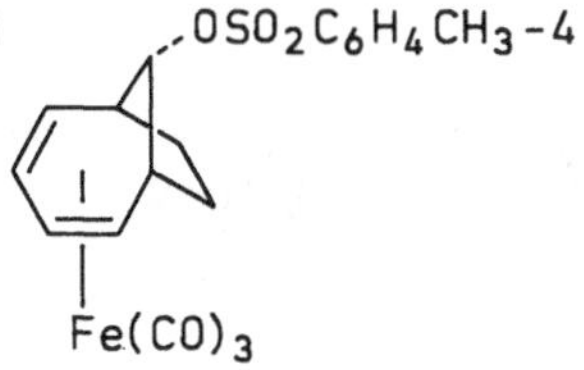 $OSO_2C_6H_4CH_3$-4, $Fe(CO)_3$	see No. 79b, p. 90	[118]
161	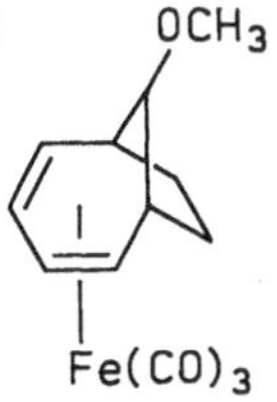 OCH_3, $Fe(CO)_3$	see No. 160, p. 98	[118]

Table 2 [continued]

No.	$^{4}LFe(CO)_3$	method of preparation (yield in %), properties and remarks	Ref.
*162	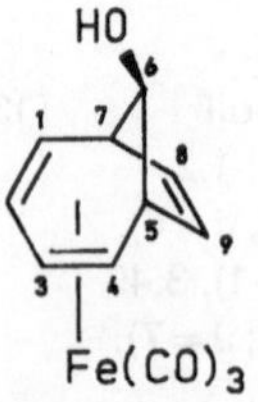	see No. 166, p. 98 pale yellow needles (from hexane) ^{13}C NMR ($CDCl_3$): 46.6 (C-5,7), 66.0 (C-1,4), 79.9 (C-6), 91.9 (C-2,3), 142.0 (C-8,9), 211.2 (CO) IR: 1984, 1992, 2053 (CO)	[120]
*163	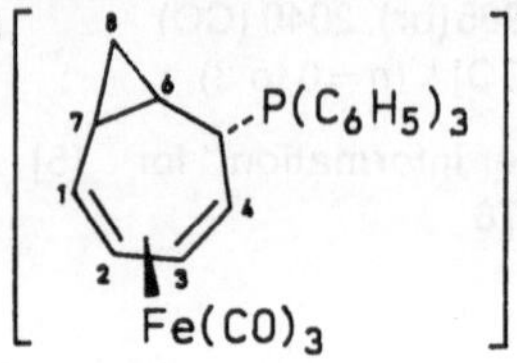 $[BF_4]^-$	^{13}C NMR (CD_3NO_2): 14.4 (C-8; J(P,C) = 9.0), 14.7 (C-7), 17.2 (C-6; J(P,C) = 5.0), 35.5 (C-5; J(P,C) = 36.0), 53.1 (C-4; J(P,C) = 7.0), 63.5 (C-1), 89.3 (C-2,3)	[128]

^{4}L is a tetrasubstituted cyclohepta-1,3-diene:

No.	$^{4}LFe(CO)_3$	method of preparation (yield in %), properties and remarks	Ref.
*164	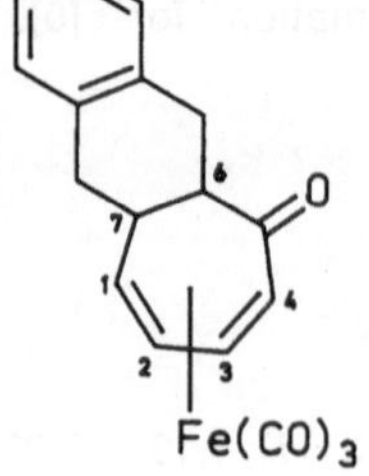	Vd (92) m.p. 121 to 122°, yellow crystals (from n-hexane) ^{1}H NMR ($CDCl_3$): 2.5 to 2.9 (m, 6H), 2.96 (d, H-4; J = 7.0), 3.30 (m, H-6), 5.14 (dd, H-2; J = 5.5, J = 8.0), 5.67 (dd, H-3; J = 5.5, J = 7.0), 7.0 to 7.2 (m, 4H) IR: 1645, 2000, 2090 mass spectrum: $[M-nCO]^+$ (n = 0 to 3)	[127]
*165a	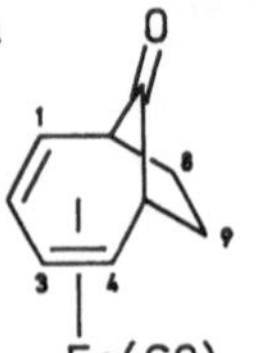	see also No. 166, p. 99 ^{13}C NMR (benzene-d_6): 27.0 (C-8,9), 49.1 (C-5,7), 67.4 (C-1,4), 90.1 (C-2,3), 210.6 (CO), C-6 not observed due to low intensity IR: 1765 (C=O); 1992 (br), 2058 (CO)	[118, 120]
b	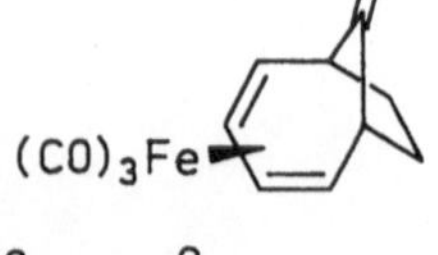 $(CO)_3Fe$	see also No. 166, p. 99 ^{13}C NMR (benzene-d_6): 26.8 (C-8,9), 43.5 (C-5,7), 62.4 (C-1,4), 87.1 (C-2,3), 204.9 (C=O), 209.9 (CO) IR: 1765 (C=O); 1986, 2005, 2063 (CO)	[118, 120]
*166a	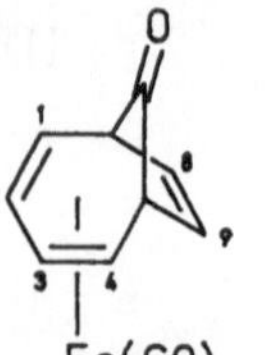	yellow crystalline platelets (from n-hexane at −30°) ^{13}C NMR (benzene-d_6): 52.6 (C-5,7), 72.8 (C-1,4), 92.4 (C-2,3), 137.7 (C-8,9), 200.7 (C=O), 210.1 (CO) IR: 1781 (C=O); 1992, 1997, 2059 (CO)	[120]

Table 2 [continued]

No.	$^4LFe(CO)_3$	method of preparation (yield in %), properties and remarks	Ref.
b	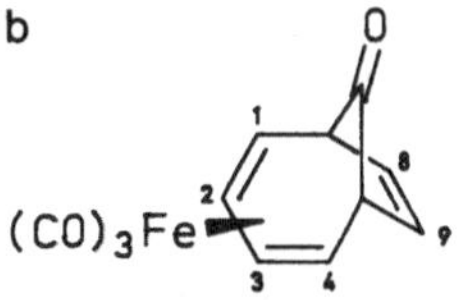	yellow needles (from hexane) ^{13}C NMR (benzene-d_6): 46.7 (C-5,7), 54.5 (C-1,4), 87.2 (C-2,3), 138.7 (C-8,9), 196.5 (C=O), 209.5 (CO) IR: 1775 (C=O); 1988, 2007, 2064 (CO)	[120]
*167	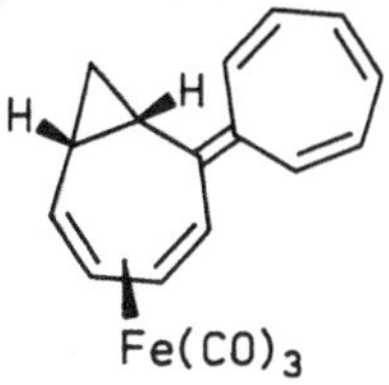	see No. 123a, p. 94 moderately stable brown oil ^1H NMR (CCl_4): 0.23 (m, 1H), 1.07 (m, 1H), 1.56 (m, 2H), 3.32 (m, 1H), 3.52 (m, 1H), 5.11 (m, 2H), 5.3 to 6.4 (m, 6H) IR (CCl_4): 1975, 1987, 2040 mass spectrum: $[M-nCO]^+$ (n=0 to 3)	[122]
168	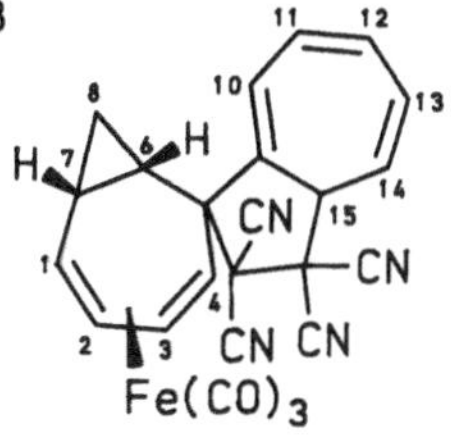	see No. 167, p. 99 m.p. 144 to 146° (dec.) ^1H NMR ($CDCl_3$): 0.65 (m, 1H), 1.25 (m, 1H), 1.35 (m, 1H), 1.86 (m, H-7), 2.59 (d, H-4; J=7.0), 3.24 (m, H-15), 3.87 (t, H-1; J=6.0), 5.18 (dd, H-2; J=5.0, J=6.0), 5.42 (dd, H-3; J=5.0, J=7.0), 5.69 (dd, H-14; J=4.0, J=9.0), 6.45 (m, H-10, 13), 6.79 (m, H-11, 12) IR (KBr): 1985, 1990, 2050 mass spectrum: $[M-nCO]^+$ (n=0 to 3), $[M-3CO-C_2(CN)_4]^+$	[122]

^4L is a pentasubstituted cyclohepta-1,3-diene:

No.	$^4LFe(CO)_3$	method of preparation (yield in %), properties and remarks	Ref.
*169	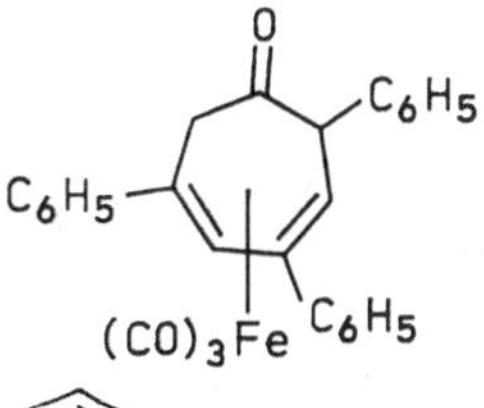	m.p. 141 to 144° (dec.), yellow needles IR: 1706 (C=O); 1968, 2036 (CO)	[110]
*170	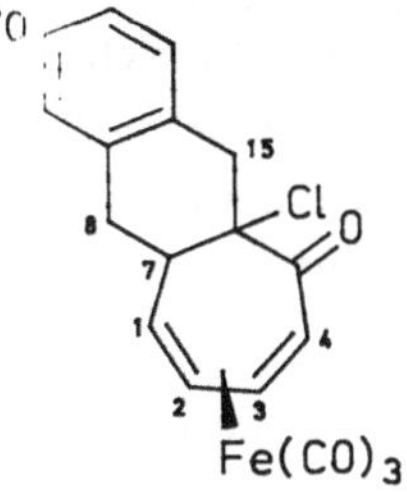	Vd (55) m.p. 129 to 130° (dec., from n-hexane) ^1H NMR (benzene-d_6): 2.16 (d, H-8; J=16.5), 2.24 (d, H-1; J=8.0); 2.84 and 3.17 (ABq, 2H-15; J(AB)=19.0), 2.84 (d, H-4; J=6.5), 3.13 (d, H-7; J=6.0), 3.37 (dd, H-8; J=6.0, J=16.5), 4.13 (dd, H-2; J=6.0, J=8.0), 4.70 (dd, H-3; J=6.0, J=6.5), 6.6 to 7.1 (m, 4H) IR: 1670, 2020, 2120 mass spectrum: $[M+n]^+$ (n=0, 2)	[127]

References on pp. 99/102

Table 2 [continued]

No.	$^4LFe(CO)_3$	method of preparation (yield in %), properties and remarks	Ref.
171	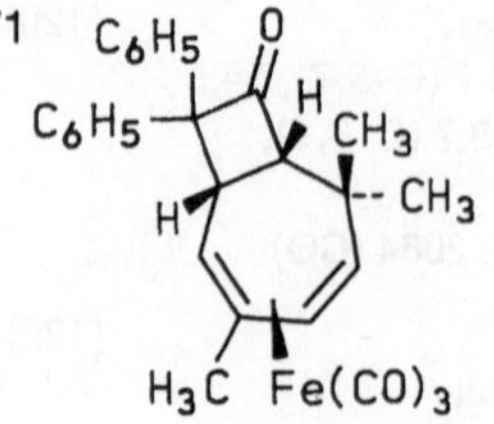	Vc (5) m.p. 141 to 142°, bright yellow crystals (from CH_2Cl_2/hexane) ^1H NMR ($CDCl_3$): 1.00 (s, CH_3), 1.30 (s, CH_3), 1.38 (s, CH_3), 2.59 (d, 1H; J=8.0), 2.92 (d, 1H; J=9.8), 3.09 (br,s, 1H), 3.83 (dd, 1H; J=3.1, J=9.8), 4.96 (d, 1H; J=8), 7.2 to 7.8 (m, 10H) IR ($CHCl_3$): 1770, 1970, 2035	[126]
*172	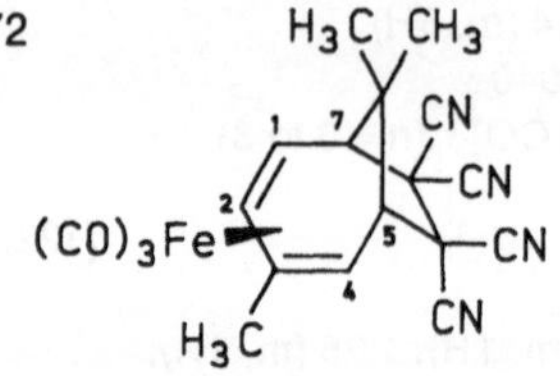	^1H NMR: 1.24 (s, CH_3), 1.42 (s, CH_3), 2.38 (s, CH_3), 2.60 (H-5), 2.87 (H-7), 3.17 (H-4), 3.77 (H-1), 5.29 (H-2)	[126]
	^4L is a heptasubstituted cyclohepta-1,3-diene:		
*173	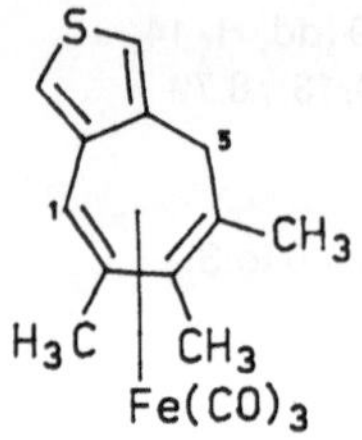	m.p. 87° (from hexane) ^1H NMR: 1.75 (s, CH_3-4), 1.80 (s, CH_3-3), 3.10 (dd, H-5; J(H-5, CH_3-4)=6.8), 5.90 (m, H-1), 6.70 (d, H-10), 6.90 (d, H-8; J(H-8, 10)=4.5), 7.00 (d, CH_3-2) IR: 1964, 1980, 2040	[119]

*Further information:

5-$HOC_7H_9Fe(CO)_3$ (Table **2**, No. **1**). The temperature dependent ^{13}C NMR spectrum shows scrambling of the three CO groups; the coalescence temperature being −16(3) °C. The scrambling proceeds in a concerted manner [75].

Attempts to obtain the free organic ligand 5-HOC_7H_9 (^4L) by oxidation with ferric ions failed (an impure sample is obtained by oxidation with $ON(CH_3)_3$). The ketone cyclohepta-2,4-dienone is obtained by oxidation with pyridinium chlorochromate in 32% yield [79].

5-$CH_3OC_7H_9Fe(CO)_3$ (Table **2**, No. **2**). The exo compound No. 2a is also obtained directly from $C_7H_9Fe(CO)_2CO_2CH_3$ on raising the temperature by spontaneous rearrangement both in the solid state and in solution ($CDCl_3$). First order kinetics are obeyed which suggests a dissociative mechanism with loss of OCH_3^- and subsequent ring addition to give No. 2a as confirmed by rearranging $C_7H_9Fe(CO)_2CO_2C_2H_5$ in $CDCl_3$ in the presence of a fivefold excess of $HOCH_3$, which also produces No. 2a [90]. Independent experiments prove that No. 2a cannot be interconverted by a fivefold excess of C_2H_5OH in $CHCl_3$ at room temperature for 15 h into compound No. 3a; only the starting compound No. 2a is recovered on work-up

[61, 90]. Oxidation with $[NH_4]_2[Ce(NO_3)_6]$ brings about rapid decomposition. Freshly sublimed $ON(CH_3)_3$ in acetone at room temperature (24 h) or in boiling C_6H_6 (3 h) liberates the organic ligand 5-$CH_3OC_7H_9$ (^{4}L) in 42% yield [79].

5-$C_2H_5OC_7H_9Fe(CO)_3$ (Table **2**, No. **3**) is also prepared directly from $C_7H_9Fe(CO)_2CO_2C_2H_5$ in hexane for 12 h. The solution is filtered through $MgSO_4$ followed by concentration and cooling. It is also formed in the reaction of $C_7H_9Fe(CO)_2CO_2CH_3$ with a fivefold excess of C_2H_5OH at room temperature for 15 h. After evaporation of solvent, solubilization in pentane, and filtration through $MgSO_4$, compound No. 3 is obtained on concentration and cooling. Compound No. 3 undergoes no change after 64 h in CH_3OH at 20 °C [61, 90]. Reaction with $[C(C_6H_5)_3]BF_4$ in CH_2Cl_2 gives $[C_7H_9Fe(CO)_3]BF_4$ [13].

5-$(CH_3)_2CHOC_7H_9Fe(CO)_3$ (Table **2**, No. **4**) is also prepared directly from $C_7H_9Fe(CO)_2CO_2$-C_3H_7-i as described for No. 3 [61, 90].

5-$C_6H_5OC_7H_9Fe(CO)_3$ (Table **2**, No. **5**) is oxidized by $ON(CH_3)_3$ in boiling C_6H_6 within 3 h to give 5-$C_6H_5OC_7H_9$ (^{4}L) in 39% yield [79].

5-$C_6H_5SC_7H_9Fe(CO)_3$ (Table **2**, No. **7**) is oxidized by freshly sublimed $ON(CH_3)_3$ in acetone at room temperature (24 h) or in boiling C_6H_6 (3 h) to give 5-$C_6H_5SC_7H_9$ (^{4}L) in 33% yield. Oxidation with $[NH_4]_2[Ce(NO_3)_6]$ brings about rapid decomposition [79].

5-$N_3C_7H_9Fe(CO)_3$ (Table **2**, No. **10**) is decomposed by attempted catalytic hydrogenation to the amine with Pd/C in CH_3OH or with $LiAlH_4$ [79].

5-$XCNC_7H_9Fe(CO)_3$ (Table **2**, Nos. **11** and **12** with X=S or Se) rearrange rapidly on exposure to air to Nos. 6 and 9. But accompanying decomposition precludes isolation of pure No. 9 [60].

5-$(CH_3)_3CNHC_7H_9Fe(CO)_3$ (Table **2**, No. **15**) is oxidized as shown for No. 7 to give 5-$(CH_3)_3CNHC_7H_9$ (^{4}L) in 72% yield [79].

5-$C_6H_5NHC_7H_9Fe(CO)_3$ (Table **2**, No. **16**). For the ^{1}H NMR spectra of both isomers given in a figure see [78]. Oxidation of both isomers with $ON(CH_3)_3$ at room temperature for 12 h gives 5-$C_6H_5NHC_7H_9$ (^{4}L) with different R_f values if chromatographed on basic alumina [78].

5-$(CH_3)_2NC_7H_9Fe(CO)_3$ (Table **2**, No. **21**) is oxidized by $ON(CH_3)_3$ as shown for No. 7 to give 5-$(CH_3)_2NC_7H_9$ (^{4}L) in 52% yield [79]. A vigorous reaction occurs with CH_3I, but the precipitate is too unstable to be recrystallized [8].

5-$C_6H_5NOC_7H_9Fe(CO)_3$ (Table **2**, No. **25**) is prepared by mixing $[C_7H_9Fe(CO)_3]BF_4$ and ONC_6H_5. Similar reaction with ONC_6D_5 gives 5-$C_6D_5NOC_7H_9Fe(CO)_3$. The ESR spectrum shows hyperfine splitting constants at 1.00 (H-3′,5′), 2.80 (H-2′,4′), 2.95 (H-5,6′), and 11.10 (NO) Gauss. The deuterated derivative shows triplets of doublets at 3.00 (H-5) and 10.88 (NO) Gauss [73].

5-$CH_3C_7H_9Fe(CO)_3$ (Table **2**, No. **26**) is prepared by the reaction of $[C_7H_9Fe(CO)_3]BF_4$ and $LiCu(CH_3)_2$ (from 0.06 mol CuI and 1.5 mol $LiCH_3$ in ether at −15 °C) in ether at room temperature [79, 124]. After 2 h the reaction mixture is poured into H_2O, extracted with petroleum ether, and chromatographed on deactivated alumina. Distillation gives compound No. 26a in 49% yield [79] or in 15 to 20% yield together with $(CO)_3FeC_7H_9C_7H_9Fe(CO)_3$ (see 2.8.1.1 in "Organoiron Compounds" C5, 1981, Table 4, No. 37, p. 51) [124]. The isomer No. 26b is obtained by refluxing XXVIII in cyclohexane for 8 h followed by chromatography on silica gel with pentane as eluent. In this latter product the CH_3 doublet at 0.48 ppm (see Table 2, p. 49) integrates for 2.6 H relative to the olefin signals and the signals between 0.8 and 1.8 ppm due to ring protons for somewhat more than 5.0 H. It is suggested that this is

 References on pp. 99/102

due to a small amount of another isomeric product [54]. Oxidation of No. 26a as described for No. 7 gives 5-$CH_3C_7H_9$ (^{4}L) in 48% yield [79]. Reaction of No. 26a with $[C(C_6H_5)_3]BF_4$ gives [6-exo-$CH_3C_7H_8Fe(CO)_3$]BF_4 [123, 124], and reaction of No. 26b with $[C(C_6H_5)_3]BF_4$ in CH_2Cl_2 gives [1-$CH_3C_7H_8Fe(CO)_3$]BF_4 [54].

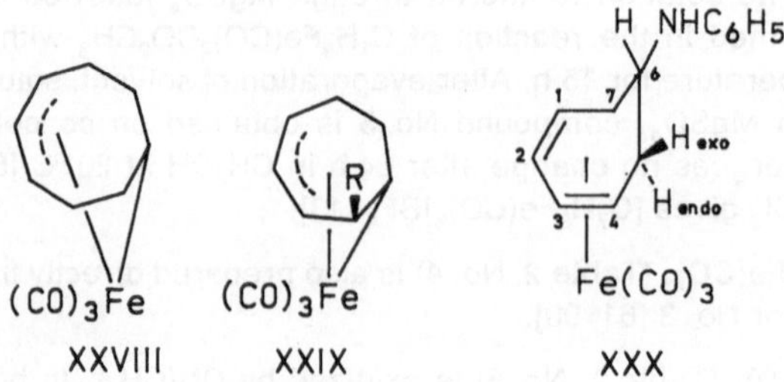

5-$(C_2H_5O_2C)_2CHC_7H_9Fe(CO)_3$ (Table **2**, No. **28**) does not react with $[C(C_6H_5)_3]BF_4$ in CH_2Cl_2 [13].

5-$C_5H_5C_7H_9Fe(CO)_3$ (Table **2**, No. **29**) is prepared by the reaction of $[C_7H_9Fe(CO)_3]BF_4$ with TlC_5H_5 in CH_2Cl_2 in the dark for 12 h followed by chromatography on alumina with 5% $CH_3CO_2C_2H_5$/light petroleum ether (b.p. 30 to 40 °C). Sublimation (some decomposition) gives a mixture of No. 29 (85%) and XXIX (R=C_5H_5, 15%) in a total yield of 55% [52].

5-$NCC_7H_9Fe(CO)_3$ (Table **2**, No. **31**) gives $[C_7H_9Fe(CO)_3]BF_4$ in the reaction with $[C(C_6H_5)_3]BF_4$ in CH_2Cl_2 [13].

6-$HOC_7H_9Fe(CO)_3$ (Table **2**, No. **32**) is prepared in 95% yield [79] from $C_7H_8Fe(CO)_3$ by reduction with B_2H_6 followed by oxidation (OH^-, H_2O_2) [79, 103]. Oxidation with $FeCl_3 \cdot 9H_2O$ in 95% C_2H_5OH/H_2O for 2 h gives 51% yield of 6-HOC_7H_9 (^{4}L) [79]. Similar oxidation with $[NH_4]_2[Ce(NO_3)_6]$ in C_2H_5OH/H_2O gives also 6-HOC_7H_9 [103], but oxidation with pyridinium chlorochromate yields cyclohepta-3,5-dienone [79].

6-$C_6H_5NHC_7H_9Fe(CO)_3$ (Table **2**, No. **33**). The endo configuration as shown in Formula XXX for this compound is assigned on the basis of the ^{1}H NMR spectrum (see figure in [78]). The high field 4-signal pattern, which integrates for two protons, is identified as an A_2 branch of an A_2B_2X system (H-5,6,7). This A_2 branch is assigned to H-5endo, 7endo of XXX (structure XXX possesses a plane of symmetry). The two large coupling constants observed, J=17 and J=12 Hz, are assigned to the geminal and H-6 spin interactions, respectively. On the basis of the magnitude of the latter J value (12 Hz) the exo configuration is assigned to H-6 in XXX. Consequently, the anilino group at C-6 must possess the endo configuration. The H-1,4,6 resonate at the same field value (δ=ca. 3.2 ppm). Irradiation at the center of the complex signal transforms the multiple high field AB pattern to a sharp 4-line A_2B_2 spectrum (J=17 Hz). This behavior corroborates the above assignments of the signals [78].

6-$CH_3O_2CCH_2CH_2C_7H_9Fe(CO)_3$ (Table **2**, No. **34**) is prepared by the photochemical reaction of $CH_3O_2CCH{=}CH_2Fe(CO)_4$, in the presence of methyl acrylate, with cyclohepta-1,3-diene. The product is also obtained using $(CH_3O_2CCH{=}CH_2)_2Fe(CO)_3$ as the starting complex in a thermal reaction [102].

[5-$C_6H_5(CH_3)_2PC_7H_9Fe(CO)_3$]$BF_4$ (Table **2**, No. **48**) is dissolved in CH_2Cl_2 and cooled to −78 °C. NaH and C_6H_5CHO (molar ratio 1:2:2) are then added. A rapid reaction occurs at about −20 °C. After hydrolysis, the organic layer is separated, dried ($MgSO_4$), and evaporated. Chromatography of the residue on silica gel and elution with pentane gives No. 58 as an inseparable 1:2 mixture of isomers in 80% yield [33], cf. [48]. Similar treatment using

$OHCC_6H_5Cr(CO)_3$ (see 1.6.1.2.1.3 in "Chrom-Organische Verbindungen", Erg.-Werk Bd. 3, 1971, Table 40, No. 32, p. 202) instead of C_6H_5CHO followed by chromatography on silica plates with a C_6H_6/pentane mixture gives two bands. The first band gives one isomer of No. 59 (m.p. 138 to 140 °C) in 57% yield and the second band the other isomer (m.p. 158 to 160 °C) in 38% yield [33]. Attempts to liberate the organic ligand ^4L by treatment with high pressures of CO at high temperatures are unsuccessful [33].

$[5-(C_6H_5)_3PC_7H_9Fe(CO)_3]BF_4 \cdot CH_2Cl_2$ (Table **2**, No. **50**) is oxidized with $[NH_4]_2[Ce(NO_3)_6]$ in CH_3OH followed by addition of $NaBF_4$ to give the uncomplexed phosphonium salt $[5-(C_6H_5)_3PC_7H_9]BF_4$ in 70 to 80% yield [128]. Compound No. 50 does not react with $N(C_2H_5)_3$ or 1,8-bis(dimethylamino)naphthalene [85].

5-exo-CH_3(5-endo-HO)$C_7H_8Fe(CO)_3$ (Table **2**, No. **51**). The steric configuration is derived from the ^1H NMR spectrum in the presence of Eu(X-D)$_3$ (X-D = 2,2,6,6-tetramethylheptanedionato) [28].

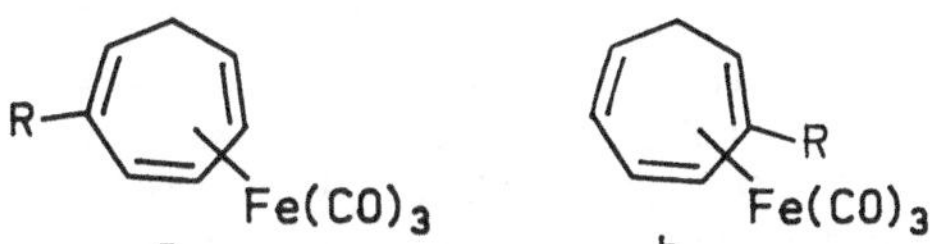

R Fe(CO)$_3$ a XXXI b R Fe(CO)$_3$ O Fe(CO)$_3$ XXXII

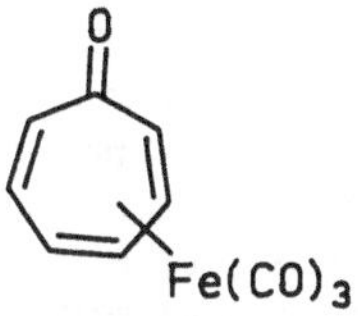

5-$CH_3O_2CCH_2$(5-HO)$C_7H_8Fe(CO)_3$ (Table **2**, No. **52**) gives $[1-CH_3O_2CCH_2C_7H_8Fe(CO)_3]BF_4$ in the reaction with $[C(C_6H_5)_3]BF_4$ [34, 37] in CH_2Cl_2 [37].

5-R(5-HO)$C_7H_8Fe(CO)_3$ (Table **2**, Nos. **53** and **54** with R = C_6H_5, $C_6H_4CH_3$-4) are dehydrated on silica gel (eluent: toluene) to give a mixture of XXXIa and b in a total yield of ca. 60% [76].

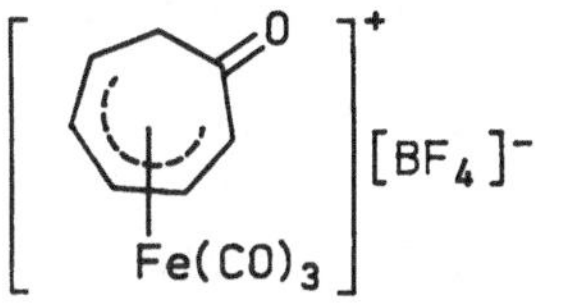

[O Fe(CO)$_3$]$^+$ $[BF_4]^-$ XXXIII O R (CO)$_3$Fe XXXIV

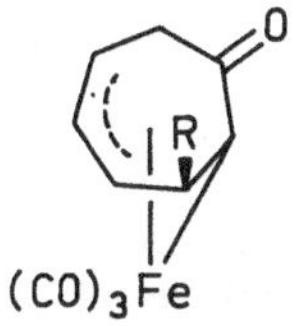

$OC_7H_8Fe(CO)_3$ (Table **2**, No. **55**) is also prepared by the reaction of XXXII with H_2/Pd (10% charcoal) [3, 43] in C_2H_5OH at 65 °C and 150 atm in an autoclave for 6 h. The reaction mixture is filtered, evaporated, and the residue is dissolved in C_6H_6. The insoluble part of the residue gives small amounts of a tetranuclear complex. Concentration of the benzene solution and addition of petroleum ether give No. 55 in 46% yield [3]. Similar reaction at 55 to 60 °C gives No. 55 and small amounts of $C_{14}H_{14}O_2Fe_2(CO)_6$ (see 2.8.1.1 in "Organoiron Compounds" C5, 1981, Table 4, No. 38, p. 51) and a trinuclear compound [3]. Reduction of XXXIII with $NaBH_4/H_2O$ covered with ether at 0 °C for 10 min, followed by extraction with ether, drying of the organic layer, and chromatography of the residue on silica gel after evaporation gives No. 55 and XXXIV (R = H) [70]. The compound No. 55 is also formed in the reaction of XXXIII with $HSi(C_2H_5)_3$. This reduction takes place with a high degree of stereospecificity as demonstrated by deuteration experiments by a stereoselective process involving cis addition of deuterium to the exo side of the ring [26, 43]. Reduction of XXXII with $HSi(C_2H_5)_3$ in CF_3CO_2H for 5 min also gives No. 55. The reaction mixture is diluted with CH_2Cl_2, then neutralized with K_2CO_3 followed by addition of H_2O. The CH_2Cl_2 layer is washed with K_2CO_3, dried ($MgSO_4$), and chromatographed with ether/hexane (2:1)

References on pp. 99/102

to give No. 55 in 75% yield [111]. Compound No. 55 is also formed in the reaction of XXXII with excess cyclopentadiene at 70 to 80 °C in an autoclave. After 3 d the mixture of products is separated by chromatography on silica gel to give XXXV, No. 55, $(C_5H_5Fe(CO)_2)_2$, and No. 133, the latter in 25% yield [74]. Heating XXXVI in C_6H_6 near reflux gives No. 55 in 25 to 30% yield [80]. Deuterated derivatives are prepared in different ways. Reaction of No. 55 with $NaOC(CH_3)_3$ in $DOC(CH_3)_3$ at room temperature for 3 h followed by acidification with dilute D_2SO_4, neutralization with K_2CO_3, extraction with ether, drying the organic layer ($MgSO_4$), and sublimation gives XXXVIIIb (R=R'=R'''=H, R''=D) in 82% (isotopical purity >92%) as yellow solid. Reduction of XXXII with $DSi(C_2H_5)_3$ as above gives XXXVIIIb (R=R'=R''=H, R'''=D). Compound XXXVIIIb (R=R'=D, R'=R'''=H) is mentioned, but no details of the preparation are given. Reduction of XXXII with D_2 (167 atm) in the presence of 10% Pd/C in hexane at 40 °C in an autoclave followed by chromatography on silica gel (70 to 325 mesh) with C_6H_6/ether (5:1) as eluent gives a rapidly decomposing product, presumably dideuterated No. 60, as the first band and XXXVIIIb (R=R'=H, R''=R'''=D) in 30% yield (purity >96%), m.p. 102 to 104 °C, subl.p. 50 °C/0.02 Torr, from the second band. This latter compound can also be obtained from No. 55 with $DSi(C_2H_5)_3$ dissolved in CF_3CO_2D as described above. Analogous reaction of No. 55 with $HSi(C_2H_5)_3$ but in D_2SO_4 gives XXXVIIIb (R=R'=R''=D, R'''=H). No details are given for the preparation of XXXVIIIb (R=R'=R'''=D, R''=H), and of XXXVIIIb (R=R'''=H, R'=R''=D). XXXVIIIb (R=R'=R''=R'''=D) is prepared from No. 55 with $DSi(C_2H_5)_3$ in D_2SO_4. The ^1H NMR spectra of these deuterated complexes XXXVIIIb are as expected and are employed to assign the ^1H NMR spectrum of No. 55 [111].

The compound No. 55 is monomeric in C_6H_6 (by cryoscopy), and shows in C_6H_6 at 25.0 °C a dipole moment of $\mu_D=4.53\pm0.06$ D, $\mu_{15\%}=4.48\pm0.07$ D [3]. For the effect of $Eu(X\text{-}D)_3$ (X-D=2,2,6,6-tetramethylheptanedionato) on the chemical shifts in the ^1H NMR spectrum see [28]. The ^{13}C NMR chemical shifts of the CO groups show at 183 K, $\delta=210.6$, 210.9 (basal CO), 214.6 (apical CO) ppm [107]. The coalescence temperature is −13(3) °C [75]. The activation parameters for the scrambling of the CO groups [107], which proceeds in a concerted manner [75, 107], are $\Delta H^{*}=10.4\pm0.5$ kcal/mol, $\Delta S^{*}=-3.0\pm2$ e.u., $\Delta G^{\ddagger}_{298}=11.3\pm0.5$ kcal/mol [107].

Electrochemical reduction affords $[OC_7H_8Fe(CO)_3]^-$ which shows hyperfine coupling with $\Delta H=8$ Gauss (very weak), g=2.0572 and with $\Delta H=8$ Gauss (strong), g=2.0572 and no evidence of spin density on the organic ligand ^4L [10]. For a description of No. 55 as a resonance hybrid of valence bond structures see [94].

The compound shows stability similar to that of XXXII [3]. Polarographic, cyclic triangular, and potential electrolytic studies of No. 55 (2×10^{-3} M) in $CH_3OCH_2CH_2OCH_3$ with $[N(C_4H_9)_4]ClO_4$ as supporting electrolyte against the reference electrode $AgClO_4$ (10^{-3} M)/Ag shows $-E^{I}_{1/2}=2.1$ V, $-E_{p,c}=2.3$ V, and $-E_{p,a}=1.8$ V, and $[OC_7H_8Fe(CO)_3]^+$ (n=1.5) is formed [10].

Protonation with CF_3CO_2H is shown to occur at the oxygen to give the expected XXXVII [26, 43, 89]. Deuteration with CF_3CO_2D or D_2SO_4 gives XXXVII with OD instead of OH [111]. Refluxing No. 55 with $P(C_6H_5)_3$ (molar ratio 1:2) in xylene gives $OC_7H_8Fe(CO)_2P(C_6H_5)_3$ [3]. Reaction with $H_2NNHC_6H_3(NO_2)_2$-2,4 [3, 37] in C_2H_5OH/H_3PO_4 at ca. 100 °C followed by concentration of the reaction mixture [3] gives No. 56 [3, 37]. Reaction under the condition of the Mannich reaction [16] ($[H_2N(CH_3)_2]Cl$/paraformaldehyde, reflux C_2H_5OH, 15 h [37]) gives No. 83 [16, 37] after cooling to 0 °C followed by addition of H_2O [37]. NaOH is added to the aqueous layer followed by extraction with ether and evaporation [37] to give No. 83 in 27 [37] or 88% yield [16]. Under the conditions of the Reformatzky reaction, No. 52 is formed with $Zn/BrCH_2CO_2CH_3$ [16, 37]. Reaction with a tenfold excess of XMgR ($R=C_6H_5$,

$C_6H_4CH_3$-4) in ether at −40 °C for 0.5 h followed by addition of aqueous $[NH_4]Cl$ until hydrolysis is complete, separation of the organic layer which is washed with H_2O, dried ($MgSO_4$), and evaporated gives Nos. 53 and 54 [76]. Conversion of No. 55 to XXXVIIIa by hydride abstraction using $[C(C_6H_5)_3]BF_4/K_2CO_3$ involves removal of the same exo hydrogen introduced in the reduction of XXXIII (see above) [26, 43, 111]. Thus, reaction of XXXVIIIb (R=R′=H, R″=R‴=D) with $[C(C_6H_5)_3]BF_4$ in CH_2Cl_2 gives LVIII (p. 92) (R′=R″=H, R=H) which on treatment with K_2CO_3 gives XXXII [111].

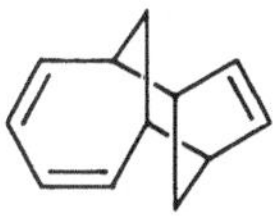

XXXV 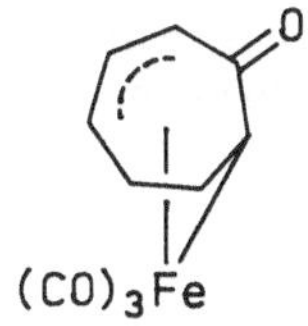 XXXVI 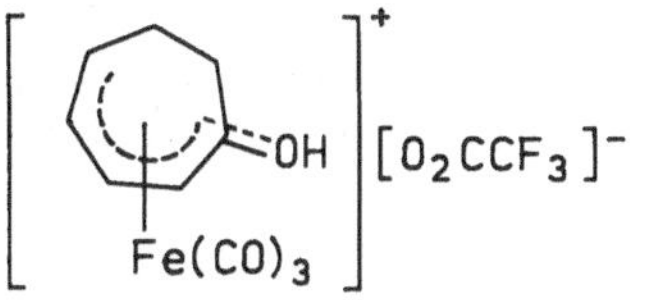 XXXVII XXXVIII

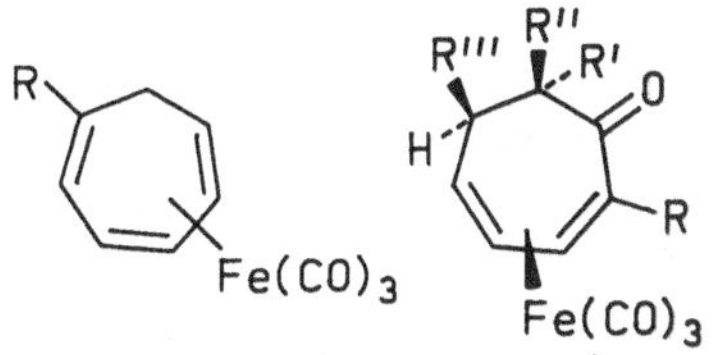

$\mathbf{CH_3CO_2CHC_7H_8Fe(CO)_3}$ (Table **2**, No. **57**). An X-ray analysis confirms the exocyclic double bond [34].

$\mathbf{OC_7H_8Fe(CO)_3}$ (Table **2**, No. **60**) is prepared by the reaction of XXXII with H_2/Pd (10% charcoal) in petroleum ether at 40 °C and 150 atm within 66 h in an autoclave followed by chromatography on silica gel with C_6H_6/ether (0 to 2 parts ether) in 20% yield. Further elution with C_6H_6/ether (5 to 50 parts ether) gives No. 55 in 61% yield [3]. For a dideuterated derivative see "Further information" for No. 55 [26, 111]. The temperature dependent ^{13}C NMR spectrum shows scrambling of the three CO groups proceeding in a concerted manner with $\Delta G^{\ddagger}_{298}$=11.3 kcal/mol [75]. The compound is very air-sensitive, even in the solid state, and gives No. 61 in the reaction with $H_2NNHC_6H_3(NO_2)_2$-2,4 in C_2H_5OH/H_3PO_4 [3].

$\mathbf{5\text{-}C_2H_5(6\text{-}CH_3CO)C_7H_8Fe(CO)_3}$ (Table **2**, No. **62**) is prepared by the reaction of XXXVIIIa (R=$COCH_3$) with C_2H_5MgBr in ether at −50 to −30 °C for 20 min followed by hydrolysis with aqueous $[NH_4]Cl$. The aqueous layer contains $[CH_3(C_2H_5)C{=}C_7H_6Fe(CO)_3]^+$, and neutralization of the acidic ethereal solution with $NaHCO_3$, followed by washing with H_2O, drying ($MgSO_4$) gives a ca. 1:6 mixture of XXXVIIIa (R=$C(C_2H_5)(CH_3)OH$, not isolated) and No. 62. Chromatography on silica gel with 5% $CH_3CO_2C_2H_5$/toluene followed by distillation gives pure No. 62 in 60% yield [38].

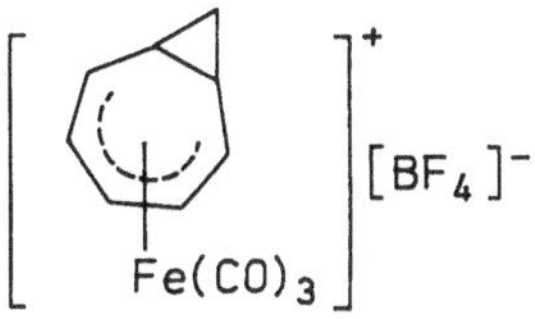

XXXIX 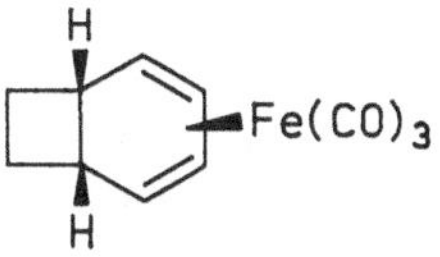 XL 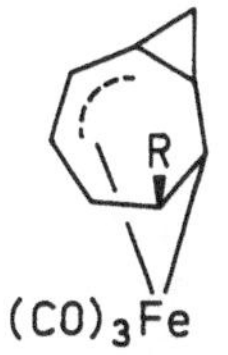 XLI 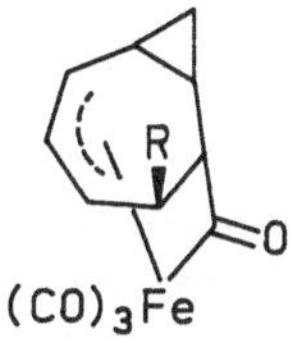 XLII 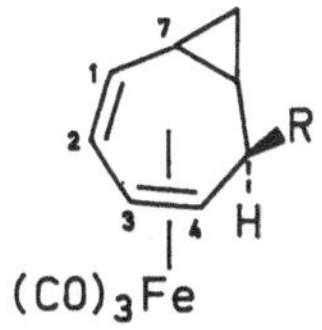 XLIII

$\mathbf{C_8H_{10}Fe(CO)_3}$ (Table **2**, No. **63**) can be also prepared by reduction of XXXIX with $NaBH_4$ [1, 2, 9, 44, 67]. Depending of the reaction conditions, a variety of compounds including $C_8H_8Fe(CO)_3$ (C_8H_8=cyclooctatetraene), $C_8H_{10}Fe(CO)_3$ (C_8H_{10}=cycloocta-1,3,5-triene), XL [1], XLI (R=H) [9], or XLII (R=H) [67] are obtained; No. 63 being the main product [1, 9, 44]. Thus, reduction in tetrahydrofuran [1, 2], cf. [44] at 25 °C [1] followed by addition of light petroleum ether (b.p. 60 to 80 °C), separation of the organic layer which is washed with H_2O, dried ($CaCl_2$) [2], and chromatographed [1, 2] on alumina (grade III) [2] gives No. 63 [1, 2, 9, 44] in 16% yield [2] as main product [9], XLI (R=H) [9] and small amounts

 References on pp. 99/102

of $C_8H_8Fe(CO)_3$ [1]. Reduction in ice-water/ether for 1 h followed by dilution with ether, washing the separated organic layer with H_2O, drying ($MgSO_4$), and evaporation (at 15 °C/15 Torr) gives a 5:1 mixture of No. 63 and XLII (R=H) in a total yield of 90% as oil. Chromatography on silica gel with petroleum ether gives No. 63 (72%), and elution with ether gives XLII (R=H) in 18% yield [40, 67]. The kinetically controlled addition of H^- to XXXIX with 3% $NaBH_4$ in ice-water gives XLIII (R=OH; No. 77), No. 63, and 18% yield of XLI (R=H) [40]. Similar reduction of XXXIX in ice-water with $NaBD_4$ gives XLIII (R=D) and XLII (R=D) in the molar ratio 4:1 [46]. Analogous reduction of perdeuterated XXXIX ($[C_8D_9Fe(CO)_3]BF_4$) with $NaBH_4$ in ice-water gives XLIV and XLV, separated by chromatography [65]. Reduction of $C_7H_8Fe(CO)_3$ with Simmons-Smith reagent according to Method Va (p. 39) gives XLVI [58, 91]. Attempts to prepare No. 63 with endo configuration failed [57, 77].

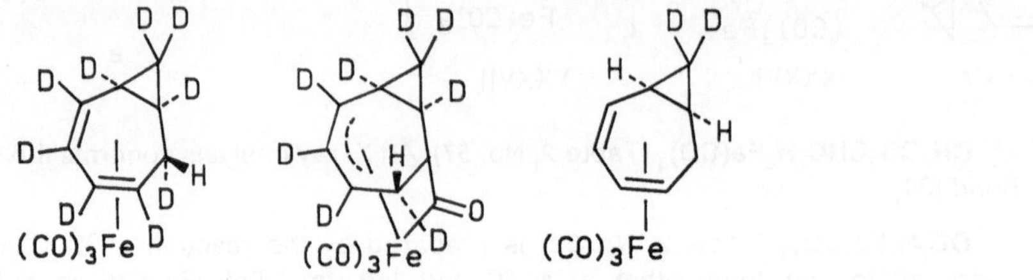

XLIV XLV XLVI

The deuterated isomer XLIII (R=D) melts at 39 °C [40]. For a figure of the ^1H NMR spectrum of No. 63 see [40]. From the appearance of 4 CO bands in hexane it is concluded that there exists an equilibrium mixture of rotamers [67].

Compound No. 63 is very soluble in organic solvents, giving solutions which are somewhat unstable in air although the solid itself is fairly stable [2].

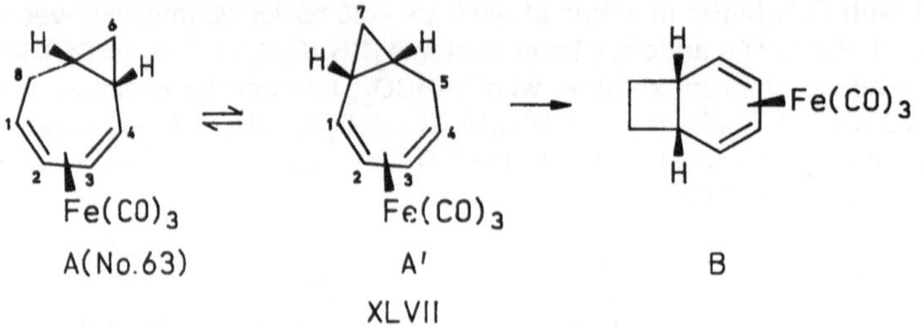

A(No.63) A′ B

XLVII

Heating compound No. 63 in C_6H_6 at 90 °C for 12 h gives an equilibrium mixture of A (No. 63) and A′ as shown in Scheme XLVII [40, 41, 77]. This degenerate valence isomerization is confirmed by using the deuterated compounds XLIII (R=D) [40, 41], and XLIV [41]. The activation energy for this migration is small [65]. Heating compound No. 63 at 100 °C [35, 47] in heptane [44] gives B (see Scheme XLVII) or in octane at 120.6±0.1 °C gives B and $C_8H_{10}Fe(CO)_3$ with C_8H_{10}=cycloocta-1,3,5-triene (see Scheme XLVII) [44]. The rates of isomerization to B in octane are shown in the following table [44]:

t in °C	k (s^{-1})	ΔG* (kcal/mol)
110±1	$(6.7\pm1)\times10^{-6}$	31.8
120.6±0.1	$(1.7\pm0.1)\times10^{-5}$	31.8
120.6±0.1	$(1.8\pm0.1)\times10^{-5}$	31.8

This isomerization follows good first-order kinetics. Chromatographic results coupled with a kinetic analysis very strongly suggest that $C_8H_{10}Fe(CO)_3$ (C_8H_{10}=cycloocta-1,3,5-triene) is an intermediate in this isomerization [44], and the mechanism involves a single-step H-1,5 shift or two sequential H-1,3 shifts [44, 100].

Oxidation of compound No. 63 with excess $Fe(H_2O)_6Cl_3$ in ether at 40 °C liberates the organic ligand ^{4}L (64%), and reaction with CO (130 atm) in a 30% pentane solution at 90 °C for 20 h gives XLVIII in 92% yield [67]. Reaction with $[C(C_6H_5)_3]BF_4$ in CH_2Cl_2 gives XXXIX [2].

OC CO Fe O — C_6H_5 C_6H_5 Fe(CO)$_3$ O

XLVIII IL L

$Cl_2CC_7H_8Fe(CO)_3$ (Table **2**, No. **65**) is oxidized with $ON(CH_3)_3$ in boiling C_6H_6 within 40 min to give the free organic ligand in 68% yield. Reaction with two equivalents of o-chloranil in CH_2Cl_2 for 1 h followed by chromatography on silica gel with C_6H_6/n-hexane (3:7) gives LXXVIa and b (see p. 97) in 19 and 27% yield [129].

$Br_2CC_7H_8Fe(CO)_3$ (Table **2**, No. **66**) crystallizes in the monoclinic space group $P2_1/c-C^5_{2h}$ with the unit cell parameters a=14.872(6), b=6.816(3), c=12.693(8) Å, β=94.26(4)°; Z=4 molecules per unit cell. The calculated density is 2.091 g/cm³. The main bond distances and angles are shown in **Fig. 8** [59]. Reaction with $LiCH_3$ in ether at −90 to −40 °C gives IL (21% yield) [23, 24, 32, 59] as the major product [24].

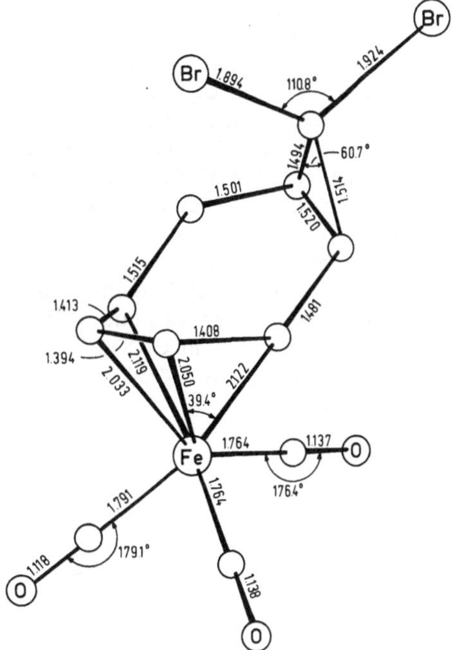

Fig. 8. Molecular structure of $Br_2CC_7H_8Fe(CO)_3$ (No. 66) [59].

References on pp. 99/102

$O(C_6H_5)_2C_2C_7H_8Fe(CO)_3$ (Table **2**, No. **67**) gives L in boiling C_6H_6 [97, 105]. This formation of L is kinetically controlled and represents a unique [2,2]-dyotropic rearrangement [105]. Attempts to chromatograph No. 67 over acid washed silica gel results in isomerization to XXXVIII a ($R=COC(C_6H_5)_2H$). Ketone XXXVIII a is also the sole product obtained by treatment of No. 67 with a catalytic amount of $HOSO_2C_6H_4CH_3$-4 in boiling benzene or by treatment of No. 67 with NaOH in CH_3OH at room temperature [97]. For the mechanism of these reactions see [97, 105], cf. [126].

$CH_3O_2CC_5H_3O_2C_7H_8Fe(CO)_3$ (Table **2**, No. **69**). Attempts to decarboxylate this compound by thermolysis are unsuccessful [109].

5-HO(7-C_6H_5NH)$C_7H_8Fe(CO)_3$ (Table **2**, No. **70**) is also obtained by reduction of VI (p. 38) with Zn/CH_3COOH followed by complexation with $Fe_2(CO)_9$ [78].

5-HO(7-CH_3CO)$C_7H_8Fe(CO)_3$ and **5-CH_3O(7-C_6H_5CO)$C_7H_8Fe(CO)_3$** (Table **2**, Nos. **71** and **72**). Both compounds give 1-$RCOC_7H_7Fe(CO)_3$ ($R=CH_3$ or C_6H_5) on treatment with silica gel in toluene. Treatment with HBF_4 gives [6-$RCOC_7H_8Fe(CO)_3$]BF_4 [42].

$OC_2H_2C_7H_8Fe(CO)_3$ (Table **2**, No. **75**). Oxidation of the isomer No. 75a with $FeCl_3 \cdot 6H_2O$ in ether at 20 °C [77] or reaction with CO [66, 77] liberates bicyclo[4.2.1]nona-2,4-dien-7-one (^{4}L) [66, 77].

$(CH_3O_2C)_2C_2C_7H_8Fe(CO)_3$ (Table **2**, No. **76**) crystallizes in the monoclinic space group P2/c-C^5_{2h} with the unit cell parameters a = 12.360, b = 9.295, c = 14.195 Å, β = 102.2°; Z = 4 molecules per unit cell [53, 114].

$HOC_8H_9Fe(CO)_3$ (Table **2**, No. **77**) is oxidized with CrO_3 in pyridine [7, 12]. The mixture is stirred for 4 h in the cold, followed by 12 h at room temperature, filtered, washed, and the combined filtrates are evaporated. Trituration of the residue with ether, filtration, and evaporation gives No. 123a in 17% yield [12].

$CH_3OC_{10}H_{11}OFe(CO)_3$ (Table **2**, No. **78**) liberates the organic ligand ^{4}L with $[NH_4]_2[Ce(NO_3)_6]$ in CH_3OH [31].

$HOC_9H_{11}Fe(CO)_3$ (Table **2**, No. **79**). Compound No. 79a is prepared from LI (R=H) and $AlCl_3$ in C_6H_6 at 10 °C within 3 h. The reaction mixture is hydrolyzed with H_2O/conc. HCl, and the aqueous phase is extracted with CH_2Cl_2. The organic layer is dried ($MgSO_4$) and chromatographed on silica gel with 10% $CH_3CO_2C_2H_5/C_6H_6$ to give LII (ca. 20%) and No. 79a (ca. 25% yield). Similar reaction of LI (R=D) gives monodeuterated compound No. 79a (see Formula LIII with R′=H, R=D), m.p. 134 °C. For the ^{1}H NMR spectrum in the presence of Eu(fod)$_3$ (fod = 1,1,1,2,2,3,3-heptafluoro-7,7-dimethyl-4,6-octanedionate) see [56]. The ^{1}H NMR spectrum of LIII (R′=H, R=D) is very similar to that of No. 79a except that the doublet at δ = 1.10 ppm now only integrates for one proton, and the doublet of doublets at δ = 1.80 ppm becomes a complex multiplet [56]. The chemical shifts in the ^{13}C NMR spectrum of No. 79c for the CO groups are not observed due to low intensity [120]. Acetylation of compound No. 79a or LIII (R′=H, R=D) with $O(COCH_3)_2$ in pyridine gives compound No. 80 or its monodeuterated derivative LIII ($R'=COCH_3$, R=D) [56]. Reaction of No. 79b with $ClSO_2C_6H_4CH_3$-4 gives No. 160 [118].

LI 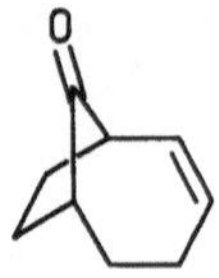 LII 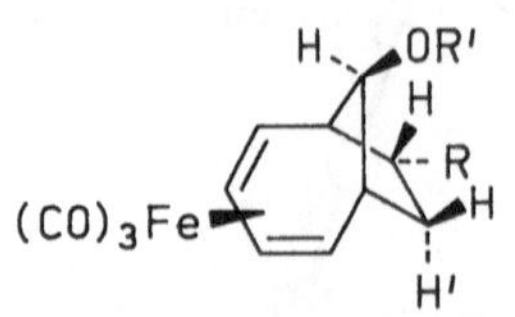 LIII 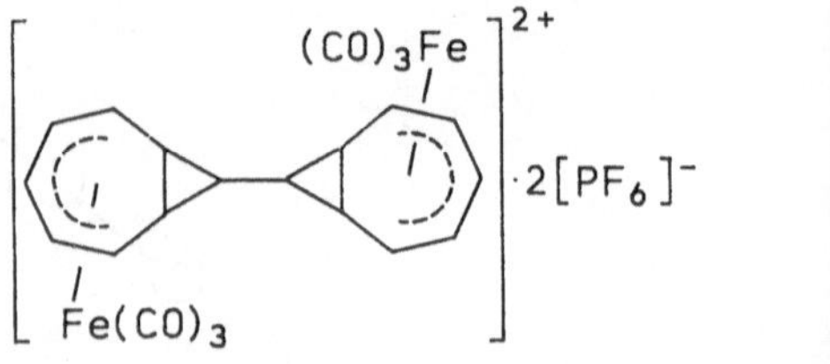 LIV

References on pp. 99/102

$C_{16}H_{16}Fe(CO)_3$ (Table **2**, No. **81**) is prepared by the reaction of LIV with $[CH_3P(C_6H_5)_3]Br$ (molar ratio 1:2) [88, 95], cf. [121] in acetone at −78 °C for 4 h [95]. After removing the solvent in vacuum the residue is dissolved in CH_2Cl_2 and chromatographed on silica gel with hexane. Elution with hexane/ether (10:1) gives No. 81 in 54% yield. The compound No. 81 crystallizes in the monoclinic space group $P2_1/c-C^5_{2h}$ (No. 14) with the unit cell parameters a = 11.882(8), b = 10.877(10), c = 12.586(10) Å, β = 112.98(6)°; Z = 4 molecules per unit cell. The measured density is 1.43 g/cm^3 (flotation), and the calculated density is 1.54 g/cm^3. The main bond distances and angles are shown in **Fig. 9**.

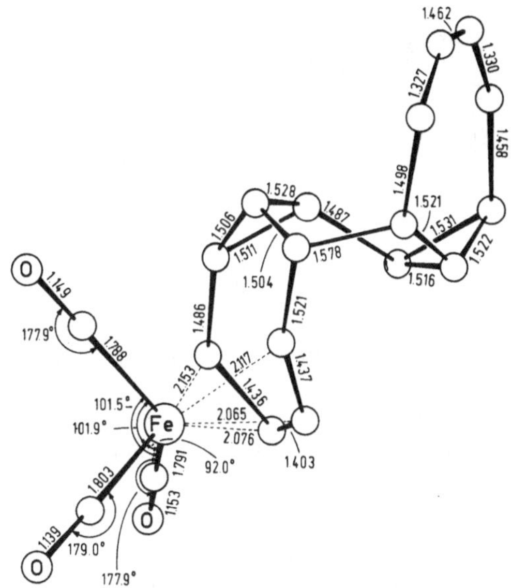

Fig. 9. Molecular structure of $C_{16}H_{16}Fe(CO)_3$ (No. 81) [95].

Oxidation of complex No. 81 with $ON(CH_3)_3 \cdot 2H_2O$ in refluxing C_6H_6 for 7 h gives LV ([4]L) in 51% yield. Refluxing with $Fe_2(CO)_9$ in n-hexane for 1.5 h gives LVI in 46% yield [95].

Fe(CO)3 Fe(CO)3 D H N H (CO)3Fe [PF6]−

LV LVI LVII

$[C_5H_5NC_8H_8DFe(CO)_3]PF_6$ (Table **2**, No. **82**). A figure of the ^1H NMR spectrum in acetone-d_6 is shown in [46]. It gives LVII in nearly quantitative yield in solution at 25 °C within 30 min [46].

$(CH_3O_2C)_2C_2HN_2C_7H_7Fe(CO)_3$ (Table **2**, No. **84**). Slow crystallization of a CH_3OH/acetone solution gives single crystals, possessing the space group $P\bar{1}-C^1_i, S^1_2$ with the unit cell parameters a = 10.683(4), b = 12.527(5), c = 7.058(2) Å, α = 94.98(3)°, β = 101.10(3)°, and γ = 114.55(3)°; Z = 2 molecules per unit cell. The measured density is 1.603 g/cm^3 (flotation

References on pp. 99/102

H_2O/KI), and the calculated density is 1.613 g/cm³. The main bond distances are shown in **Fig. 10**. The cycloheptadiene ring has a tub conformation in the complex. The H-5 atom occupies the endo configuration with respect to the Fe. The dihedral angle H-5, C-5, C-4, H-4 is 71.4°.

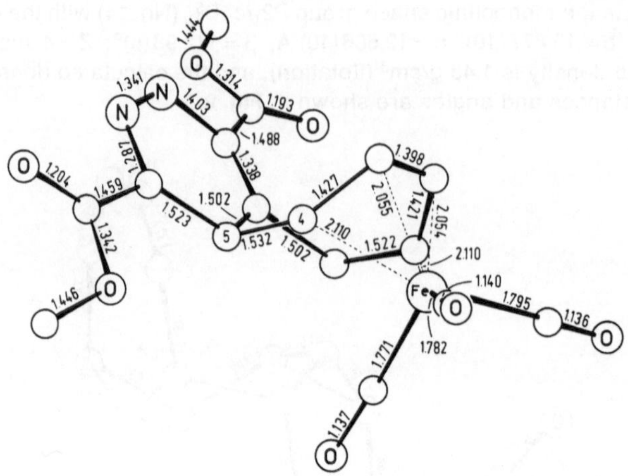

Fig. 10. Molecular structure of $(CH_3O_2C)_2C_2HN_2C_7H_7Fe(CO)_3$ (No. 84) [109].

Decomplexation reaction with $ON(CH_3)_3$ in C_6H_6 at room temperature for 1 h gives the organic ligand ⁴L in 33% yield [109].

$HOC_7H_7OFe(CO)_3$ (Table **2**, No. **93**). Compound XXXII (p. 85) is dissolved in CH_2Cl_2 and extracted with conc. H_2SO_4. After 15 min the solution is poured onto excess $NaHCO_3$, and the resulting paste is diluted with CH_2Cl_2. Addition of H_2O and CH_2Cl_2 followed by extraction with CH_2Cl_2 and drying gives No. 93 (71% yield) after chromatographic work-up. Analogous reaction of XXXII with D_2SO_4 and K_2CO_3/H_2O gives LXXVII (p. 97) in 70% yield [111]. A "hydroxy-cycloheptadienon-eisen-tricarbonyl" is occasionally obtained in the high pressure hydrogenation reaction of XXXII (p. 85) in C_2H_5OH, cf. Nos. 55 and 60 [3]. This compound possibly possesses the structure of No. 93. Reaction of LXXVII (p. 97) with $[C(C_6H_5)_3]BF_4$ gives LXXVIII (p. 97) in 78% yield [111].

$CH_3OC_7H_7OFe(CO)_3$ (Table **2**, No. **94**). The trideuterated compound No. 94 (deuterated at C-4,6, cf. Table 2, p. 65) can be prepared from LVIII (R=R′=R″=D) and Na_2CO_3/CH_3OH at 0 °C (cf. Method VIIb, p. 43) [27, 106]. Quenching a solution of XXXII (p. 85) in liquid SO_2/$HOSO_2F$ with cold Na_2CO_3/CH_3OH gives No. 94 (>99% yield). Reaction of No. 94 with HBF_4 in $O(COCH_3)_2$/ether gives XXIII (p. 41) [27]. Similar reaction of the trideuterated compound with a little excess of HBF_4 in $O(COCH_3)_2$ gives LVIII [106].

R″ R′ O R $[BF_4]^-$ $Fe(CO)_3$ — LVIII

D D O CN D $(CO)_3Fe$ — LIX

CRR′ $Fe(CO)_3$ — LX

$NCC_7H_7OFe(CO)_3$ (Table **2**, No. **98**). Reaction of LVIII (R=R′=R″=D) with aqueous NaCN as described in Method VIIb (p. 43) gives trideuterated compound No. 98 (deuterated at C-4,6, cf. Table 2, p. 66) and LIX. The ^1H NMR spectrum ($CDCl_3$) of the trideuterated derivative shows the expected shifts at $\delta=3.27$ (dt, H-1), 3.68 (br, s, H-7), 5.8 (dd, H-2), and 6.03 (dd, H-3) ppm [70].

$HO(RCH{=})C_7H_7Fe(CO)_3$ (Table **2**, Nos. **99** to **103** with R=H, CH_3, $CH_2C_6H_5$, C_6H_5, $C_6H_4CH_3$-4). The two protons at C-6 in the ^1H NMR spectra of Nos. 99, 100, and 103 (R=H, CH_3, $C_6H_4CH_3$-4) give rise to an AB quartet, which contains further coupling. The chemical shift difference is influenced to a large extent by the nature of the trans-substituted R (defined with respect to the $Fe(CO)_3$ group and the largest substituent). Using these data it is possible to obtain an excellent correlation: trans H, CH_3, and $C_6H_4CH_3$-4 groups give large splittings, while in the cis position they have little effect as shown by comparison of No. 103a and b [49]. Abstraction of H_2O from Nos. 101 and 102 (R=$CH_2C_6H_5$, C_6H_5) gives LX (R=$CH_2C_6H_5$, C_6H_5; R′=H) [29].

$HO(R_2C{=})C_7H_7Fe(CO)_3$ (Table **2**, Nos. **104** and **105** with R=CH_3, C_6H_5). For the ^1H NMR spectrum of No. 104 compare Nos. 99, 100, and 103 (before) [49]. Dehydratation of both compounds gives $RR'C{=}C_7H_6Fe(CO)_3$ (cf. LX with R=R′=CH_3 or C_6H_5) [29].

$HO(4\text{-}CH_3C_6H_4(CH_3)C{=})C_7H_7Fe(CO)_3$ (Table **2**, No. **106**). For the ^1H NMR spectrum compare Nos. 99, 100, and 103 (above) [49].

$5\text{-}RR'C{=}(7\text{-}CH_3O)C_7H_7Fe(CO)_3$ (Table **2**, Nos. **108** to **115** with R=H, R′=CH_3, $CH_2C_6H_5$, C_6H_5, $C_6H_4CH_3$-4; R=R′=CH_3; R=R′=C_6H_5; R=CH_3, R′=C_2H_5, R′=$C_6H_4CH_3$-4) are isomerically pure. They show no splitting of the OCH_3 resonance upon addition of europium shift reagent. The OCH_3 resonance moves rapidly on varying the europium concentration. All these compounds are easily demethoxylated on silica gel [38] to give the appropriate LX [29, 38].

$5\text{-}HO(4,7,7\text{-}(CH_3)_3)C_7H_6Fe(CO)_3$ (Table **2**, No. **117**) can be gas-chromatographed at temperatures up to 150 °C with no apparent thermal decomposition. Teflon columns employing E-301 methyl silicone gum as the stationary phase give best results, provided that precautions are taken to exclude air and prevent prolonged exposure to light [36]. The complex is refluxed in C_6H_6 containing $HOSO_2C_6H_4CH_3$-4 for 1.5 h, stirred with anhydrous Na_2CO_3, and filtered. Removal of the solvent gives LXXXII, p. 98 (70% yield). Attempts to effect the elimination under milder conditions, using $OPCl_3$ or $OSCl_2$ in pyridine, failed [126].

$4\text{-}R(7\text{-}CH_3O)C_7H_6OFe(CO)_3$ (Table **2**, Nos. **118** to **120** with R=Cl, CH_3, C_6H_5) are treated with 40% aqueous HBF_4 in $O(COCH_3)_2$ to give LXI with R=H, CH_3, C_6H_5, respectively [63].

LXI LXII LXIII LXIV

$CH_2C_7H_6OFe(CO)_3$ (Table **2**, No. **123**). For the preparation of the dideuterated exo compound No. 123a (Formula LXII) see "Further information" for No. 128. This compound LXII shows in the ^1H NMR spectrum ($CDCl_3$) chemical shifts at $\delta=1.56$ (d, H-6; J=7.5), 2.00

(dd, H-7; J=7.5, J=7.5), 3.11 (d, H-4; J=7.5), 3.62 (dd, H-1; J=7.5, J=7.5), 5.46 (m, H-2 or H-3), 5.81 (m, H-2 or H-3) ppm [69]. Irradiation of No. 123a in CH_3CO_2H gives LXIII in 90% yield [81]. Oxidation with $[NH_4]_2[Ce(NO_3)_6]$ [7] in the presence of Na_2HPO_4 in acetone at 0 °C for 3 h 45 min [12] liberates the organic ligand ^{4}L [7, 12] (68.7% yield [12]). ^{4}L is also obtained in >90% yield with $ON(CH_3)_3$ [62]. Treatment of No. 123a with CF_3CO_2H, H_2SO_4, or D_2SO_4 at −10 to +5 °C gives LXIV (R=R′=H) [89]. Treatment of No. 123a with LXXXI (p. 98) gives No. 167 (78% yield) [122].

$RR'CC_7H_6OFe(CO)_3$ (Table **2**, Nos. **124** and **125** with R=H, R′=CH_3 or R=R′=CH_3). Oxidation of a mixture of Nos. 124a (16%) and b (84%) with $ON(CH_3)_3$ gives LXV in a total yield of 55%. Similar oxidation of No. 125 gives XII with R=CH_3 (see p. 39) in >90% yield [62]. Treatment of Nos. 124 or 125 with CF_3CO_2H, H_2SO_4, or D_2SO_4 at −10 to +5 °C gives LXIV with R=H, R′=CH_3 or R=R′=CH_3, respectively [89].

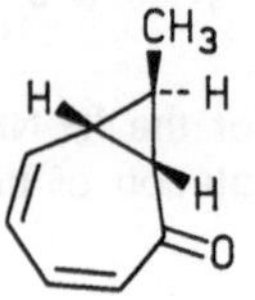

LXV 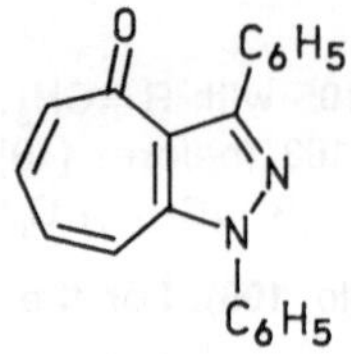 LXVI 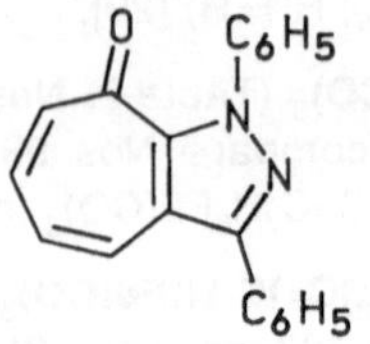 LXVII 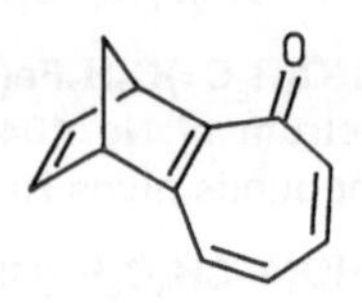 LXVIII

$RCNOC_7H_6OFe(CO)_3$ (Table **2**, Nos. **126** and **127** with R=C_6H_5 or $C_6H_2(CH_3)_3$-2,4,6). Oxidation of Nos. 126 or 127 with $[NH_4]_2[Ce(NO_3)_6]$ in acetone liberates the organic ligands ^{4}L in 50 and 86% yield, respectively [72].

$RR'CN_2C_7H_6Fe(CO)_3$ (Table **2**, Nos. **128** to **130** with R=R′=H; R=H, R′=CH_3 and R=R′=CH_3). The dideuterated compound No. 128 with R=R′=D is prepared as in Method Vd (pp. 40/1) but with excess CD_2N_2. This compound shows in the ^{1}H NMR spectrum ($CDCl_3$) chemical shifts at δ=2.56 (d, H-6; J=10.0), 3.16 (d, H-4; J=7.5), 3.40 (dd, H-1; J=4.0, J=7.5), 5.40 to 5.90 (m, H-2,3,7) ppm. IR spectrum (cyclohexane): 1651 (C=O); 1998, 2008, 2066 cm^{-1}. Heating compound No. 128 or its deuterated derivative in toluene at 125 °C for 15 min gives No. 123a or its deuterated derivative in 91% yield [69]. Heating No. 128 to 80 °C or No. 130 to 110 °C gives N_2 and Nos. 123a or 125 in quantitative yield. Refluxing the mixture of isomers No. 129 in C_6H_6 gives N_2 and both isomers of No. 124, which are separated nicely by chromatography on silica gel. Refluxing pure isomer No. 129a in C_6H_6 gives N_2, Nos. 124a (7%) and b (93%). Refluxing pure isomer No. 129b in C_6H_6 gives N_2, Nos. 124a (40%) and b (60%). Reaction of No. 130 with $ON(CH_3)_3$ in boiling C_6H_6 gives XII with R=CH_3 (see p. 39) and the free organic ligand ^{4}L (47%) [62].

$(C_6H_5)_2CN_2C_7H_6OFe(CO)_3$ (Table **2**, Nos. **131** and **132**). The mixture of regioisomers No. 131 and 132 obtained by Method Vd (pp. 40/1) shows in the ^{1}H NMR spectrum ($CDCl_3$) chemical shifts at δ=2.85 to 3.30 (m, H-1,4), 4.01 (d, H-7; J(H-6,7)=9.3), 5.15 to 5.80 (m, H-2,3,6) ppm. Competitive reaction of tropone and XVII (R=H, p. 40) in the ratio 3:1 with C_6H_5CCl=$NNHC_6H_5$/excess $N(C_2H_5)_3$ in C_6H_6 in the dark at room temperature for 24 h followed by filtration, evaporation of the filtrate, and chromatography of the residue gives No. 131 and its regioisomer No. 132 (total yield 67%; ratio 2.7:1), LXVI (25%), and traces which were not identified. Oxidation of the mixture of products obtained by Method Vd with $[NH_4]_2[Ce(NO_3)_6]/Na_2HPO_4$ in acetone at room temperature for 48 h gives LXVI (53%) and LXVII (3.0%) in the ratio 17:1. Similar oxidation of the mixture of products at 0 °C for 1.5 h followed by chromatography gives LXVII and LXVI as well as a mixture of their irontricarbonyl complexes [71].

$C_5H_6C_7H_6OFe(CO)_3$ (Table **2**, No. **133**) is unchanged after 3 h reflux in C_6H_6 and gives a retro Diels-Alder reaction at 140 °C. The compound No. 133 gives a mixture of products on oxidation with $FeCl_3$, $[NH_4]_2[Ce(NO_3)_6]$, or CrO_3. Oxidation with excess $ON(CH_3)_3$ in boiling C_6H_6 gives LXIX in 50% yield beside products of a retro Diels-Alder reaction. Repetition of this reaction in ether at room temperature for 6 h (60% transformation) liberates the organic ligand ^4L (55%) and a small amount of LXIX together with LXVIII. The yield of the latter product increases with prolonged reaction time and derives very probably from ^4L. LXIX also derives from the free ligand ^4L. Catalytic reduction with H_2 on PtO_2/C_2H_5OH gives No. 134 which does not give a retro Diels-Alder reaction [74].

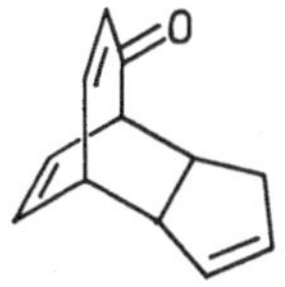

LXIX 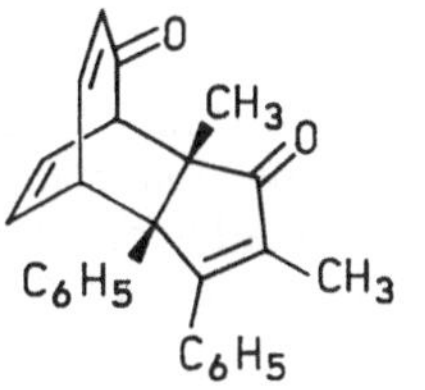 LXX 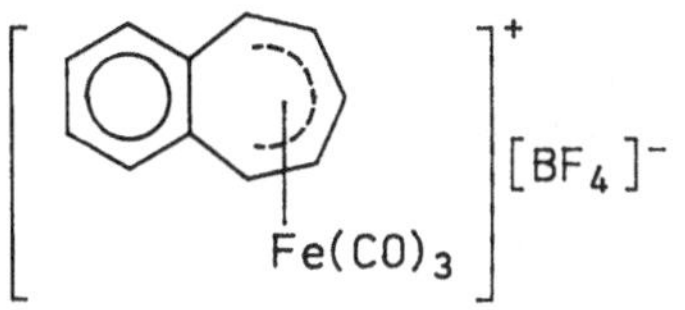 LXXI

$(CH_3)_2(C_6H_5)_2C_5OC_7H_6OFe(CO)_3$ (Table **2**, No. **135**) appears to be in thermal equilibrium with XVII (R=H, p. 40) and 2,5-dimethyl-3,4-diphenyl-cyclopenta-2,4-dienone (A) at 95 °C. Oxidation with $[NH_4]_2[Ce(NO_3)_6]$ in acetone at 0 °C gives LXX (53%) and A (29%, isolated as its dimer) smoothly via the liberated organic ligand ^4L. The formation of the intermediate adduct ^4L is considered most probable on the basis of the following evidence. On addition of $[NH_4]_2[Ce(NO_3)_6]$ to No. 135, thin layer chromatography monitoring reveals the formation of LXX and A with a third spot which represents the dominant reaction product at the end of the oxidant addition. On standing, the spot of the unknown compound fades while the spots of LXX and A are enhanced. Compound ^4L is transformed into LXX through a [3.3] sigmatropic rearrangement and into A by a retro Diels-Alder reaction. Thus, the formation of LXX demonstrates in an unmistakable way the endo structure of No. 135 as shown in Table 2, p. 74 [72].

$C_4H_4C_7H_6Fe(CO)_3$ (Table **2**, No. **138**) gives LXXI in the reaction with $[C(C_6H_5)_3]BF_4$ in CH_2Cl_2 [14].

$C_7H_6O_2Fe(CO)_3$ (Table **2**, No. **139**) gives No. 140 in the reaction with $H_2NNHC_6H_3(NO_2)_2$-2,4. Reaction of No. 139 with $[C_6H_7Fe(CO)_3]BF_4$ in H_2O/C_2H_5OH at 100 °C for 15 min gives LXXIII [15].

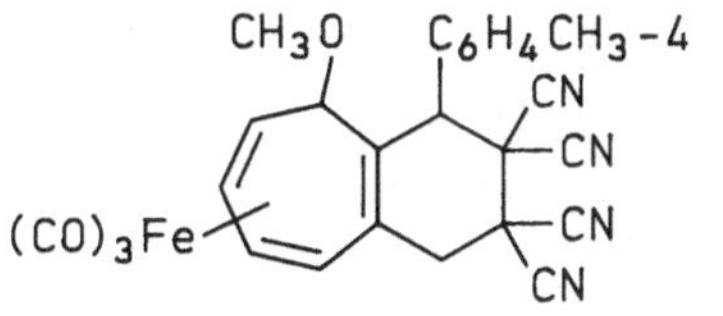

a LXXII b

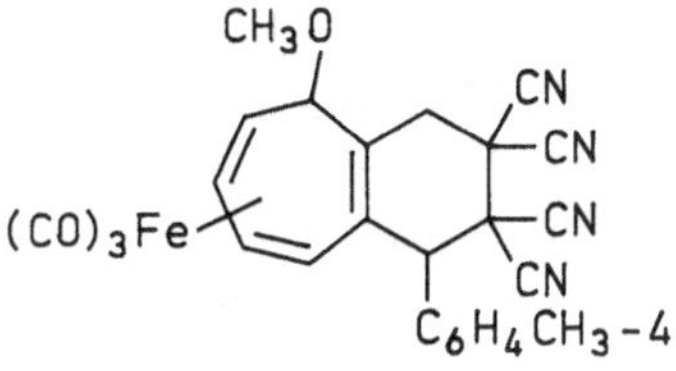

$(CH_3)_2C{=}C_7H_6OFe(CO)_3$ (Table **2**, No. **141**). Refluxing with $ON(CH_3)_3$ in CH_2Cl_2 for 30 min liberates the organic ligand ^4L (75% yield). Dehydrogenation with an excess of MnO_2 in refluxed CH_2Cl_2 leads to LXXIV (50% yield) within 2 h [82].

References on pp. 99/102

4,7,7-$(CH_3)_3C_7H_5OFe(CO)_3$ (Table **2**, No. **142**). $Fe_2(CO)_9$ is gradually added during 4 h to stirred 2,6,6-trimethylcyclohepta-2,4-dienone (eucarvone = ^{4}L) at 60 °C. Stirring is continued at 60 °C for another 10 h. Pentane is added, and the precipitate is filtered off. The solvent and excess $Fe(CO)_5$ are removed at room temperature, and the residue distilled to give No. 142 in 37% yield. A solution of compound No. 142 in dry ether is added dropwise to a slurry of $LiAlH_4$ in ether cooled in a dry ice-acetone mixture. The mixture is stirred for 1 h, then treated with $CH_3CO_2C_2H_5$ followed by 10% aqueous HCl. The aqueous solution is extracted with ether, and the ethereal solutions are washed with brine and dried over Na_2SO_4. Removal of the solvent and chromatography on Kieselgel 60 with toluene/$CHCl_3$ gives No. 117 (40% yield) [126]. The compound can be gas-chromatographed as described for No. 117 [36].

6-$CH_3COC_{10}H_{11}N_2OFe(CO)_3$ (Table **2**, No. **143**) loses N_2 almost instantaneously at 85 °C. Heating of No. 143 in CH_2Cl_2 at 30 °C for 30 min leads to LXXV (98% yield). Heating of No. 143 in C_2H_5OH at 78 °C for 1 h followed by addition of K_2CO_3 and stirring for another 30 min gives No. 141 (88% yield) after chromatography on SiO_2 with hexane/ether [82].

4-$CH_3C_6H_4NC_7H_5C_2HN_2(CO_2CH_3)_2Fe(CO)_3$ (Table **2**, No. **145**). In the ^{1}H NMR spectrum (see Table 2, p. 76), two stereoisomers (concerning the relationship of the $Fe(CO)_3$ and the 4-$CH_3C_6H_4$ group) are observed. Decomplexation with $ON(CH_3)_3$ in C_6H_6 at room temperature for 30 min liberates the organic ligand ^{4}L in 34% yield [109].

$CH_3O(4\text{-}CH_3C_6H_4CH{=})C_7H_5CH_2Fe(CO)_3$ (Table **2**, No. **146**). The ^{1}H NMR spectrum contains two OCH_3 resonances (cf. Table 2, p. 76) which suggests the formation of two isomers No. 146a and b. The presence of two exocyclic double bonds in No. 146 with the 4-$CH_3C_6H_4$ group trans to the second methylene group is substantiated by the ready reaction with $(NC)_2C{=}C(CN)_2$ in C_6H_6 at 80 °C within 3 h to give the two isomers LXXII [38].

6-$CH_3CO[(CH_3)_2C{=}]C_7H_5OFe(CO)_3$ (Table **2**, No. **147**) is prepared by heating LXXV in C_6H_6 at 80 °C for 2 h (98% yield). Refluxing No. 147 in ethanolic K_2CO_3 for 30 min gives No. 141 in 90% yield [82].

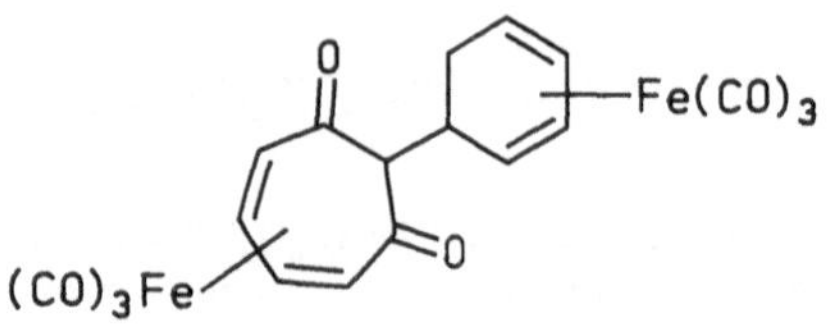

LXXIII 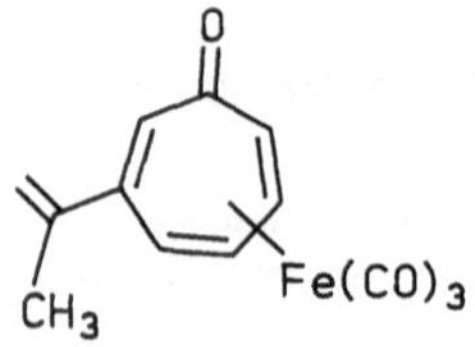 LXXIV 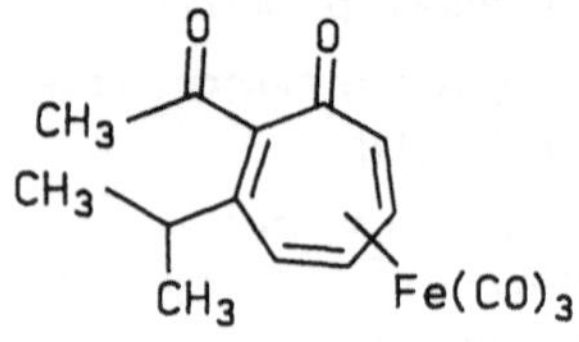 LXXV

1,3-$(CH_3)_2C_7H_4C_2H_2SFe(CO)_3$ (Table **2**, No. **149**) is photolyzed by sunlight within 3 d to give X (49%) and XI (51%), see p. 39 for the formulas [83].

$C_7H_nF_{10-n}Fe(CO)_3$ (Table **2**, Nos. **152** to **154** with n = 0 to 2). Compound No. 154 crystallizes in the orthorhombic space group Pnma-D_{2h}^{16} with the unit cell parameters a = 11.305 ± 0.010, b = 12.110 ± 0.010, c = 9.075 ± 0.010 Å; Z = 4 molecules per unit cell. The calculated density is 2.212 g/cm^3. The main bond distances and angles are shown in **Fig. 11**. The complex has exact C_s symmetry [50].

Compound No. 154 is recovered unchanged from a sealed tube at 200 °C after 8 h. Pyrolysis of No. 152 at 440 °C using a flow system yields p-$C_6F_4H_2$ (61%) and m-$C_6F_4H_2$ (39%). Similar pyrolysis of No. 153 gives C_6F_5H, and compound No. 154 yields C_6F_6, C_2F_4, and SiF_4 (from the glass) [51].

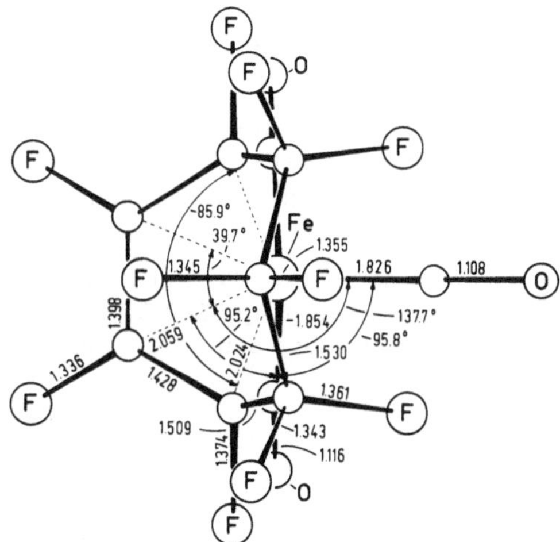

Fig. 11. Molecular structure of $C_7F_{10}Fe(CO)_3$ (No. 154) [50].

$C_2H_5O_2CCHC_7H_8Fe(CO)_3$ (Table **2**, No. **156**). A solution of $N_2CHCO_2C_2H_5$ in $C_7H_8Fe(CO)_3$ (see Formula XXXI, p. 85, with R=H) is added dropwise with vigorous stirring during 15 min to $C_7H_8Fe(CO)_3$ containing a soluble Cu^{II} chelate of salicylaldehyde-aniline Schiff's base as catalyst at 75 °C. When the evolution of N_2 ceases the mixture is heated for an additional 15 min and then diluted with hexane. The solution is filtered and then chromatographed on Kieselgel 60 with hexane. The first fraction contains unreacted $C_7H_8Fe(CO)_3$ followed by $(CO)_3FeC_7H_7C_7H_7Fe(CO)_3$ (see 2.8.1.1 in "Organoiron Compounds" C5, 1981, Table 4, No. 39, p. 51). Elution with toluene and distillation gives No. 156 in a yield of 40% [130].

a LXXVI b LXXVII

5,7-$(CH_3)_2C_7H_8Fe(CO)_3$ (Table **2**, No. **159**). Treatment of LXXIX with $LiCu(CH_3)_2$ affords an inseparable 3:1 mixture of No. 159 and LXXX in only 15 to 20% yield [123, 124].

LXXVIII LXXIX LXXX

4-$CH_3C_6H_4O_2SOC_9H_{11}Fe(CO)_3$ (Table **2**, No. **160**). The rates of solvolysis of this complex and its free ligand are measured. Based on orbital symmetry considerations, it is predicted

References on pp. 99/102

that complexation increases the rate of solvolysis. The measured rate of methanolysis of No. 160 at 65 °C is 28 times faster than that of the ligand. This methanolysis gives No. 161 [118].

$HOC_9H_9Fe(CO)_3$ (Table **2**, No. **162**) is dissolved in n-hexane, and a 5% Pd/C catalyst is added. The solution is flushed vigorously with H_2 and then stirred for 2 d under a constant pressure of 1 atm H_2. After filtering over cotton wool and cooling to −30 °C, No. 79c precipitates (85 to 90% yield) [120].

$[exo\text{-}(C_6H_5)_3PC_8H_9Fe(CO)_3]BF_4$ (Table **2**, No. **163**) is prepared by reaction of $P(C_6H_5)_3$ with XXXIX (p. 87). The complex is oxidized by $[NH_4]_2[Ce(NO_3)_6]$ in CH_3OH. Addition of $NaBF_4$ precipitates the uncomplexed phosphonium salt $[exo\text{-}(C_6H_5)_3PC_8H_9]BF_4$ in 70 to 80% yield [128].

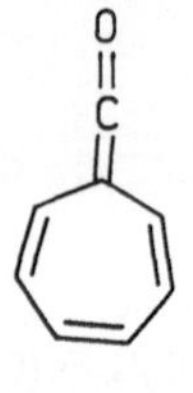

LXXXI 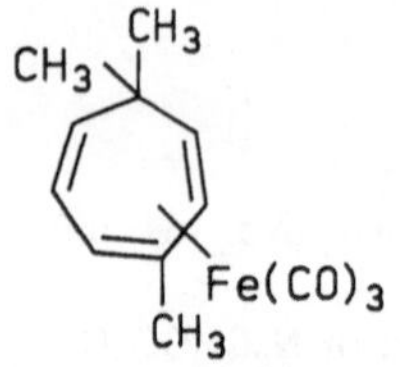 LXXXII 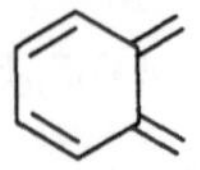 LXXXIII 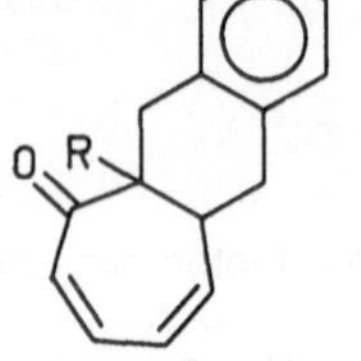 LXXXIV 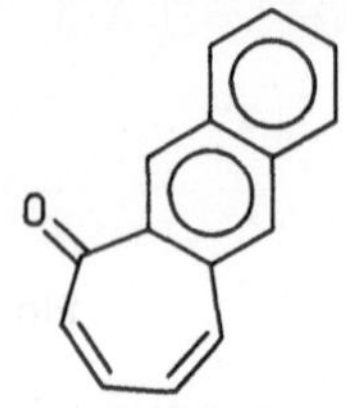 LXXXV

$C_8H_8C_7H_6OFe(CO)_3$ (Table **2**, No. **164**) is treated with o-chloranil in refluxing CH_2Cl_2 to give LXXXIV (R=H) in 50% yield. Treatment with $ON(CH_3)_3$ gives LXXXV along with LXXXIV (R=H) and its isomers [127].

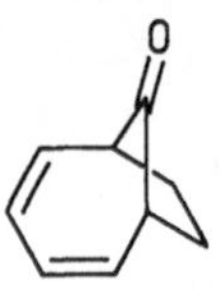

LXXXVI 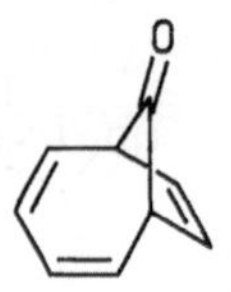 LXXXVII 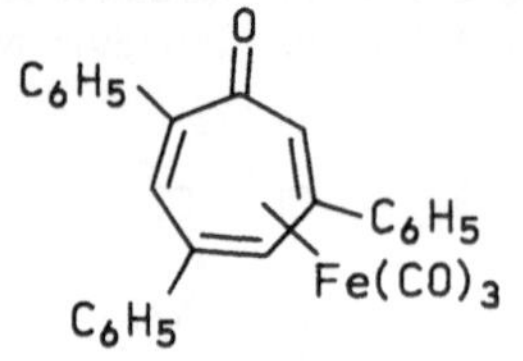 a LXXXVIII b

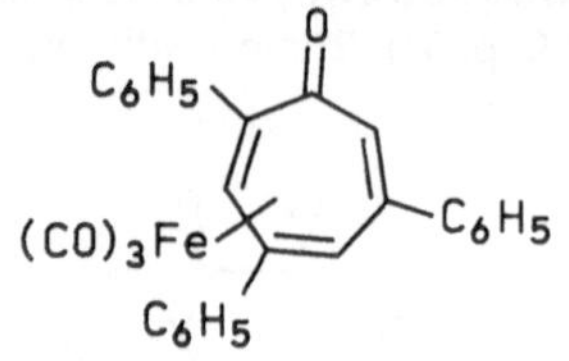

$OC_9H_{10}Fe(CO)_3$ (Table **2**, No. **165**). Depending on the solvent, either No. 165a or b or a mixture of the two is formed when LXXXVI reacts with $Fe_2(CO)_9$. To identify the isomers, a structural determination of No. 165b is performed using X-ray techniques. The reduction of No. 165a with $LiAlH_4$ or $NaBH_4$ gives only No. 79c. Compound No. 79b can be obtained in a 45:55% mixture with No. 79c via reduction with aluminium isopropoxide. Separation of the epimeric alcohols No. 79b and c requires reverse phase high pressure liquid chromatography [118].

$OC_9H_8Fe(CO)_3$ (Table **2**, No. **166**). LXXXVII is stirred with an excess of $Fe_2(CO)_9$ in CH_3OH at 30 °C for 2 d. Evaporation of the solvent and $Fe(CO)_5$ under vacuum yields a solid residue which is extracted with ether and filtered. Chromatography on Al_2O_3 (activity grade III) gives No. 166a as the major fraction (42% yield). Heating LXXXVII with an excess $Fe_2(CO)_9$ in C_6H_6 at 60 °C for 2 h, followed by removal of the solvent and $Fe(CO)_5$ under vacuum, extracting the residue with hot hexane, filtration, and cooling to −30 °C gives No. 166b in 55% yield [117, 120], cf. [115]. Reduction of No. 166a with $NaBH_4$ in CH_3OH/H_2O followed by removal of the solvent and dilution of the residue with H_2O gives No. 162 after extraction with ether (85% yield). Similar attempted reduction of No. 166b failed. Both isomers No. 166a and b are reduced by H_2. Thus, No. 166a is dissolved in n-hexane, and a 5% Pd/C catalyst is added. The solution is flushed vigorously with H_2 and then stirred for 2 d under a constant pressure of 1 atm H_2. After filtering over cotton wool and cooling to −30 °C, compound No. 165a is produced. Similar reaction of No. 166b gives No. 165b. The yields are 85 to 90% [120].

$C_7H_6C_8H_8Fe(CO)_3$ (Table **2**, No. **167**) reacts with $(NC)_2C{=}C(CN)_2$ to give No. 168 in 60% yield [122].

1,3,5-$(C_6H_5)_3C_7H_5OFe(CO)_3$ (Table **2**, No. **169**). Reduction of LXXXVIIIa or b with H_2 (150 atm) in the presence of Pd/C catalyst in C_6H_6 at 50 °C for 12 h gives, upon chromatography on silica gel and elution with benzene, oily products and a complex, probably No. 169, which is extremely sensitive to air and light [110].

$C_8H_8(Cl)C_7H_5OFe(CO)_3$ (Table **2**, No. **170**) is treated with o-chloranil to give LXXXIV (R= Cl) in 33% yield [127].

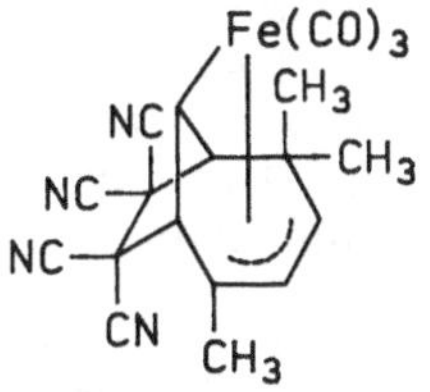

Fe(CO)3 CH3 NC CH3 NC NC CN CH3 CH3 S H CH3 CH3

LXXXIX 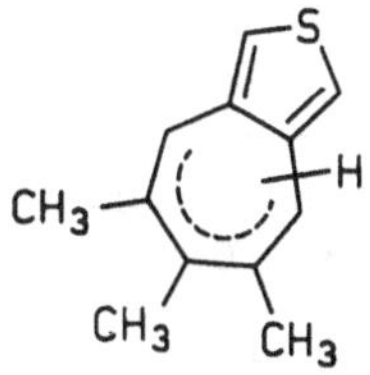 XC

3,6,6-$(CH_3)_3C_7H_5C_6N_4Fe(CO)_3$ (Table **2**, No. **172**) is prepared from LXXXII and $(NC)_2C{=}C(CN)_2$ in CH_2Cl_2 at room temperature within 10 min. The elemental analysis corresponds to a 1:1 adduct. However, the ^1H NMR spectrum reveals the presence of a 4:1 mixture of isomers, identified as LXXXIX and No. 172 [126].

2,3,4-$(CH_3)_3C_7H_3C_2H_2SFe(CO)_3$ (Table **2**, No. **173**) is prepared by refluxing XC with $Fe_2(CO)_9$ in C_6H_6 for 12 h. The solid formed during the reaction is filtered. The filtrate is evaporated and the residue is chromatographed on silica gel using pentane as eluent (yield 12%) [119].

References:

[1] A Davison, W. McFarlane, G. Wilkinson (Chem. Ind. [London] **1962** 820/1). – [2] A. Davison, W. McFarlane, L. Pratt, G. Wilkinson (J. Chem. Soc. **1962** 4821/9). – [3] E. Weiss, W. Hübel, J. Nielsen, A. Gerondal, J. Vandervalle (Chem. Ber. **95** [1962] 1179/85). – [4] N.G. Connelly, A.R. Lucy, M.W. Whiteley (J. Chem. Soc. Dalton Trans. **1983** 111/5). – [5] Z. Goldschmidt, S. Antebi, I. Goldberg (J. Organometal. Chem. **260** [1984] 105/13).

[6] G.R. John, L.A.P. Kane-Maguire (J. Chem. Soc. Dalton Trans. **1979** 873/8). – [7] J.D. Holmes, R. Pettit (J. Am. Chem. Soc. **85** [1963] 2531/2). – [8] F.M. Chaudhari, P.L. Pauson (J. Organometal. Chem. **5** [1966] 73/8). – [9] C.E. Keller (Diss. Univ. Texas 1966 from [64]). – [10] R.E. Dessy, R.B. King, H. Waldrop (J. Am. Chem. Soc. **88** [1966] 5112/7).

[11] M.A. Haas, J.M. Wilson (unpublished results from [17]). – [12] L.A. Paquette, O. Cox (J. Am. Chem. Soc. **89** [1967] 5633/9). – [13] M.A. Hashmi, J.D. Munro, P.L. Pauson, J.M. Williamson (J. Chem. Soc. A **1967** 240/3). – [14] D.J. Bertelli, J.M. Viebrock (Inorg. Chem. **7** [1968] 1240/2). – [15] A.J. Birch, P.E. Cross, J. Lewis, D.A. White, S.B. Wild (J. Chem. Soc. A **1968** 332/40).

[16] J. Lewis, A.W. Parkins (Chem. Commun. **1968** 1194/5). – [17] M.I. Bruce (Advan. Organometal. Chem. **6** [1968] 273/333, 306/12). – [18] R.B. King (Appl. Spectrosc. **23** [1969] 536/46). – [19] G.R. Knox, A. Pryde (J. Organometal. Chem. **18** [1969] 169/73). – [20] B.F.G. Johnson, J. Lewis, G.L.P. Randall (Chem. Commun. **1969** 1273/4).

[21] A.W. Parkins (Diss. Univ. London 1969 from [76]). – [22] B.L. Kane (Diss. Univ. Birmingham 1970 from [51]). – [23] P.J. van Vuuren, R.J. Fletterick, J. Meinwald, R.E. Hughes (Chem. Commun. **1970** 883/4). – [24] P.J. van Vuuren (Diss. Cornell Univ. 1970; Diss. Abstr. Intern. B **31** [1971] 7201/2). – [25] J. Lewis, M. Schiavon (unpublished results from [30]).

[26] G.C. Farrant, G.T. Rodeheaver, D.F. Hunt (5th Intern. Conf. Organometal. Chem., Moscow 1971, Vol. 1, Abstr. No. 20). – [27] A. Eisenstadt, S. Winstein (Tetrahedron Letters **1971** 613/6). – [28] M.I. Foreman, D.G. Leppard (J. Organometal. Chem. **31** [1971] C31/C33). – [29] B.F.G. Johnson, J. Lewis, P. McArdle, G.L.P. Randall (Chem. Commun. **1971** 177/8). – [30] L.A.P. Kane-Maguire (J. Chem. Soc. A **1971** 1602/6).

[31] B.F.G. Johnson, J. Lewis, G.L.P. Randall (J. Chem. Soc. A **1971** 422/9). – [32] P.J. van Vuuren, R.J. Fletterick, J. Meinwald, R.E. Hughes (J. Am. Chem. Soc. **93** [1971] 4394/9). – [33] P. Hackett, B.F.G. Johnson, J. Lewis, G. Jaouen (J. Chem. Soc. Dalton Trans. **1982** 1247/51). – [34] R. Mason (unpublished results from [37]). – [35] M.S. Brookhart, N.M. Lippman, B.F. Lewis (Abstr. Papers 163rd Natl. Meeting Am. Chem. Soc., Boston 1972, ORGN 16).

[36] E.J. Forbes, M.K. Sultan, P.C. Uden (Anal. Letters **5** [1972] 927/30). – [37] R.J.H. Cowles, B.F.G. Johnson, J. Lewis, A.W. Parkins (J. Chem. Soc. Dalton Trans. **1972** 1768/75). – [38] B.F.G. Johnson, J. Lewis, P. McArdle, G.L.P. Randall (J. Chem. Soc. Dalton Trans. **1972** 2076/83). – [39] A.M. Brodie, B.F.G. Johnson, J. Lewis (J. Chem. Soc. Dalton Trans. **1973** 1997/2002). – [40] R. Aumann (Angew. Chem. **85** [1973] 628/9).

[41] R. Aumann (unpublished results from [40]). – [42] B.F.G. Johnson, J. Lewis, P. McArdle, G.L.P. Randall (J. Chem. Soc. Dalton Trans. **1972** 456/62). – [43] G.T. Rodeheaver (Diss. Univ. Virginia 1973; Diss. Abstr. Intern. B **34** [1973] 1433). – [44] M. Brookhart, R.E. Dedmond, B.F. Lewis (J. Organometal. Chem. **72** [1974] 239/45). – [45] R. Aumann (J. Organometal. Chem. **77** [1974] C33/C36).

[46] R. Aumann (J. Organometal. Chem. **78** [1974] C31/C34). – [47] R. Aumann (unpublished results from [44]). – [48] G. Jaouen, B.F.G. Johnson, J. Lewis (J. Organometal. Chem. **231** [1982] C21/C24). – [49] B.F.G. Johnson, J. Lewis, P. McArdle, J.R. Norton (J. Chem. Soc. Dalton Trans. **1974** 1253/8). – [50] P. Dodman, T.A. Hamor (J. Chem. Soc. Dalton Trans. **1974** 1010/3).

[51] P. Dodman, J.C. Tatlow (J. Organometal. Chem. **67** [1974] 87/92). – [52] R. Edwards, J.A.S. Howell, B.F.G. Johnson, J. Lewis (J. Chem. Soc. Dalton Trans. **1974** 2105/12). – [53] R.E. Davis, T.A. Dodds, T.H. Hseu, J.C. Wagnon, T. Devon, J. Tancrede, J.S. McKennis, R. Pettit (J. Am. Chem. Soc. **96** [1974] 7562/3). – [54] A.J. Deeming, S.S. Ullah, A.J.P. Domingos, B.F.G. Johnson, J. Lewis (J. Chem. Soc. Dalton Trans. **1974** 2093/104). – [55] R. Pettit (Organotransition-Metal Chem. Proc. 1st. Japan.-Am. Semin., Honolulu 1974 [1975], pp. 157/67; C.A. **85** [1976] No. 176360).

[56] B.F.G. Johnson, J. Lewis, D.J. Thompson, B. Heil (J. Chem. Soc. Dalton Trans. **1975** 567/71). – [57] K.E. Hine, B.F.G. Johnson, J. Lewis (J. Chem. Soc. Chem. Commun. **1975** 81/2). – [58] D.L. Reger, A. Gabrielli (J. Am. Chem. Soc. **97** [1975] 4421/2). – [59] P. Skarstad, P.J. van Vuuren, J. Meinwald, R.E. Hughes (J. Chem. Soc. Perkin Trans. II **1975** 88/92). – [60] D.A. Brown, N.J. Fitzpatrick, W.K. Glass, P.K. Sayal (J. Organometal. Chem. **234** [1982] C52/C54).

[61] D.A. Brown, S.K. Chawla, W.K. Glass, F.M. Hussein (Inorg. Chem. **21** [1982] 2726/32). — [62] M. Franck-Neumann, D. Martina (Tetrahedron Letters **1975** 1759/62). — [63] A. Eisenstadt (J. Organometal. Chem. **97** [1975] 443/51). — [64] R. Pettit, L.W. Haynes (Carbonium Ions **5** [1976] 2263/302). — [65] R. Aumann (Chem. Ber. **109** [1976] 168/73).

[66] R. Aumann, J. Knecht (unpublished results from [67]). — [67] R. Aumann, J. Knecht (Chem. Ber. **109** [1976] 174/9). — [68] D.A. Brown, S.K. Chawla, W.K. Glass (Inorg. Chim. Acta **19** [1976] L31/L32). — [69] B.F.G. Johnson, J. Lewis, D. Wege (J. Chem. Soc. Dalton Trans. **1976** 1874/80). — [70] A. Eisenstadt (J. Organometal. Chem. **113** [1976] 147/56).

[71] M. Bonadeo, C. de Micheli, R. Gandolfi (J. Chem. Soc. Perkin Trans. I **1977** 939/44). — [72] M. Bonadeo, R. Gandolfi, C. de Micheli (Gazz. Chim. Ital. **107** [1977] 577/8). — [73] M. Cais, P. Ashkenazi, J. Gottlieb (Rev. Roumaine Chim. **22** [1977] 545/9). — [74] M. Franck-Neumann, D. Martina (Tetrahedron Letters **1977** 2293/6). — [75] F.A. Cotton, B.E. Hanson (Israel J. Chem. **15** [1977] 165/73).

[76] D. Cunningham, P. McArdle, H. Sherlock, B.F.G. Johnson, J. Lewis (J. Chem. Soc. Dalton Trans. **1977** 2340/4). — [77] R. Aumann, J. Knecht (Chem. Ber. **111** [1978] 3927/31). — [78] Y. Becker, A. Eisenstadt, Y. Shvo (J. Organometal. Chem. **155** [1978] 63/75). — [79] B.Y. Shu, E.R. Biehl, P.C. Reeves (Syn. Commun. **8** [1978] 523/31). — [80] B.F.G. Johnson, K.D. Karlin, J. Lewis (J. Organometal. Chem. **145** [1978] C23/C25).

[81] M. Franck-Neumann, D. Martina, F. Brion (Angew. Chem. **90** [1978] 736/7). — [82] M. Franck-Neumann, F. Brion, D. Martina (Tetrahedron Letters **1978** 5033/6). — [83] M. El Borai, R. Guilard, P. Fournari, Y. Dusausoy, J. Protas (J. Organometal. Chem. **148** [1978] 285/301). — [84] G.A. Taylor (J. Chem. Soc. Perkin Trans. I **1979** 1716/9). — [85] J. Evans, D.V. Howe, B.F.G. Johnson, J. Lewis (J. Organometal. Chem. **61** [1973] C48/C50).

[86] A.M. Gabrielli (Diss. Univ. South Carolina 1978; Diss. Abstr. Intern. B **39** [1979] 5911). — [87] Z. Goldschmidt, Y. Bakal (J. Organometal. Chem. **168** [1979] 215/25). — [88] G. Deganello (Transition Metal Complexes of Cyclic Polyolefins, Academic, London 1979, p. 224 from [95]). — [89] R.F. Childs, A. Varadarajan (J. Organometal. Chem. **184** [1980] C28/C32). — [90] D.A. Brown, W.K. Glass, F.H. Hussein (J. Organometal. Chem. **186** [1980] C58/C60).

[91] D.L. Reger, R. Gabrielli (J. Organometal. Chem. **187** [1980] 243/52). — [92] M. El-Borai, M.F. Abdel-Megeed (Phosphorus Sulfur **9** [1980] 165/73). — [93] L.A.P. Kane-Maguire, T.I. Odiaka, S. Turgoose, P.A. Williams (J. Organometal. Chem. **188** [1980] C5/C9). — [94] W.C. Herndon (J. Am. Chem. Soc. **102** [1980] 1538/41). — [95] N.G. Connelly, R.L. Kelly, M.D. Kitchen, R.M. Mills, R.F.D. Stansfield, M.W. Whiteley, S.M. Whiting, P. Woodward (J. Chem. Soc. Dalton Trans. **1981** 1317/26).

[96] T.I. Odiaka, L.A.P. Kane-Maguire (J. Chem. Soc. Dalton Trans. **1981** 1162/8). — [97] Z. Goldschmidt, S. Antebi (J. Organometal. Chem. **206** [1981] C1/C3). — [98] L.A.P. Kane-Maguire, T.I. Odiaka, S. Turgoose, P.A. Williams (J. Chem. Soc. Dalton Trans. **1981** 2489/95). — [99] L.A.P. Kane-Maguire, T.I. Odiaka, P.A. Williams (J. Chem. Soc. Dalton Trans. **1981** 200/4). — [100] K.J. Karel, M. Brookhart, R. Aumann (J. Am. Chem. Soc. **103** [1981] 2695/8).

[101] J.G. Atton, L.A.P. Kane-Maguire (J. Chem. Soc. Dalton Trans. **1982** 1491/8). — [102] R. Benn, F.-W. Grevels, R. Schrader (22nd Intern. Conf. Coord. Chem., Budapest 1982, Vol. 1, p. 324). — [103] C.H. Mauldin, E.R. Biehl, P.C. Reeves (Tetrahedron Letters **1972** 2955/8). — [104] G.R. John, L.A.P. Kane-Maguire (J. Chem. Soc. Dalton Trans. **1979** 873/8). — [105] Z. Goldschmidt, S. Antebi (Tetrahedron Letters **1978** 271/4).

[106] A. Eisenstadt, J.M. Guss, R. Mason (J. Organometal. Chem. **80** [1974] 245/56). — [107] L. Kruczynski, J. Takats (Inorg. Chem. **15** [1976] 3140/7). — [108] S.K. Chopra, G. Moran,

P. McArdle (J. Organometal. Chem. **214** [1981] C36/C38). – [109] T. Ban, K. Nagai, Y. Miyamoto, K. Harano, M. Yasuda, K. Kanematsu (J. Org. Chem. **47** [1982] 110/6). – [110] E.H. Braye, W. Hübel (J. Organometal. Chem. **3** [1965] 25/37).

[111] D.F. Hunt, G.C. Farrant, G.T. Rodeheaver (J. Organometal. Chem. **38** [1972] 349/65). – [112] P. McArdle, H. Sherlock (Proc. 16th Intern. Conf. Coord. Chem., Dublin 1974, Abstr. No. R 89). – [113] R. Pettit, J.S. McKennis, W. Slegeir, W.H. Starnes, T. Devon, R. Case, J.C. Wagnon, L. Breuer, J. Wristers (Ann. N.Y. Acad. Sci. **239** [1974] 22/31). – [114] T.A. Dodds, R.E. Davis (unpublished results from [53]). – [115] A. Salzer, W. von Philipsborn (unpublished results from [116]).

[116] A. Salzer (8th Intern. Conf. Organometal. Chem., Kyoto 1977, p. 144). – [117] A. Salzer (9th Intern. Conf. Organometal. Chem., Dijon 1979, Abstr. No. B 70). – [118] W.G.L. Aalbersberg (Diss. Univ. California 1980; Diss. Abstr. Intern. B **41** [1981] 3029/30). – [119] M. El Borai, A. Akelah, M.A. Hassen (Egypt. J. Chem. **24** [1981] 91/6). – [120] A. Salzer, W. von Philipsborn (J. Organometal. Chem. **170** [1979] 63/70).

[121] N.G. Connelly, A.R. Lucy, M.W. Whiteley (Proc. 10th Intern. Conf. Organometal. Chem., Toronto 1981, p. 216). – [122] N. Morita, T. Asao (Chem. Letters **1982** 1575/8). – [123] A.J. Pearson, S.L. Kole, B. Chen (J. Am. Chem. Soc. **105** [1983] 4483/4). – [124] A.J. Pearson (Pure Appl. Chem. **55** [1983] 1767/79). – [125] J.G. Atton, L.A.P. Kane-Maguire (J. Organometal. Chem. **246** [1983] C23/C26).

[126] Z. Goldschmidt, S. Antebi (J. Organometal. Chem. **259** [1983] 119/25). – [127] K. Hayakawa, N. Hori, K. Kanematsu (Chem. Pharm. Bull. [Tokyo] **31** [1983] 1809/11). – [128] A. Salzer, A. Hafner (Helv. Chim. Acta **66** [1983] 1774/85). – [129] T. Ishizu, K. Harano, N. Hori, M. Yasuda, K. Kanematsu (Tetrahedron **39** [1983] 1281/9).

1.4.1.4.1.2.4.2.3 ^{4}L is Cycloheptatriene

1.4.1.4.1.2.4.2.3.1 $C_7H_8Fe(CO)_3$

$C_7H_8Fe(CO)_3$ (C_7H_8=cycloheptatriene). The complex, which was originally formulated as $C_7H_8Fe(CO)_2$ [1, 115], is shown to be $C_7H_8Fe(CO)_3$ [3, 5, 7, 12, 21].

Preparation and Formation

The compound was first prepared by refluxing a 1:1 mixture of $Fe(CO)_5$ and cycloheptatriene without a solvent [1, 3, 5, 8, 21, 37, 115] or by heating these components in a sealed tube at 100 to 150 °C [14]. Refluxing a ca. 2:3 mixture for 21 h followed by evaporation gives a yield of 52% [115]. Heating the mixture of components at 110 °C for 3 d gives $C_7H_8Fe(CO)_3$ in 20% yield [37], and for 7 d gives $C_7H_8Fe(CO)_3$, which is purified by chromatography on alumina with light petroleum ether as eluent [5]. Heating a 1:1 mixture of $Fe(CO)_5$ and C_7H_8 at 135 °C for 1 d gives $C_7H_8Fe(CO)_3$ and $C_7H_{10}Fe(CO)_3$ (C_7H_{10}=cyclohepta-1,3-diene) [3, 8, 21] in the ratio 19:1; after 5 d the ratio is 2:1 [3]. Repeated recrystallization of the product obtained after 1 to 2 d from hexane at −78 °C gives pure $C_7H_8Fe(CO)_3$ [3]. Similar reaction at 135 to 140 °C for 7 d gives $C_7H_8Fe(CO)_3$ and $C_7H_{10}Fe(CO)_3$ (ratio ca. 1:1) and additionally $(C_7H_8)_2Fe_3(CO)_9$ [5]. Refluxing $Fe(CO)_5$ and C_7H_8 in nonane (ca. 118 to 129 °C) followed by filtration, evaporation, and chromatography of the residue on alumina with petroleum ether (b.p. 37 to 43 °C) gives $C_7H_8Fe(CO)_3$. $C_7H_8Fe(CO)_3$ is also the sole product with excess C_7H_8 or $Fe(CO)_5$ under various conditions (refluxing for 6 h, standing for 2 d and refluxing again for 8 h, or reaction at 100 °C with 1 atm CO, or reaction at 150 °C with slight CO pressure) [9]. Refluxing $Fe(CO)_5$ and C_7H_8 in methylcyclohexane for 15 h [26], 48 h [64], or 3 d [67] followed by evaporation, distillation, and chromatography on neutral Al_2O_3 (Woelm, activity II) with pentane gives $C_7H_8Fe(CO)_3$ in 18% yield [26]; no purification is necessary for spectral applications after a reaction time of 3 d [67]. Refluxing $Fe(CO)_5$ and C_7H_8 in ethylcyclohexane gives $C_7H_8Fe(CO)_3$ after distillation [16]. Irradiation of a solution of $Fe(CO)_5$ and C_7H_8 in C_6H_6 at reflux temperature [35, 105, 106] for 36 h [106] or 48 h [35] followed by evaporation and distillation gives $C_7H_8Fe(CO)_3$ in 68% yield [106]. This photochemical reaction is better than the thermal procedures above [105]. Irradiation of $Fe(CO)_5$ with excess C_7H_8 in C_6H_6 for 12 h followed by filtration through Celite, concentration of the filtrate, and chromatography on alumina (activity II) with hexane gives $C_7H_8Fe(CO)_3$ in ca. 12% yield [116]. $C_7H_8Fe(CO)_3$ can also be prepared by the reaction of C_7H_8 with $Fe_2(CO)_9$ [11, 14, 21, 33] in a sealed tube at 100 to 150 °C [14] or in boiling pentane within 18 h [33] together with $C_7H_8Fe_2(CO)_6$ (see 2.6.1.2.3 in "Organoiron Compounds" C5, 1981, Table 2, No. 13, p. 17) [11, 33]. The reaction in pentane gives small amounts of $C_7H_8Fe(CO)_3$, distilled from the residue and 8.6% $C_7H_8Fe_2(CO)_6$, separated by chromatography [33]. The reaction of $Fe_3(CO)_{12}$ with C_7H_8 [1, 14, 21, 32] proceeds more readily than that with $Fe(CO)_5$ [1] and yields in boiling C_6H_6 almost exclusively $C_7H_8Fe(CO)_3$ [32]. The reaction of I (R=CH_3, R'=R''=H) with C_7H_8 (ratio 1:2) in C_6H_6 at 60 °C [94] or in toluene at 70 °C [103] for 72 h [94], followed by chromatography on basic alumina (activity III) with hexane [94] gives $C_7H_8Fe(CO)_3$ in 50% yield [94, 103]. Similar reaction of I (R=R'=R''=H; R=C_6H_5, R'=H, R''=H or CH_3) with C_7H_8 (ratio 1:10 to 1:200) in toluene at 90 °C ensures first-order conditions. For the kinetics and mechanism of this reaction see [88, 103]. The reaction of I (R=CH_3, R'=R''=H) with C_7H_8 in the presence of C_6H_8 (molar ratio 1:12:6) at 60 °C in C_6H_6 gives $C_7H_8Fe(CO)_3$ and $C_6H_8Fe(CO)_3$ showing a reactivity ratio of C_7H_8:C_6H_8 of ca. 0.003 (corrected assuming a C_6H_8:C_8H_8 reactivity ratio of 16:1) [94]. Addition of excess free benzylideneacetone to the reaction mixture of I (R=CH_3, R'=R''=H) in C_6H_6 shows the rate of $Fe(CO)_3$ transfer and gives insight to the detailed mechanism of the trapping reaction [81, 94]. The experimental results at 67 °C (reaction time 24 h) are shown in the following table [94]:

I (R=CH_3, R'=R''=H):C_7H_8:benzylideneacetone (molar ratio)			$C_7H_8Fe(CO)_3$ (yield in %)
1	1.5	0	50
1	1.5	2	20
1	1.5	5	13
1	1.5	10	<1

$C_7H_8Fe(CO)_3$ is formed in the reaction of Na[$C_7H_7Fe(CO)_3$] with H_2O [60], dilute acids, or C_2H_5OH [68]. Similar reaction of Li[$C_7H_7Fe(CO)_3$] with D_2O gives XXI (p. 110), containing 4% $C_7H_8Fe(CO)_3$ (by mass spectrometry) [29]. $C_7H_8Fe(CO)_3$ is very likely formed by the reaction of the very unstable $C_7H_7Fe(CO)_2CH_2CH{=}CH_2$ intermediate with moisture during chromatography [60, 61, 102]. Direct reaction of K[$C_7H_7Fe(CO)_3$] occurs with $ClSiR_3$, $ClCO_2R$, and $[C_7H_7M(CO)_3]^-$ (M=Cr, Mo, W) yielding variable yields of $C_7H_8Fe(CO)_3$ depending upon the reagents. Chromatography on Florisil gives $C_7H_8Fe(CO)_3$ and trace amounts of two other compounds which are not further investigated [61]. Dropwise addition of Na[$C_7H_7Fe(CO)_3$] in tetrahydrofuran to a solution of [$C_6H_7Fe(CO)_3$]BF_4 in tetrahydrofuran followed by evaporation and chromatographic work-up on Al_2O_3 with n-hexane gives $C_7H_8Fe(CO)_3$ in the first fraction, $(OC)_3FeC_6H_7$-$C_7H_7Fe(CO)_3$ together with small amounts of byproducts in the second fraction, and finally $C_6H_8Fe(CO)_3$, $(C_6H_7Fe(CO)_3)_2$, and $(C_7H_7Fe(CO)_3)_2$ as main products in the third fraction [79]. Dissolving II in organic solvents regenerates $C_7H_8Fe(CO)_3$ [76]. At 90.5 °C in C_6H_6 solution III a or b cleanly rearranges to $C_7H_8Fe(CO)_3$ in a first-order reaction [111]. Refluxing [$C_7H_9Fe(CO)_3$]BF_4 with $N(CH_3)_3$ in acetone for 2 h gives $C_7H_8Fe(CO)_3$ in quantitative yield. Refluxing [$C_7H_9Fe(CO)_3$]BF_4 with dicyclohexylethylamine in acetone for 3 h

References on pp. 113/6

followed by evaporation of the reaction mixture and extraction of the residue with ligroin gives $C_7H_8Fe(CO)_3$ in 83% yield and $(C_7H_9Fe(CO)_2)_2$ in 8% yield after extraction with CH_2Cl_2. Analogous reaction in boiling isopropanol gives quantitatively $C_7H_8Fe(CO)_3$ and traces of $C_7H_{10}Fe(CO)_3$ [19]. Reaction of $[C_7H_9Fe(CO)_3]BF_4$ with $HN(CH_3)C_6H_5$ (ratio ca. 1:7) in CH_2Cl_2 at room temperature for 20 min gives 5-exo-$C_6H_5(CH_3)NC_7H_9Fe(CO)_3$ and $C_7H_8Fe(CO)_3$. Similarly, $(CH_3)_2NC_6H_5$ gives only $C_7H_8Fe(CO)_3$ as a product [78]. Refluxing $[C_7H_9Fe(CO)_3]BF_4$ with pyridine in acetone for 3 h gives $C_7H_8Fe(CO)_3$ in only 25% yield [19]. For the preparation of a deuterated compound see "Chemical Behavior" (pp. 110/1).

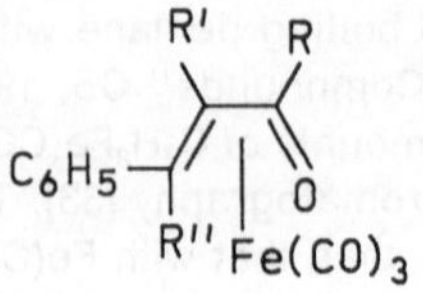

R' R (CO)$_3$Fe—SO$_2$ H H

C_6H_5 O Fe(CO)$_3$ Fe(CO)$_3$

R'' Fe(CO)$_3$ H H

I 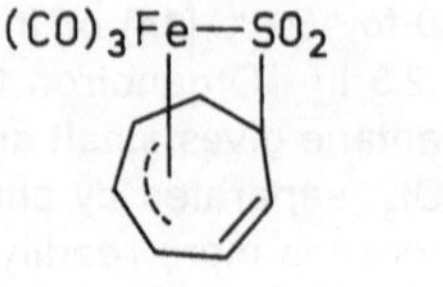 II 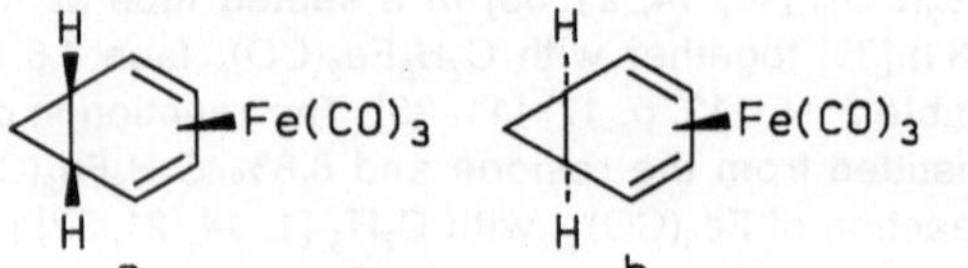 a III b

Physical Properties

$C_7H_8Fe(CO)_3$ melts at ca. −2 °C (from hexane at −78 °C) [3] or at ca. 5 °C [5], a ca. 1:1 mixture of $C_7H_8Fe(CO)_3$ and $C_7H_{10}Fe(CO)_3$ melts at 23 °C [5]. The compound boils at 50 °C/1 Torr [9], 53 °C/0.02 Torr [26], 55 to 65 °C/0.05 Torr [35, 106], 60 °C/0.5 Torr [3], 66 to 70 °C/0.15 Torr [115], 70 °C/0.4 Torr [1, 5, 16]. At room temperature it is a mobile [1] yellow [1, 49, 102], orange-yellow [5], orange [3, 9, 35, 106], or orange-red [115] liquid [1, 3, 5, 16, 115] or oil [9, 35, 106]. The compound is monomeric in solution [1], and exhibits a dipole moment of $\mu=2.35\pm0.11$ D [2] or $\mu=2.43\pm0.04$ D in C_6H_6 at 25±0.1 °C [5]. Refractive index: $n_D^{20}=1.6277$ [5]. The ^1H NMR spectra (for assignments see Formula IV) in different solvents at ambient temperatures (if not mentioned otherwise) are shown in the following table (δ in ppm):

solvent	H-1	H-2	H-3	H-4	H-5	H-6	H-7	Ref.
benzene-d_6	2.89	4.59 to 4.75		2.72	5.68	5.03	1.89 (exo), 2.12 (endo)	[87, 116]
benzene-d_6	2.87	4.66 to 5.14		2.87	5.61	4.66 to 5.14	2.12	[29]
benzene-d_6 (30 to 80 °C)	2.883 (t)	4.688 (t)	4.694	2.687 (d)	5.573 (d)	4.933	1.873, 2.083	[26]
toluene-d_8	2.89 (m)	4.66 (m)		2.72 (m)	5.68 (ddd)	5.03 (m)	1.89 (endo), 2.12 (exo)	[112]
CCl_4	3.30 (d)	5.13 (dt)		2.96 (t)	5.24	5.155 (d)	2.27, 238	[5]
CS_2	3.2	5.2		3.2	5.8	5.3	2.4, 3.2	[3]

The spectrum in benzene-d_6 consists of an 8 spin system of the type ABMNPQXY [26]. The coupling constants (for assignments see Formula IV) are shown in the following table (J in Hz):

assignment	coupling constant, from		
	[26]	[112, 116]	[5]
J(H-1,2)	7.76	–	7.7
J(H-1,3)	1.55	–	–
J(H-1,6)	0.83	0.9 or 1.4	–
J(H-1,7exo)	2.91 or 4.54	3.4 or 4.3	–
J(H-1,7endo)	2.91 or 4.54	3.4 or 4.3	–
J(H-2,3)	4.96	–	4.74
J(H-2,4)	1.10	–	1.5
J(H-2,5)	0.0	–	–
J(H-2,7)	−0.82	–	–
J(H-3,4)	7.78	–	7.6
J(H-3,5)	0.0	–	–
J(H-3,6)	0.0	–	–
J(H-4,5)	7.94	7.9	7.7
J(H-4,6)	1.37	0.9 or 1.4	–
J(H-5,6)	10.62	10.6	10.6
J(H-5,7)	−1.89, −2.82	1.9, 2.6	2.2
J(H-6,7)	3.43, 4.54	3.5, 4.2	3.8
J(H-7exo, 7endo)	−21.95	22.0	–

The exo deuterated compound shows in benzene-d_6 the same ^1H NMR spectrum as $C_7H_8Fe(CO)_3$ except that the shift at $\delta = 1.89$ ppm has disappeared [116]. The ^1H NMR spectrum of $C_7H_8Fe(CO)_3$ shows no temperature dependence [22], no line broadening due to 1,3 iron shifts at temperatures up to 100 °C (ΔG^{*} > ca. 20 kcal/mol, −120 °C) [38, 40, 114], and is therefore nonfluxional at room temperature [15, 80]. In an effort to determine the activation energy for the 1,3 iron shifts (see Formula IV), spin transfer experiments are performed at several temperatures as shown in the following table [83, 87, 112]:

t in °C	M(0)/M(∞)	T_1 (s)	$k \cdot 10^2$ (s^{-1})	ΔG^{*} (kcal/mol)
61	1.27	14.5	1.90	22.2
75	1.80	17.5	4.55	22.4
86	2.52	10.7	14.2	22.5
90	4.50	10.7	32.7	22.1

At 90 °C a dramatic decrease (78%) is measured in the integrated intensity of the signal due to H-6 when H-1 is saturated. Data for M(0)/M(∞) are summarized in the table along with T_1 values measured for H-6. Also included are the rate constants calculated from $k = [(M(0)/M(\infty)) - 1]/T_{1A}$ where T_{1A} is the spin-lattice relaxation time of the proton at site A, k is the rate constant in a two-site, equal-population system, $A \rightleftharpoons B$, M(0) is the normal equilibrium magnetization of the proton A, and M(∞) is the equilibrium magnetization of the proton A after saturating the proton at site B. Experimentally M(0)/M(∞) is determined by comparing the signal areas of proton A with and without saturation of proton B. Although the averaging process corresponds to a net 1,3 iron shift, two mechanisms for iron migration seem feasible as shown in IV. A norcaradiene intermediate is attractive in that only 1,2 iron shifts are involved and is also compatible with the averaged free energy of activation of 22.3 kcal/mol [83, 87, 112], cf. [124]. All intermolecular exchange mechanisms involving dissociation of the complex can be eliminated since there is no scrambling of the exo

 References on pp. 113/6

and endo substituents observed in 7-exo-$HC_7D_7Fe(CO)_3$ even after several days at 72 °C [112].

(CO)₃Fe IV

The ^{13}C NMR spectrum ($CDCl_3$) [76, 95] shows chemical shifts at δ=30.70 (C-7), 55.92 (C-1 or 4), 60.26 (C-1 or 4), 88.28 (C-2 or 3), 93.39 (C-2 or 3), 125.43 (C-6), 128.41 (C-5), 211.32 (CO) ppm [95], and in acetone-d_6 at 183 K δ=30.9 (C-7), 56.2 (C-1 or 4), 60.7 (C-1 or 4), 88.6 (C-2 or 3), 93.7 (C-2 or 3), 125.6 (C-5 or 6), 128.6 (C-5 or 6), 210.6 (CO basal), 210.9 (CO basal), 214.6 (CO apical) ppm (for assignments see Formula IV). At 280 K the shifts for the CO groups are averaged to δ=211.9 ppm [67, 117]. The scrambling of the CO groups proceeds in a concerted manner [67, 77, 117]. The activation parameters for this scrambling are ΔH^{*}=10.8 kcal/mol, ΔS^{*}=−0.1 e.u. [67], $\Delta G^{\ddagger}_{298}$=10.4 kcal/mol [67, 77], and E_a=11.6(3) kcal/mol [77, 117]. $C_7H_8Fe(CO)_3$ is shown to be fluxional by use of the Forsén-Hoffman spin saturation transfer method in the ^{13}C NMR spectrum (see ^1H NMR spectrum). This method yields an approximate rate constant of 0.2 s^{-1} at 94 °C, which corresponds to a ΔG^{*} of ca. 23 kcal/mol (significant decomposition). Again two mechanisms, as shown in IV, are discussed, but no decision is given between these two possibilities. An activitation energy of ca. 16 kcal/mol could be attributed to the formation of the norcaradiene complex [73].

The ^{57}Fe NMR spectrum in C_6H_6 (0.83 M) at 300 K shows a chemical shift referred to neat $Fe(CO)_5$ (δ=0 ppm) of δ=169.4±4 ppm [109].

The IR spectrum of the pure liquid [3, 5, 9] shows bands at 704, 804, 834, 864, 893, 933, 943, 1000, 1153, 1189, 1214, 1346, 1356, 1404, 1419, 1446, 1471, 1665, 1668, 1975, 1989, 2050, 2796, 2841, 2872, 2921, 2992, and 3026 cm^{-1} [5]. The following bands are reported for the ν(CO) region: 1975, 1989, 2050 cm^{-1} in Nujol/Hostaflon [17], 1976, 1986, 2052 cm^{-1} in cyclohexane [2], 1974.7, 1987.1, 2049.8 cm^{-1} in n-hexadecane [10, 14], 1970, 1983, 2047 cm^{-1} in decalin [53], 1970, 2048 cm^{-1} in CH_2Cl_2 [82, 96], and 1974, 2048 cm^{-1} in tetrahydrofuran [49, 100]. The 7-exo deuterated compound (neat) lacks the absorption at 2795 cm^{-1} observed for $C_7H_8Fe(CO)_3$ [29]. The stretching force constant for the CO group is k=16.23 mdyn/Å [104] which corresponds to a binding energy of E(C1s)=292.79 eV [75, 104] and E(O1s)=539.09 eV [104]. The data indicate that transfer of electron density to the CO is accompanied by weakening of the C-O bonds and that in back-bonding to CO groups, more charge is transferred to the C atoms than to the O atoms [104]. The He(I) photoelectron spectra show ionization energies at 7.76 and 8.39 (Fe), 8.78 and 10.23 ($1a_2$), 11.10 ($1b_1$), and 11.82 (CO) eV [69, 123]. The experimental data can be satisfactorily explained by a localization energy approximation of the Hückel type [18]. For extended molecular orbital calculations to probe the geometry and electronic structure of $C_7H_8Fe(CO)_3$ see [110, 112]. The HOMO coefficients calculated for C-1, C-4, and C-6 (see Formula IV) are −0.098, −0.515, and +0.454. Frontier orbital theory predicts that 4,6-addition is the preferred cycloaddition mode for $(NC)_2C=C(CN)_2$ [110]. Molecular orbital calculations for $[C_7H_7Fe(CO)_3]^-$ predict interesting consequences for $C_7H_8Fe(CO)_3$ [89]. For the application of topology and group theory to $C_7H_8Fe(CO)_3$ see [24, 72].

Chemical Reactions

$C_7H_8Fe(CO)_3$ decomposes in air on prolonged storage [1, 3, 5] and is sensitive to light [3]. Pyrolysis at 200 to 350 °C gives C_7H_8 (95.54 to 97.38%) and cycloheptadiene (0.62 to 1.55%); at 400 to 450 °C the products are C_7H_8 (90.05 to 94.93%), C_7H_{10} (1.32 to 3.00%), and toluene (0.31 to 2.18%). Other products of this pyrolysis are CO, CO_2, H_2, and decomposition products of C_7H_8 in low yield. The pyrolytic Fe film is contaminated by carbon [85]. The mass spectrum shows the ion $[M-CO]^+$ [3]. $C_7H_8Fe(CO)_3$ forms molecular anions in high abundance in its negative ion mass spectrum in apparent violation of the rare-gas rule [99]. Gas phase protonation reactions are investigated between a series of Brønsted acid reagent ions as shown in the following table (mass spectra obtained with ionization source temperature in the range 100 to 110 °C) [101]:

ions	reagent gas			
	H_2	CH_4	$i\text{-}C_4H_{10}$	NH_3
$[M+H]^+$	13	8	53	96
$[M]^{+\cdot}$	0.1	0.1	0	0
$[M-H]^+$	0.6	0.4	4	0
$[M+H-CO]^+$	29	38	4	0
$[M-CO]^{+\cdot}$	11	19	0	1
$[M-H-CO]^+$	0.4	0.3	0	0
$[M+H-2CO]^+$	42	24	11	3
$[M-2CO]^{+\cdot}$	2	2	1	0
$[M-H-2CO]^{+\cdot}$	0.3	0.2	1	0
$[M+H-3CO]^{+\cdot}$	3	2	9	0.5
$[M-3CO]^{+\cdot}$	5	2	2	0
$[M-H-3CO]^+$	0.7	0.8	3	0
$[C_7H_9]^+$	2	3	5	0
$[C_7H_8]^{+\cdot}$	0	0	0	0
$[C_7H_7]^+$	2	1	7	0

For a discussion of these protonation reactions and mass spectra in terms of the relative acid strengths of the protonating reagent ions see [101]. Irradiation of $C_7H_8Fe(CO)_3$ in tetrahydrofuran (^2D) at −78 °C gives Va and/or b as intermediates (not isolated), which after removal of the light source and addition of RC≡CR give VI (n=2 and/or 3, m=0 and/or 1 intermediate) and finally VII (for $R=CO_2CH_3$), or VIII (for $R=C_6H_5$) in 25% yield [52, 54, 55].

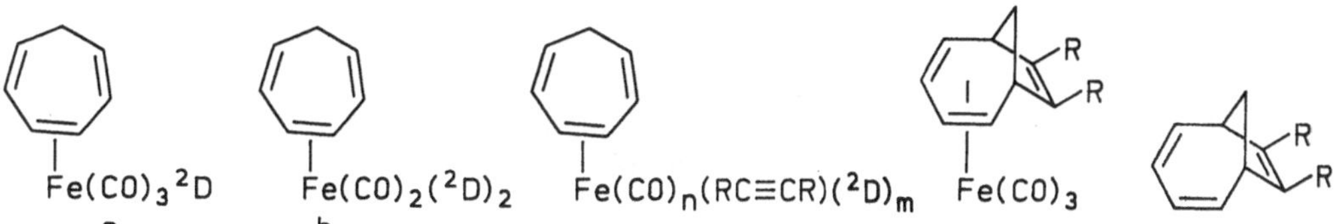

a V b VI VII VIII

Reactions with Elements and Inorganic Compounds. Hydrogenation with H_2/Raney Ni at 130 °C in C_2H_5OH at a pressure of 100 atm gives cycloheptane (40%). Halogens give a

References on pp. 113/6

rapid reaction, but no apparent change is observed with Zn/CH_3CO_2H or HCl. Concentrated HCl in $C_2H_5OH/CHCl_3$ (1:1) in the presence of air gives $[C_7H_9Fe(CO)_3]FeCl_4$ within 6 d. Similar reaction with HBr in light petroleum ether gives $[C_7H_9Fe(CO)_3]Br$ [5]. Reaction with conc. H_2SO_4 [4] or D_2SO_4 [4, 116] gives stable solutions of $[C_7H_9Fe(CO)_3]^+$ or $[C_7H_8DFe(CO)_3]^+$ [4, 116]. Attempted protonation with $HOSO_2F$ gives a dimeric binuclear species which arises by Fe protonation and its formation is inhibited by the presence of electron donating ligands such as PR_3 [97]. Protonation with 40% aqueous [64] HBF_4 [8, 36, 64] in propionic anhydride [36] or $O(COCH_3)_2$ at 0 °C [64, 125, 126] gives $[C_7H_9Fe(CO)_3]BF_4$. Similar dropwise addition of $C_7H_8Fe(CO)_3$ in CF_3CO_2D to 40% HBF_4 in propionic anhydride at 0 °C gives IX [35]. Oxidation with NOX (X=BF_4 or PF_4) in CH_2Cl_2 or with $AgBF_4$ gives $[C_7H_9Fe(CO)_3]X$ (X=BF_4, PF_6) [70]. Reaction with $AgNO_3$ gives a pale yellow insoluble adduct which decomposes quite quickly, the Ag^+ being reduced to Ag^0 [5]. Immediate decomposition is observed with OsO_4 [5]. Reacting $C_7H_8Fe(CO)_3$ with $WCl_6/C_2H_5AlCl_2/C_2H_5OH$ in C_6H_6 in a sealed tube for 24 h, addition of more catalyst, and further reaction for 17 h gives cis-$(C_7H_9Fe(CO)_2)_2$ and $C_7H_9Fe(CO)_2Cl$. Similar reaction with WCl_6 in toluene/LiC_4H_9-n in hexane in a sealed tube with toluene as solvent at 80 °C for 17 h gives $C_{8.7}H_{10.7}Fe(CO)_3$ (see Formula X) [106]. Dissolving $C_7H_8Fe(CO)_3$ in liquid SO_2 gives XI [76]. $C_7H_8Fe(CO)_3$ does not give CO insertion products with $AlCl_3$ in C_6H_6 at room temperature within 16 h [59], but gives $C_7H_8Fe(CO)_3AlX_3$ (X=Cl, Br) with AlX_3 in CH_2Cl_2 [82, 96]. Dropwise addition of $C_7H_8Fe(CO)_3$ in tetrahydrofuran to a slurry of KH in tetrahydrofuran (^2D) at 25 °C gives $K[C_7H_7Fe(CO)_3]\cdot{}^2D$ [49, 93, 98]. Reaction with B_2H_6 followed by OH^- and H_2O_2 gives 6-$HOC_7H_9Fe(CO)_3$ (Formula XII) [36, 84]. Reduction with $C_4H_8OBH_3$ (C_4H_8O=tetrahydrofuran) in tetrahydrofuran for 1 h gives XIII. How such a molecule would form in the reaction is not clear [91, 100].

IX X XI

Reaction with Organic Electrophiles. $C_7H_8Fe(CO)_3$ gives XIV in the reaction with $[C_7H_7]BF_4$ [8, 44], but excess $C_7H_8Fe(CO)_3$ gives XV (R=C_7H_7) with $[C_7H_7]PF_6$ within 24 h and not XIV [65]. Addition of $[C(C_6H_5)_3]BF_4$ in CH_2Cl_2 gives XVI (R=$C(C_6H_5)_3$, X=BF_4) in 96% yield [3]. Similar reaction with $[C_3(C_6H_5)_3]BF_4$ gives low yields (6%) of XVI (R=$C_3(C_6H_5)_3$, X=BF_4) [121]. Reaction with $RCOF/BF_3$ (R=CH_3, C_6H_5) in CH_2Cl_2 at −80 °C for 10 min gives XVI (R=$COCH_3$, COC_6H_5, X=BF_4) [35]. Similar reaction with $CH_3COCl/AlCl_3$ in CH_2Cl_2 at

XII XIII XIV

References on pp. 113/6

0 °C for 10 min followed by addition of H_2O gives XVII (R = $COCH_3$) from the organic phase and XVI (R = $COCH_3$, X = PF_6) from the aqueous phase after addition of $[NH_4]PF_6$ [25, 35]. Refluxing $C_7H_8Fe(CO)_3$ with CH_2I_2/Zn/Cu in ether gives a product described as XVIII [35], but a repetition of this reaction in the presence of one crystal of I_2 gives XIX (R = R′ = H) only, and no XVIII can be detected [56, 100]. Analogous reaction with CD_2I_2/Zn/Cu gives XIX (R = R′ = D) [100]. The copper catalyzed (Cu^{II} chelate of salicylaldehyde-aniline Schiff's base) reaction with $N_2CHCO_2C_2H_5$ at ca. 50 °C gives XIX (R = H, R′ = $CO_2C_2H_5$) and XVIII; no reaction occurs at room temperature. However, the analogous reaction with $N_2C(C_6H_5)COC_6H_5$ gives XXXIII (p. 112) [127]. Reaction with $[N_2C_6H_4R'\text{-}4]BF_4$ in acetone gives 4-$R'C_6H_4N_2C_7H_7Fe(CO)_3$ (R′ = F or NO_2) [119] or at −23 °C yellow solutions are obtained from which unstable yellow solids are isolable on addition of ether. These complexes are formulated as XVI (R = $N_2C_6H_4R'$-4 with R′ = F, OCH_3, or NO_2, X = BF_4) but are not fully characterized (loss of N_2 precludes good elemental analyses) [122].

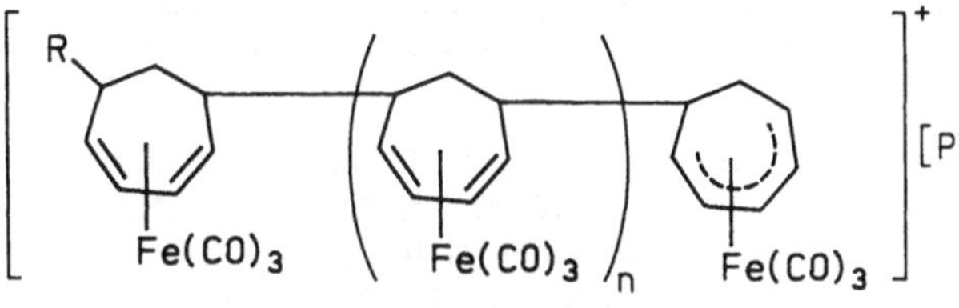

R, $Fe(CO)_3$, $Fe(CO)_3$, n, $Fe(CO)_3$, $[PF_6]^-$, +

XV

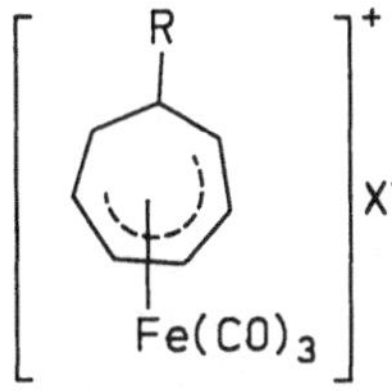

R, $Fe(CO)_3$, X^-, +

XVI

A 15-fold excess of $C_7H_8Fe(CO)_3$ with XVI (R = XX, X = PF_6) in CH_2Cl_2 gives XV with R = XX. Similar reaction of a 100-fold excess of $C_7H_8Fe(CO)_3$ with XVI (R = H, X = PF_6) in CH_2Cl_2 in a sealed tube gives XV (R = H, X = PF_6; n < 30) within 3 weeks [65]. Formylation by the Vilsmeyer-Haack reaction ($OHCN(CH_3)_2/OPCl_3$ [25, 37, 41], 0 °C, 1 h [37]) gives XVII (R = CHO) [25, 37, 41].

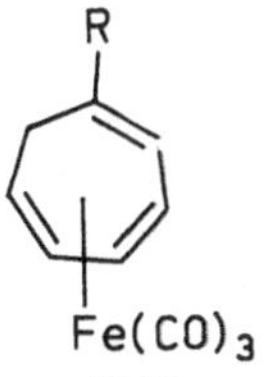

R, $Fe(CO)_3$

XVII

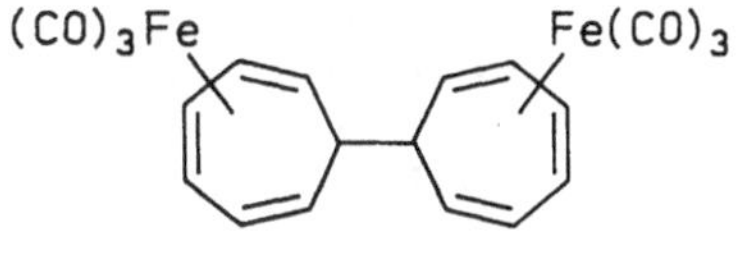

$(CO)_3Fe$, $Fe(CO)_3$

XVIII

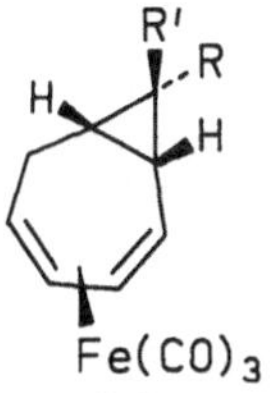

R′, R, H, H, $Fe(CO)_3$

XIX

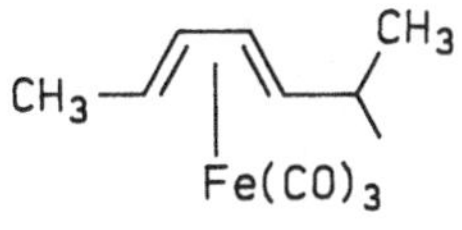

CH_3, CH_3, $Fe(CO)_3$

XX

Reaction with Nucleophiles. Reaction of $C_7H_8Fe(CO)_3$ with PF_3 gives $C_7H_8Fe(CO)_n(PF_3)_{3-n}$ (n = 0 to 2) [28]. Reaction with excess $P(OCH_3)_3$ at 100 to 110 °C for 3 d gives $(CO)_2Fe(P(OCH_3)_3)_3$ [14] and not cis-$(CO)_3Fe(P(OCH_3)_3)_2$ as claimed in [10]. Reaction with $P(C_2H_5)_3$ gives trans-$(CO)_3Fe(P(C_2H_5)_3)_2$ and $C_7H_8Fe(CO)_2P(C_2H_5)_3$ [10, 14], but prolonged reaction with PR_3 (R = OCH_3 or C_2H_5) at elevated temperatures does not give $C_7H_8Fe(CO)_2PR_3$ obtained under milder conditions, but almost certainly $C_7H_{10}Fe(CO)_2PR_3$ (C_7H_{10} = cyclohepta-1,3-diene) [19]. A mixture of $P(OCH_3)_3/P(C_2H_5)_3$ at 50 °C gives $C_7H_8Fe(CO)_2P(OCH_3)_3$ and $C_7H_8Fe(CO)_2P(C_2H_5)_3$ and at 70 °C a mixture of complexes $(CO)_2Fe(P(C_2H_5)_3)_n(P(OCH_3)_3)_{3-n}$ (n = 1, 2) [14]. Photochemical reaction with $P(C_6H_5)_3$ in C_6H_6 for 18 h yields only $C_7H_8Fe(CO)_2P(C_6H_5)_3$; no trace of $C_7H_8Fe(P(C_6H_5)_3)_2CO$ can be isolated even after prolonged irradiation (36 h) using an excess of $P(C_6H_5)_3$ [19]. Irradiation with $P(C_6H_5)_3$ at −78 °C gives $(CO)_2Fe(P(C_6H_5)_3)_3$ [55]. Refluxing with $P(C_6H_5)_3$ in ethylcyclohexane for 7 h gives $(CO)_3Fe(P(C_6H_5)_3)_2$ and $(CO)_2Fe(P(C_6H_5)_3)_3$ as a mixture of isomers [14, 74, 115]. Reaction

References on pp. 113/6

with $P(C_6H_5)_3$ in decalin at 154 °C involves both first- and second-order processes. The pseudo first-order constants for this reaction at 153.6 °C are shown in the following table [53]:

$[P(C_6H_5)_3]$ in mol/L	$k_{obs} \cdot 10^3$ in s^{-1}
0.0348	3.69
0.0552	4.06
0.0797	4.29
0.1065	4.41
0.1300	4.70

The rate constant for $C_7H_8Fe(CO)_3 + P(C_6H_5)_3 \rightarrow C_7H_8Fe(CO)_2P(C_6H_5)_3 + CO$ is $k_1 = (3.42 \pm 0.16) \times 10^{-4}\ s^{-1}$ and for $C_7H_8Fe(CO)_3 + P(C_6H_5)_3 \rightarrow$ trans-$(CO)_3Fe(P(C_6H_5)_3)_2 + C_7H_8$ is $k_2 = (9.9 \pm 2.0) \times 10^{-4}\ L \cdot mol^{-1} \cdot s^{-1}$ [53]. Refluxing $P(C_6H_4F\text{-}4)_3$ in 2,2,5-trimethylhexane for 7 h [27] gives $(CO)_2Fe(P(C_6H_4F\text{-}4)_3)_3$ [27, 74] and $(CO)_3Fe(P(C_6H_4F\text{-}4)_3)_2$ [27]. Reaction with $^4D = P(C_6H_5)_2(CH_2)_2P(C_6H_5)_2$ gives $(CO)_3Fe^4D$ [10].

Reaction with $NaN(Si(CH_3)_3)_2$ gives $Na[C_7H_7Fe(CO)_3]$ (see 1.3.1.4.2.2 in "Eisen-Organische Verbindungen" B5, 1978, Table 17, No. 12, p. 74) [20, 63, 66, 93, 98] and $HN(Si(CH_3)_3)_2$ [63, 66]. Refluxing with $NaOC_2H_5/[As(C_6H_5)_4]Cl$ in acetic ester for 10 h gives $[As(C_6H_5)_4][C_7H_7Fe(CO)_3]$. This reaction is better than the above reaction with $NaN(Si(CH_3)_3)_2$ [86]. Similarly KOC_4H_9-t removes the exo proton in $C_7H_8Fe(CO)_3$ under various conditions in tetrahydrofuran to give $K[C_7H_7Fe(CO)_3]$ [49, 60, 61, 93, 98, 102]. Thus, dropwise addition of $C_7H_8Fe(CO)_3$ in tetrahydrofuran to a slurry of KOC_4H_9-t in tetrahydrofuran at 25 °C [49, 102], or addition of solid KOC_4H_9-t to a stirred tetrahydrofuran solution of $C_7H_8Fe(CO)_3$ gives $K[C_7H_7Fe(CO)_3]$ with comparable results [102] (compare reaction with KH, p. 108). Addition of $C_7H_8Fe(CO)_3$ in hexane to KOC_4H_9-t in t-C_4H_9OH followed by addition of $CHCl_3$ in hexane at 0 °C gives XIX (R = R′ = Cl) [92]. Reaction of $C_7H_8Fe(CO)_3$ with LiC_4H_9-n [29, 45 to 47, 49, 71, 93, 98] in tetrahydrofuran [29, 45, 46, 49, 71, 93, 98] at −78 [29] or −60 °C [49, 71] gives $Li[C_7H_7Fe(CO)_3]$. Reaction of XXI with LiC_4H_9-t followed by addition of H_2O gives $C_7H_8Fe(CO)_3$, $C_7H_7DFe(CO)_3$, $C_7H_6D_2Fe(CO)_3$, and $C_7H_5D_3Fe(CO)_3$ in 76, 10, 6, and 9% yield (by mass spectrometry) [29]. XXI is also obtained by the reaction of $C_7H_8Fe(CO)_3$ with $NaOCH_3/CH_3OD$ at room temperature for 1 h [29] or 5.5 h [116]. The reaction is quenched with saturated $[NH_4]Cl/H_2O$, extracted with ether, dried (K_2CO_3), and the mixture is concentrated. The residual oil is distilled under vacuum to give 70% yield of the product [116], consisting of 6% $C_7H_8Fe(CO)_3$, 93% XXI, and 1% $C_7H_6D_2Fe(CO)_3$ (by mass spectrometry) [29]. The stereoselective deuterium exchange of $C_7H_8Fe(CO)_3$ occurs with CH_3OD with a faster rate than does XXI in CH_3OH [47]. Treatment of IX with excess $HN(C_2H_5)_2$ in acetone followed by addition of HCl until pH 2 was reached, extraction with ether/brine, drying of the organic layer ($MgSO_4$), and chromatography on alumina with light petroleum ether gives a ca. 1:1 mixture of $C_7H_8Fe(CO)_3$ and XXI as a yellow oil [35]. Attempts to specifically incorporate deuterium at C-5 or C-6 (compare XXI) in $C_7H_8Fe(CO)_3$ failed [50]. Thermal isomerization of XXI in

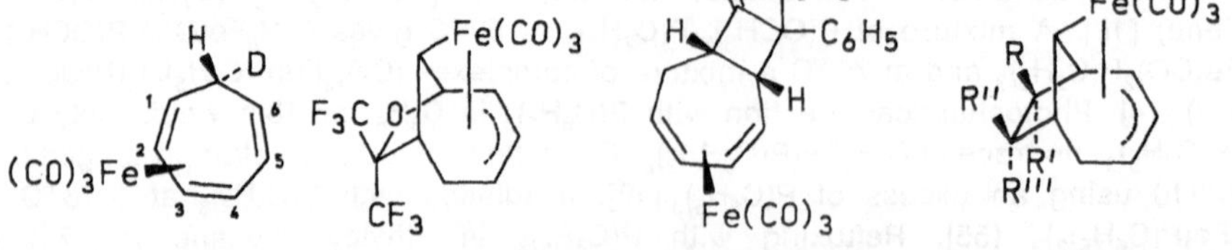

XXI XXII XXIII XXIV

References on pp. 113/6

benzene-d_6 for 1 h at 52 °C followed by 2 h at 72 °C gives a ca. 25% incorporation of H into the exo site of XXI; scrambling increases to ca. 60% after an additional 6 h at 72 °C [83, 116]. The above reactions with XXI indicate stereoselectivity for the deuterium or proton abstraction [29]. Reaction of $C_7H_8Fe(CO)_3$ with LiC_6H_5 [5, 29] or $BrMgC_6H_5$ [5] in tetrahydrofuran at −78 °C [29] gives $M[C_7H_7Fe(CO)_3]$ (M=Li [5, 29], BrMg [5]).

Reactions with Olefins and Analogous Compounds. $C_7H_8Fe(CO)_3$ displays no activity as a diene in a purely thermal Diels-Alder reaction [55]. Thus, the following reactions with, e.g., $O{=}C(CF_3)_2$, $(NC)_2C{=}C(CF_3)_2$, trans-$CF_3(NC)C{=}C(CN)CF_3$, or $(NC)_2C{=}C(CN)_2$ proceed via the initial formation of transient dipolar species with electrophilic attack exo to the Fe [48]. Frontier orbital calculations predict that 1,3-addition is the preferred cycloaddition mode for $(NC)_2C{=}C(CN)_2$ to $C_7H_8Fe(CO)_3$ [110]. Reaction with excess $O{=}C(CF_3)_3$ condensed on $C_7H_8Fe(CO)_3$ at −196 °C in hexane in a Carius tube at room temperature for 2 d [48] gives a 1:1 adduct [30, 31] as shown in Formula XXII ($R{=}CF_3$) [48]. Reaction of $C_7H_8Fe(CO)_3$ with $O{=}C{=}C(C_6H_5)_2$ at room temperature gives XXIII; whereas in refluxing C_6H_6, XXIV (R, R′=O, R″=R‴=C_6H_5) is obtained [90]. Reaction with XXV in C_6H_6 gives XXVI and similar reaction with XXVII gives XXVIII. The rate constants for the reaction of $C_7H_8Fe(CO)_3$ with XXVII in different solvents at 34.1 °C are shown in the following table:

solvent	$[C_7H_8Fe(CO)_3]$ / [XXVII]	$k \cdot 10^3$ (±3%) (s^{-1})	$k(C_7H_8Fe(CO)_3)$ / $k(C_2H_5O_2CNC_6H_6Fe(CO)_3)$
benzene	100	16.6	–
chlorobenzene	100	34.5	27.3
chlorobenzene	10	32.8	–
chlorobenzene	5	35.7	–
dichloroethane	100	73.4	–
acetonitrile	5	37.0	–

The value $k(C_7H_8Fe(CO)_3)/k(C_2H_5O_2CNC_6H_6Fe(CO)_3)=27.3$ shows that $C_7H_8Fe(CO)_3$ is more reactive than the N-(ethoxycarbonyl)azepine complex [113]. Reaction with $(NC)_2C{=}C(CN)_2$ [30, 37, 42, 48, 107, 110, 118] in CH_2Cl_2 [48] or C_6H_6 [37] at room temperature [30, 37, 42, 48, 107, 110] gives XXIV with R=R′=R″=R‴=CN. The second-order rate constant for this addition of $(NC)_2C{=}C(CN)_2$ at 25 °C is $k=62.2\pm0.9$ in CH_2Cl_2 and $k=234\pm6$ in CH_3NO_2 $dm^3 \cdot mol^{-1} \cdot s^{-1}$. The observed pseudo-first-order rate constants at 25 °C are shown in the following table [107]:

solvent	$[(NC)_2C{=}C(CN)_2]$ mmol/dm^3	k_{obs} in s^{-1}
CH_2Cl_2	5.0	2.78×10^{-1}
	10.0	5.68×10^{-1}
	15.0	8.95×10^{-1}
	20.0	11.85×10^{-1}
	25.1	15.3×10^{-1}
CH_3NO_2	2.16	0.46
	4.02	0.90
	6.12	1.44
	7.99	1.87
	9.99	2.27

References on pp. 113/6

Reaction of $C_7H_8Fe(CO)_3$ with $(NC)_2C=C(CF_3)_2$ [30, 48] in hexane at room temperature [48] gives a 1:1 adduct [30] as shown in XXIV with R=R′=CF_3 and R″=R‴=CN [48]. Reaction with trans-$CF_3(NC)C=C(CN)CF_3$ in CH_2Cl_2 at room temperature [30, 31, 48] gives a 1:1 adduct [30, 31] consisting of XXIV (R=R″=CN, R′=R‴=CF_3) and XXIV (R=R″=CF_3, R′=R‴=CN) in the ratio 1:1.5 [48]; reaction with a 1:3 mixture of cis- and trans-$CF_3(NC)C=C(CN)CF_3$ gives the above mentioned two isomers of XXIV and in one set XXIV with R=R″=CN, R′=R‴=CF_3 in the ratio 3:1 which implies that in the formation of the adducts there is no change in the stereochemistry of the reacting olefin [48]. Similar reactions with fluoroolefins [34] such as $CF_2=CF_2$ and $CF_3CF=CF_2$ [42] gives 1:1 adducts [34] as π-allylic species, in which the fluoroolefin formally links the Fe and the organic ligand [42]. Reaction of $C_7H_8Fe(CO)_3$ with XXIX in C_6H_6 gives XXX [113]; and refluxing with C_8H_8 in ethylcyclohexane for 4.5 h gives $C_8H_8Fe(CO)_3$ as the sole product [115]. Irradiation of $C_7H_8Fe(CO)_3$ in tetrahydrofuran at −78 °C, and subsequent addition of cis-CH_3O_2C-HC=CH-CO_2CH_3 gives XXXIV [55].

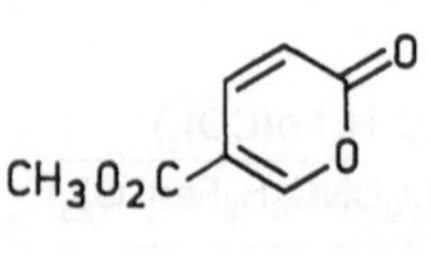

XXV 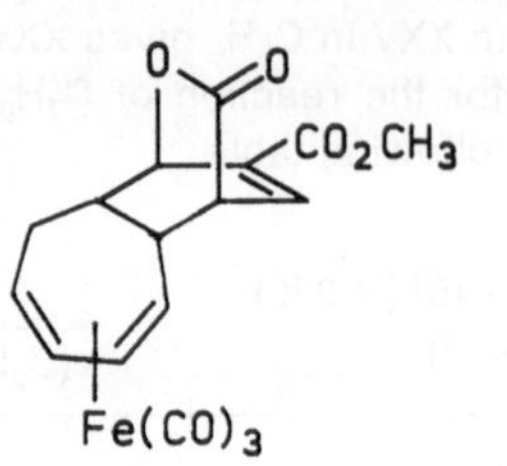 XXVI 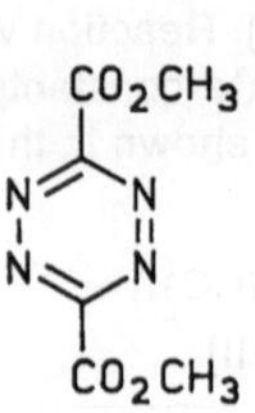 XXVII 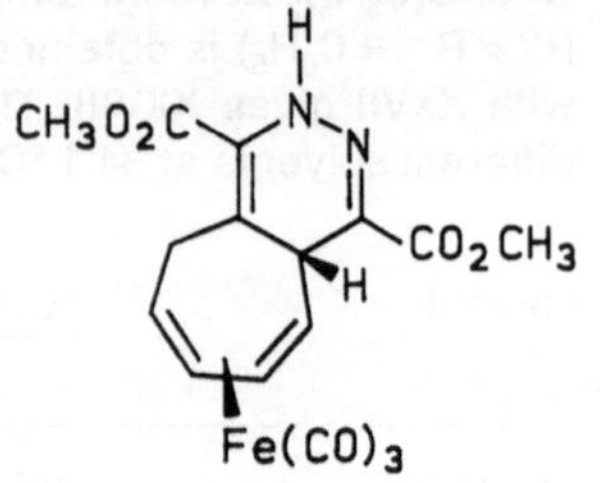 XXVIII

Reactions with Alkynes. Irradiation of $C_7H_8Fe(CO)_3$ with RC≡CR in tetrahydrofuran at 0 °C gives VII (R=CO_2CH_3) or VIII (R=C_6H_5) [51, 52, 54, 55]. Similar irradiation with $CF_3C≡CCF_3$ in hexane gives XXXI [62].

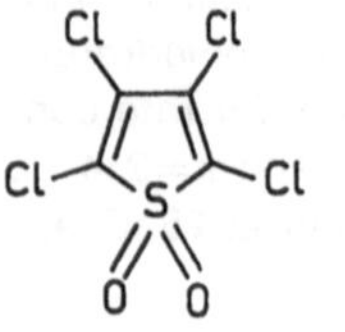

XXIX 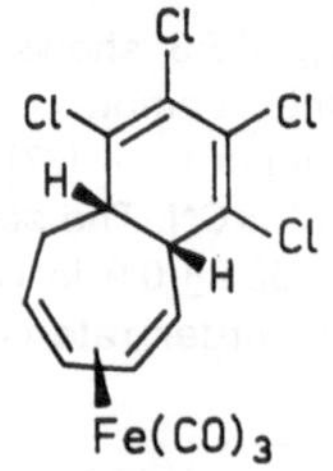 XXX 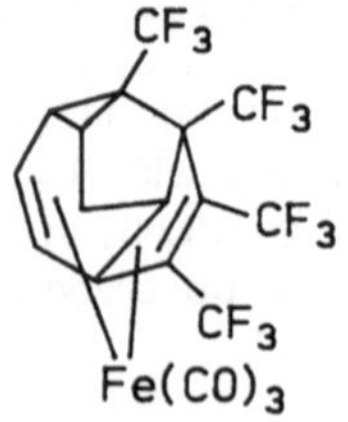 XXXI 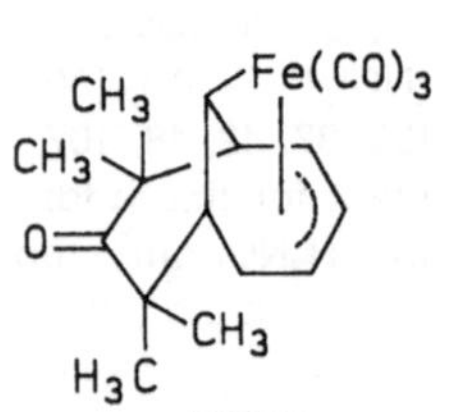 XXXII

Other Reactions. Attempted oxidation of $C_7H_8Fe(CO)_3$ with N-bromosuccinimide gives extensive decomposition and immediate decomposition is also observed with perphthalic

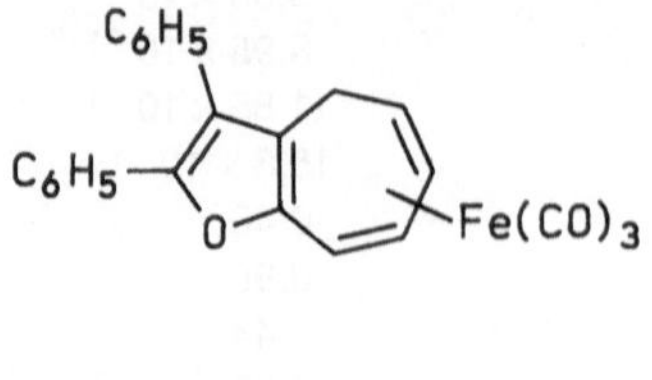

XXXIII 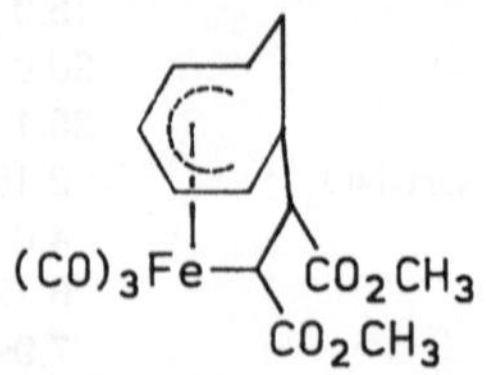 XXXIV

acid [5]. Reaction with CF_3CO_2H in CD_2Cl_2 at −78 °C gives $[C_7H_9Fe(CO)_3]^+$, and analogous deuteration with CF_3CO_2D gives XVI (R = D) with exo addition of the proton (or deuterium) [116]. Picryl azide gives a product for which an aziridine or an imine structure can be proposed from the ^1H NMR data. X-ray analysis confirms the imine structure [58]. Reaction with 50% aqueous $NaOH/[C_6H_5CH_2N(C_2H_5)_3]Cl$ in C_6H_6 followed by slow addition of $CHCl_3$ gives XIX (R = R′ = Cl) [57]. Stirring $C_7H_8Fe(CO)_3$ with $Fe_2(CO)_9$ and $(CH_3)_2CBrCOC(CH_3)_2Br$ in C_6H_6 at 50 °C for 8 h yields XXXII [108, 120].

Catalytic Properties and Uses. $C_7H_8Fe(CO)_3$ is proposed as starting compound for metal plating [9, 85] or iron coating for glass [9, 85], quartz, or halite [85], and as antiknock agent in petroleum hydrocarbons, or as an additive to fuels and oils [9]. An aqueous suspension of $C_7H_8Fe(CO)_3$ and Tween-80 (sorbimacrogololeate) exhibits fungicidal activity as well as an aerosol formulation from $C_7H_8Fe(CO)_3$, acetone, and CCl_2F_2 at 140.6 atm; effectiveness is proved by the Agar-plate technique [6].

References:

[1] R. Burton, M.L.H. Green, E.W. Abel, G. Wilkinson (Chem. Ind. [London] **1958** 1592). – [2] R.D. Fischer (Diss. Univ. München 1961, pp. 1/99). – [3] H.J. Dauben, D.J. Bertelli (J. Am. Chem. Soc. **83** [1961] 497/8). – [4] A. Davison, W. McFarlane, L. Pratt, G. Wilkinson (Chem. Ind. [London] **1961** 553). – [5] R. Burton, L. Pratt, G. Wilkinson (J. Chem. Soc. **1961** 594/602).

[6] W.E. Burt, Ethyl Corp. (U.S. 3123524 [1961/64]). – [7] E. Weiss, W. Hübel, J. Nielsen, A. Gerondal, J. Vandervalle (Chem. Ber. **95** [1962] 1179/85). – [8] D.J. Bertelli (Diss. Univ. Washington, Seattle, 1961; Diss. Abstr. **22** [1962] 3850). – [9] K.G. Ihrmann, T.H. Coffield, Ethyl Corp. (U.S. 3037999 [1962]). – [10] A. Reckziegel, M. Bigorgne (Compt. Rend. **258** [1964] 4065/8).

[11] G.F. Emerson, J.E. Mahler, R. Pettit, R. Collins (J. Am. Chem. Soc. **86** [1964] 3590/1). – [12] H.J. Dauben (unpublished results from [13]). – [13] R. Pettit, G.F. Emerson (Advan. Organometal. Chem. **1** [1964] 1/46). – [14] A. Reckziegel, M. Bigorgne (J. Organometal. Chem. **3** [1965] 341/54). – [15] H.W. Whitlock, Y.N. Chuah (J. Am. Chem. Soc. **87** [1965] 3605/8).

[16] R.B. King (Organometal. Syn. **1** [1965] 127/8). – [17] E.O. Fischer, H. Rühle (Z. Anorg. Allgem. Chem. **341** [1965] 137/45). – [18] B.J. Nicholson (J. Am. Chem. Soc. **88** [1966] 5156/65). – [19] F.M. Chaudhari, P.L. Pauson (J. Organometal. Chem. **5** [1966] 73/8). – [20] H. Behrens, M. Moll, W. Popp, V. Schneider, K.H. Trummer (22nd Intern. Conf. Coord. Chem., Budapest 1982, Vol. 1, p. 103).

[21] D.A. White (Organometal. Chem. Rev. A **3** [1968] 497/554, 516/7). – [22] R. Wenzl (Diss. Univ. Köln 1969 from [23]). – [23] R. Aumann, H. Averbeck, C. Krüger (Chem. Ber. **108** [1975] 3336/48). – [24] R.B. King (J. Am. Chem. Soc. **91** [1969] 7217/23). – [25] B.F.G. Johnson, J. Lewis, G.L.P. Randall (J. Chem. Soc. D **1969** 1273/4).

[26] R. Wenzl (Diss. Univ. Köln 1969, pp. 1/136). – [27] F.T. Delbeke, G.P. Van der Kelen (J. Organometal. Chem. **21** [1970] 155/8). – [28] J.D. Warren (Diss. Florida State Univ. 1970; Diss. Abstr. Intern. B **31** [1971] 5245/6). – [29] H. Maltz, B.A. Kelly (J. Chem. Soc. D **1971** 1390/1). – [30] M. Green, S. Tolson, J. Weaver, D.C. Wood, P. Woodward (J. Chem. Soc. D **1971** 222/3).

[31] M. Green, S. Tolson (unpublished results from [30]). – [32] F.A. Cotton, C.R. Reich (unpublished results from [33]). – [33] F.A. Cotton, B.G. DeBoer, T.J. Marks (J. Am. Chem. Soc. **93** [1971] 5069/75). – [34] A. Bond, M. Green, B. Lewis, S.F.W. Lowrie (J. Chem. Soc.

D **1971** 1230/1). – [35] B.F.G. Johnson, J. Lewis, P. McArdle, G.L.P. Randall (J. Chem. Soc. Dalton Trans. **1972** 456/62).

[36] C.H. Mauldin, E.R. Biehl, P.C. Reeves (Tetrahedron Letters **1972** 2955/8). – [37] D.J. Ehntholt, R.C. Kerber (J. Organometal. Chem. **38** [1972] 139/45). – [38] M. Brookhart (unpublished results from [39]). – [39] M. Brookhart, E.R. Davis, D.L. Harris (J. Am. Chem. Soc. **94** [1972] 7853/8). – [40] A. Davison, M. Brookhart (unpublished results from [39]).

[41] G.T. Rodeheaver (Diss. Univ. Virginia 1973; Diss. Abstr. Intern. B **34** [1973] 1433). – [42] M. Green, S. Tolson (unpublished results from [43]). – [43] M. Green, B. Lewis (J. Chem. Soc. Chem. Commun. **1973** 114/5). – [44] P. McArdle (J. Chem. Soc. Chem. Commun. **1973** 482/3). – [45] L. Kruczynski, LiShingMan, J. Takats (6th Intern. Conf. Organometal. Chem., Amherst, Mass., 1973, Abstr. No. 6).

[46] J. Takats (unpublished results from [45]). – [47] B.A. Kelly (Diss. Univ. Boston 1973; Diss. Abstr. Intern. B **34** [1973] 1031). – [48] M. Green, S. Heathcock, D.C. Wood (J. Chem. Soc. Dalton Trans. **1973** 1564/9). – [49] L.K.K. LiShingMan, J.G.A. Reuvers, J. Takats, G. Deganello (Organometallics **2** [1983] 28/39). – [50] P. McArdle, H. Sherlock (Proc. 16th Intern. Conf. Coord. Chem., Dublin 1974, Abstr. No. R 89).

[51] R. Pettit (Organotransition Met. Chem. Proc. 1st Japan. Am. Semin., Honolulu 1974 [1975], pp. 157/67; C.A. **85** [1976] No. 176360). – [52] R. Pettit, J.S. McKennis, W. Slegeir, W.H. Starnes, T. Devon, R. Case, J.C. Wagnon, L. Breuer, J. Wristers (Ann. N.Y. Acad. Sci. **239** [1974] 22/31). – [53] B.F.G. Johnson, J. Lewis, M.V. Twigg (J. Chem. Soc. Dalton Trans. **1974** 2546/9). – [54] T.J. Devon (Diss. Univ. Texas 1974; Diss. Abstr. Intern. B **35** [1975] 3825). – [55] R.E. Davis, T.A. Dodds, T.H. Hseu, J.C. Wagnon, T. Devon, J. Tancrede, J.S. McKennis, R. Pettit (J. Am. Chem. Soc. **96** [1974] 7562/3).

[56] D.L. Reger, A. Gabrielli (J. Am. Chem. Soc. **97** [1975] 4421/2). – [57] T. Ishizu, K. Harano, N. Mori, M. Yasuda, K. Kanematsu (Tetrahedron Letters **39** [1983] 1281/9). – [58] V.N. Narasimhachari (Diss. Univ. Texas 1975; Diss. Abstr. Intern. B **36** [1975] 745). – [59] B.F.G. Johnson, J. Lewis, D.J. Thompson, B. Heil (J. Chem. Soc. Dalton Trans. **1975** 567/71). – [60] G. Deganello, T. Boschi, L. Toniolo (7th Intern. Conf. Organometal. Chem., Venice 1975, Abstr. No. 40).

[61] G. Deganello, T. Boschi, L. Toniolo (J. Organometal. Chem. **97** [1975] C46/C48). – [62] M. Bottrill, R. Goddard, M. Green, R.P. Hughes, M.K. Lloyd, B. Lewis, P. Woodward (J. Chem. Soc. Chem. Commun. **1975** 253/5). – [63] H. Behrens, K. Geibel, R. Kellner, M. Moll. P. Würstl (7th Intern. Conf. Organometal. Chem., Venice 1975, Abstr. No. 30). – [64] G.A. Olah, S.H. Yu, G. Liang (J. Org. Chem. **41** [1976] 2383/6). – [65] P. McArdle, H. Sherlock (J. Organometal. Chem. **116** [1976] C23/C24).

[66] M. Moll, H. Behrens, R. Kellner, H. Knöchel, P. Würstl (Z. Naturforsch. **31b** [1976] 1019/20). – [67] L. Kruczynski, J. Takats (Inorg. Chem. **15** [1976] 3140/7). – [68] H. Behrens, K. Geibel, R. Kellner, H. Knöchel, M. Moll, E. Sepp (Z. Naturforsch. **31b** [1976] 1021/2). – [69] J.C. Green, P. Powell, J. van Tilborg (J. Chem. Soc. Dalton Trans. **1976** 1974/6). – [70] N.G. Connelly, R.L. Kelly (J. Organometal. Chem. **120** [1976] C16/C17).

[71] M.J. Bennett, J.L. Pratt, K.A. Simpson, L.K.K. LiShingMan, J. Takats (J. Am. Chem. Soc. **98** [1976] 4810/7). – [72] R.B. King (Israel J. Chem. **15** [1976/77] 181/9). – [73] B.E. Mann (J. Organometal. Chem. **141** [1977] C33/C34). – [74] G.P. Sollott, D.L. Daughdrill, W.R. Peterson (J. Organometal. Chem. **133** [1977] 347/57). – [75] W.L. Jolly, S.C. Avanzino, R.R. Rietz (Inorg. Chem. **16** [1977] 964/6).

[76] D. Cunningham, P. McArdle, H. Sherlock, B.F.G. Johnson, J. Lewis (J. Chem. Soc. Dalton Trans. **1977** 2340/4). – [77] F.A. Cotton, B.E. Hanson (Israel J. Chem. **15** [1977]

165/73). – [78] D.A. Brown, S.K. Chawla, W.K. Glass, F.M. Hussein (Inorg. Chem. **21** [1982] 2726/32). – [79] M. Moll, P. Würstl, H. Behrens, P. Merbach (Z. Naturforsch. **33b** [1978] 1304/8). – [80] S. Litman, A. Gedanken, Z. Goldschmidt, Y. Bakal (J. Chem. Soc. Chem. Commun. **1978** 983/4).

[81] G.O. Nelson (Diss. Univ. North Carolina 1977; Diss. Abstr. Intern. B **39** [1978] 237/8). – [82] K.D. Karlin, B.F.G. Johnson, J. Lewis (J. Organometal. Chem. **160** [1978] C21/C23). – [83] K.J. Karel (Diss. Univ Princeton 1978; Diss. Abstr. Intern. B **39** [1979] 4353). – [84] B.Y. Shu, E.R. Biehl, P.C. Reeves (Syn. Commun. **8** [1978] 523/31). – [85] I.A. Shatkovskii, N.V. Perevozchikova, B.S. Kaverin, Yu.A. Sorokin (Khim. Elementoorg. Soedin. [Gor'kiy] **1978** 58/9).

[86] E. Sepp, A. Pürzer, G. Thiele, H. Behrens (Z. Naturforsch. **33b** [1978] 261/4). – [87] K.J. Karel, M. Brookhart (J. Am. Chem. Soc. **100** [1978] 1619/21). – [88] J.A.S. Howell, P.M. Burkinshaw (J. Organometal. Chem. **152** [1978] C5/C8). – [89] P. Hofmann (Z. Naturforsch. B **33** [1978] 251/60). – [90] Z. Goldschmidt, S. Antebi (Tetrahedron Letters **1978** 271/4).

[91] A.M. Gabrielli (Diss. Univ. South Carolina 1978; Diss. Abstr. Intern. B **39** [1979] 5911). – [92] G.A. Taylor (J. Chem. Soc. Perkin Trans. I **1979** 1716/9). – [93] J.G.A. Reuvers, J. Takats (J. Organometal. Chem. **175** [1979] C13/C16). – [94] M. Brookhart, G.O. Nelson (J. Organometal. Chem. **164** [1979] 193/202). – [95] Z. Goldschmidt, Y. Bakal (J. Organometal. Chem. **168** [1979] 215/25).

[96] B.F.G. Johnson, K.D. Karlin, J. Lewis (J. Organometal. Chem. **174** [1979] C29/C31). – [97] J. McGavock Crockett (Diss. Univ. North Carolina 1979; Diss. Abstr. Intern. B **40** [1979] 2189). – [98] C.S. Day, V.W. Day, G. Deganello, L.K.K. LiShingMan, J.G.A. Reuvers, J. Takats (unpublished results from [93]). – [99] M.R. Blake, J.L. Garnett, I.K. Gregor, S.B. Wild (J. Organometal. Chem. **178** [1979] C37/C42). – [100] D.L. Reger, A. Gabrielli (J. Organometal. Chem. **187** [1980] 243/52).

[101] M.R. Blake, J.L. Garnett, I.K. Gregor, D. Nelson (J. Organometal. Chem. **188** [1980] 203/10). – [102] M. Airoldi, G. Deganello, G. Dia, P. Sacco, J. Takats (Inorg. Chim. Acta **41** [1980] 171/8). – [103] P.M. Burkinshaw, D.T. Dixon, J.A.S. Howell (J. Chem. Soc. Dalton Trans. **1980** 999/1004). – [104] S.C. Avanzino, A.A. Bakke, H.-W. Chen, C.J. Donahue, W.L. Jolly, T.H. Lee, A.J. Ricco (Inorg. Chem. **19** [1980] 1931/6). – [105] G.L.P. Randall (unpublished results from [106]).

[106] H.A. Bockmeulen, A.W. Parkins (J. Chem. Soc. Dalton Trans. **1981** 262/6). – [107] S.K. Chopra, M.J. Hynes, P. McArdle (J. Chem. Soc. Dalton Trans. **1981** 586/9). – [108] T. Ishizu, K. Harano, M. Yasuda, K. Kanematsu (J. Org. Chem. **46** [1981] 3630/4). – [109] T. Jenny, W. von Philipsborn, J. Kronenbitter, A. Schwenk (J. Organometal. Chem. **205** [1981] 211/22). – [110] S.K. Chopra, G. Moran, P. McArdle (J. Organometal. Chem. **214** [1981] C36/C38).

[111] W. Grimme, H.G. Köser (J. Am. Chem. Soc. **103** [1981] 5919/20). – [112] K.J. Karel, T.A. Albright, M. Brookhart (Organometallics **1** [1982] 419/30). – [113] T. Ban, K. Nagai, Y. Miyamoto, K. Harano, M. Yasuda, K. Kanematsu (J. Org. Chem. **47** [1982] 110/6). – [114] H. Behrens, M. Moll, W. Popp, V. Schneider, K.H. Trummer (22nd Intern. Conf. Coord. Chem., Budapest 1982, Vol. 1, p. 103). – [115] T.A. Manuel, F.G.A. Stone (J. Am. Chem. Soc. **82** [1960] 366/72).

[116] M. Brookhart, K.J. Karel, L.E. Nance (J. Organometal. Chem. **140** [1977] 203/10). – [117] L. Kruczynski, J. Takats (J. Am. Chem. Soc. **96** [1974] 932/4). – [118] D.J. Ehntholt (Diss. State Univ. New York 1971; Diss. Abstr. Intern. B **32** [1972] 4486). – [119] N.G. Connelly, A.R. Lucy, M.W. Whiteley (J. Chem. Soc. Chem. Commun. **1979** 985/6). – [120] T. Ishizu, K. Harano, M. Yasuda, K. Kanematsu (Tetrahedron Letters **22** [1981] 1601/4).

[121] K. Broadley, N.G. Connelly, J.A.K. Howard, W. Risse (J. Organometal. Chem. **221** [1981] C29/C32). – [122] N.G. Connelly, A.R. Lucy, J.B. Sheridan (J. Chem. Soc. Dalton Trans. **1983** 1465/8). – [123] I.L. Fragalà, J. Takats, M.A. Zerbo (Organometallics **2** [1983] 1502/4). – [124] Z. Goldschmidt, Y. Bakal, D. Hoffer, A. Eisenstadt, P.F. Lindley (J. Organometal. Chem. **258** [1983] 53/62). – [125] A.J. Pearson (Pure Appl. Chem. **55** [1983] 1767/79).

[126] A. Salzer, A. Hafner (Helv. Chim. Acta **66** [1983] 1774/85). – [127] Z. Goldschmidt, S. Antebi, I. Goldberg (J. Organometal. Chem. **260** [1984] 105/13).

1.4.1.4.1.2.4.2.3.2 $C_7H_8Fe(CO)_3AlX_3$

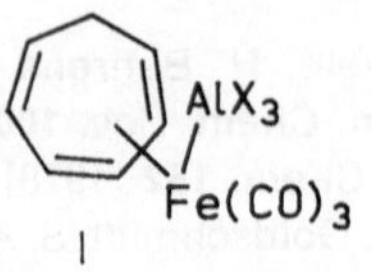

AlX_3

$Fe(CO)_3$

I

$C_7H_8Fe(CO)_3AlX_3$ (see Formula I with X=Cl, Br) are prepared from $C_7H_8Fe(CO)_3$ and AlX_3 in CH_2Cl_2. The compound with X=Cl shows in the IR spectrum in CH_2Cl_2 bands at 2069 and 2114 (νCO) cm^{-1}, and the compound with X=Br shows bands at 2067 and 2114 (νCO) cm^{-1}. Hydrolysis of the compound affords only about 45% recovery of $C_7H_8Fe(CO)_3$ [1, 2].

References:

[1] K.D. Karlin, B.F.G. Johnson, J. Lewis (J. Organometal. Chem. **160** [1978] C21/C23). – [2] B.F.G. Johnson, K.D. Karlin, J. Lewis (J. Organometal. Chem. **174** [1979] C29/C31).

1.4.1.4.1.2.4.2.4 ^{4}L is a Substituted Cycloheptatriene

1.4.1.4.1.2.4.2.4.1 ^{4}L is a Monosubstituted Cycloheptatriene

The complexes $RC_7H_7Fe(CO)_3$ described in this section are arranged in Table 3 as follows. Compounds of the type 1-$RC_7H_7Fe(CO)_3$ (Nos. 1 to 21, 54 to 59) are described first, followed by the 2-$RC_7H_7Fe(CO)_3$ (Nos. 22 to 26), and 3-$RC_7H_7Fe(CO)_3$ complexes (Nos. 27, 28). Next, 7-$RC_7H_7Fe(CO)_3$ complexes (Nos. 29 to 51, 60) are described followed by two complexes $RC_7H_7Fe(CO)_3$ (Nos. 52, 53) with unknown position of substitution. Most of the $RC_7H_7Fe(CO)_3$ complexes described in Table 3, pp. 119/32, can be prepared by the following methods:

Method I: $Fe(CO)_5$ and 2,4-$(C_2H_5)_2C_6H_3C_7H_7$ are refluxed in pentylbutanoate for 2 h. Filtration and evaporation followed by chromatography on alumina gives No. 52 [2].

Method II: $Fe_2(CO)_9$ reacts with an organic substrate in ether or benzene.

a. $Fe_2(CO)_9$ and I are refluxed in ether for 60 h. The reaction mixture is filtered, evaporated, and the residue is extracted with petroleum ether. Chromatography on neutral alumina (activity III) with petroleum ether gives $C_8H_8Fe(CO)_3$, trans-$C_8H_8Fe_2(CO)_6$ (see 2.8.1.1 in "Organoiron Compounds" C5, 1981, Table 4, No. 25, p. 49), and II (see 2.6.1.2.3 in "Organoiron Compounds" C5, 1981, Table 2, No. 15, p. 18). Elution with petroleum ether/C_6H_6 (1:1) gives $C_8H_8Fe_2(CO)_5$ (see 2.8.2.2 in "Organoiron Compounds" C5, 1981, Table 7, No. 1, p. 114) and No. 14. Further elution with petroleum ether/C_6H_6 (1:2) gives III, IV (see 2.6.1.2.4 in "Organoiron Compounds" C5, 1981, Table 3, No. 2, p. 28), and a large amount of polymer. Under conditions where no II is isolated

References on pp. 143/4

(reflux in C_6H_6 or toluene), the yield of No. 14 greatly increases and becomes the principal product [9]. Similarly, excess $Fe_2(CO)_9$ reacts with 7-$HOCH_2C_7H_7$ in ether at room temperature for 17 h followed by 4 h at reflux and another 2 d at room temperature. The reaction mixture is filtered through infusorial earth, and the filtrate is evaporated. Distillation of the residue at reduced pressure gives phenylsubstituted products, V, and a liquid which may be No. 40 [19]. Similar reaction of $Fe_2(CO)_9$ with 7-$CH_3OC_7H_7$ gives No. 30 [3, 4] and $CH_3OC_7H_7Fe_2(CO)_6$ (see 2.6.1.2.3 in "Organoiron Compounds" C5, 1981, Table 2, No. 14, p. 17) [4].

b. Stirring $Fe_2(CO)_9$ with 7-$C_6H_5C_7H_7$ in C_6H_6 gives No. 45 [30].

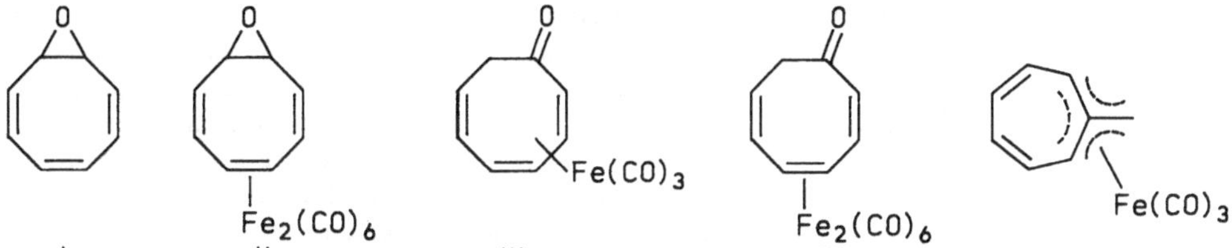

O, $Fe_2(CO)_6$, $Fe(CO)_3$, $Fe_2(CO)_6$, $Fe(CO)_3$

I II III IV V

Method III: $Fe_3(CO)_{12}$ and I are refluxed in C_6H_6 to give No. 14 and small amounts of complexes of cyclooctatetraene and cyclooctatrienone [9].

Method IV: $^3LFe(CO)_3$ complexes such as $^3LFe(CO)_3$ (^3L = benzylideneacetone) or VI are the starting compounds.

a. $^3LFe(CO)_3$ (^3L = benzylideneacetone) reacts with $CH_3OC_7H_7$ to give No. 30 [47].

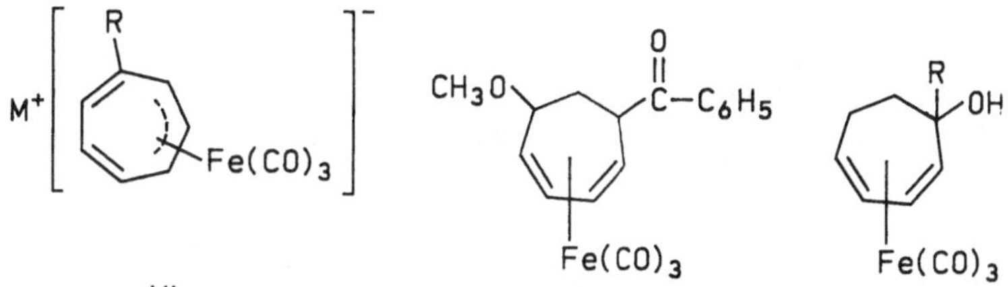

M^+ [R, $Fe(CO)_3$]$^-$ — CH_3O, $C(=O)-C_6H_5$, $Fe(CO)_3$ — R, OH, $Fe(CO)_3$

VI VII VIII

b. VI (M = Li, Na, or K, R = H) in tetrahydrofuran reacts with $ClSi(CH_3)_3$ to give No. 33 [23, 25, 27, 42, 43, 52], with $ClSi(C_2H_5)_3$ to give No. 34 [25, 26, 42, 43], with $ClGe(CH_3)_3$ to give No. 35 [25, 27, 42, 43, 52], with $ClGe(C_2H_5)_3$ to give No. 36 [25, 26, 42, 43], with $BrGe(C_6H_5)_3$ to give No. 37 [29, 42, 43, 52], with $ClSn(CH_3)_3$ to give an inseparable mixture of No. 38 and $C_7H_7Fe(CO)_2Sn(CH_3)_3$ [27], with $XCH_2CH{=}CH_2$ to give No. 41 [20, 23] and with $ClCO_2R'$ (R' = CH_3, C_2H_5) to give Nos. 43 and 44, respectively [23, 25, 26, 43]. Thus, VI (M = K, R = H) in tetrahydrofuran is added dropwise over a period of 2 h to a stirred solution of $ClSi(CH_3)_3$ in tetrahydrofuran. The mixture is stirred for an additional 4 h to ensure completion of the reaction. The reaction mixture is filtered, the solvent removed, and the residue is stirred for 4 h at room temperature in hexane. Removal of hexane gives a solid which is Soxhlet extracted with pentane for 5 h. Cooling the mixture to room temperature followed by filtration and drying gives No. 33. No traces of $(CH_3)_3SiOSi(CH_3)_3$ are encountered;

References on pp. 143/4

however, various amounts of $(C_7H_7Fe(CO)_3)_2$ are also obtained and separated by chromatography on alumina (activity II). Similar reaction with $ClGe(CH_3)_3$ gives No. 35 and various amounts of $(C_7H_7Fe(CO)_3)_2$, separated as before. Similar reaction with $BrGe(C_6H_5)_3$ gives No. 37 and various amounts of $(C_6H_5)_3GeOGe(C_6H_5)_3$. Only fractional crystallization from hexane or hexane washing and Soxhlet extraction, described above, proves successful in obtaining pure No. 37 [52]. Nucleophilic attack of H_2O on VI (M=Li, R=H) gives No. 29 [20].

c. Dropwise addition of $[C_7H_7Cr(CO)_3]BF_4$ (see 1.6.1.8 in "Chrom-Organische Verbindungen", Erg.-Wk. Bd. 3, 1971, Table 57, No. 14, p. 306 Cr) in tetrahydrofuran to VI (M=K, R=H) in tetrahydrofuran gives a precipitate after 1 h. The reaction mixture is evaporated and the residue is extracted several times with a CH_2Cl_2/ether (1:1) mixture. The extracts are concentrated to give some orange-red crystalline material which is filtered off and washed with ether. The ether washings are added to the filtrate, concentrated, and chromatographed on alumina. Elution with petroleum ether gives traces of $C_7H_8Fe(CO)_3$ and a tiny amount of $(C_7H_7Fe(CO)_3)_2$. Elution with ether gives an orange-red solid which is combined with the previous crop and recrystallized to give No. 48 [47], cf. [23, 24, 28, 43]. Nos. 49 and 50 are prepared by analogous route and work-up [23, 24, 28, 43, 47] in variable yields depending on the reagent [24].

d. Addition of VI (M=K, R=H) in tetrahydrofuran to $[C_7H_7]BF_4$ in tetrahydrofuran followed by stirring at room temperature for 4 h, evaporation, and extraction of the residue with petroleum ether gives three bands by chromatography on alumina (activity III). The first and second band contain small amounts of $C_7H_8Fe(CO)_3$ and $(C_7H_7Fe(CO)_3)_2$, respectively, while the third band gives, after usual work-up, No. 47 [47].

e. Deuteration (or reprotonation) of VI (M=K, R=$Ge(C_6H_5)_3$) gives two isomeric compounds, No. 1 and 27, isolated in a 1:2 ratio by repeated fractional crystallization. More than 80% deuterium incorporation occurs exclusively at the 7-exo position [43].

Method V: $^4LFe(CO)_3$ complexes with ^4L=cyclohepta-1,3,5-triene, tropone, or cyclooctatetraene, and the complexes VII and VIII are the starting compounds.

a. $C_7H_8Fe(CO)_3$ reacts with $OHCN(CH_3)_2/OPCl_3$ [8, 13, 17, 50] at 0 °C for 1 h [13]. The reaction mixture is poured into H_2O and stirred at 20 °C for 15 h. Extraction with ether, washing the organic layer with H_2O, drying ($MgSO_4$), and chromatography on silica gel with 5% $CH_3CO_2C_2H_5$/toluene gives No. 14 [13]. Similar reaction of $C_7H_8Fe(CO)_3$ with $CH_3COCl/AlCl_3$ [8, 13] in CH_2Cl_2 at 0 °C for 10 min, followed by hydrolysis with ice-water, and work-up as before gives No. 15 [13]. Addition of 15% aqueous $[NH_4]PF_6$ to the aqueous solution gives XVb (R=H, R′=$COCH_3$, X=PF_6, p. 133) [13].

b. $^4LFe(CO)_3$ (^4L=tropone) in CH_3OH reacts with $HSC(CH_3)_2H$ and $(C_2H_5)_2OBF_3$ at room temperature for 1 h. Evaporation followed by chromatography of the residue on neutral alumina (Merck, aluminium oxide G) gives No. 31 (R_f= 0.36). Similar reaction with HSC_6H_5 (carefully freed from $C_6H_5SSC_6H_5$) for 75 min leaves, after evaporation, an oily residue which on addition of a little ether changes to crystals [46].

References on pp. 143/4

c. $^4LFe(CO)_3$ ($^4L = C_8H_8$) with $[C_7H_7]BF_4$ in acetone at −23 °C gives an orange solution from which crystals of No. 42 are obtained. The yield increases in the presence of pyridine [48, 58]. After 7 h the reaction mixture is evaporated, and the residue extracted with ether. Chromatography of the concentrated extract on alumina with n-hexane gives a yellow band which is eluted with n-hexane/ether (10:1). Concentration and cooling to −78 °C gives No. 42 [58].

d. VII is placed on top of a column of silica gel made up in toluene. After 6 h at 20 °C a deep yellow band is eluted with 10% $CH_3CO_2C_2H_5$/toluene. Evaporation gives a golden oil. Recrystallization from pentane gives pure No. 16 [13]. Similar dehydration of VIII ($R = C_6H_6$ or $C_6H_4CH_3$-4) gives a mixture of the two isomers, No. 25a and b (ratio 1:1) [33] and No. 26a and b (ratio 1.6:1) [22, 33], respectively, in ca. 60% total yield [33].

Method VI: $[^5LFe(CO)_3]BF_4$ ($^5L = C_7H_7$) or complexes IX and X react with bases.

a. $[C_7H_7Fe(CO)_3]BF_4$ [3] or IX [5, 7] react with H_2O to give No. 29 [3] or No. 17b [5, 7].

b. X is treated with $NaOCH_3$ and silica gel to give No. 47 [49].

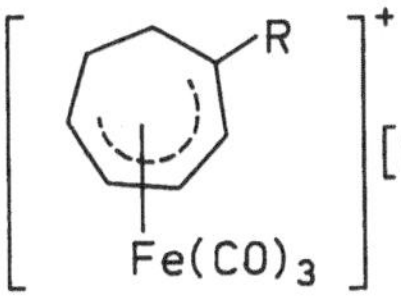

$[RC_7H_6Fe(CO)_3]^+[BF_4]^-$ — IX

$[(C_7H_7)C_7H_6Fe(CO)_3]^+[BF_4]^-$ — X

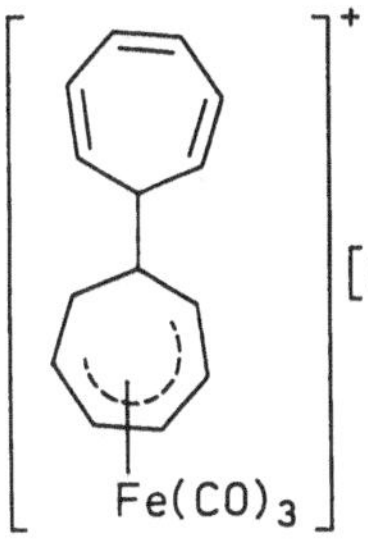

Method VII: Compound II (see 2.6.1.2.3 in "Organoiron Compounds" C5, 1981, Table 2, No. 15, p. 18) is refluxed in toluene for 3 h. The reaction mixture is evaporated, and the residue is extracted with petroleum ether. Chromatography on neutral alumina (activity III) with petroleum ether gives $C_8H_8Fe(CO)_3$, and elution with petroleum ether/C_6H_6 gives $C_8H_8Fe_2(CO)_5$ (see 2.8.2.2 in "Organoiron Compounds" C5, 1981, Table 7, No. 1, p. 114), trans-$C_8H_8Fe_2(CO)_6$ (see 2.8.1.1 in "Organoiron Compounds" C5, 1981, Table 4, No. 25, p. 49), and No. 14 [9].

Table 3
Compounds of the Type $RC_7H_7Fe(CO)_3$.
Further information on numbers preceded by an asterisk is given at the end of the table, pp. 132/43.
For abbreviations and dimensions see p. VIII.

No.	$RC_7H_7Fe(CO)_3$	method of preparation (yield in %), properties and remarks	Ref.
compounds of the type 1-$RC_7H_7Fe(CO)_3$ (continued with No. 54, p. 131):			
1	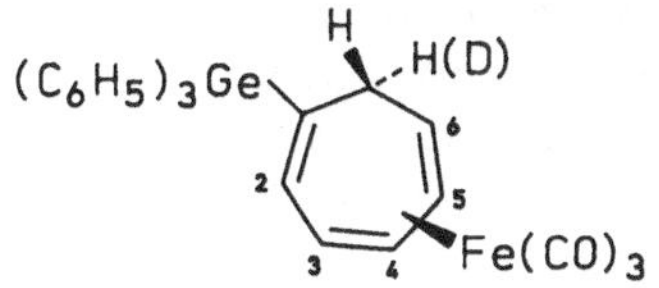 $(C_6H_5)_3Ge$, H, H(D) on C-1; ring positions 2–6; $Fe(CO)_3$	IVe ^1H NMR (CS_2): 2.42 (H-7), 3.01 (H-3), 3.28 (H-6), 5.15 to 5.30 (H-4, 5), 6.10 (H-2)	[43]

References on pp. 143/4

Table 3 [continued]

No.	$RC_7H_7Fe(CO)_3$	method of preparation (yield in %), properties and remarks	Ref.
*2	$(CH_3)_2CH$– [ring] $Fe(CO)_3$	—	[1]
3	$CH_2=CH$– [ring] $Fe(CO)_3$	see No. 9, p. 133	[8]
*4	CH_3O_2C–CH=CH– [ring] $Fe(CO)_3$	see No. 14, p. 134 m.p. 101° (from hexane), orange crystals ^{1}H NMR ($CDCl_3$): 2.45 (ABX, H-7; J(H-7 exo, 7 endo) = 20), 3.12 (ddd, H-3; J(H-3,5) = 3), 3.45 (m, H-6; J(H-6,7) = 5), 3.71 (s, CH_3), 5.38 (m, H-4,5; J(H-3,4) = 6), 5.63 (d, H-9; J(H-8,9) = 16), 6.32 (d, H-2; J(H-2,3) = 8), 7.06 (d, H-8) ^{13}C NMR ($CDCl_3$): 29.05 (C-7), 48.95 (C-3 or 6), 51.37 (CH_3), 55.48 (C-3 or 6); 88.63, 93.69 (C-4,5); 113.09 (C-9), 132.26 (C-1), 141.41 (C-2), 147.29 (C-8), 210.14 (CO) IR (Nujol): 1705, 1950, 1980, 2020 mass spectrum: $[M-nCO]^+$ (n = 1 to 3)	[44, 57]
*5	$C_2H_5O_2C$–CH=CH– [ring] $Fe(CO)_3$	see No. 14, p. 134 yellow oil which cannot be crystallized ^{1}H NMR ($CDCl_3$): 1.26 (t, CH_3; J = 7), 2.43 (ABX, H-7; J(H-7 exo, 7 endo) = 20), 3.12 (ddd, H-3; J(H-3,5) = 3), 3.45 (m, H-6; J(H-6,7) = 5), 4.16 (q, CH_2), 5.38 (m, H-4,5; J(H-3,4) = 6), 5.64 (d, H-9; J(H-8,9) = 16), 6.32 (d, H-2; J(H-2,3) = 8), 7.04 (d, H-8) ^{13}C NMR ($CDCl_3$): 14.34 (CH_3), 29.04 (C-7); 55.53, 59.01 (C-3,6); 60.12 (CH_2); 88.55, 93.67 (C-4,5), 113.51 (C-9), 132.31 (C-1), 141.23 (C-2), 147.02 (C-8), 210.19 (CO) IR (neat): 1690, 1950, 1990, 2045 mass spectrum: $[M-nCO]^+$ (n = 0 to 3)	[44]

Table 3 [continued]

No.	$RC_7H_7Fe(CO)_3$	method of preparation (yield in %), properties and remarks	Ref.
*6	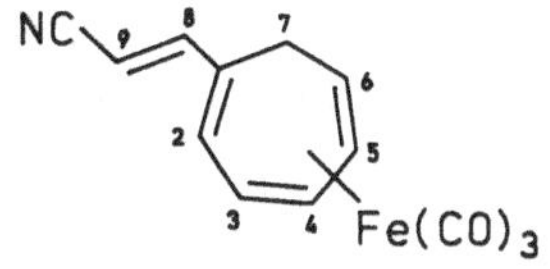 NC, $Fe(CO)_3$	see No. 14, p. 134 m.p. 106° (from hexane), orange crystals ^1H NMR ($CDCl_3$): 2.36 (ABX, H-7; J(H-7 exo, 7 endo) = 20), 3.09 (ddd, H-3; J(H-3,5) = 3), 3.44 (m, H-6; J(H-6,7) = 5), 5.01 (d, H-9; J(H-8,9) = 16), 5.40 (m, H-4,5; J(H-3,4) = 6), 6.28 (d, H-2; J(H-2,3) = 8), 6.72 (d, H-8) ^{13}C NMR ($CDCl_3$): 28.31 (C-7); 54.84, 58.82 (C-3,6); 88.67, 90.63 (C-4,5); 94.05 (C-9), 119.06 (CN), 131.64 (C-1), 142.69 (C-2), 152.40 (C-8), 209.84 (CO) IR (Nujol): 1960, 1990, 2025, 2204 mass spectrum: $[M-CO]^+$ and other fragments given	[44]
*7a	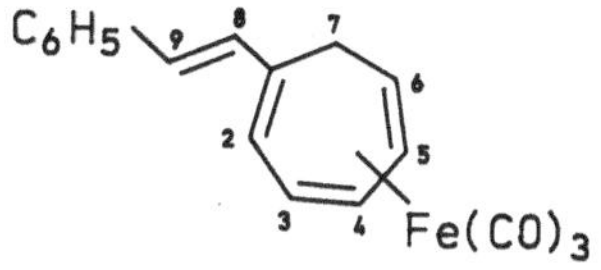 C_6H_5, $Fe(CO)_3$	see No. 14, p. 134 m.p. 143° (from hexane), orange-yellow crystals ^1H NMR ($CDCl_3$): 2.60 (ABX, H-7; J(H-7 exo, 7 endo) = 20), 3.19 (ddd, H-3; J(H-3,5) = 3), 3.49 (m, H-6; J(H-6,7) = 5), 5.35 (m, H-4,5; J(H-3,4) = 6), 6.02 (d, H-2; J(H-2,3) = 8), 6.44 (ABq, H-8,9; J(H-8,9) = 16), 7.3 (m, C_6H_5) ^{13}C NMR ($CDCl_3$): 29.48 (C-7); 56.92, 59.19 (C-3,6); 88.18, 92.82 (C-4,5); 124.38 (C-8); 126.15, 127.0, 128.65 (C_6H_5); 131.64 (C-9), 132.74 (C-2), 134.27 (C-1), 137.81 (C_6H_5), 210.75 (CO) IR (Nujol): 1975, 1990, 2050 mass spectrum: $[M-nCO]^+$ (n = 1 to 3)	[44]
b	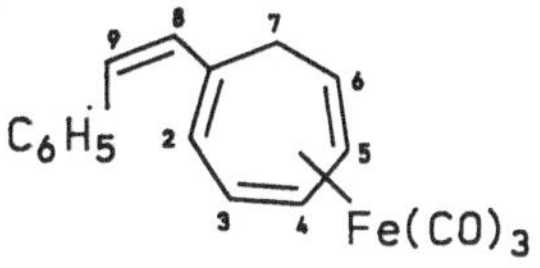 C_6H_5, $Fe(CO)_3$	see No. 14, p. 134 yellow oil ^1H NMR ($CDCl_3$): 2.24 (ABX, H-7; J(H-7 exo, 7 endo) = 20), 3.04 (ddd, H-3; J(H-3,5) = 3), 3.20 (m, H-6; J(H-6,7) = 5), 5.21 (m, H-4,5; J(H-3,4) = 6), 5.92 (d, H-2; J(H-2,3) = 8), 6.04 (ABq, H-8,9; J(H-8,9) = 12), 7.2 (m, C_6H_5) IR (neat): 1970, 1990, 2045	[44]

References on pp. 143/4

Table 3 [continued]

No.	$RC_7H_7Fe(CO)_3$	method of preparation (yield in %), properties and remarks	Ref.
*8	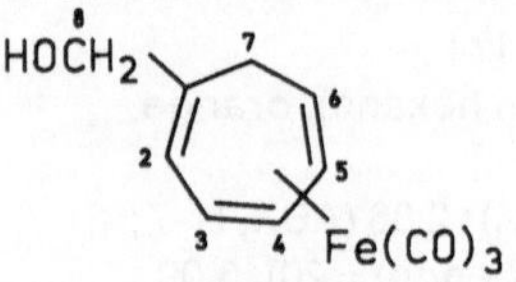 $HOCH_2$–$C_7H_7Fe(CO)_3$ (positions 2–8 numbered)	see No. 14, p. 133 subl. p. 70°/0.05 Torr, yellow volatile solid ^{1}H NMR (CS_2): 2.2 (s and m, H-7, OH), 3.0 (t, H-3; J=8), 3.3 (m, H-6), 3.6 (H-8), 5.25 (m, H-4,5), 5.7 (d, H-2; J=8) IR (liquid film): 3300 (OH) IR (CCl_4): 1973, 1983, 2048 (CO)	[8, 13, 17]
*9	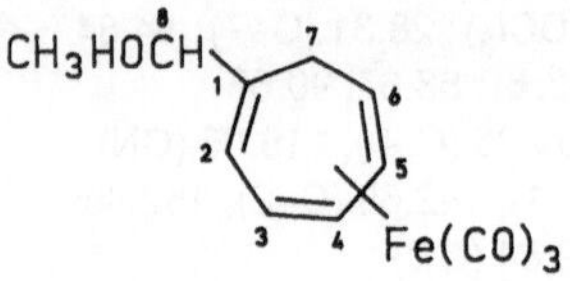 CH_3HOCH–$C_7H_7Fe(CO)_3$ (positions 1–8 numbered)	see Nos. 14, pp. 133/4, and 15, p. 135 b.p. 100°/0.05 Torr, yellow oil ^{1}H NMR ($CDCl_3$): 1.05 (d, CH_3; J=6), 2.2 (m, H-7), 2.5 (s, OH), 3.0 (t, H-3; J=8), 3.3 (m, H-6), 3.8 (q, H-8; J=6), 5.2 (m, H-4,5), 5.7 (d, H-2; J=8) IR (liquid film): 3350 (OH) IR (CCl_4): 1982, 1985, 2053	[8, 13, 14, 44]
*10	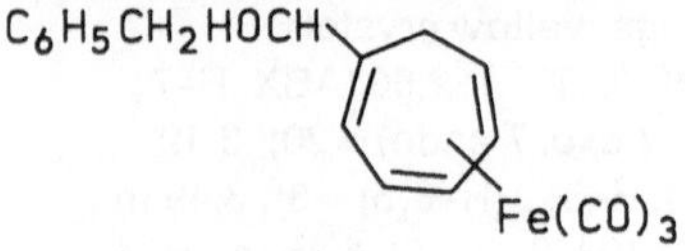 $C_6H_5CH_2HOCH$–$C_7H_7Fe(CO)_3$	–	[10]
*11	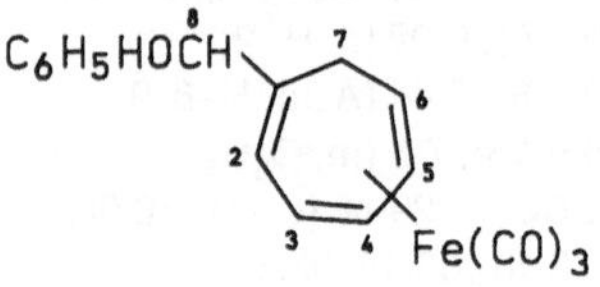 C_6H_5HOCH–$C_7H_7Fe(CO)_3$ (positions 2–8 numbered)	see Nos. 14, p. 133, and 16, p. 136 pale yellow nonvolatile oil ^{1}H NMR (CS_2): 2.0 (m, H-7), 3.0 (m, OH, H-3,6), 4.58 (s, H-8), 5.06 (m, H-4,5), 5.70 (d, H-2; J=8), 7.06 (m, C_6H_5) IR (liquid film): 3390 (OH) IR ($CHCl_3$): 1980, 1990, 2055 (CO)	[13, 14]
12	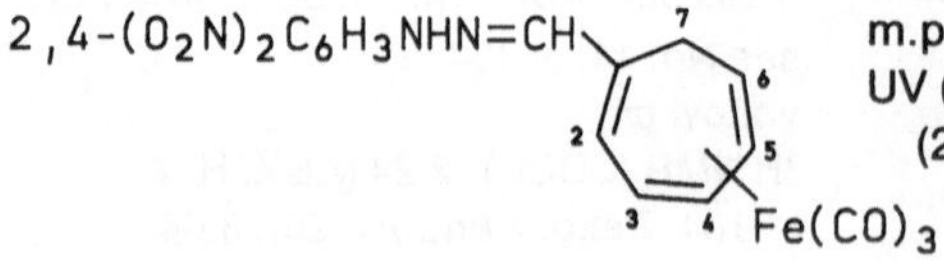 $2,4\text{-}(O_2N)_2C_6H_3NHN{=}CH$–$C_7H_7Fe(CO)_3$ (positions 2–7 numbered)	see No. 14, p. 133 m.p. 180° (dec.) UV (CH_2Cl_2): $\lambda_{max}(\varepsilon)=23800$ (28.8×10^3) cm^{-1}	[8, 13, 21]
13	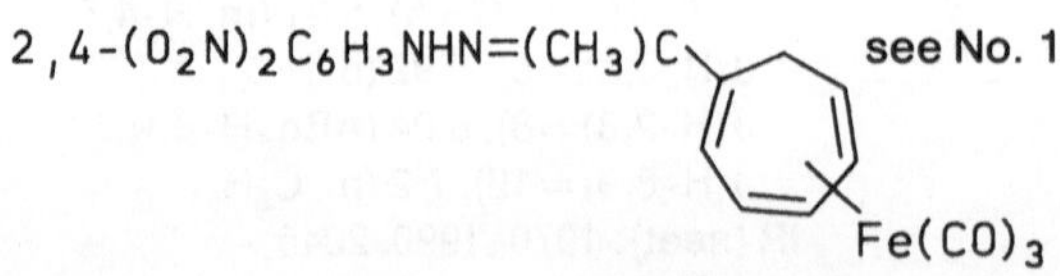 $2,4\text{-}(O_2N)_2C_6H_3NHN{=}(CH_3)C$–$C_7H_7Fe(CO)_3$	see No. 15, p. 135	[8]

References on pp. 143/4

Table 3 [continued]

No.	$RC_7H_7Fe(CO)_3$	method of preparation (yield in %), properties and remarks	Ref.
*14	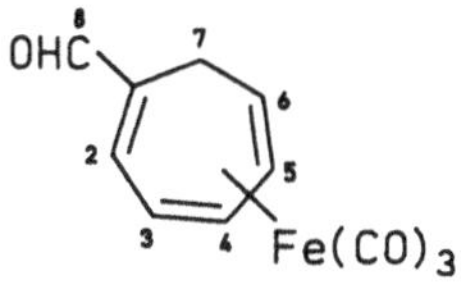 OHC(8)-substituted $C_7H_7Fe(CO)_3$ (ring positions 2–7)	IIa (1.5), III (19), Va (70), VII (17) m.p. 88 to 89°, m.p. 88 to 90°, subl. p. 70°/0.01 Torr, subl.p. 70°/ 0.02 Torr, yellow volatile solid, orange-yellow solid, yellow crystals or yellow powder (from petroleum ether) ^1H NMR (benzene-d_6): 1.62 to 2.95 (m, H-3, 6, 7), 4.47 (m, H-4, 5), 6.07 (d, H-2; J=8), 9.03 (s, H-8) ^1H NMR (CS_2): 2.43 (m, H-7), 3.18 (t, H-3; J=8.5), 3.55 (m, H-6), 5.50 (m, H-4, 5), 6.80 (d, H-2; J=8.5), 9.02 (s, H-8) ^{13}C NMR ($CDCl_3$): 26.85 (C-7), 53.85 (C-6), 60.42 (C-3), 89.23 (C-5), 94.61 (C-4), 136.90 (C-1), 154.0 (C-2), 192.33 (C-8), 209.72 (CO) IR (cyclohexane): 1640 (C=C), 1682 (C=O); 1990, 1995, 2050 (CO) IR (CCl_4): 1644 (C=C), 1684 (C=O); 1999, 2004, 2044 (CO) mass spectrum: $[M-nCO]^+$ (n=0 to 3), $[C_7H_nFe]^+$ (n=6, 7)	[8, 9, 13, 17, 44, 45]
*15	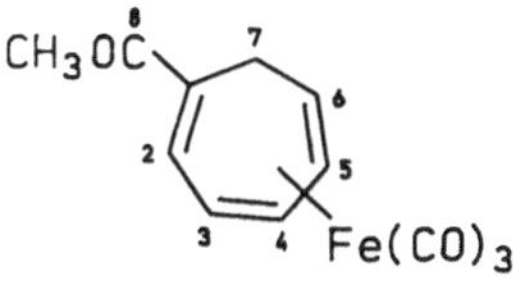 CH_3OC(8)-substituted $C_7H_7Fe(CO)_3$	Va (24) m.p. 88 to 89°, yellow oil or solid ^1H NMR (CS_2): 2.04 (s, CH_3), 2.45 (m, H-7), 3.02 (t, H-3; J=8.0), 3.46 (m, H-6), 5.35 (m, H-4, 5), 6.94 (d, H-2; J=8.0) IR (CCl_4): 1663 (C=O); 1983, 1995, 2060 (CO)	[8, 13]
*16	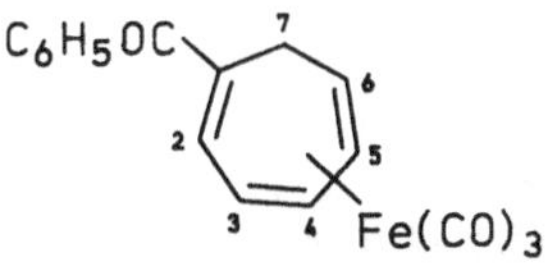 C_6H_5OC-substituted $C_7H_7Fe(CO)_3$	Vd (80) m.p. 95° (from pentane) ^1H NMR (CS_2): 2.48 (m, H-7), 3.04 (t, H-3; J=8.0), 3.47 (m, H-6), 5.34 (m, H-4, 5), 6.74 (d, H-2; J=8.0), 7.32 (m, C_6H_5) IR (cyclohexane): 1630 (C=O); 1980, 1993, 2060 (CO)	[13]
*17a	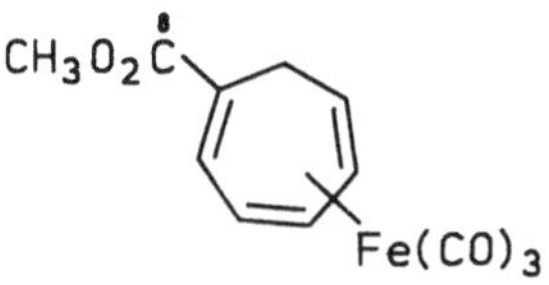 CH_3O_2C(8)-substituted $C_7H_7Fe(CO)_3$	see No. 43, p. 141 ^{13}C NMR: 170 (C-8) IR: 1643 (C=C), 1709 (C=O)	[25, 26]

Table 3 [continued]

No.	$RC_7H_7Fe(CO)_3$	method of preparation (yield in %), properties and remarks	Ref.
b	CH_3O_2C–; $(CO)_3Fe$	VIa IR: 1610 (C=C)	[5, 7]
*18	$C_2H_5O_2C$ (C-8); $Fe(CO)_3$	see No. 44, p. 141 ^{13}C NMR: 170 (C-8) IR: 1643 (C=C), 1709 (C=O)	[25, 26]
*19	$HO(CH_3)_2C$; $Fe(CO)_3$	–	[10, 14]
*20	$HO(C_6H_5)_2C$; $Fe(CO)_3$	–	[10]
*21	C_6H_5; $Fe(CO)_3$	–	[36, 39]

compounds of the type 2-$RC_7H_7Fe(CO)_3$:

No.	$RC_7H_7Fe(CO)_3$	method of preparation (yield in %), properties and remarks	Ref.
*22	C_2H_5; $Fe(CO)_3$	–	[1]
*23	CH_3OC; $Fe(CO)_3$ (positions 1, 3, 4, 5, 6, 7)	see No. 15, p. 135 orange oil ^1H NMR ($CDCl_3$): 2.22 (s, CH_3), 2.54 (m, H-7), 3.23 (m, H-6), 3.93 (m, H-3), 5.31 (m, H-4, 5), 6.37 (m, H-1) IR (n-heptane): 1674 (C=O); 1984, 1993, 2058 (CO)	[33]
*24	$Fe(CO)_3$	–	[1]

Table 3 [continued]

No.	$RC_7H_7Fe(CO)_3$	method of preparation (yield in %), properties and remarks	Ref.
*25a	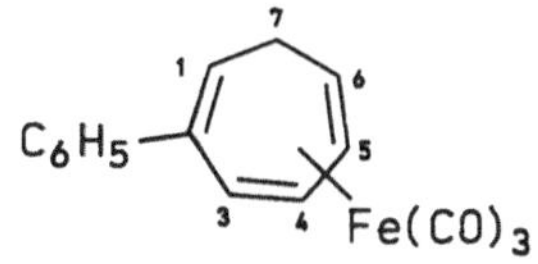 C_6H_5; $Fe(CO)_3$	Vd, see No. 21, p. 136 ^1H NMR ($CDCl_3$): 2.46 (H-7), 3.25 (m, H-3,6), 5.18 (H-1,4,5), 7.28 (m, C_6H_5) IR (liquid film): 1980, 2050	[33, 36]
b	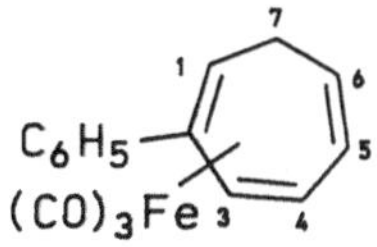 C_6H_5; $(CO)_3Fe$	Vd ^1H NMR ($CDCl_3$): 2.46 (H-7), 2.92 (t, H-4; J=8), 3.65 (m, H-1), 5.18 (H-3), 5.5 (d of d, H-5; J=2, J=8), 5.8 (m, H-6), 7.28 (m, C_6H_5) IR (liquid film): 1980, 2050	[33]
*26a	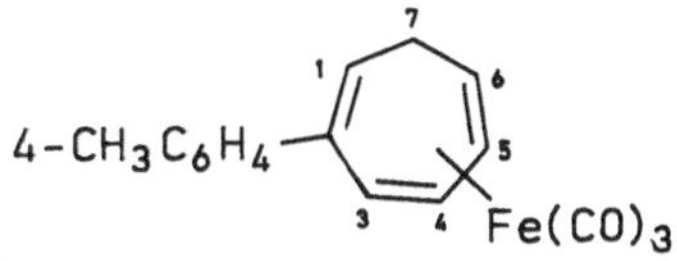 4-$CH_3C_6H_4$; $Fe(CO)_3$	Vd ^1H NMR ($CDCl_3$): 2.32 (s, CH_3), 2.55 (m, H-7), 3.34 (m, H-3,6), 5.30 (m, H-1,4,5), 7.20 (m, C_6H_4) IR (liquid film): 1980, 2050	[22, 33]
b	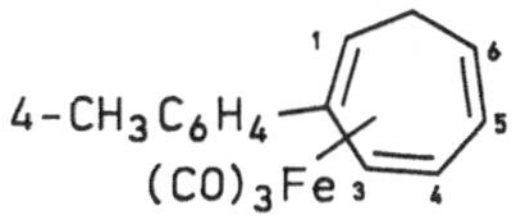 4-$CH_3C_6H_4$; $(CO)_3Fe$	Vd ^1H NMR ($CDCl_3$): 2.38 (s, CH_3), 2.46 (m, H-7), 3.0 (t, H-4; J=8), 3.76 (m, H-1), 5.30 (m, H-3), 5.6 (d of d, H-5; J=2, J=8), 5.86 (m, H-6), 7.20 (m, C_6H_4) ^1H NMR (liquid SO_2): 2.32 (s, CH_3), 2.48 (m, H-7), 3.1 (t, H-4; J=7), 3.8 (m, H-1), 5.0 to 5.4 (m, H-3), 5.68 (d of d, H-5; J=8, J=2), 5.9 (m, H-6), 7.2 (m, C_6H_4) IR (liquid film): 1980, 2050	[22, 33]

compounds of the type 3-$RC_7H_7Fe(CO)_3$:

No.	$RC_7H_7Fe(CO)_3$	method of preparation (yield in %), properties and remarks	Ref.
27	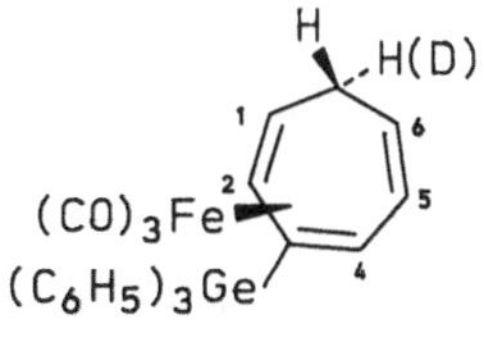 H; H(D); $(CO)_3Fe$; $(C_6H_5)_3Ge$	IVe ^1H NMR (CS_2): 2.40 (H-7), 2.94 (H-4), 3.28 (H-1), 4.82 (H-2), 5.15 (H-6), 5.75 (H-5)	[43]
*28	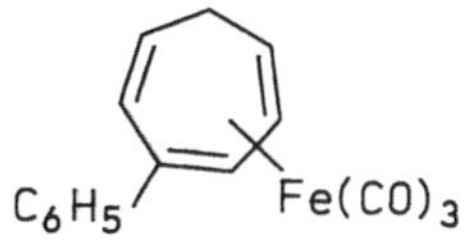 C_6H_5; $Fe(CO)_3$	see No. 21, p. 136	[36, 39]

References on pp. 143/4

Table 3 [continued]

No.	$RC_7H_7Fe(CO)_3$	method of preparation (yield in %), properties and remarks	Ref.
compounds of the type 7-$RC_7H_7Fe(CO)_3$ (continued with No. 60):			
*29	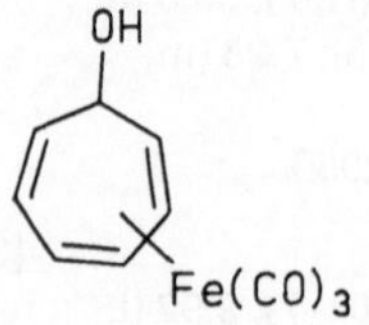 OH; Fe(CO)$_3$	IVb, VIa	[3, 20]
*30	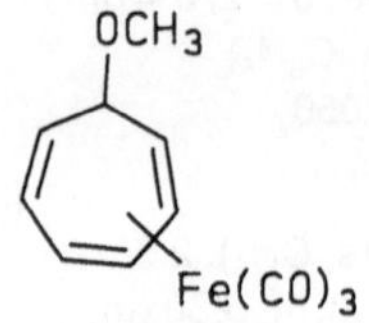 OCH$_3$; Fe(CO)$_3$	IIa, IVa	[3, 4, 47]
31	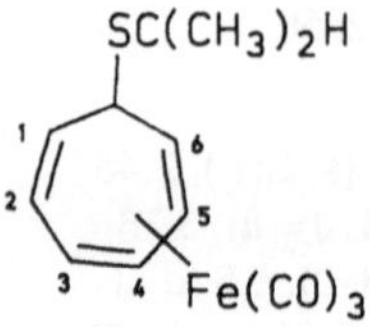 SC(CH$_3$)$_2$H; Fe(CO)$_3$	Vb (34) orange-yellow semisolid, particularly labile, decomposing to large extent during isolation to give tars ^{1}H NMR (benzene-d_6): 1.05 (dd, 2CH_3; J=6, J=1), 1.9 (t, H-7; J=12), 2.4 (m, H-2), 2.6 (sept, CH; J=6), 3.0 (dd, H-3 or 6; J=7, J=1), 3.25 (dd, H-3 or 6; J=7.2, J=1), 3.65 (ddd, H-1; J=12, J=5, J=1), 4.65 (pseudo t of d, H-4 or 5; J=5), 4.85 (pseudo t of d, H-4 or 5; J=5) IR (Nujol): 1637 (C=C); 1980, 2030 (CO)	[46]
*32	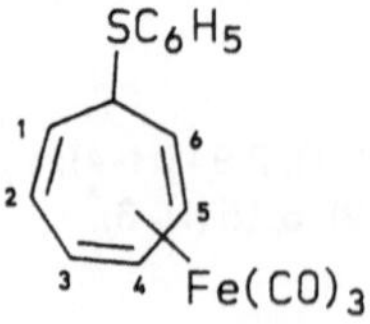 SC$_6$H$_5$; Fe(CO)$_3$	Vb (82) m.p. 121° (from C_6H_6), yellow crystals ^{1}H NMR (benzene-d_6): 1.9 (t, H-7; J=12), 2.5 (complex m, H-2), 2.95 (dd, H-3 or 6; J=7.5, J=1), 3.15 (dd, H-3 or 6; J=7.5, J=1), 3.95 (ddd, H-1; J=12, J=5, J=1), 4.65 to 4.95 (complex m, H-4,5), 6.9 to 7.3 (complex m, C_6H_5) IR ($CHCl_3$): 1660 (C=C); 1995, 2060 (CO)	[46]

References on pp. 143/4

Table 3 [continued]

No.	$RC_7H_7Fe(CO)_3$	method of preparation (yield in %), properties and remarks	Ref.
*33	H, $Si(CH_3)_3$; 1, 2, 3, 4, 5, 6; $Fe(CO)_3$	IVb (50) subl. p. 50°/10^{-3} Torr, yellow crystals (from pentane at −78 °) ^{1}H NMR (toluene): −0.10 (CH_3), 1.68 (H-7), 2.97 (H-3), 3.36 (H-6), 4.37 to 4.48 (H-4, 5), 5.10 (H-1), 5.56 (H-2) ^{13}C NMR (toluene): −2.5 (CH_3), 35.5 (C-7), 59.6 (C-3), 67.0 (C-6), 84.8 (C-5), 91.4 (C-4), 126.0 (C-2), 128.3 (C-1), 213.3 (CO) IR (cyclohexane): 1969, 1983, 2044 (CO) mass spectrum: $[M-nCO]^+$ (n = 1 to 3), $[C_nH_nFe]^+$ (n = 5 to 7), $[C_7H_7Si(CH_3)_3]^+$, $[C_7H_7]^+$, $[Si(CH_3)_3]^+$	[23, 25, 27, 42, 43, 52]
*34	H, $Si(C_2H_5)_3$; $Fe(CO)_3$	IVb	[25, 26, 42, 43]
*35	H, $Ge(CH_3)_3$; 1, 2, 3, 4, 5, 6; $Fe(CO)_3$	IVb (50) subl. p. 50°/10^{-3} Torr, yellow crystals (from pentane at −78 °) ^{1}H NMR (toluene): −0.10 (CH_3), 1.81 (H-7), 2.96 (H-3), 3.41 (H-6), 4.34 to 4.59 (H-4, 5), 5.02 (H-1), 5.51 (H-2) ^{13}C NMR (n-decane): −2.6 (CH_3), 36.4 (C-7), 59.2 (C-3), 67.4 (C-6), 84.4 (C-5), 91.7 (C-4), 125.6 (C-2), 128.5 (C-1), 212.4 (CO) IR (cyclohexane): 1968, 1982, 2043 (CO) mass spectrum: $[M-nCO]^+$, $[M-nCO-Ge(CH_3)_3]^+$ (n = 0 to 3), $[C_7H_7Fe]^+$, $[C_7H_7Ge(CH_3)_3]^+$, $[C_7H_7]^+$, $[Ge(CH_3)_3]^+$	[25, 27, 42, 43, 52]
*36	H, $Ge(C_2H_5)_3$; $Fe(CO)_3$	IVb	[25, 26, 42, 43]

References on pp. 143/4

Table 3 [continued]

No.	$RC_7H_7Fe(CO)_3$	method of preparation (yield in %), properties and remarks	Ref.
*37	R = $Ge(C_6H_5)_3$ (H, $Ge(C_6H_5)_3$ on C-7; ring positions 1–6; $Fe(CO)_3$)	IVb (65) bright yellow solid (from hexane), yellow crystals ^1H NMR (toluene): 2.80 (H-3), 2.94 (H-7), 3.75 (H-6), 4.06 to 4.42 (H-4, 5), 5.26 to 5.56 (H-1, 2), 7.25 to 7.50 (C_6H_5) ^{13}C NMR (toluene): 36.2 (C-7), 58.6 (C-3), 65.8 (C-6), 85.4 (C-5), 91.8 (C-4), 127.0 to 136.4 (C-1, 2, C_6H_5), 212.5 (CO) IR (cyclohexane): 1965, 1978, 2046 (CO) mass spectrum: $[M-nCO]^+$ (n=0 to 3), $[M-3CO-nC_6H_6]^+$ (n=1, 2), $[C_7H_7Fe]^+$, $[C_7H_7Ge(C_6H_5)_3]^+$, $[C_7H_7]^+$, $[Ge(C_6H_5)_3]^+$	[29, 42, 43, 52]
38	R = $Sn(CH_3)_3$ ($Fe(CO)_3$)	IVb	[27]
*39	R = CH_3 (H, CH_3; $Fe(CO)_3$)	–	[1, 37]
40	R = CH_2OH ($Fe(CO)_3$)	IIa orange liquid ^1H NMR ($CDCl_3$): 2.2 (m, 4H), 3.3 (m, 4H), 5.4 (m, 2H) IR (film): 874, 1029, 1054, 1075, 1346, 1413, 1456, 1600, 1678; 1970, 2050 (CO); 2874, 3288	[19]
41	R = $CH_2CH{=}CH_2$ ($Fe(CO)_3$)	IVb	[20, 23]

References on pp. 143/4

Table 3 [continued]

No.	$RC_7H_7Fe(CO)_3$	method of preparation (yield in %), properties and remarks	Ref.
*42	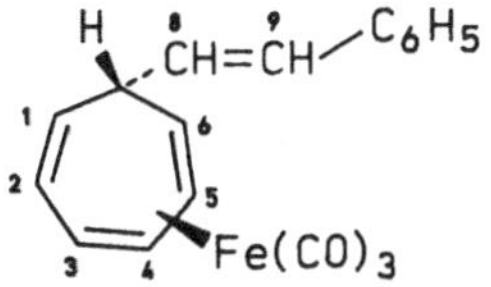 H, $CH=CH-C_6H_5$ (positions 8, 9); ring positions 1–6; $Fe(CO)_3$	Vc (32 to 41) m.p. 64 to 66°, pale yellow crystals, thin long yellow needles (from hexane at −20°) ^{1}H NMR ($CDCl_3$): 3.05 (dddd, H-3), 3.25 (m, H-6,7), 5.13 (ddd, H-1), 5.36 (m, H-4,5), 5.85 (dd, H-8), 5.86 (ddd, H-2), 6.43 (d, H-9; J(H-8,9) = 16), 7.30 (m, C_6H_5) IR (hexane): 1974, 1986, 2046 (CO) mass spectrum: $[M]^+$	[48, 58]
*43	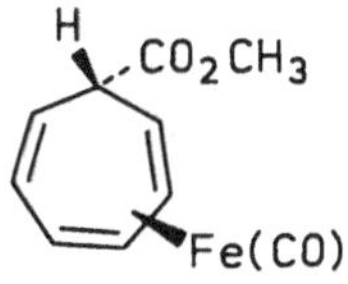 H, CO_2CH_3; $Fe(CO)_3$	IVb unstable ^{13}C NMR: 170 (CO_2) IR: 1655 (C=C), 1745 (C=O)	[23, 25, 26, 43]
*44	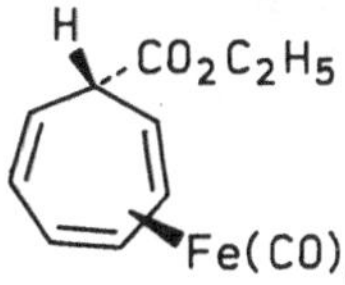 H, $CO_2C_2H_5$; $Fe(CO)_3$	IVb unstable ^{13}C NMR: 170 (CO_2) IR: 1655 (C=C), 1745 (C=O)	[23, 25, 26]
*45	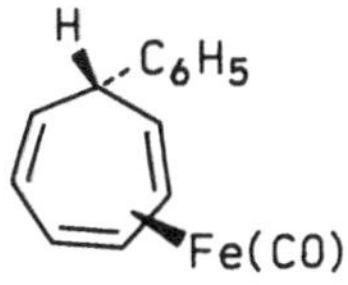 H, C_6H_5; $Fe(CO)_3$	IIb, compare No. 21, p. 136 m.p. 69.5 to 71.5° (from light petroleum ether), yellow blades	[30, 31, 36, 39]
46	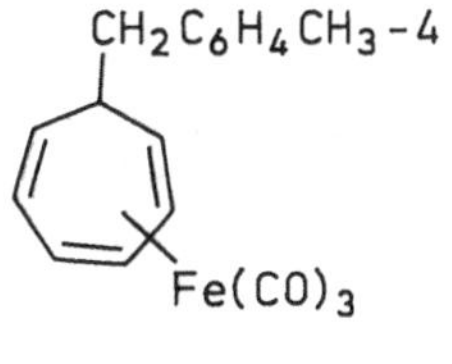 $CH_2C_6H_4CH_3$-4; $Fe(CO)_3$	^{1}H NMR (liquid SO_2): normal	[33]
*47	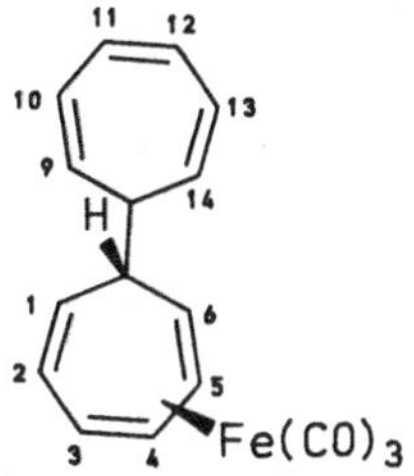 H, cycloheptatrienyl (positions 9–14); ring positions 1–6; $Fe(CO)_3$	IVd (25), VIb (60) waxy orange-yellow compound, yellow oil ^{1}H NMR (CS_2): 1.8 (q, H-8 or 7; J = 6), 3.0 (m, H-3,6,7 or 8), 5.1 to 6.3 (m, H-1,2,4,5,9,10,13,14), 6.6 (t, H-11,12; J = 4) ^{1}H NMR ($CDCl_3$): 1.82 (q, H-7 or 8; J = 6.0), 3.0 (H-8 or 7), 3.05 (H-3,6), 5.2 (H-1), 5.3 (H-9,14),	[18, 41, 47, 49]

References on pp. 143/4

Table 3 [continued]

No.	$RC_7H_7Fe(CO)_3$	method of preparation (yield in %), properties and remarks	Ref.
		5.38 (H-4,5), 5.92 (H-2), 6.22 (H-10,13), 6.63 (H-11,12) IR (CCl_4): 1976, 1988, 2050 mass spectrum: $[M-nCO]^+$ (n=1 to 3)	
48	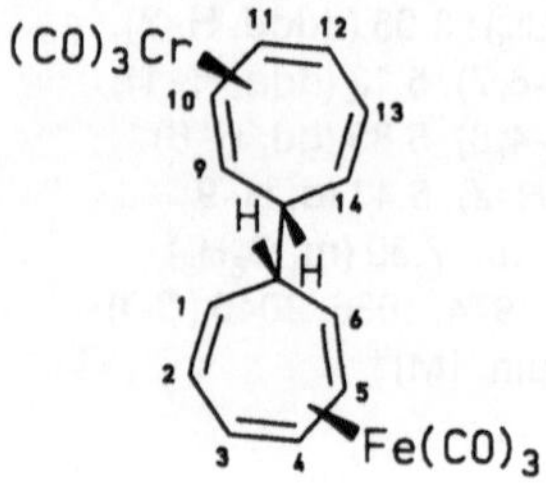 $(CO)_3Cr$ (η-10–13 coordinated C₇ ring) linked at C-14/C-7 to C₇ ring with Fe(CO)₃ at C-3–6	IVc (75) m.p. 146 to 149° (from ether/petroleum ether), orange-red solid ^1H NMR ($CDCl_3$): 0.88 (H-7), 2.52 (H-8), 2.96 (H-3,6), 3.57 (H-14), 3.77 (H-9), 4.7 (H-1, overlapped), 4.88 (H-10,13), 5.29 (H-4,5), 5.78 (H-2), 5.94 (H-11,12) IR: 1893, 1920, 1978 br, 2058 (CO) mass spectrum: $[M]^+$	[23, 24, 28, 43, 47]
*49	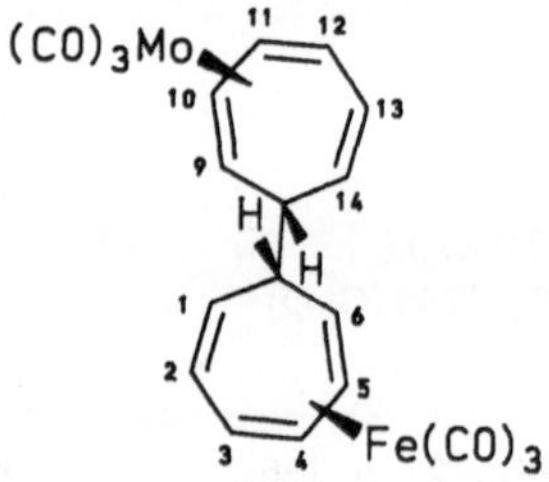 $(CO)_3Mo$ analogue, Fe(CO)₃	IVc (77) m.p. 126 to 130° (from ether/petroleum ether), red solid ^1H NMR ($CDCl_3$): 1.36 (H-7), 2.74 (H-8), 3.05 (H-3,6), 3.82 (H-14), 4.03 (H-9), 4.9 (H-1, overlapped), 5.00 (H-10,13), 5.34 (H-4,5), 5.84 (H-2), 6.02 (H-11,12) IR: 1898, 1922, 1988 br, 2056 (CO) mass spectrum: $[M]^+$	[23, 24, 28, 43, 47]
50	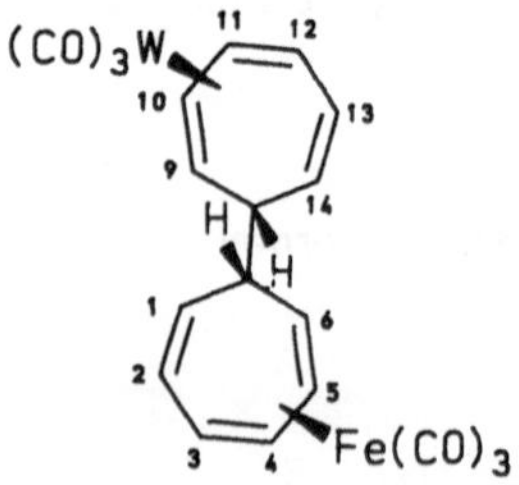 $(CO)_3W$ analogue, Fe(CO)₃	IVc (62) m.p. 124 to 126° (from ether/petroleum ether), wine-red solid ^1H NMR ($CDCl_3$, 60 MHz): 1.37 (H-7), 2.77 (H-8), 2.97 (H-3,6), 3.73 (H-14), 3.94 (H-9), 4.8 (H-1, overlapped), 4.82 (H-10,13), 5.27 (H-4,5), 5.80 (H-2), 5.95 (H-11,12) ^1H NMR ($CDCl_3$, 400 MHz): 1.34 (H-7; J(H-7,8)=8.5, J(H-1,7)=J(H-6,7)=4.6), 2.78 (H-8; J(H-8,9)=J(H-8,14)=8.5), 2.96 (H-3; J(H-2,3)=7.8), 3.03 (H-6), 3.74 (H-14), 3.98 (H-9), 4.83 (H-10 or 13), 4.96 (H-1, 10, or 13, overlapped; J(H-1,2)=10.6), 5.29 (H-4,5), 5.81 (H-2; J(H-2,7)=1.5), 5.99 (H-11,12) IR: 1890, 1918, 1985 br, 2055 (CO) mass spectrum: $[M]^+$	[23, 24, 28, 43, 47]

References on pp. 143/4

Table 3 [continued]

No.	$RC_7H_7Fe(CO)_3$	method of preparation (yield in %), properties and remarks	Ref.
*51	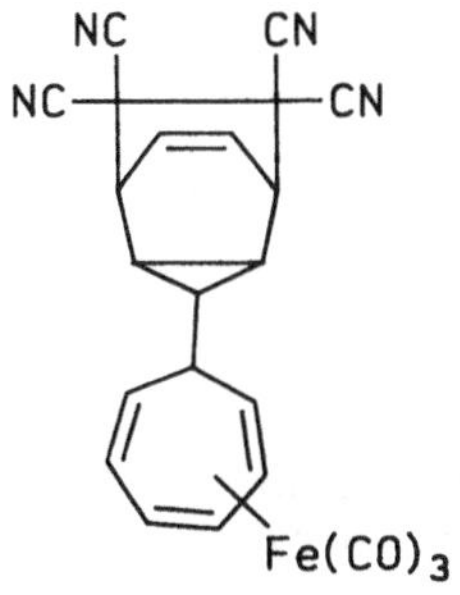 NC, NC, CN, CN … Fe(CO)$_3$	see No. 47, p. 141 off-white solid IR ($CHCl_3$): 1983, 2054 (CO)	[18, 49]
other compounds:			
*52	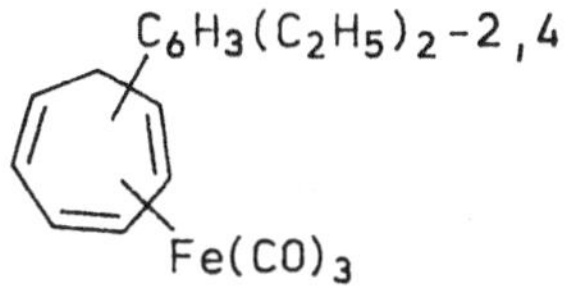 $C_6H_3(C_2H_5)_2$-2,4 … Fe(CO)$_3$	I (good yield)	[2]
*53	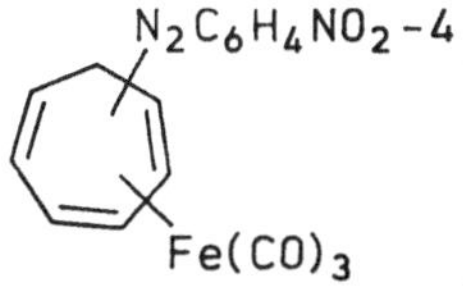 $N_2C_6H_4NO_2$-4 … Fe(CO)$_3$	–	[53]
supplements:			
compounds of the type 1-$RC_7H_7Fe(CO)_3$:			
54	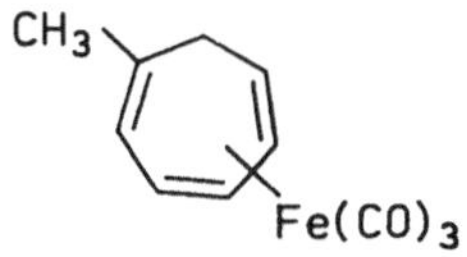 CH_3 … Fe(CO)$_3$	see No. 39, p. 140	[37]
*55	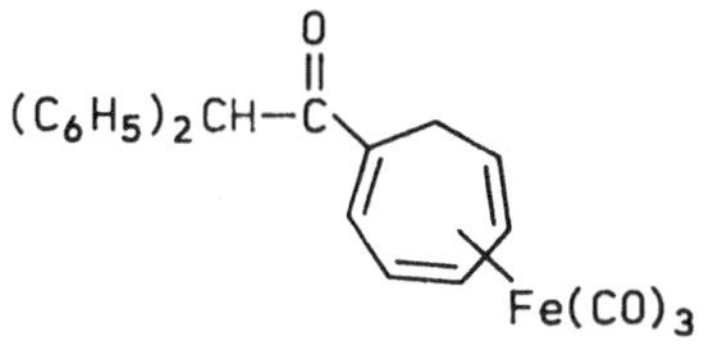 $(C_6H_5)_2CH$–C(=O) … Fe(CO)$_3$	m.p. 160 to 161° (from CH_2Cl_2/hexane) ^{1}H NMR ($CDCl_3$): 2.60 (2H, ABX; J=3, J=22), 2.99 (t, 1H; J=8), 3.49 (m, 1H), 5.34 (m, 2H), 5.61 (s, 1H), 6.84 to 7.52 (m, 11H) IR (Nujol): 1640 (C=O); 1955, 1975, 2020 (CO)	[54]

References on pp. 143/4

Table 3 [continued]

No.	$RC_7H_7Fe(CO)_3$	method of preparation (yield in %), properties and remarks	Ref.
*56	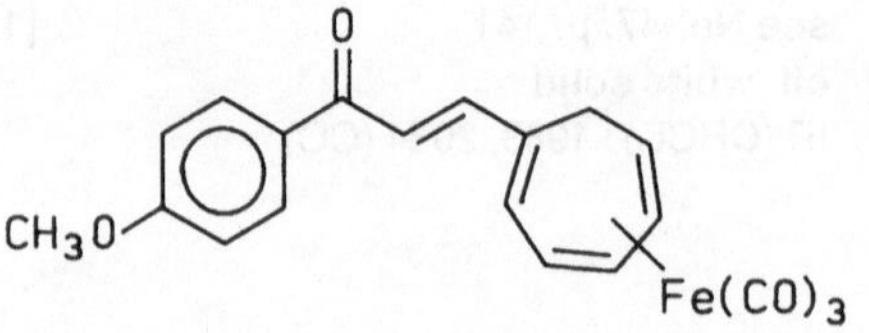 O; CH_3O; $Fe(CO)_3$	see No. 14, p. 134 m.p. 147° (from CH_2Cl_2/hexane) ^1H NMR ($CDCl_3$): 2.52 (dd, 2H; J=5, J=20), 3.13 (t, 1H; J=9), 3.43 (m, 1H), 3.80 (s, OCH_3), 5.40 (m, 2H), 6.38 (d, 1H; J=8), 6.7 (d, 1H; J=16), 7.18 (d, 1H; J=16); 6.87, 7.87 (m, C_6H_4) IR (Nujol): 1640, 1975, 2045 mass spectrum: $[M-nCO]^+$ (n=1 to 3)	[57]
*57	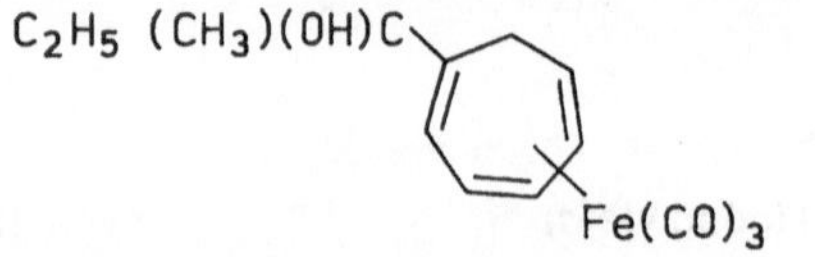 $C_2H_5(CH_3)(OH)C$; $Fe(CO)_3$	see No. 15, p. 135 (≈10%), not isolated	[14]
*58	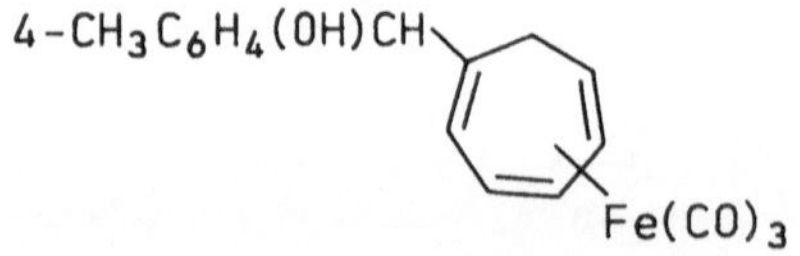 $4\text{-}CH_3C_6H_4(OH)CH$; $Fe(CO)_3$	see No. 14, p. 134 (nearly quantitative), not isolated	[14]
*59	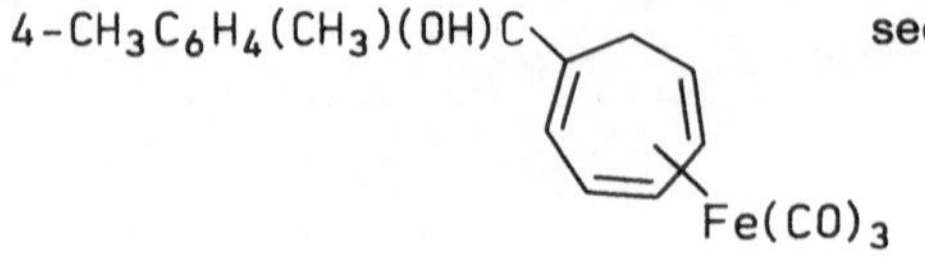 $4\text{-}CH_3C_6H_4(CH_3)(OH)C$; $Fe(CO)_3$	see No. 15, p. 135, not isolated	[14]
compound of the type $7\text{-}RC_7H_7Fe(CO)_3$:			
*60	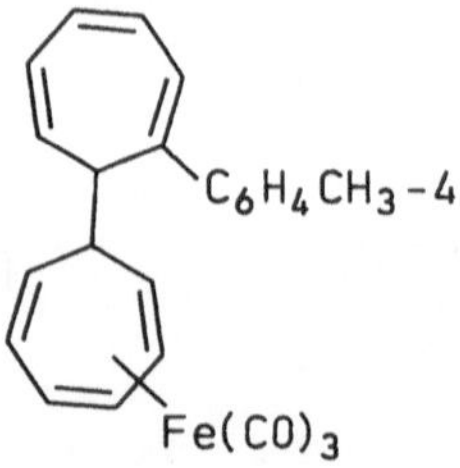 $C_6H_4CH_3\text{-}4$; $Fe(CO)_3$	–	[59]

*Further information:

1-$(CH_3)_2CHC_7H_7Fe(CO)_3$ (Table **3**, No. **2**). The colloidal formulation obtained from No. 2 with kerosine, H_2O, and mannitan monooleate in a colloid mill shows fungicidal effectiveness proved by the Agar-plate technique [1].

1-RCH=CHC$_7$H$_7$Fe(CO)$_3$ (Table **3**, Nos. **4** to **7** with R=CO_2CH_3, $CO_2C_2H_5$, CN, C_6H_5). Complex No. 4 is heated in toluene at 90 to 100 °C for 48 h to give XXIV (R=CO_2CH_3, p. 137) in 27% yield. Analogous thermal reaction of No. 7 at 100 °C gives the metal-hydrogen shift isomer XXV ($R=C_6H_5$, p. 142) and XXIV ($R=C_6H_5$) in 25 and 12% yield, respectively. Refluxing Nos. 4 or 7 with $Fe_2(CO)_9$ in toluene for 17 h gives XXIV ($R=CO_2CH_3$) in 57% or XXIV ($R=C_6H_5$) in 64% yield [57].

Mixing equal amounts of freshly sublimed $(NC)_2C{=}C(CN)_2$ and the corresponding complex No. 4 to 7 in CH_2Cl_2 gives XI ($R=CO_2CH_3$, $CO_2C_2H_5$, CN, C_6H_5). Similar reaction of the appropriate complexes No. 4 to 7 with N-phenyltriazolinedione gives XII ($R=CO_2CH_3$, $CO_2C_2H_5$, CN, C_6H_5) [44].

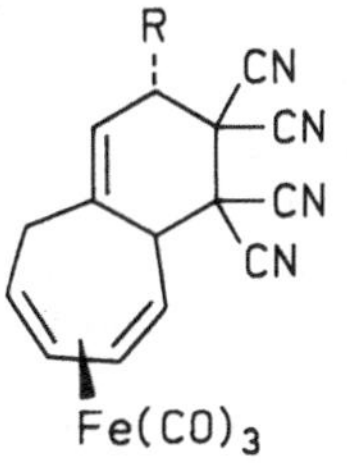

XI 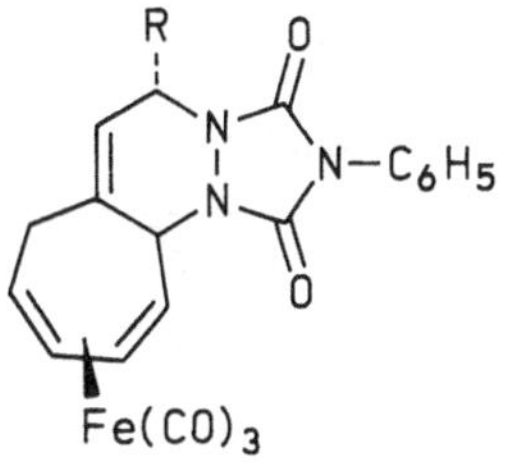 XII 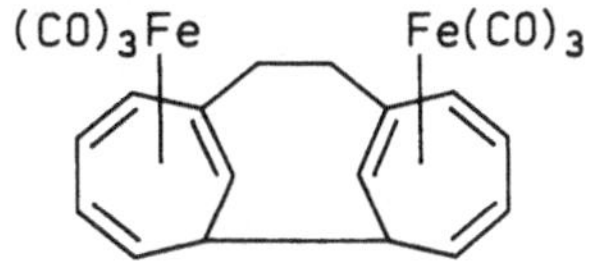 XIII

1-HOCH$_2$C$_7$H$_7$Fe(CO)$_3$ (Table **3**, No. **8**) in toluene is placed on top of a silica gel column. After 15 h toluene elutes XIII (60%) and XIV ($R=R'=H$) in ca. 1% [14]. Reaction with 65% HPF_6/H_2O [8, 13, 14] or $[C(C_6H_5)_3]BF_4$ [8, 17] in ether [13] gives XVb ($R=R'=H$, $X=PF_6$ or BF_4) [8, 13, 14, 17].

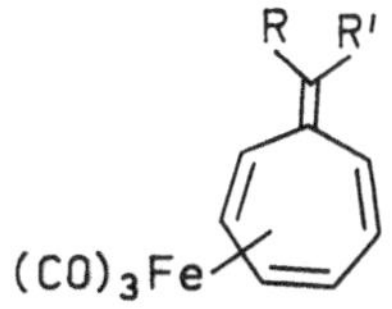

XIV 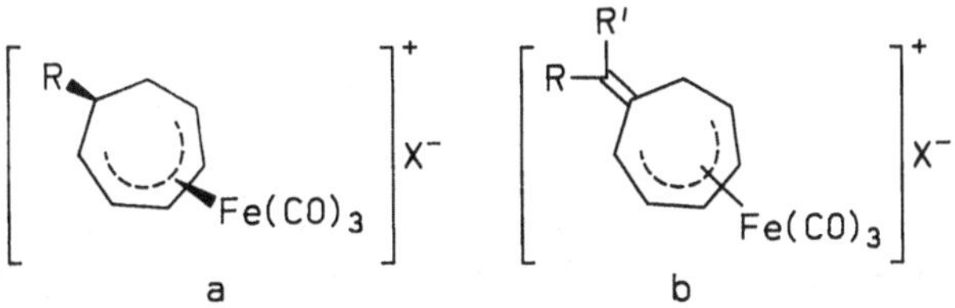 XV (a, b)

1-RHOHCC$_7$H$_7$Fe(CO)$_3$ (Table **3**, Nos. **9** to **11**) give XIV ($R=CH_3$, $CH_2C_6H_5$, C_6H_5) by dehydration on silica gel [10, 13, 14]; No. 9 is reported to give also No. 3 [8]. Reaction of Nos. 9 to 11 with 65% aqueous HPF_6 in ether gives XVb ($R=H$, $R'=CH_3$, $CH_2C_6H_5$, C_6H_5) [10, 13, 14].

1-OHCC$_7$H$_7$Fe(CO)$_3$ (Table **3**, No. **14**) is also formed in the reaction of XVb ($R=H$, $R'=OH$) with CH_3OH or CH_2N_2 [13]. Compound No. 14 gives a normal ^1H NMR spectrum in liquid SO_2 solution [33]. The methane chemical ionization mass spectrum is discussed in [16]. In contrast to many other complexes attempts to remove the $OHCC_7H_7$ ligand are unsuccessful. Reduction with $NaBH_4$ in C_2H_5OH at 20 °C [8, 13] for 45 min [13], pouring the reaction mixture into H_2O, extracting with ether, drying the organic layer ($MgSO_4$), and chromatography on silica gel with 5% $CH_3CO_2C_2H_5$/toluene gives No. 8 in 91% yield [13], cf. [17]. Reaction with 60% aqueous HPF_6 in ether gives XVb ($R=H$, $R'=OH$, $X=PF_6$) [13], cf. [50]. Reaction with $H_2NNHC_6H_3(NO_2)_2$-2,4 gives No. 12 [8, 13, 21], and no reaction is observed with SO_2 [33]. Compound No. 14 reacts with C_6H_5MgBr (ratio 1:4) in ether at −30 °C for 15 min. A saturated aqueous $[NH_4]Cl$ solution is added, and the mixture is warmed to 20 °C. The organic layer is washed with aqueous $NaHCO_3$, dried ($MgSO_4$), and evaporated to give No. 11 (90% yield)

 References on pp. 143/4

[13], cf. [14]. Similar treatment with CH_3MgI gives No. 9 [8]. Similar treatment with 4-$CH_3C_6H_4MgBr$ at $-50\,°C$ for 20 min, hydrolysis with aqueous $[NH_4]Cl$, and drying the ethereal layer with $MgSO_4$ gives No. 58, which is not isolated, in nearly quantitative yield. Treatment of the solution with HPF_6 gives XVb (R=H, R′=$C_6H_4CH_3$-4, X=PF_6) [14]. Treatment of No. 14 with $HC(OCH_3)_3$/2 drops conc. HCl in CH_3OH at room temperature for 24 h gives a 3:1 mixture of XIV with R=OCH_3, R′=H and XIV with R=H, R′=OCH_3. Similar results are obtained with $HC(OC_2H_5)_3$ [45]. Treatment with $(C_6H_5)_3P$=CH_2 in tetrahydrofuran at 0 °C for 2 h gives XVb with R=H, R′=CH_3. Compound No. 14 reacts with $(C_6H_5)_3P$=CHC_6H_5 in tetrahydrofuran at 0 °C within 4 h. The reaction mixture is quenched with H_2O, extracted with ether, and the organic layer is washed with H_2O, dried ($MgSO_4$), and evaporated. Chromatography of the residue on silica gel with hexane affords first No. 7b (8% yield), and No. 7a (24%) elutes second. Similar reaction with $(C_6H_5)_3P$=$CHCO_2C_2H_5$ gives No. 5 (37%), and very slow reaction with $(C_6H_5)_3P$=$CHCO_2CH_3$ in CH_2Cl_2 at room temperature [44, 57] for 3 d followed by removal of the solvent and chromatography of the residue on Kieselgel 60 with hexane [57] gives No. 4 (50%) [44, 57]. Similar reaction with $(C_6H_5)_3P$=$CHCOC_6H_4OCH_3$-4 at room temperature for two weeks gives No. 56 in 18% yield [57]. Reaction with $(C_6H_5)_3P$=CHCN in C_6H_6 at 40 to 50 °C for one month followed by removal of the solvent and chromatography as before gives No. 6 (21%) [44]. $(NC)_2C$=$C(CN)_2$ in CH_2Cl_2 gives XVI in a slow reaction [32]. The observed rate constants for the addition of $(NC)_2C$=$C(CN)_2$ in CH_2Cl_2 at 25 °C are shown in the following table [49].

$[(NC)_2C$=$C(CN)_2]$ in mmol/L	k_{obs} in s^{-1}
10.0	1.58×10^{-3}
20.0	2.05×10^{-3}
30.0	2.40×10^{-3}
40.0	2.93×10^{-3}

A plot of k_{obs} against $[(NC)_2C$=$C(CN)_2]$ gives a good straight line with a large positive intercept. The presence of the intercept suggests that the system contains a rate-determining reaction and does not depend upon the concentration of $(NC)_2C$=$C(CN)_2$. A rate-determining radical closure could be excluded by ESR. When the reaction was carried out in an NMR tube, it became clear that three times the molar quantity of $(NC)_2C$=$C(CN)_2$ was necessary to remove all of No. 14, and the new absorptions are due to XVII (R=H). XVI is formed in a much slower reaction.

No. 14 + $(CN)_2C$=$C(CN)_2$ $\underset{k_{-1}}{\overset{k_1}{\rightleftharpoons}}$ XVII

k'_2 → XVI

References on pp. 143/4

The slope and intercept of the plot of k_{obs} versus $[(NC)_2C{=}C(CN)_2]$ is assigned to k_1 and k_{-1}, respectively. k_{-1} was also determined by addition of 4-isopropyl-1-methylcyclohexa-1,3-diene to XVII to remove $(NC)_2C{=}C(CN)_2$ (direct IR). The rate constants and equilibrium constants in CH_2Cl_2 (unless otherwise stated) by the various methods are summarized in the following table [49]:

method	$k_1 \cdot 10^2$ in $dm^3 \cdot mol^{-1} \cdot s^{-1}$	$k_{-1} \cdot 10^3$ in s^{-1}	k	$k_2' \pm 20\%$ in $dm^3 \cdot mol^{-1} \cdot s^{-1}$
UV-visible	4.40 ± 0.23	1.14 ± 0.06	38.2 ± 3.0	—
IR	4.72 ± 0.18	1.25 ± 0.08	37.8 ± 2.9	9.4×10^{-4}
direct IR	—	1.32 ± 0.07	29.0	—
NMR (in acetone-d_6)	—	—	10.5	1.06×10^{-2}

Using the kinetic data, a pure sample of XVII (R=H) is obtained after 0.5 h in CH_2Cl_2 if the solution is frozen with liquid N_2 and the volatiles are removed rapidly under vacuum [49].

1-$CH_3OCC_7H_7Fe(CO)_3$ (Table **3**, No. **15**) is also obtained by the reaction of XVb (R=H, R′=$COCH_3$, X=PF_6) with $NaOCH_3/CH_3OH$ [8] or $N(C_2H_5)_3$ [13]. Compound No. 15 gives a normal ^1H NMR spectrum in liquid SO_2 solution [33] and shows in the presence of $Eu(fod)_3$ (fod=6,6,7,7,8,8,8-heptafluoro-2,2-dimethyloctane-3,5-dionate) chemical shifts at $\delta = 0.0$ (CH_3), 0.1 (H-7), 1.4 (H-7), 4.0 (H-2), 7.8 (H-3), 8.2 (H-6), 8.6 (H-4), and 8.9 (H-5) ppm [33]. The compound No. 15 reacts with Na in CH_3OH at 25 °C for 0.5 h. The reaction mixture is poured into ether/H_2O (1:2) and treated with dilute HCl. The organic layer is washed with $NaHCO_3$ and H_2O, dried ($MgSO_4$), and evaporated. Chromatography of the residue on silica gel with 10% $CH_3CO_2C_2H_5$/toluene gives No. 23 in 80% yield [33]. Reduction with $NaBH_4$ [8, 13] in C_2H_5OH at 20 °C for 45 min followed by work-up as described with No. 14 gives No. 9 in 80% yield [13]. Reaction with $H_2NNHC_6H_3(NO_2)_2$-2,4 gives No. 13 [8], and no reaction is observed with SO_2 [33]. Protonation with 60% HPF_6 in ether gives an immediate precipitate of a yellow salt. On continued stirring, the endo acyl complex XVa (R=CH_3CO, X=PF_6, p. 133) is formed. The yellow salt is probably the protonated oxygen intermediate XVb (R=CH_3, R′=OH, X=PF_6, p. 133) formed prior to formation of XVa. XVa probably arises from exo proton attack on XVb [13]. The compound No. 15 reacts with C_2H_5MgBr in ether at −50 °C followed by stirring at −30 °C for 20 min. The reaction mixture is hydrolyzed with aqueous $[NH_4]Cl$, the ethereal layer is washed with H_2O, dried ($MgSO_4$), and evaporated to give a yellow oil (No. 57). Addition of HPF_6 in ether to the oily residue gives XVb (R′=CH_3, R=C_2H_5, X=PF_6). Neutralization of the acidic solution with $NaHCO_3$, washing with H_2O, drying ($MgSO_4$), and chromatography on silica gel with 5% $CH_3CO_2C_2H_5$/toluene gives XVIII [14]. Reaction of No. 15 in ether at −50 °C with 4-$CH_3C_6H_4MgBr$ for 20 min followed by hydrolyzation with $[NH_4]Cl$ and dried ($MgSO_4$), gives No. 59 which is not isolated. Acidification with HPF_6 gives XVb (R=CH_3, R′=4-$CH_3C_6H_4$, X=PF_6, p. 133) in 20% yield. Reaction with $(NC)_2C{=}C(CN)_2$ gives XVII (R=CH_3) [33, 49]. The observed rate constants for this 1,3-addition in CH_2Cl_2 at 25 °C are shown in the following table [49]:

$[(NC)_2C{=}C(CN)_2]$ in mmol/dm^3	k_{obs} in s^{-1}
6.7	1.26×10^{-3}
10.9	1.94×10^{-3}
15.1	2.77×10^{-3}
20.3	3.51×10^{-3}
27.0	4.63×10^{-3}

References on pp. 143/4

A plot of k_{obs} against $[(NC)_2C{=}C(CN)_2]$ gives a good straight line through the origin. From this, the second-order rate constant for this addition is obtained as $k_2 = (1.66 \pm 0.04) \times 10^{-1}\ dm^3 \cdot mol^{-1} \cdot s^{-1}$ [49].

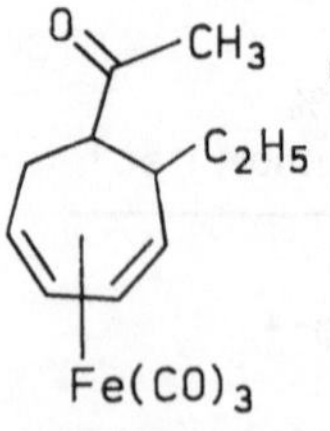

XVIII 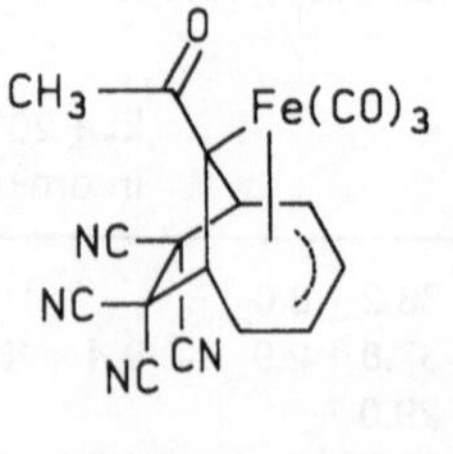 XIX 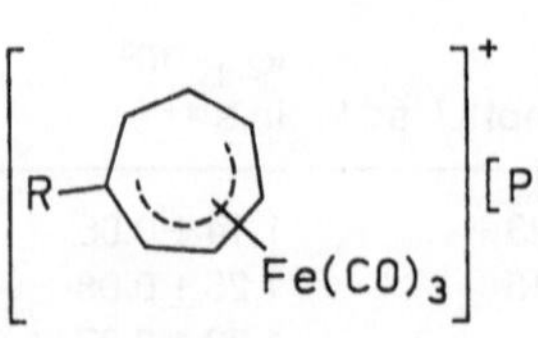 XX 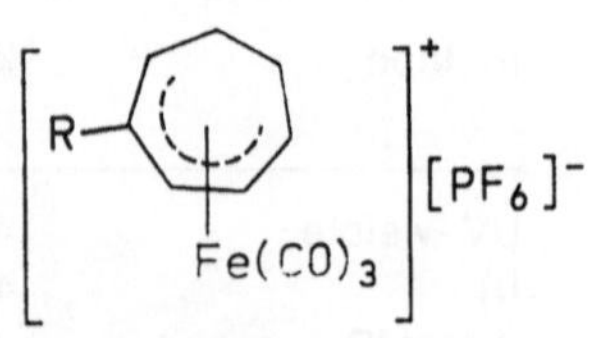 XXI

1-$C_6H_5OCC_7H_7Fe(CO)_3$ (Table **3**, No. **16**) gives No. 11 on reduction with $NaBH_4$ [13]. Treatment with $HC(OCH_3)_3$/2 drops conc. HCl in CH_3OH at room temperature for 24 h gives a 4:1 mixture of XIV with R=OCH_3, R'=C_6H_5 and XIV with R=C_6H_5, R'=OCH_3 [45].

1-$CH_3O_2CC_7H_7Fe(CO)_3$ (Table **3**, No. **17**). Reaction of isomer No. 17a with $NaN(Si(CH_3)_3)_2$ in C_6H_6 gives $Na[CH_3O_2CC_7H_7Fe(CO)_3]$ and $HN(Si(CH_3)_3)_2$ [25, 26]. Protonation of isomer No. 17b with HBF_4 gives $[1\text{-}CH_3O_2CC_7H_8Fe(CO)_3]BF_4$ [5], and similar reaction of No. 17b with concentrated H_2SO_4 gives $[1\text{-}CH_3O_2CC_7H_8Fe(CO)_3]^+$ [7], cf. [11].

1-$C_2H_5O_2CC_7H_7Fe(CO)_3$ (Table **3**, No. **18**). Reaction of No. 18 with $NaN(Si(CH_3)_3)_2$ in C_6H_6 gives $Na[C_2H_5O_2CC_7H_7Fe(CO)_3]$ and $HN(Si(CH_3)_3)_2$ [25, 26].

1-$HOR_2CC_7H_7Fe(CO)_3$ (Table **3**, Nos. **19** and **20** with R=CH_3 or C_6H_5) give XVb with R=R'=CH_3 [10, 14] or C_6H_5 [10] on protonation. Dehydration on silica gel gives XIV with R=R'=CH_3 [10, 14] or C_6H_5 [10], respectively [10, 14].

1-$C_6H_5C_7H_7Fe(CO)_3$ (Table **3**, No. **21**). Nos. 21, 25, and 28 are prepared by thermal rearrangement of 7-$C_6H_5C_7H_7Fe(CO)_3$ and isolated as pure samples by thin layer chromatography or high performance liquid chromatography (No. 25) [36, 39]. However, thin layer techniques offer inadequate resolution for a detailed analysis of the products of the rearrangement. Thus, high-performance liquid chromatography is used to analyze the thermal rearrangement of 1,2,3- and 7-$C_6H_5C_7H_7Fe(CO)_3$ (Nos. 21, 25, 28, and 45) in isooctane at 90 °C. The results are shown in the following table [36]:

starting isomer	h	ratios obtained (%)		
		No. 21	No. 25	No. 28
1-$C_6H_5C_7H_7Fe(CO)_3$ (No. 21)	47	48.2	16.5	35.3
	69	48.1	17.0	34.8
2-$C_6H_5C_7H_7Fe(CO)_3$ (No. 25)	21	38	25	37
3-$C_6H_5C_7H_7Fe(CO)_3$ (No. 28)	46	48.9	16.6	34.5
	69	48.0	17.1	34.9
7-$C_6H_5C_7H_7Fe(CO)_3$ (No. 45)	43.5	50.0	14	36

The ratios indicate that in each case an equilibrium is obtained [36].

2-$C_2H_5C_7H_7Fe(CO)_3$ (Table **3**, No. **22**) is mixed with Attaclay and Nacconol. Stirring the resulting powder with H_2O gives a suspension which shows fungicidal effectiveness as proved by the Agar-plate technique [1].

References on pp. 143/4

2-$CH_3OCC_7H_7Fe(CO)_3$ (Table **3**, No. **23**) shows in the ^1H NMR spectrum in the presence of Eu(fod)$_3$ (fod = 6,6,7,7,8,8,8-heptafluoro-2,2-dimethyloctane-3,5-dionate) chemical shifts at $\delta = 0$ (CH_3), 5.0 (H-3), 8.5 (H-7), 9.0 (H-6), and 9.1 (H-4,5) ppm. Protonation with CF_3CO_2H in $CDCl_3$ followed by pouring the solution into HPF_6 in ether gives [2-$CH_3OCC_7H_8Fe(CO)_3$]PF_6. The compound gives normal ^1H NMR spectra in liquid SO_2 indicating that no reaction takes place [33]. Reaction with $(NC)_2C{=}C(CN)_2$ gives XIX [33, 49]. The rate constants for this 1,3-addition have been measured. The observed rate constants in CH_2Cl_2 at 25 °C are shown in the following table [43]:

$[(NC)_2C{=}C(CN)_2]$ in mmol/dm^3	k_{obs} in s^{-1}
6.7	1.26×10^{-3}
10.9	1.87×10^{-3}
15.1	2.65×10^{-3}
20.3	3.28×10^{-3}
27.0	4.45×10^{-3}

A plot of k_{obs} against $[(NC)_2C{=}C(CN)_2]$ gives a good straight line through the origin. From this, the second-order rate constant for this addition is $k_2 = (1.56 \pm 0.05) \times 10^{-1}$ dm$^3 \cdot$ mol$^{-1} \cdot$ s^{-1} [49].

1-$C_5H_5C_7H_7Fe(CO)_3$ (Table **3**, No. **24**) is mixed with cyclohexanone at 25 °C for 2 min followed by dilution with kerosene. The resulting mixture is suitable for application as a fungicide as proved by the Agar-plate technique [1].

1-$RC_7H_7Fe(CO)_3$ (Table **3**, Nos. **25** and **26** with R = C_6H_5, $C_6H_4CH_3$-4). Protonation of the isomers No. 25a and b (ratio 1:1) with CF_3CO_2H in $CDCl_3$ followed by addition of HPF_6 in ether leads to a 1:1 mixture of XX and XXI (R = C_6H_5) [33]. Similar protonation of the isomers No. 26a and b (ratio 1.6:1) [22, 33] leads to 1.6:1 mixture [33] of XX and XXI [22, 33]. Thus, 26a and b are protonated at the uncoordinated double bond. Reaction of the 1.6:1 mixture of the isomers No. 26a and b with liquid SO_2 gives XXII, and isomer No. 26b is recovered [33].

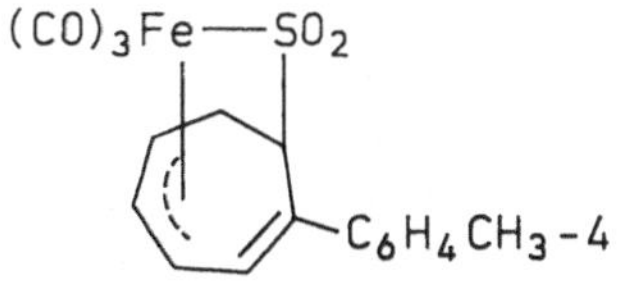

(CO)$_3$Fe—SO$_2$ ... C$_6$H$_4$CH$_3$-4

XXII

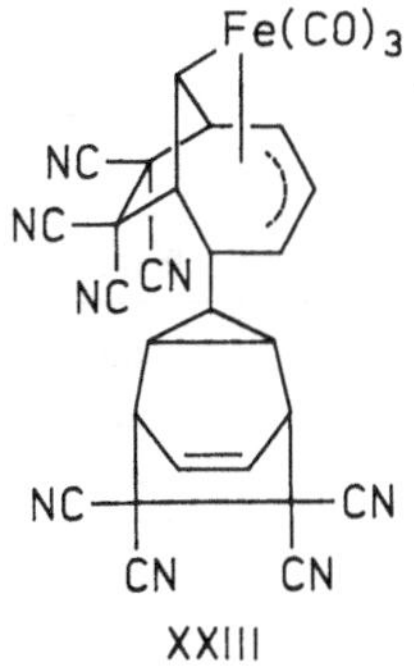

Fe(CO)$_3$ NC NC NC CN NC CN CN CN

XXIII

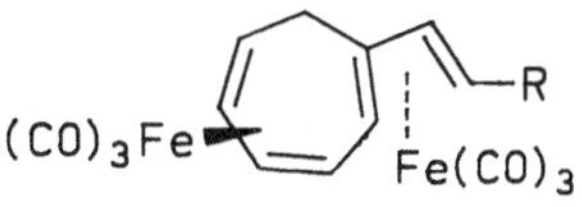

(CO)$_3$Fe R Fe(CO)$_3$

XXIV

3-$C_6H_5C_7H_7Fe(CO)_3$ (Table **3**, No. **28**) gives with kerosene, H_2O, and Triton X-100 a dispersion and with Fuller's earth a dust formulation, both showing fungicidal effectiveness as proved by the Agar-plate technique [1].

7-$HOC_7H_7Fe(CO)_3$ (Table **3**, No. **29**). Provided that precautions are taken to exclude air and prevent prolonged exposure to light, compound No. 29 can be gas chromatographed

References on pp. 143/4

at temperatures up to 150 °C with no apparent thermal decomposition. Teflon columns employing E-301 methyl silicone gum as stationary phase give best results [15]. Protonation with HBF_4 gives $[C_7H_7Fe(CO)_3]BF_4$ [3].

7-$CH_3OC_7H_7Fe(CO)_3$ (Table **3**, No. **30**) reacts with HBF_4 [3, 47] in ether [47] to give $[C_7H_7Fe(CO)_3]BF_4$ [3, 47].

7-$C_6H_5SC_7H_7Fe(CO)_3$ (Table **3**, No. **32**) reacts with $[C(C_6H_5)_3]BF_4$ in CH_3CN to give $[C_7H_7Fe(CO)_3]BF_4$ on addition of ether [46].

7-$R_3MC_7H_7Fe(CO)_3$ (Table **3**, Nos. **33** to **37** with M=Si, R=CH_3, C_2H_5 and M=Ge, R=CH_3, C_2H_5, C_6H_5). The compounds No. 33, 35, and 37 exhibit surprisingly facile oscillation of the $Fe(CO)_3$ moiety (compare XXX, p. 143) as probed by variable temperature ^1H NMR spectroscopy [27, 29, 37, 52] in $O(C_6H_5)_2$ [52] at 40 and 130 °C [27, 52], ^{13}C NMR spectroscopy [27, 29, 52] in toluene-d_8 at 33 [27, 52], 50, 70 [52], 80 [27, 52], 90, 100 [52], and 127 °C [27, 52] (here the high-temperature limiting spectrum has not been reached [27]), and ^{13}C NMR Forsén-Hoffmann spin saturation transfer technique at 40, 45, 50, 55, 60, and 70 °C [52]. For these spectra see figures in [52]. The chemical shift differences Δν, coalescence temperatures T_c, and ΔG^* values derived from the ^1H and ^{13}C NMR spectra are shown in the following table [52]:

compound No.	Δν (Hz)	T_c (°C)	ΔG^* at T_c (kJ/mol)	Ref.
^1H NMR spectrum:				
33a	145	105±5	75±3	[52]
35a	150	95±5	73±3	[27, 29, 34, 51, 52]
37a	155	100±5	74±3	[51, 52]
^{13}C NMR spectrum:				
33a	719	115±5	72±3	[52]
35a	735	115±5	72±3	[52]
37a	750	115±5	73±3	[52]

The spin saturation transfer data for No. 37 and rate constants k_{ex} are shown in the following table [52]:

T (°C)	$T_{(4)}$ (s)	$\frac{M_{z(4)}(\infty)}{M_{z(4)}(0)-M_{z(4)}(\infty)}$	k_{ex} (s^{-1})
40	0.868	2.69	0.428
45	0.994	1.44	0.699
55	1.309	0.44	1.754
60	1.462	0.20	3.367

The lifetime at C-3, $\tau_{(3)}$ is given by $\tau_{(4)}=T_{1(4)}\,M_{z(4)}(\infty)/(M_{z(4)}(0)-M_{z(4)}(\infty))$ where $M_{z(4)}(0)$ is the normal equilibrium magnetization of C-4 in the absence of irradiation at C-3, and $T_{1(4)}$ is the spin lattice relaxation time of C-4 and is measured by the (180°-τ-90°-(sample)-T)$_n$ pulse sequence with $T \geq 5T_1$. From the lifetime, the rate of leaving site C-4, i.e., the rate for the exchange between C-4 and C-3 is given by $k_{ex}=k_{3,4}=1/\tau_{(4)}=[(M_{z(4)}(0)/M_{z(4)}(\infty))-1]/T_{1(4)}$. From these spin saturation transfer data the activation parameters $\Delta H^*=84(\pm4)$ kJ/mol, $\Delta S^*=18(\pm2)\ J\cdot K^{-1}\cdot mol^{-1}$, and $\Delta G^*_{393}=77(\pm4)$ kJ/mol are derived [52]. The values of ΔG^* are approximately 75 kJ/mol for all three compounds

[52], cf. [29]. Possible pathways for the rearrangement are explored [27, 37, 52]; and on the basis of ^{13}CO incorporation and $P(C_6H_5)_3$ substitution studies, the molecular structure of No. 37 (see below), the remarkable lowering of ΔG^{+} relative to $C_7H_8Fe(CO)_3$ (see pp. 105/6), and comparison with activation barriers in other substituted $C_7H_8Fe(CO)_3$ derivatives, it is concluded that direct 4, 6 iron shift is most appropriate in describing the mechanism of fluxionality [52].

He(I) and He(II) photoelectron spectra of Nos. 33 and 35 are studied in order to identify possible electronic factors responsible for the variation in fluxional behavior. The ionization energies (in eV) and assignments are shown in the following table [55]:

compound No.	Fe 3d	assignments σ(M—C) (M=Si, Ge)	π_1	π_2	π_3
33	7.96, 8.56	10.13	10.90	11.20	11.87
35	7.62, 8.35	9.75	10.40	10.92	11.81

These data are compared with those of $C_7H_8Fe(CO)_3$. Spectral evidence indicates that there is a rupture in conjugation between the free double bond and the bound butadiene moiety and a stronger Fe-diene interaction in the substituted compounds relative to the unsubstituted molecules. These effects appear to be sufficient to cause a significant lowering of the activation energy for 1, 3 iron shifts in Nos. 33 and 35 [55].

X-ray analysis of No. 35 [38] and No. 37 [52] establishes the exo position of GeR_3 (R=CH_3, C_6H_5). The compound No. 37 crystallizes in the monoclinic space group $P2_1/n\text{-}C^5_{2h}$ with

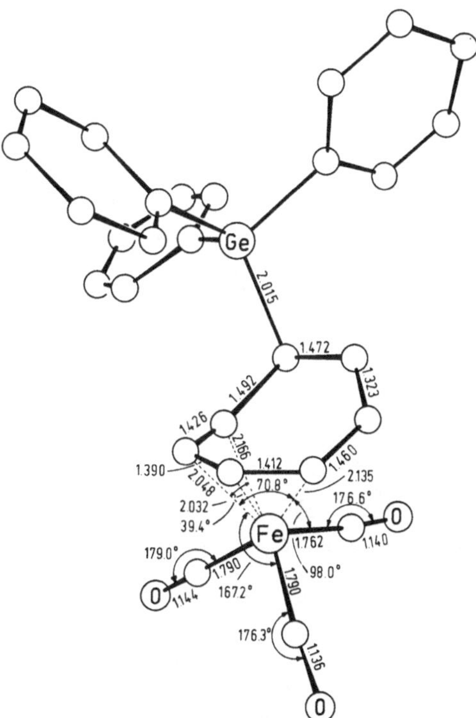

Fig. 12. Molecular structure of $7\text{-}(C_6H_5)_3GeC_7H_7Fe(CO)_3$ (No. 37) [52].

References on pp. 143/4

the unit cell parameters a = 7.0699(7), b = 28.634(2), c = 12.154(2) Å, β = 103.26(1)°; Z = 4 molecules per unit cell. The calculated density is 1.48 g/cm³. The main bond distances and angles are shown in **Fig. 12**, p. 139 [52].

Molecular orbital calculations for $[C_7H_7Fe(CO)_3]^-$ predict interesting consequences for the fluxional behavior of compounds like Nos. 33 and 35 [40].

The compounds No. 33, 35, and 37 are soluble in common organic solvents giving moderately air-sensitive solutions. However, the solid complexes can be handled in air for several days without apparent decomposition, and they can be stored at room temperature under N_2. Hydrogen migrations in No. 35 are seen to occur above 100 °C [52]. Heating Nos. 33 or 37 in toluene for 24 h under an atmosphere of ^{13}CO [52], or No. 35 at 115 °C under an atmosphere of ^{13}CO for 1 h [27] gives no enrichment in ^{13}CO [27, 52]. In this regard it is noted that heating No. 37 in the presence of two equivalents of $P(C_6H_5)_3$ for 3 d in C_6H_6 at 80 °C does not result in any reaction, refluxing in toluene at 112 °C for 3 d gives only an approximate 10% yield of $(CO)_3Fe(P(C_6H_5)_3)_2$ [52]. Reaction of Nos. 34 and 36 with $NaN(Si(CH_3)_3)_2$ gives $Na[C_7H_7Fe(CO)_3]$ and $(C_2H_5)_3MN(Si(CH_3)_3)_2$ (M = Si or Ge) [26]. Deprotonation of No. 37 gives $[(C_6H_5)_3GeC_7H_6Fe(CO)_3]^-$ [29, 43].

7-$CH_3C_7H_7Fe(CO)_3$ (Table **3**, No. **39**). The thermal isomerization of 7-exo-$CH_3C_7H_7Fe(CO)_3$ (No. 39) is examined by ^1H NMR spectroscopy. Although the rate of isomerization is comparable to that of the deuterium-labelled complexes $C_7H_8Fe(CO)_3$ (see pp. 110/1), the initial formation of 1-$CH_3C_7H_7Fe(CO)_3$ (No. 54) indicates that substituent site preferences of the CH_3 group play an important role in determining both, the kinetically and thermodynamically favored products [37]. Mixing 7-$CH_3C_7H_7Fe(CO)_3$ with petrolatum gives a fine paste which shows fungicidal effectiveness as proved by the Agar-plate technique [1].

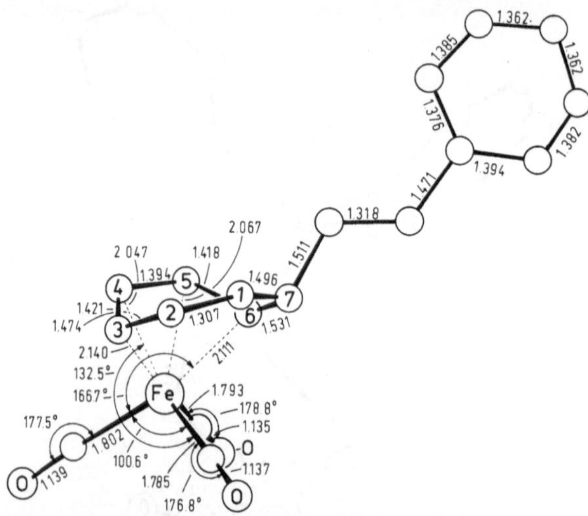

Fig. 13. Molecular structure of 7-C_6H_5CH=$CHC_7H_7Fe(CO)_3$ (No. 42) [48, 58].

7-C_6H_5CH=$CHC_7H_7Fe(CO)_3$ (Table **3**, No. **42**). The unexpected identity of this compound is established by an X-ray diffraction study. Compound No. 42 crystallizes in the monoclinic space group C2/c-C^6_{2h} with the unit cell parameters a = 25.035(8), b = 6.195(2), c = 20.938(12) Å, β = 102.69(4)°; Z = 8 molecules per unit cell. The molecular structure is shown in **Fig. 13** [48, 58]. The C_7 ring comprises two coplanar sections, the first consisting of the four atoms which are bonded to the Fe atom (C(3) to C(6)), the second of the atom sequence

C(1)-C(3), C(6), and C(7). This latter plane makes a dihedral angle of 97° with the plane of the styryl moiety (so that the two ring systems are approximately mutually perpendicular), while the interplanar angle in the C_7 ring is 138°. The main bond distances and angles are shown in Fig. 13. The compound is soluble in all common organic solvents, and the solutions slowly decompose in air. Reaction of No. 42 in ether with a 40% aqueous solution of HBF_4 gives XXVI. Reaction with $(NC)_2C{=}C(CN)_2$ in CH_2Cl_2 gives XXVII [58].

7-$RO_2CC_7H_7Fe(CO)_3$ (Table **3**, Nos. **43** and **44** with R = CH_3, C_2H_5) react with bases to give Nos. 17a and 18, respectively. Reaction with $NaN(Si(CH_3)_3)_2$ in C_6H_6 gives $Na[RO_2CC_7H_6$-$Fe(CO)_3]$ (R = CH_3, C_2H_5) and $HN(Si(CH_3)_3)_2$ [25, 26].

7-$C_6H_5C_7H_7Fe(CO)_3$ (Table **3**, No. **45**) crystallizes in the orthorhombic space group $Pca2_1$-C^5_{2v} with the unit cell parameters a = 12.48(1), b = 14.90(2), c = 7.53(1) Å; Z = 4 molecules per unit cell. The measured density (flotation with KI/H_2O and little detergent) is 1.42 g/cm³ and the calculated density is 1.46 g/cm³. The main bond distances and angles are shown in **Fig. 14** [31].

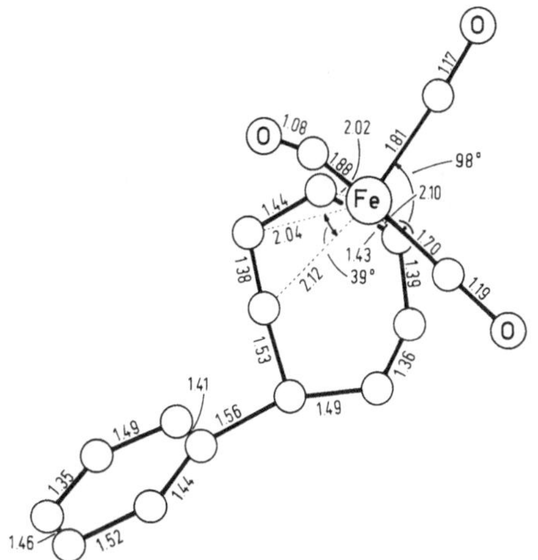

Fig. 14. Molecular structure of 7-$C_6H_5C_7H_7Fe(CO)_3$ (No. 45) [31].

7-$C_7H_7C_7H_7Fe(CO)_3$ (Table **3**, No. **47**) is also prepared by treatment of $(CO)_3$-$FeC_7H_7C_7H_7M(CO)_3$ (M = Fe (see 2.8.1.1 in "Organoiron Compounds" C5, 1981, Table 4, No. 39, p. 51), Cr (No. 48), Mo (No. 49), W (No. 50) with $N(CH_2CH_2NH_2)_3$ [28, 47]. Thus, treatment of No. 49 with $N(CH_2CH_2NH_2)_3$ (ratio 1:1.5) in CH_2Cl_2 at room temperature for 24 h, filtration of the reaction mixture, evaporation of the filtrate, and extraction of the residue with petroleum ether, followed by chromatography on alumina with petroleum ether, gives No. 47 [47]. Compound No. 47 is also obtained together with $Fe_3(CO)_{12}$ and Fe by the reaction of the solvate of the radical anion $[^3LFe(CO)_3]^-$ (^3L = benzylideneacetone) with the cation $[C_7H_7]^+$ in 85% yield [41]. Compound No. 47 reacts with $^3LFe(CO)_3$ (^3L = benzylideneacetone) [28, 47] in C_6H_6 at 65 °C for 48 h to give $(C_7H_7Fe(CO)_3)_2$ (see 2.8.1.1 in "Organoiron Compounds" C5, 1981, Table 4, No. 39, p. 51) [47]. Reaction with $(NC)_2C{=}C(CN)_2$ [18, 49] in CH_2Cl_2 at 20 °C for 0.5 h [49] gives No. 51 [18, 49] in 80% yield [18]. Further reaction with $(NC)_2C{=}C(CN)_2$ gives XXIII [18, 49]. Kinetic experiments confirm the stepwise addition. The measured rate constants in CH_2Cl_2 at 25 °C for both steps are shown in the following table [49]:

References on pp. 143/4

$[(NC)_2C{=}C(CN)_2]$ (mmol/dm^3)	fast step: k_{obs} (s^{-1})	slow step: k_{obs} (s^{-1})
5.0	0.17	1.35×10^{-3}
10.0	0.32	2.48×10^{-3}
15.0	0.52	—
16.8	—	3.75×10^{-3}
20.0	0.72	4.47×10^{-3}
25.1	0.88	5.95×10^{-3}

The second-order rate constant for the first step (fast) is $k_2 = (3.63 \pm 0.11) \times 10^1$ $dm^3 \cdot mol^{-1} \cdot s^{-1}$ and for the second step (slow) $k_2 = (2.2 \pm 0.01) \times 10^{-1}$ at 25 °C. The rates are sufficiently different to explain the synthetic results [49].

7-$(CO)_3MoC_7H_7C_7H_7Fe(CO)_3$ (Table **3**, No. **49**) gives No. 47 on treatment with $N(CH_2CH_2NH_2)_3$ [47].

7-$(NC)_4C_9H_7C_7H_7Fe(CO)_3$ (Table **3**, No. **51**) gives XXIII in the reaction with excess $(NC)_2C{=}C(CN)_2$ [18], cf. [49].

?-(2,4-$C_2H_5)_2C_6H_3C_7H_7Fe(CO)_3$ (Table **3**, No. **52**) is proposed as an additive to fuels and oils, as antiknock in petroleum hydrocarbons, and for metal plating applications [2].

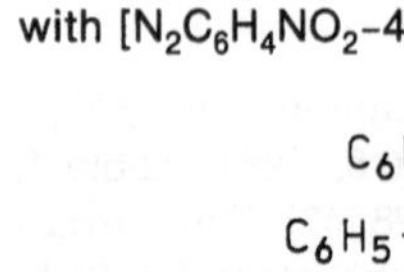

XXV 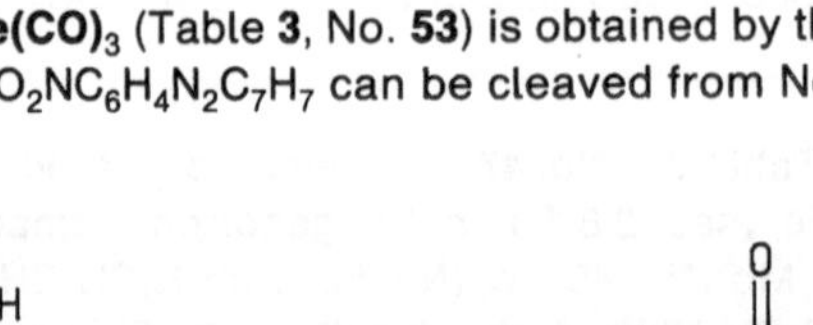 XXVI 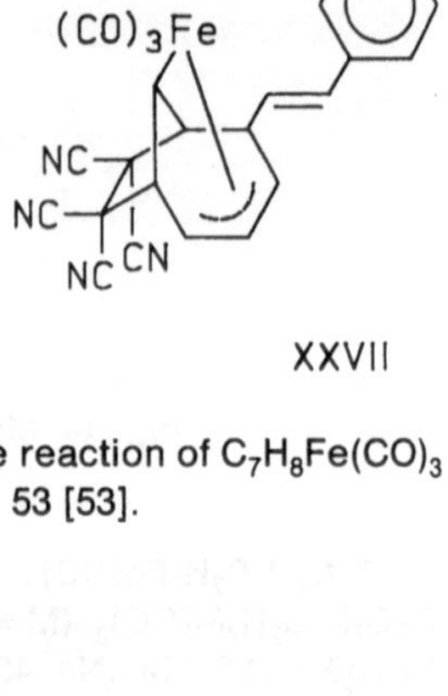 XXVII

?-(4-$O_2NC_6H_4N_2)C_7H_7Fe(CO)_3$ (Table **3**, No. **53**) is obtained by the reaction of $C_7H_8Fe(CO)_3$ with $[N_2C_6H_4NO_2\text{-}4]BF_4$; 4-$O_2NC_6H_4N_2C_7H_7$ can be cleaved from No. 53 [53].

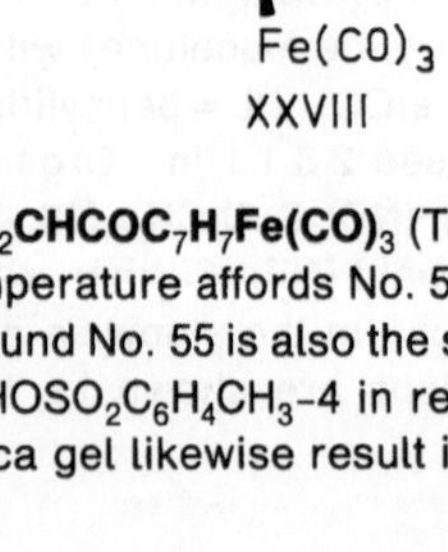

XXVIII 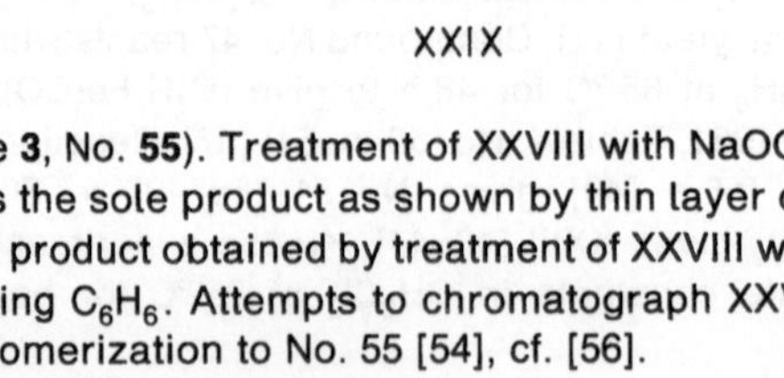 XXIX

1-$(C_6H_5)_2CHCOC_7H_7Fe(CO)_3$ (Table **3**, No. **55**). Treatment of XXVIII with $NaOCH_3$ in CH_3OH at room temperature affords No. 55 as the sole product as shown by thin layer chromatography. Compound No. 55 is also the sole product obtained by treatment of XXVIII with a catalytic amount of $HOSO_2C_6H_4CH_3$-4 in refluxing C_6H_6. Attempts to chromatograph XXVIII over acid washed silica gel likewise result in isomerization to No. 55 [54], cf. [56].

References on pp. 143/4

1-(4-$CH_3OC_6H_4COCH{=}CH)C_7H_7Fe(CO)_3$ (Table **3**, No. **56**) is heated in toluene at 100 °C for 48 h to give XXIV ($R{=}COC_6H_4OCH_3$-4, p. 137) in 30% yield. Refluxing with $Fe_2(CO)_9$ in toluene for 17 h gives XXIV ($R{=}COC_6H_4OCH_3$-4) in 37% yield and its isomeric XXIX [57].

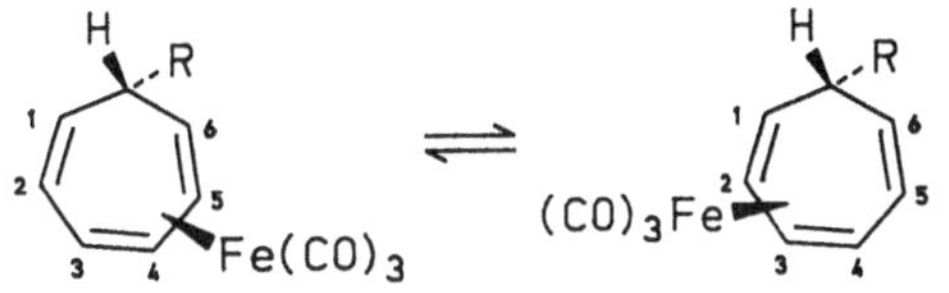

XXX

1-RR′(HO)$CC_7H_7Fe(CO)_3$ (Table **3**, Nos. **57** to **59** with $R{=}CH_3$, $R'{=}C_2H_5$; R=H, $R'{=}C_6H_4CH_3$-4; $R{=}CH_3$, $R'{=}C_6H_4CH_3$-4) can be dehydrated on silica gel to give XIV ($R{=}CH_3$, $R'{=}C_2H_5$; R=H, $R'{=}C_6H_4CH_3$-4; $R{=}CH_3$, $R'{=}C_6H_4CH_3$-4, p. 133) [14].

7-(1-(4-$CH_3)C_6H_4C_7H_6)C_7H_7Fe(CO)_3$ (Table **3**, No. **60**) is prepared from $Fe(CO)_5$ and the free organic ligand [4]L in boiling tetrahydrofuran within 20 h and purified by chromatography. The position of the $Fe(CO)_3$ is at one of the C_7 rings (no details given). The compound decomposes above 400 °C to give iron mirrors. The compound is proposed as a fuel additive or antiknock [59].

References:

[1] W.E. Burt, Ethyl Corp. (U.S. 3123524 [1961/64]). – [2] K.G. Ihrmann, T.H. Coffield, Ethyl Corp. (U.S. 3037999 [1962]). – [3] J.E. Mahler, D.A.K. Jones, R. Pettit (J. Am. Chem. Soc. **86** [1964] 3589/90). – [4] G.F. Emerson, J.E. Mahler, R. Pettit, R. Collins (J. Am. Chem. Soc. **86** [1964] 3590/1). – [5] A.W. Parkins (unpublished results from [6]).

[6] D.A. White (Organometal. Chem. Rev. A **3** [1968] 497/554, 509/19). – [7] J. Lewis, A.W. Parkins (Chem. Commun. **1968** 1194/5). – [8] B.F.G. Johnson, J. Lewis, G.L.P. Randall (J. Chem. Soc. D **1969** 1273/4). – [9] H. Maltz, G. Deganello (J. Organometal. Chem. **27** [1971] 383/7). – [10] B.F.G. Johnson, J. Lewis, P. McArdle, G.L.P. Randall (Chem. Commun. **1971** 177/8).

[11] R. Mason (unpublished results from [12]). – [12] R.J.H. Cowles, B.F.G. Johnson, J. Lewis, A.W. Parkins (J. Chem. Soc. Dalton Trans. **1972** 1768/71). – [13] B.F.G. Johnson, J. Lewis, P. McArdle, G.L.P. Randall (J. Chem. Soc. Dalton Trans. **1972** 456/62). – [14] B.F.G. Johnson, J. Lewis, P. McArdle, G.L.P. Randall (J. Chem. Soc. Dalton Trans. **1972** 2076/83). – [15] E.J. Forbes, M.K. Sultan, P.C. Uden (Anal. Letters **5** [1972] 927/30).

[16] D.F. Hunt, J.W. Russell, R.L. Torian (J. Organometal. Chem. **43** [1972] 175/83). – [17] G.T. Rodeheaver (Diss. Univ. Virginia 1973; Diss. Abstr. Intern. B **34** [1973] 1433). – [18] P. McArdle (J. Chem. Soc. Chem. Commun. **1973** 482/3). – [19] R.C. Kerber, D.J. Ehntholt (J. Am. Chem. Soc. **95** [1973] 2927/34). – [20] B.A. Kelly (Diss. Boston Coll. 1973; Diss. Abstr. Intern. B **34** [1973] 1031).

[21] A.M. Brodie, B.F.G. Johnson, J. Lewis (J. Chem. Soc. Dalton Trans. **1973** 1997/2002). – [22] P. McArdle, H. Sherlock (Proc. 16th Intern. Conf. Coord. Chem., Dublin 1974, Abstr. No. R 89). – [23] G. Deganello, T. Boschi, L. Toniolo (7th Intern. Conf. Organometal. Chem., Venice 1975, Abstr. No. 40). – [24] G. Deganello, T. Boschi, L. Toniolo (J. Organometal. Chem. **97** [1975] C46/C48). – [25] H. Behrens, K. Geibel, R. Kellner, M. Moll, P. Würstl (7th Intern. Conf. Organometal. Chem., Venice 1975, Abstr. No. 30).

[26] H. Behrens, K. Geibel, R. Kellner, H. Knöchel, M. Moll, E. Sepp (Z. Naturforsch. **31** [1976] 1021/2). – [27] L.K.K. LiShingMan, J. Takats (J. Organometal. Chem. **117** [1976] C104/C110). – [28] G. Deganello, T. Boschi (Chim. Ind. [Milan] **58** [1976] 654). – [29] J.

Takats, J.G. Reuvers (Abstr. Papers Joint Conf. Chem. Inst. Can. Am. Chem. Soc. **2** [1977] C No. 90). – [30] P.L. Pauson (unpublished results from [31]).

[31] J.A.D. Jeffreys, C. Metters (J. Chem. Soc. Dalton Trans. **1977** 729/33). – [32] Z. Goldschmidt, Y. Bakal (Tetrahedron Letters **1977** 955/8). – [33] D. Cunningham, P. McArdle, H. Sherlock, B.F.G. Johnson, J. Lewis (J. Chem. Soc. Dalton Trans. **1977** 2340/4). – [34] J. Takats (unpublished results from [35]). – [35] K.J. Karel, M. Brookhart (J. Am. Chem. Soc. **100** [1978] 1619/21).

[36] A. Pryde (J. Chromatog. **152** [1978] 123/9). – [37] K.J. Karel (Diss. Princeton Univ. 1978; Diss. Abstr. Intern. B **39** [1979] 4353). – [38] V.W. Day (unpublished results from [47]). – [39] M.I. Foreman, G.R. Knox, D.G. Leppard, P.L. Pauson, A. Pryde, P.J. Walker, W.E. Watts (unpublished results from [36]). – [40] P. Hofmann (Z. Naturforsch. **33b** [1978] 251/60).

[41] N. El Murr, M. Riveccie, P. Dixneuf (J. Chem. Soc. Chem. Commun. **1978** 552/3). – [42] C.S. Day, V.W. Day, G. Deganello, L.K.K. LiShingMan, J.G.A. Reuvers, J. Takats (unpublished results from [43]). – [43] J.G.A. Reuvers, J. Takats (J. Organometal. Chem. **175** [1979] C13/C16). – [44] Z. Goldschmidt, Y. Bakal (J. Organometal. Chem. **168** [1979] 215/25). – [45] Z. Goldschmidt, Y. Bakal (J. Organometal. Chem. **179** [1979] 197/204).

[46] M. Cavazza, G. Morganti, F. Pietra (J. Org. Chem. **45** [1980] 2001/4). – [47] M. Airoldi, G. Deganello, G. Dia, P. Saccone, J. Takats (Inorg. Chim. Acta **41** [1980] 171/8). – [48] K. Broadley, N.G. Connelly, R.M. Mills, M.W. Whiteley, P. Woodward (J. Chem. Soc. Chem. Commun. **1981** 19/20). – [49] S.K. Chopra, M.J. Hynes, P. McArdle (J. Chem. Soc. Dalton Trans. **1981** 586/9). – [50] P. Hackett, B.F.G. Johnson, J. Lewis (J. Chem. Soc. Dalton Trans. **1982** 1253/9).

[51] K.J. Karel, T.A. Albright, M. Brookhart (Organometallics **1** [1982] 419/30). – [52] L.K.K. LiShingMan, J.G.A. Reuvers, J. Takats, G. Deganello (Organometallics **2** [1983] 28/39). – [53] N.G. Connelly, A.R. Lucy, M.W. Whiteley (J. Chem. Soc. Chem. Commun. **1979** 985/6). – [54] Z. Goldschmidt, S. Antebi (J. Organometal. Chem. **206** [1981] C1/C3). – [55] I.L. Fragalà, J. Takats, M.A. Zerbo (Organometallics **2** [1983] 1502/4).

[56] Z. Goldschmidt, S. Antebi (J. Organometal. Chem. **259** [1983] 119/25). – [57] Z. Goldschmidt, Y. Bakal, D. Hoffer, A. Eisenstadt, P.F. Lindley (J. Organometal. Chem. **258** [1983] 53/62). – [58] K. Broadley, N.G. Connelly, R.M. Mills, M.W. Whiteley, P. Woodward (J. Chem. Soc. Dalton Trans. **1984** 683/8). – [59] G. Wilkinson (Brit. 894852 [1962]).

1.4.1.4.1.2.4.2.4.2 ^{4}L is a Disubstituted Cycloheptatriene

These complexes are arranged in Table 4, pp. 148/63, as follows:

Complexes of the type 1-R(7-R′) $C_7H_6Fe(CO)_3$ are described as Nos. 1 to 12, 53, 54 followed by 7-R(7-R′)$C_7H_6Fe(CO)_3$ complexes as Nos. 13 to 25. Complexes of type $X{=}C_7H_6Fe(CO)_3$ with X=O or NR are arranged as Nos. 26 to 34 and 56, followed by complexes of type $RCH{=}C_7H_6Fe(CO)_3$ as Nos. 35 to 40. The compounds No. 41 to 51 are of the type $RR'C{=}C_7H_6Fe(CO)_3$, No. 52 is an ionic complex, and No. 55 is of the type 1-R(2-R′)$C_7H_6Fe(CO)_3$. Many of the complexes in the following Table 4 can be prepared by the following methods:

Method I: $Fe(CO)_5$ reacts with an organic substrate.

a. $Fe(CO)_5$ reacts photochemically with the organic ligand ^{4}L. Thus, irradiation of $Fe(CO)_5$ with 7-phenyl-7-carbomethoxycycloheptatriene (as a ca. 2:1 mixture of norcaradiene to cycloheptatriene isomers) in C_6H_6 [60, 81] for 22 h, followed by work-up as in Method IVa, gives No. 18 [81]. Irradiation with tropone in toluene gives No. 26 [56]. Similarly, $Fe(CO)_5$ and 7-methyl-7-carbomethoxy-norcaradiene are photolyzed in C_6H_6 for 8 to 12 h. After quick filtration of the reaction mixture through a pad of Celite and alumina, the solvent is removed, and the residue is chromatographed by medium pressure liquid chromatography using hexane as eluent. The major isomer No. 14 (ca. 40%) is isolated from the first yellow band, and a second band eluted with hexane/C_6H_6 (2:1) gives the minor isomer No. 14 (21%). The last fraction contains 7-CH_3O_2C(7-CH_3)$C_7H_6Fe_2(CO)_6$ [81].

b. $Fe(CO)_5$ and the organic ligand ^4L are refluxed in tetrahydrofuran for 20 h. Filtration and evaporation of the filtrate followed by chromatography gives No. 5 [8].

c. $Fe(CO)_5$ reacts with HC≡CH (11 to 21 atm) in petroleum ether or C_6H_6 in an N_2 atmosphere (21 to 31 atm) at 60 to 70 °C to give scarcely soluble dark brown complexes (2%), $OC_5H_4Fe(CO)_3$ (see 1.4.1.4.1.2.2.2.3.1 in "Organoiron Compounds" B7, 1981, p. 145), No. 26, $C_4H_4Fe_2(CO)_6$ (see 2.4.1.2.2.1 in "Organoiron Compounds" C3, 1980, Table 3, No. 1, p. 28), $2OC_5H_4Fe(CO)_3 \cdot HO$-$C_6H_4OH$-p (see 1.4.1.4.1.2.2.2.3.1 in "Organoiron Compounds" B7, 1981, p. 149) and tropone in small yields. The products are separated by chromatography on acidic or neutral Al_2O_3 (Woelm) [10].

Method II: $Fe_2(CO)_9$ reacts with an organic substrate.

a. $Fe_2(CO)_9$ and the organic ligand ^4L are refluxed in an inert solvent or without a solvent. Thus, excess $Fe_2(CO)_9$ reacts with tropone to give No. 26a [10, 15, 19, 27, 38, 82] in high yield [19] up to 65% [38] and $OC_7H_6Fe_2(CO)_6$ (see Formula I) [82]. Reaction in C_6H_6 in a sealed tube [10, 15] at 60 °C [10] or 75 to 80 °C [15] for 6 h [10, 15] followed by filtration, evaporation of the filtrate, and chromatography of the residue [10, 15] on neutral [15] Al_2O_3 [10, 15] with pentane [15] or $CH_3CO_2C_2H_5$/ether [10] gives No. 26a [10, 15] in 14 [10] or 19% yield [15]. Refluxing the components in ether in the dark for 12 h gives No. 26a in 62% yield after recrystallization from hexane at −78 °C [27]. Refluxing in C_6H_6 followed by chromatography on silica gel gives OC_7H_6-$Fe_2(CO)_9$ (see Formula I) in ca. 15% yield and No. 26a (ca. 55%) [82]. Similarly, $Fe_2(CO)_9$ and azaheptafulvenes II (R = C_6H_4R' with R' = Cl, Br, OCH_3, CH_3, cyclo-C_7H_6) are refluxed in ether for 5 h and then left overnight. Filtration of the mixture and evaporation of the solvent affords an oily residue which is crystallized from petroleum ether to give pure compounds No. 28 to 32 as mixtures of (E) and (Z) isomers (one isomer prevails over the other by a factor of 1.5). Compound No. 30 is purified by chromatography on silica gel (Merck GF_{254}) eluting with cyclohexane/$CH_3CO_2C_2H_5$ or cyclohexane/C_6H_6. Attempted analogous reaction with II (R = CH_3) fails [71]. Similar reaction with

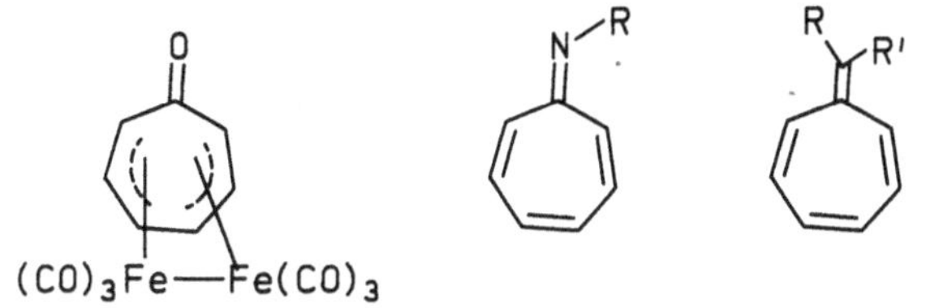

O

$(CO)_3Fe$—$Fe(CO)_3$

I

N—R

II

R, R'

III

References on pp. 177/9

III ($R=R'=N(CH_3)_2$) in tetrahydrofuran at room temperature for 20 h, followed by filtration and evaporation gives a dark red oily residue which crystallizes at 0 °C. Chromatography on silica gel with acetone gives No. 43 [78].

b. $Fe_2(CO)_9$ reacts with HC≡CH (21 to 25 atm) at 20 to 25 °C to give No. 26b [1, 2, 9] in 30% yield [2]. Similar reaction in petroleum ether at ca. 20 °C gives $Fe(CO)_5$, No. 26b (main product), traces of $C_4H_4Fe_2(CO)_6$ (see 2.4.1.2.2.1 in "Organoiron Compounds" C3, 1980, Table 3, No. 1, p. 28), and $C_{10}H_{10}Fe_2(CO)_6$ (see 2.8.1.1 in "Organoiron Compounds" C5, 1981, Table 4, No. 27, p. 50), and a compound of the composition $(C_2H_2)_{2.7}Fe(CO)_{3.9}$. Reaction with 15 to 17 atm HC≡CH at 0 to 90 °C gives a maximum yield of No. 26b (22%) at 20 to 25 °C together with $(C_2H_2)_{2.7}Fe(CO)_{3.9}$ (40%). Reaction with 25 atm HC≡CH at 20 to 25 °C gives No. 26b in 28% yield [10]. The product mixture is chromatographed on acidic [9, 10], neutral [10], or basic [9] Al_2O_3 (Woelm) [9, 10] with $CH_3CO_2C_2H_5$, $CH_3CO_2C_2H_5/C_6H_6$, or $CH_3CO_2C_2H_5$/ether as eluents [9]. Elution with petroleum ether separates $Fe(CO)_5$, further elution with C_6H_6 gives $C_4H_4Fe_2(CO)_6$ and $C_{10}H_{10}Fe_2(CO)_6$, and $CH_3CO_2C_2H_5$ elutes No. 26b [10].

Method III: $Fe_3(CO)_{12}$ reacts with an organic substrate.

a. Reaction of $Fe_3(CO)_{12}$ with tropone at 80 °C gives No. 26a and b together with $C_4H_2O_3Fe(CO)_4$ (see 1.2.1.2.3.1.4.2 in "Eisen-Organische Verbindungen" B4, 1978, Table 29, No. 8, p. 303), a yellow compound (m.p. 152 to 153 °C) and a dark brown complex in small quantities (not identified) [1]. Similar reaction in petroleum ether at 90 °C for 3 h gives No. 26 in 5% yield [10]. Refluxing in C_6H_6 for 7 h followed by chromatography on alumina with C_6H_6 and sublimation gives No. 26 in 7.4% yield [11].

b. $Fe_3(CO)_{12}$ reacts with HC≡CH at atmospheric pressure [1, 10], but higher pressures give better yields [1]. At 15 to 23 atm and 45 to 50 °C scarcely soluble $(C_2H_2)_{1.9}Fe(CO)_{3.5}$ is isolated in 50 to 60% yield together with $Fe(CO)_5$ [10], $C_4H_4Fe_2(CO)_6$ (see 2.4.1.2.2.1 in "Organoiron Compounds" C3, 1980, Table 3, No. 1, p. 28), IV, $C_6H_6Fe_2(CO)_6$ (see 2.6.1.2.2 in "Organoiron Compounds" C5, 1981, Table 1, No. 1, p. 9), modification a and b of No. 26 [1, 10], $C_{10}H_{10}Fe_2(CO)_6$ (see 2.8.1.1 in "Organoiron Compounds" C5, 1981, Table 4, No. 27, p. 50), $C_{10}H_{10}Fe_3(CO)_{10}$, $C_4H_2O_3Fe(CO)_4$ (see 1.2.1.2.3.1.4.2 in "Eisen-Organische Verbindungen" B4, 1978, Table 29, No. 8, p. 303), $C_6H_6Fe_2(CO)_7$, $C_6H_6Fe_3(CO)_8$, and $(C_2H_2)_4Fe(CO)_3$ in a total yield of 22% [10]. The products are chromatographed as described in Method IIb. Elution with $CH_3CO_2C_2H_5$/ether gives Nos. 26a and b and $C_{10}H_{10}Fe_3(CO)_{10}$, elution with CH_3OH gives $C_4H_2O_3Fe(CO)_4$ (0.2%), and elution with C_6H_6/petroleum ether gives C_9H_8O-$Fe(CO)_3$ together with small amounts of $(C_2H_2)_{1.9}Fe(CO)_{3.5}$ [10]. No. 26 is formed in better yields by Method IIb [1].

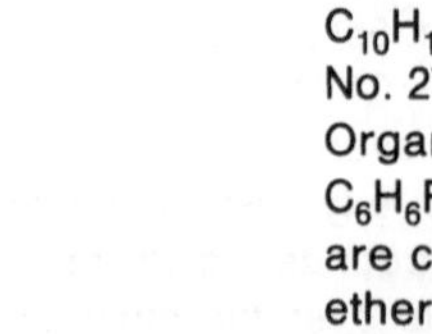

$(CO)_3Fe$ … $Fe(CO)_3$ IV

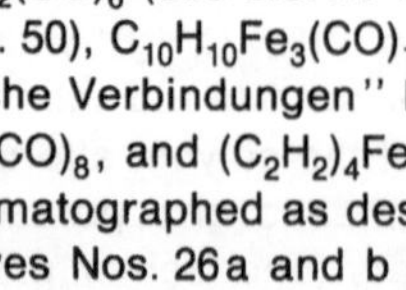

R … $Fe(CO)_3$ V

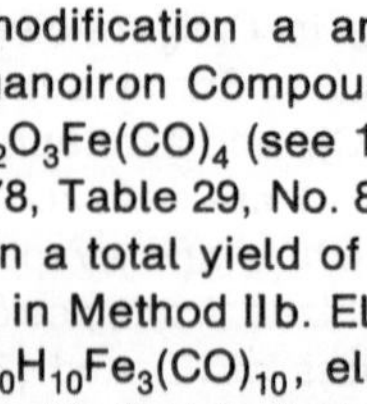

CH_2OH … $Fe(CO)_3$ VI

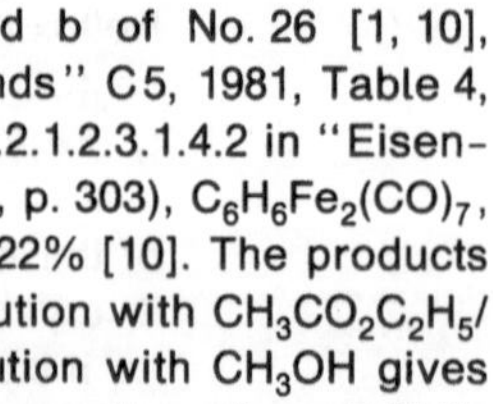

R' R … $Fe(CO)_3$ X^- VII

Method IV: $^3LFe(CO)_3$ (^3L = benzylideneacetone) or Na[$^3LFe(CO)_3$] ($^3L = CH_3O_2CC_7H_6$) react with an organic substrate.

a. $^3LFe(CO)_3$ reacts with III (R = R′ = CH_3) in toluene at 50 °C for 6 h. Chromatography on silica gel gives No. 44. Analogous reaction with III (R = R′ = C_6H_5) gives No. 45 [29]. Similar reaction of $^3LFe(CO)_3$ with 7-phenyl-7-carbomethoxycycloheptatriene (as a ca. 2:1 mixture of norcaradiene to cycloheptatriene) [60, 81] in C_6H_6 at 52 to 53 °C for 72 h gives No. 18 [81]. The reaction mixture is filtered, and the filtrate is refluxed for 24 h. Filtration, removal of the solvent, and chromatography on Florisil with C_6H_6 gives pure No. 18 [81].

b. Na[$CH_3O_2CC_7H_6Fe(CO)_3$] reacts with $ClCO_2CH_3$ to give No. 4 [43].

Method V: Complexes of the type $^4LFe(CO)_3$ as the starting compounds.

a. To a solution of V (R = CHO) in CH_3OH is added $HC(OCH_3)_3$ and two drops of conc. HCl. After 24 h at room temperature H_2O is added and the mixture is extracted with ether. The organic layer is washed with 5% aqueous Na_2CO_3, then H_2O, and dried ($MgSO_4$). Removal of the solvent and chromatography of the residue on silica gel with hexane affords an uncrystallizable oily 3:1 mixture of Nos. 35a and b in 58% yield [66, 79]. No. 36 is obtained similarly from V (R = CHO) and $HC(OC_2H_5)_3$. Similar reaction of V (R = COC_6H_5) and $HC(OCH_3)_3$ gives a 4:1 mixture of the two isomers No. 48a and b in 60% yield [66].

b. V (R = CHO) reacts with $(C_6H_5)_3P{=}CH_2$ (from [$(C_6H_5)_3PCH_3$]Br and LiC_4H_9 in hexane at 0 °C) in tetrahydrofuran for 2 h. The mixture is quenched with H_2O and extracted with ether. The organic layer is washed with H_2O, dried ($MgSO_4$), and the solvent is removed. The residue is chromatographed on silica gel with hexane to give No. 37 as an 1:1 mixture of isomers [65].

c. Dehydration of V (R = $CHOHCH_2C_6H_5$) on silica gel gives No. 38 [23], and dehydration of V (R = $CHOHC_6H_5$) gives a mixture of isomers No. 39a and b [23]. No. 39b is the major isomer [81]. Similarly, VI in toluene is placed on top of a silica gel column; after 15 h, toluene elutes XII (R = R′ = H, p. 164) and No. 41 [26].

d. VIII (p. 164) reacts with excess $LiAlH_4$ in ether at −78 °C. The reaction mixture is allowed to warm to room temperature, and H_2O is added followed by dilute HCl (1:4). The organic layer is washed with dilute NaCl, dried, and chromatographed on basic alumina to give No. 1 [33].

e. $^4LFe(CO)_3$ (^4L = methylcyclooctatetraene) and [C_7H_7]BF_4 in acetone at −23 °C give an orange solution from which No. 2 may be obtained [73, 88].

Method VI: The salt VII (R = H, R′ = CH_3) reacts with $N(C_2H_5)_3$ in acetone. The reaction mixture is extracted with ether/H_2O, and the ethereal layer is washed with H_2O, dried ($MgSO_4$), and distilled to give No. 37 [26]. Similar reaction of VII (R = H, R′ = $CH_2C_6H_5$ [23] or R = CH_3, R′ = C_2H_5 [26]) gives Nos. 38 [23] and 49 [26]. Similar reaction of VII (R = R′ = H, X = BF_4) in CH_2Cl_2 followed by evaporation and chromatography on silica gel with pentane or CS_2 gives No. 41 [24, 30]. VII (R = H, R′ = C_6H_5) reacts with excess $N(C_2H_5)_3$ in acetone. All volatiles are evaporated in vacuum, and the residue is extracted with toluene and chromatographed on alumina with toluene to give a mixture of isomers No. 39a and b in 96% yield [26]. Isomer No. 39b is the major isomer [81]. Similar reaction of VII (R = H, R′ = $C_6H_4CH_3$-4 or R = CH_3, R′ = $C_6H_4CH_3$-4) gives a mixture of isomers No. 40 or an 1:1 mixture of isomers No. 51 [26].

References on pp. 177/9

Table 4
Compounds of the Type $RR'C_7H_6Fe(CO)_3$.
Further information on numbers preceded by an asterisk is given at the end of the table, pp. 164/77.
For abbreviations and dimensions see p. VIII.

No.	$RR'C_7H_6Fe(CO)_3$	method of preparation (yield in %), properties and remarks	Ref.
compounds of the type 1-R(7-R')$C_7H_6Fe(CO)_3$ (continued with No. 53):			
*1	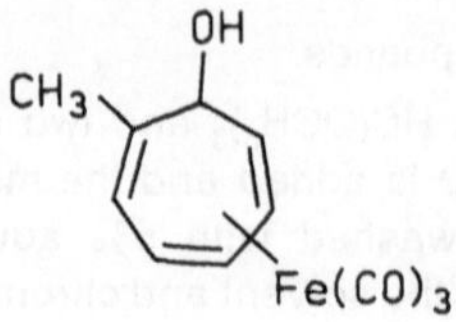	Vd (64) yellow oil ^{1}H NMR ($CDCl_3$): 1.32 (s, 1H), 1.68 (d, 3H), 3.1 (d, 1H), 3.55 (m, 2H), 5.4 (quint, 2H), 5.65 (dd, 1H; J=9)	[33]
2	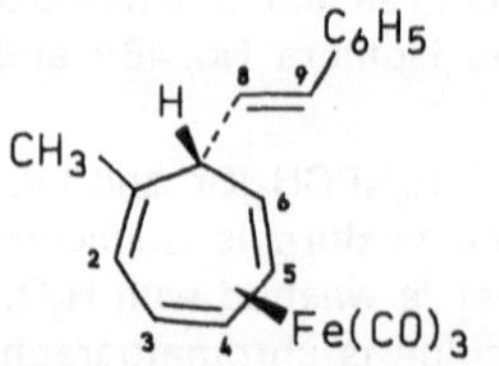	Ve (5) yellow crystals m.p. 60 to 64° ^{1}H NMR ($CDCl_3$): 1.49 (dd, CH_3; J(H-2, CH_3)=1.4, J(H-7, CH_3)=0.8), 3.02 (m, H-3, CH_3; J(H-6,7)=5, J(H-2,3)=8, J(H-7,8)=9), 3.22 (m, H-6), 5.32 (m, H-4,5), 5.72 (m, H-2,8), 6.45 (d, H-9; J(H-8,9)=15), 7.30 (m, C_6H_5) ^{13}C NMR ($CDCl_3$): 22.74 (CH_3), 50.56 (C-7); 56.80, 65.25 (C-3,6), 86.24, 94.10 (C-4,5); 124.64, 126.14, 127.24, 127.76, 128.54, 131.34, 134.71, 134.91, 137.24 (C-2,3,8,9, C_6H_5); 210.92 (CO) IR (n-hexane): 1974, 1986, 2048 (CO) mass spectrum: $[M-CO]^+$	[73, 88]
3		soluble in all common organic solvents see No. 6, p. 164 m.p. 175° (dec.), yellow crystals (from acetone/hexane) ^{1}H NMR (acetone-d_6): 3.26 (m, H-3,6), 5.44 (d, H-7; J=5), 6.18 (m, H-4,5), 7.4 (m, H-2, C_6H_5), 9.10 (s, CHO) ^{13}C NMR (acetone-d_6): 53.5 (C-6), 57.2 (C-7), 58.5 (C-3), 92.2 (C-5), 99.4 (C-4); 126.7, 128.5, 129.6, 133.4 (C_6H_5); 132.4 (C-1); 154.2, 155.0 (C=O); 161.7 (C-2), 191.5 (CHO), 209.9 ($Fe(CO)_3$) IR (Nujol): 1685, 1760, 1980, 1995, 2055 mass spectrum: $[M-C_8H_6N_3O_2-nCO]^+$ (n=1 to 3)	[66]

Table 4 [continued]

No.	$RR'C_7H_6Fe(CO)_3$	method of preparation (yield in %), properties and remarks	Ref.
4	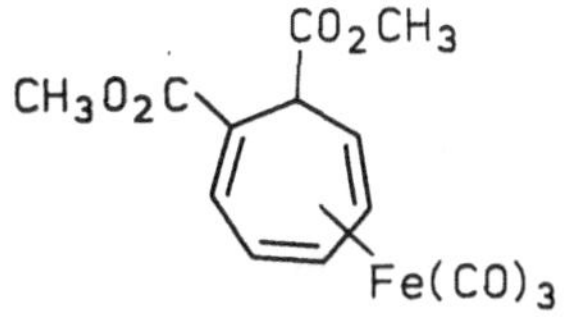	IVb IR: 1648 (C=C); 1712, 1744 (C=O) mass spectrum: $[M-nCO]^+$ (n=0 to 3), $[M-O(CH_3)_2-nCO]^+$ (n=1 to 3)	[43]
*5	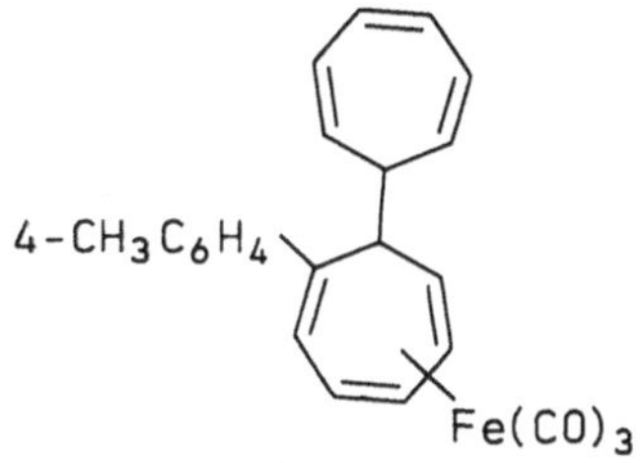	Ib	[8]
*6	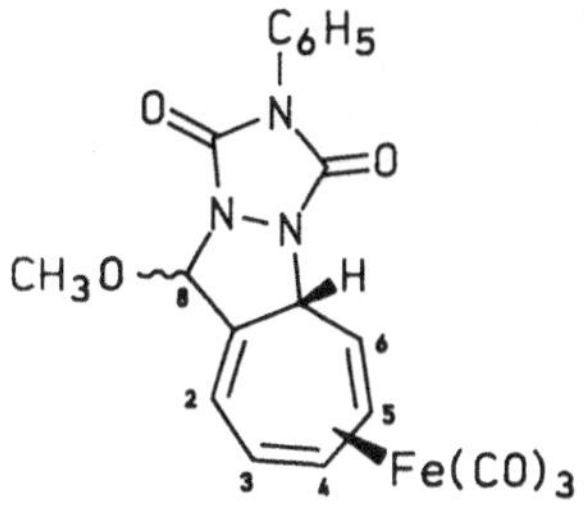	see No. 35, p. 172 m.p. 125° (dec.), yellow crystals (from CH_2Cl_2/hexane) ^1H NMR ($CDCl_3$): 2.64 (t, H-3; J=7), 3.34 (s, OCH_3), 3.54 (m, H-6), 5.06 (s, H-8), 5.14 (m, H-7), 5.66 (m, H-4, 5), 5.98 (dd, H-2; J=7, J=2), 7.4 (m, C_6H_5) IR (Nujol): 1710, 1770, 2005, 2040 mass spectrum: $[M-C_8H_5N_3O_2-nCO]^+$ (n=0 to 3)	[66]
*7a	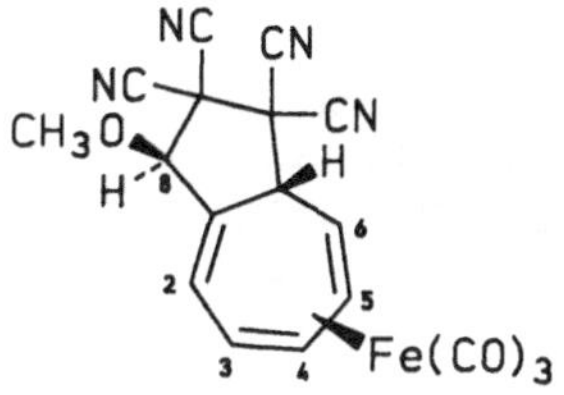	see No. 35, p. 172 a 4:5 mixture of Nos. 7a and b: m.p. 128° (dec., from CH_2Cl_2/hexane) ^1H NMR ($CDCl_3$): 2.97 (m, H-3, 6), 3.68 (s, H-7, OCH_3), 4.33 (s, H-8), 5.74 (m, H-4, 5), 6.33 (ddd, H-2; J=8, J=2, J=2) IR (Nujol): 1998, 2005, 2055 IR (CH_2Cl_2): 1997.8, 2064.0 (CO) mass spectrum: $[M-(NC)_2C=C(CN)_2-nCO]^+$ (n=0 to 3)	[66, 79]
b	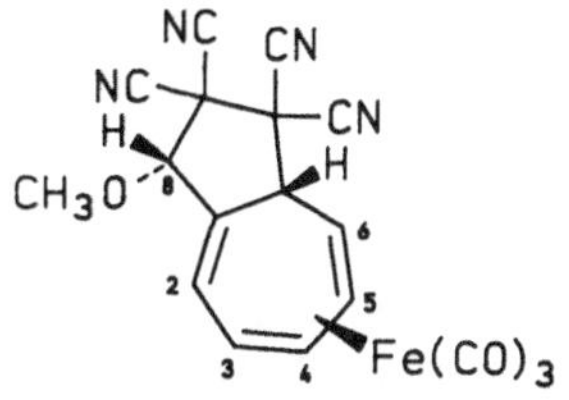	see No. 35, p. 172 a 4:5 mixture of Nos. 7a and b: m.p. 128° (dec., from CH_2Cl_2/hexane) ^1H NMR ($CDCl_3$): 2.97 (m, H-3, 6), 3.46 (s, OCH_3), 3.90 (m, H-7), 4.62 (br, s, H-8), 5.74 (m, H-4, 5), 6.33 (ddd, H-2; J=8, J=2, J=2) IR (Nujol): 1998, 2005, 2055 IR (CH_2Cl_2): 1997.8, 2064.0 (CO) mass spectrum: $[M-(NC)_2C=C(CN)_2-nCO]^+$ (n=0 to 3)	[66, 79]

Table 4 [continued]

No.	$RR'C_7H_6Fe(CO)_3$	method of preparation (yield in %), properties and remarks	Ref.
8a	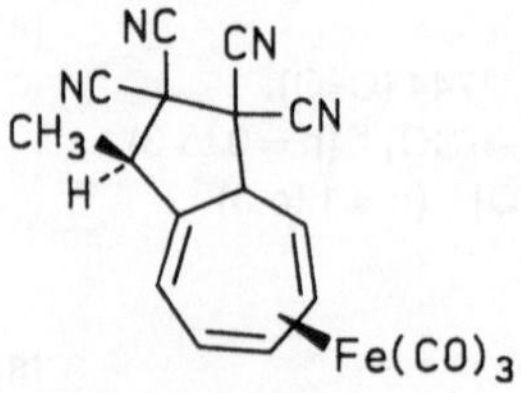	see No. 37, p. 173 mixture of isomers No. 8a and b: m.p. 139° (dec., from CH_2Cl_2/hexane) ^1H NMR ($CDCl_3$): 1.37, 1.47 (pair of d, CH_3, ratio ca. 1:3), 2.80 to 3.50 (m, 3H), 3.70 (m, 1H), 5.66 (m, 2H), 5.94 (m, 1H) IR (Nujol): 1985, 1990, 2025 mass spectrum: $[M-nCO]^+$ (n=0 to 3)	[59, 65]
b	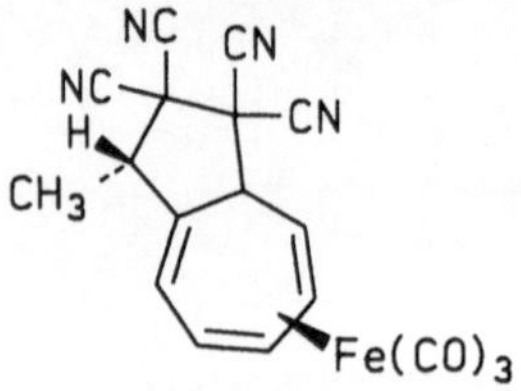		
*9	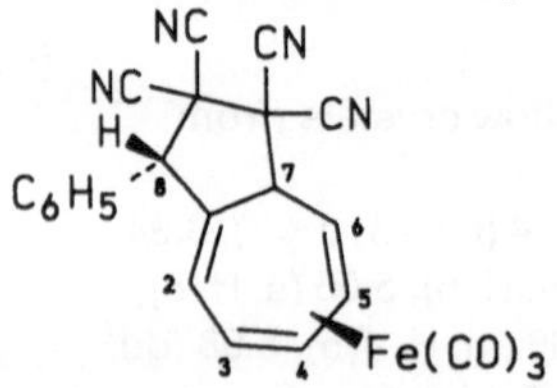	see No. 39, p. 173 ^1H NMR (acetone-d_6): 3.15 (t, H-3; J=8), 3.34 (m, H-6), 4.18 (m, H-7), 4.99 (t, H-8; J(H-7,8)=2.5), 5.72 (m, H-2), 6.01 (m, H-4,5), 7.28 (m, C_6H_5) ^{13}C NMR (acetone-d_6): 48.2 (C-6), 50.9 (C-3), 56.8 (C-7), 59.5 (C-8), 90.70 (C-5), 95.6 (C-4), 134.8 (C-2); 129.6, 130.4 (C_6H_5) IR (CH_2Cl_2): 1995.7, 2061.5 (CO) UV (CH_3NO_2): $\lambda_{max}(\varepsilon)=420$ (29200) nm	[59, 79]
10a	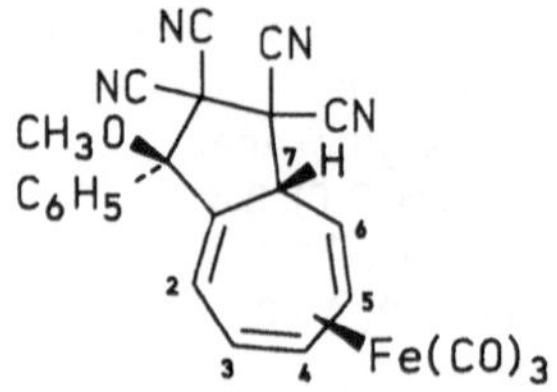	see No. 48, p. 175 mixture of isomers No. 10a and b: m.p. 146° (dec., from CH_2Cl_2/hexane) ^1H NMR (acetone-d_6): 3.16 (s, OCH_3), 3.26 (m, H-3), 3.45 (m, H-6), 4.43 (m, H-7), 6.08 (m, H-4,5), 6.38 (dd, H-2; J=8, J=3), 7.56 (m, C_6H_5) IR (Nujol): 1995, 2003, 2068 mass spectrum: $[M-(NC)_2C{=}C(CN)_2-nCO]^+$ (n=0 to 3)	[66]
b	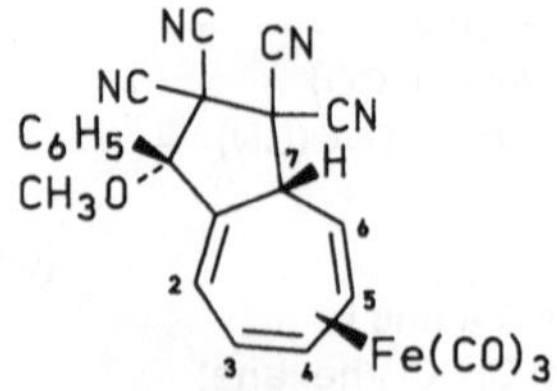		
11	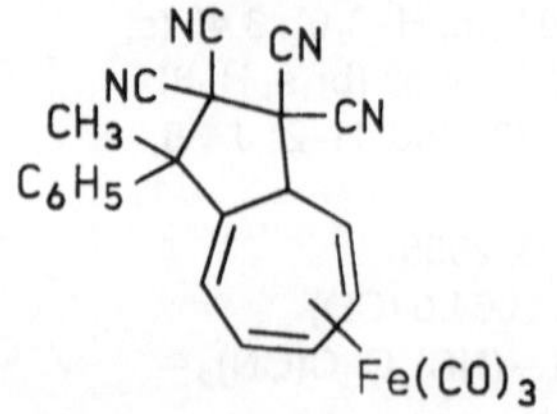	see No. 50, p. 176 ^1H NMR (acetone-d_6): 2.32 (s, CH_3), 3.15 (t, H-3; J=8), 3.34 (m, H-6), 4.18 (m, H-7), 5.72 (m, H-2), 6.01 (m, H-4,5), 7.28 (m, C_6H_5)	[59]

Table 4 [continued]

No.	$RR'C_7H_6Fe(CO)_3$	method of preparation (yield in %), properties and remarks	Ref.
*12	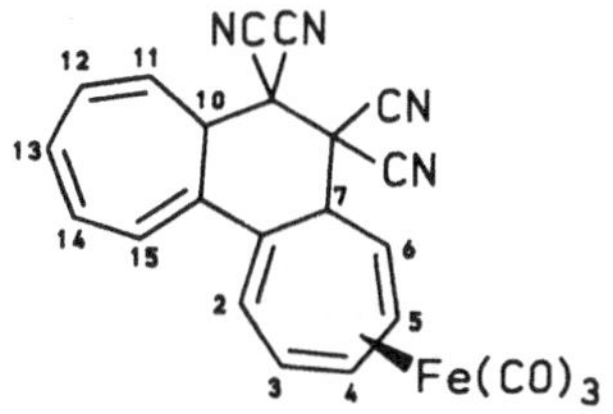	see No. 46, p. 175 m.p. 182 to 183° (dec.), yellow crystals ^{1}H NMR ($CDCl_3$): 2.16 (d, H-10; J=7.0), 2.99 (m, H-3), 3.30 (m, H-6), 3.63 (d, H-7; J=4.0), 5.46 to 5.72 (m, H-4,5,11), 6.20 (d, H-2; J=8.0), 6.25 to 6.5 (m, H-12,15), 6.79 (m, H-13,14) IR (KBr): 1940, 1996, 2060 (CO); 2240 (CN) UV (C_2H_5OH): λ_{max}(log ε)=352 (4.08) nm mass spectrum: $[M-nCO]^+$ (n=0 to 3)	[82]
compounds of the type 7-R(7-R')$C_7H_6Fe(CO)_3$:			
*13	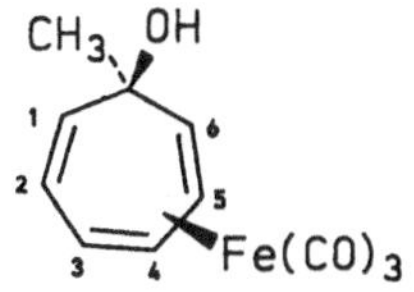	see No. 26, p. 170 m.p. 55 to 56° (from hexane), pale yellow crystals ^{1}H NMR ($CDCl_3$): 1.35 (s, 3H), 2.95 (br, t, 1H), 3.33 (dd, 1H; J=7.5), 5.03 (ddd, 1H), 5.33 (dt, 1H), 5.7 (dd, 1H; J=11, J=8.5) ^{1}H NMR ($CDCl_3$): 1.42 (CH_3), 1.85 (OH), 3.07 (t, H-3; J(H-3,4)=8.5), 3.46 (d, H-6; J(H-1,6)=1.1), 5.20 (td, H-1; J(H-1,2)=10.5), 5.48 (m, H-4,5; J(H-5,6)=8.2), 5.85 (dd, H-2; J(H-2,3)=8.5) ^{13}C NMR ($CDCl_3$): 32.4 (CH_3), 54.1 (C-3), 68.6 (C-7), 72.6 (C-6), 83.7 (C-5), 94.9 (C-4), 127.5 (C-2 or 1), 132.3 (C-1 or 2), 210.4 (CO) IR (hexane): 1985, 2048 (CO) IR ($CDCl_3$): 1100, 1264, 1410; 3590 (OH) mass spectrum: $[M-nCO]^+$ (n=0 to 3)	[33, 67]
*14	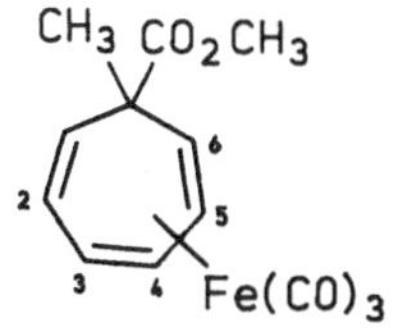	Ia major isomer (ca. 40% yield): yellow oil ^{1}H NMR ($C_6D_5CD_2H$): 1.45 (s, CH_3), 2.79 (tt, H-3; J(H-3,4)=J(H-2,3)=7.5), 3.60 (s, OCH_3), 3.69 (dt, H-6; J(H-5,6)=8), 5.03 (m, H-4,5; J(H-3,5)=J(H-1,3)=1.1), 5.51 (ddd, H-1; J(H-1,6)=2), 5.80 (dd, H-2; J(H-1,2)=11) ^{13}C NMR (toluene-d_8): 31.4 (CH_3), 48.2 (C-7), 52.3 (OCH_3); 55.0, 67.2 (C-3,6); 85.7, 95.6 (C-4,5); 128.6, 130.0 (C-1,2); 174.7 (C=O); 210.9 (CO)	[81]

References on pp. 177/9

Table 4 [continued]

No.	$RR'C_7H_6Fe(CO)_3$	method of preparation (yield in %), properties and remarks	Ref.
		IR (cyclohexane): 1740, 1970, 2050 minor isomer (ca. 21% yield): ^{1}H NMR (CD_2HNO_2): 1.26 (s, CH_3), 3.12 (dt, H-6; J(H-5,6) = 7), 3.18 (m, H-3; J(H-2,3) = 8), 3.61 (s, OCH_3), 5.04 (ddd, H-1; J(H-1,6) = 2), 5.50 (m, H-4,5), 5.90 (dd, H-2; J(H-1,2) = 11, J(H-1,3) = 1) IR (cyclohexane): 1750, 1983, 1993, 2049	
*15	$(CH_3)_2CH$, OH; Fe(CO)$_3$	see No. 26, p. 170 ^{1}H NMR ($CDCl_3$): 0.83 (d, CH_3; J = 7.5), 0.96 (d, CH_3; J = 7.5), 1.73 (OH), 1.80 (m, CH), 2.93 (t, H-3), 3.16 (dd, H-6), 5.04 (dd, H-1), 5.50 (m, H-4,5), 5.90 (dd, H-2) ^{13}C NMR ($CDCl_3$): 17.0 (CH_3), 18.0 (CH_3), 39.1 (CH), 54.3 (C-3), 67.8 (C-6), 74.2 (C-7), 85.1 (C-5), 93.4 (C-4); 129.6, 130.7 (C-1,2); 210.6 (CO) IR: 1986 ± 3, 2048 ± 4 (CO)	[67]
*16	cyclopropyl, OH; Fe(CO)$_3$	see No. 26, p. 170 ^{1}H NMR ($CDCl_3$): 0.38 (m, H-9, 10), 0.88 (m, H-8), 1.57 (OH), 3.00 (t, H-3), 3.43 (dt, H-6), 4.97 (dd, H-1), 5.47 (m, H-4, 5), 5.88 (dd, H-2) ^{13}C NMR ($CDCl_3$): 1.5, 1.7 (C-9, 10); 24.2 (C-8), 54.0 (C-3), 70.1 (C-7), 71.6 (C-6), 83.8 (C-5), 94.4 (C-4); 129.30, 129.38, (C-1,2); 210.5 (CO)	[67]
*17	C_6H_5, OH; Fe(CO)$_3$	see No. 26, p. 170 solid ^{1}H NMR ($CDCl_3$): 2.13 (OH), 3.12 (br, t, H-3), 3.44 (d, H-6), 5.23 (dd and m, H-1,5), 5.59 (m, H-4), 6.05 (dd, H-2), 7.31 (C_6H_5) ^{13}C NMR ($CDCl_3$): 54.0 (C-3), 73.1 (C-7), 73.2 (C-6), 83.7 (C-5), 94.6 (C-4), 124.8 (C-β), 127.1 (C-δ), 128.4 (C-γ); 128.9, 131.3 (C-1,2), 147.2 (C-α), 210.2 (CO)	[67]
*18	C_6H_5, CO_2CH_3; Fe(CO)$_3$	Ia, IVa (ca. 40) m.p. 110 to 112° yellow needles (from ether/pentane at −20°) ^{1}H NMR ($C_6D_5CD_2H$): 2.77 (tt, H-3; J(H-2,3) = 7.4), 3.08 (ddd, H-6;	[60, 64, 81]

References on pp. 177/9

Table 4 [continued]

No.	$RR'C_7H_6Fe(CO)_3$	method of preparation (yield in %), properties and remarks	Ref.
		J(H-4,6) = 1.5, J(H-1,6) = 2.2), 3.59 (s, CH_3), 4.76 (ddd, H-5; J(H-5,6) = 7.9, J(H-4,5) = 4.7), 4.93 (ddd, H-4; J(H-3,5) = 1.5, J(H-3,4) = 7.3), 5.89 (ddd, H-1; J(H-1,3) = 1.6), 6.07 (dd, H-2; J(H-1,2) = 11), 7.32 (m, C_6H_5) ^{13}C NMR ($CDCl_3$): 53.4 (CH_3); 52.5, 57.7, 67.0 (C-3,6,7); 85.1, 94.7 (C-4,5); 125.5, 126.7, 127.1, 128.7, 130.8, 144.8 (C-1,2, C_6H_5); 172.2 (C=O); 209.5 (CO) IR (cyclohexane): 1742, 1980, 1995, 2050	
*19	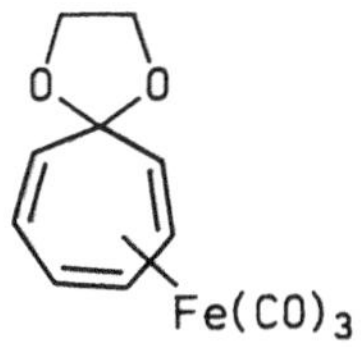 [structure: O, O, $Fe(CO)_3$]	–	[34]
*20	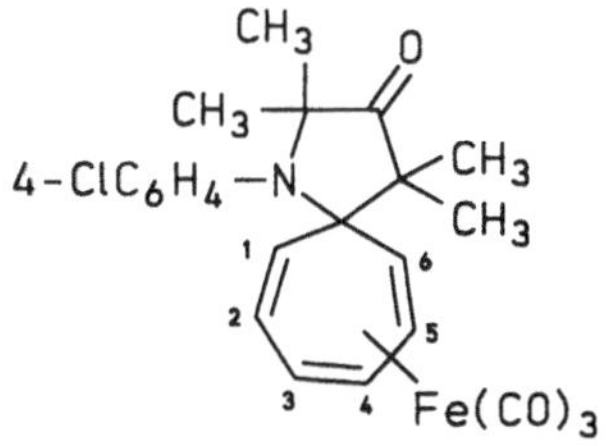 [structure: CH_3, O, CH_3, 4-ClC_6H_4–N, CH_3, CH_3, 1, 2, 3, 4, 5, 6, $Fe(CO)_3$]	see No. 28, p. 172 m.p. 185 to 187° (from n-hexane), pale yellow crystals ^1H NMR ($CDCl_3$): 0.80, 1.13, 1.18, 1.48 (s, 4CH_3); 2.92 to 3.20 (m, H-3,6), 5.24 to 5.44 (m, H-1,4,5), 5.96 (dd, H-2; J(H-2,3) = 11, J(H-1,2) = 8), 7.08 to 7.36 (m, C_6H_4) IR (Nujol): 1743 (C=O); 1980, 1990, 2080 (CO) mass spectrum: $[M-nCO]^+$ (n = 1, 3)	[75]
*21	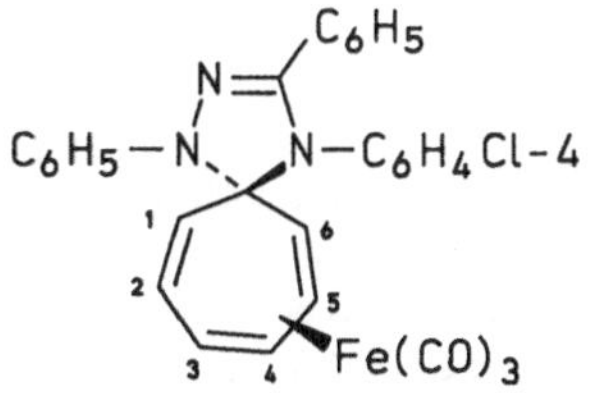 [structure: C_6H_5, N, C_6H_5–N, N–C_6H_4Cl-4, 1, 2, 3, 4, 5, 6, $Fe(CO)_3$]	see No. 28, p. 172 m.p. 135 to 138° (dec.), orange prisms (from CH_3CN) ^1H NMR ($CDCl_3$): 2.75 (br, t, H-3; J(H-2,3) = J(H-3,4) = 8.0), 3.05 (dd, H-6; J(H-5,6) = 8.0, J(H-1,6) = 3.0), 4.80 to 5.35 (m, H-4,5), 5.53 (dd, H-1; J(H-1,2) = 11.0), 6.20 (dd, H-2) IR (Nujol): 1985, 2020, 2060 (CO)	[71]
*22	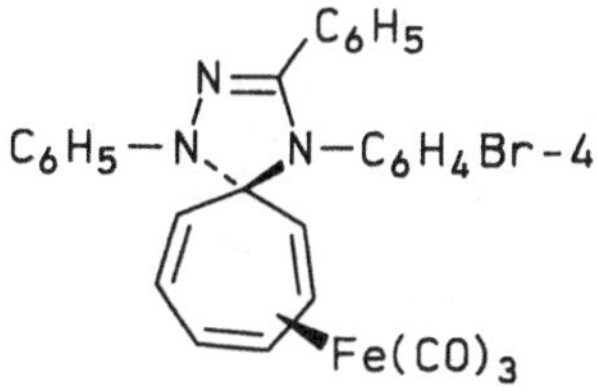 [structure: C_6H_5, N, C_6H_5–N, N–C_6H_4Br-4, $Fe(CO)_3$]	see No. 29, p. 172 m.p. 159 to 160° (dec.), orange prisms (from CH_3CN)	[71]

References on pp. 177/9

Table 4 [continued]

No.	$RR'C_7H_6Fe(CO)_3$	method of preparation (yield in %), properties and remarks	Ref.
*23	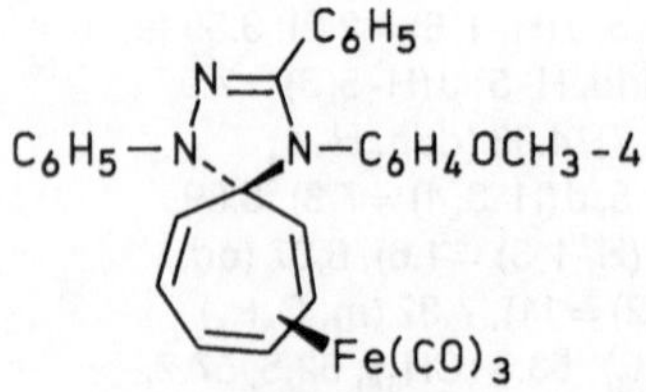	see No. 30, p. 172 m.p. 129 to 131° (dec.), orange prisms (from cyclohexane)	[71]
*24	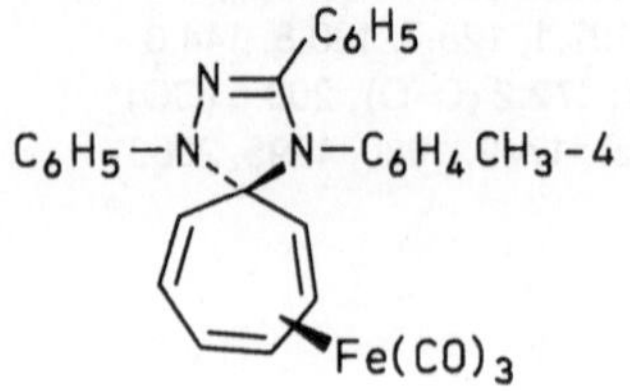	see No. 31, p. 172 m.p. 123 to 125° (dec.), orange prisms (from petroleum ether)	[71]
*25	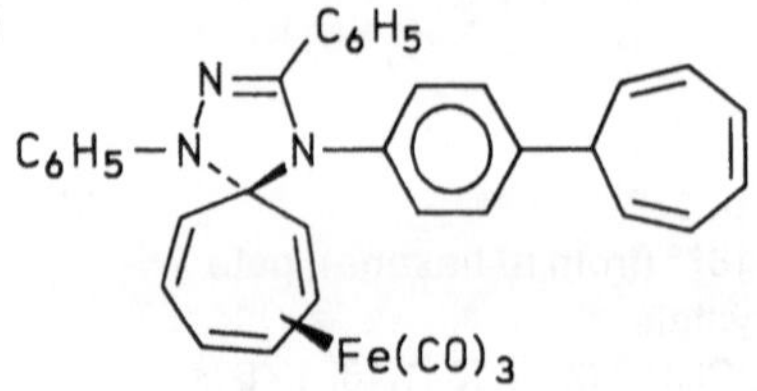	see No. 32, p. 172 m.p. 161 to 162° (dec.), yellow prisms (from CH_3CN)	[71]
compounds of the type $X{=}C_7H_6Fe(CO)_3$ (X=O, NR) (continued with No. 56):			
*26	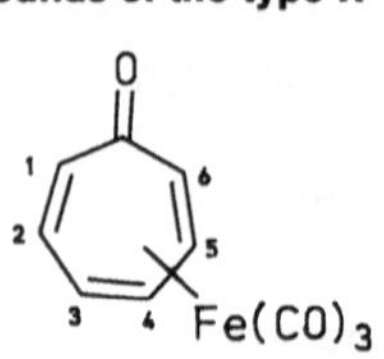	I a (85), I c (traces), II a (14 to 65), II b (22 to 30), III a (5 to 7.4), III b (4)	[1, 2, 9 to 11, 15, 19, 27, 38, 56, 82]
		modification a: m.p. 63.5 to 64.5° (from petroleum ether, ether/petroleum ether, or C_6H_6/petroleum ether), m.p. 65 to 66° (from n-hexane), sublimes at 0.01 Torr, red or orange-red prisms	[1, 9, 10, 12, 15]
		modification b: m.p. 76.5 to 77.5° (from hexane at −78°), m.p. 78 to 79°, m.p. 83 to 84° (from C_6H_6, or from C_6H_6 or ether with small amounts of petroleum ether, or from $CH_3CO_2C_2H_5$), subl.p. 40°/vacuum, subl.p. 46°/0.2 Torr, subl.p. 60°/0.1 Torr, orange crystals or solid	[1, 9 to 12, 27, 82]

Table 4 [continued]

No.	$RR'C_7H_6Fe(CO)_3$	method of preparation (yield in %), properties and remarks	Ref.
		the spectra are identical for both modifications a and b: ^{1}H NMR ($CDCl_3$): 2.723 (d, H-3; J(H-1,3) = 1.0, J(H-3,4) = 1.25, J(H-3,6) = 0.05), 3.157 (d, H-6; J(H-1,6) = 2.27), 5.033 (s, H-1; J(H-1,2) = 10.85), 6.377 (q, H-4,5; J(H-1,4) = 0.0, J(H-4,6) = 1.28, J(H-4,5) = 4.78, J(H-5,6) = 7.70), 6.562 (s, H-2; J(H-2,3) = 8.0, J(H-2,4) = J(H-2,5) = 0.0) ^{13}C NMR (benzene-d_6): 51.8 (C-3; J(C-3,4) = 41.0), 62.1 (C-6; J(C-6,7) = 50.5), 91.1 (C-5; J(C-5,6) = 38.7), 95.4 (C-4; J(C-4,5) = 45.2), 122.4 (C-1; J(C-1,2) = 65.7), 146.6 (C-2; J(C-2,3) = 51.2), 197.4 (C-7; J(C-1,7) = 55.0), 207.8 (CO) ^{13}C NMR ($CDCl_3$ at 243 K): 50.8, 60.4 (C-3,6); 91.3, 96.0 (C-4,5); 121.8 (C-1), 148.8 (C-2), 199.4 (C-7); 203.5, 204.7 (basal CO); 213.5 (apical CO) ^{13}C NMR ($CDCl_3$ at 223 K): 52.9, 62.3 (C-3,6); 92.5, 97.5 (C-4,5); 122.7, 148.8 (C-1,2); 197.9 (C-7); 205.4, 206.5 (basal CO); 216.0 (apical CO) ^{13}C NMR (CD_2Cl_2 at 213 K): 204.2, 205.4 (basal CO; J(^{57}Fe, ^{13}CO) = 24.5 ± 0.2); 214.3 (apical CO; J(^{57}Fe, ^{13}CO) = 31.5 ± 0.2) ^{57}Fe NMR (C_6H_6, relative to $Fe(CO)_5$): 849.3 ± 1.5 ^{57}Fe-γ (78 K): δ = 0.021 (^{57}Co), Δ = 0.149 ^{57}Fe-γ (−196°): δ = 0.13 (Fe), Δ = 1.44 (± 0.05) IR (KBr): 1637 (C=O); 1992, 2008, 2066 (CO) IR (CCl_4): 1613 (C=C); 1637 (C=O); 1992, 2008, 2066 (CO) mass spectrum: $[M]^+$	[1, 9, 11, 13, 15, 19, 27, 35, 46, 51, 53, 76, 77]
*27	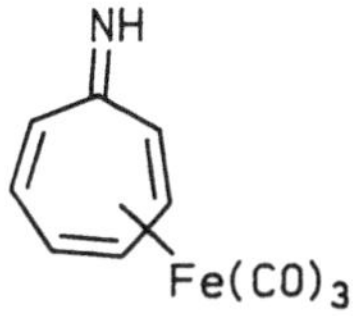 NH Fe(CO)$_3$	–	[41]

References on pp. 177/9

Table 4 [continued]

No.	$RR'C_7H_6Fe(CO)_3$	method of preparation (yield in %), properties and remarks	Ref.
*28	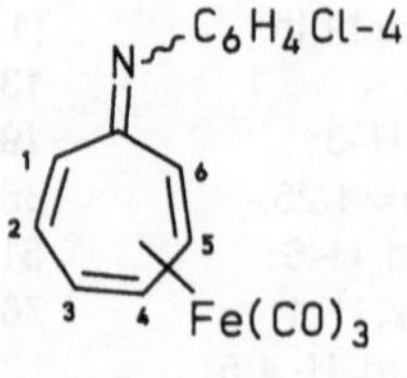 $N\sim C_6H_4Cl$-4 $Fe(CO)_3$	II a (70) m.p. 109 to 111° (from petroleum ether), orange prisms **major isomer:** ^1H NMR ($CDCl_3$): 2.75 (m, H-3), 3.72 (m, H-6), 5.15 (m, H-1), 5.70 to 6.40 (complex, H-2, 4, 5), 6.56 (AA′BB′ system, C_6H_4) **minor isomer:** ^1H NMR ($CDCl_3$): 2.90 (m, H-3), 3.24 (m, H-6), 5.45 (m, H-1), 5.70 to 6.40 (complex, H-2, 4, 5), 6.95 (AA′BB′ system, C_6H_4) **both isomers:** IR: 1990, 2700 (CO)	[71]
*29	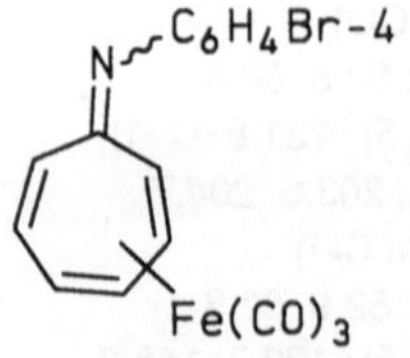 $N\sim C_6H_4Br$-4 $Fe(CO)_3$	II a (74) m.p. 119 to 121° (from petroleum ether), orange prisms	[71]
*30	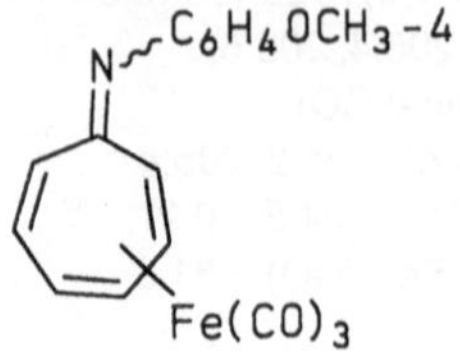 $N\sim C_6H_4OCH_3$-4 $Fe(CO)_3$	II a (50)	[71]
*31	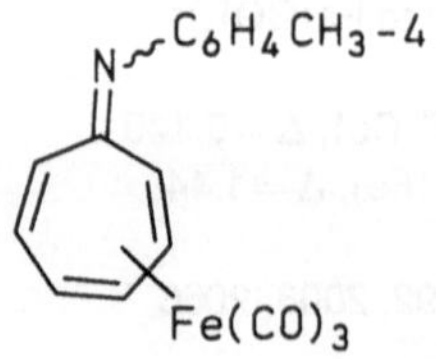 $N\sim C_6H_4CH_3$-4 $Fe(CO)_3$	II a (63) m.p. 96 to 98° (from petroleum ether), orange prisms	[71]
*32	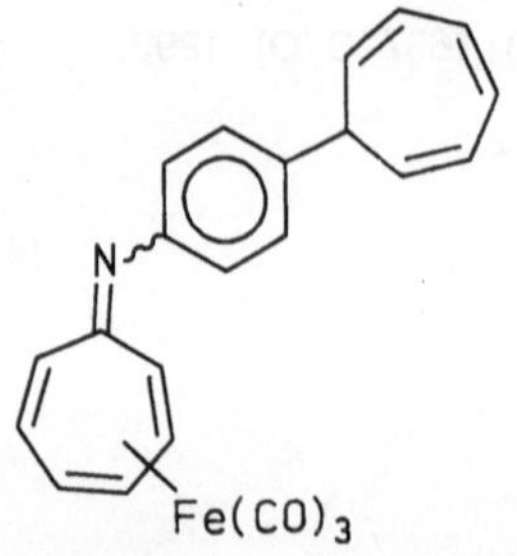 $Fe(CO)_3$	II a (75) m.p. 112 to 113° (from cyclohexane), orange prisms	[71]

References on pp. 177/9

Table 4 [continued]

No.	$RR'C_7H_6Fe(CO)_3$	method of preparation (yield in %), properties and remarks	Ref.
33	$NNHC_6H_5$ … $Fe(CO)_3$	see No. 26, p. 170 dec. ca. 130°, red oil which yields in C_2H_5OH/H_2O yellow crystals, air-sensitive IR (KBr): 1980, 2062 (CO)	[9]
34	$NNHC_6H_3(NO_2)_2$-2,4 … $Fe(CO)_3$	see No. 26, p. 170 dec. 190 to 200° (from C_6H_6/petroleum ether), red crystals IR (KBr): 1988, 2000, 2062 (CO)	[9]

compounds of the type $RHC{=}C_7H_6Fe(CO)_3$:

No.	compound	method of preparation (yield in %), properties and remarks	Ref.
*35a	H, OCH_3 (C-8); ring positions 1–6; $Fe(CO)_3$	Va uncrystallizable oil (3:1 mixture with No. 35b) ^{1}H NMR ($CDCl_3$): 3.22 (m, H-3), 3.7 (s, OCH_3), 4.22 (br, d, H-6; J=8), 5.1 to 6.0 (H-1, 2, 4, 5), 6.19 (s, H-8) ^{13}C NMR ($CDCl_3$): 56.2 (t, C-6; J(C, H)=151), 58.6 (d, C-3; J(C, H)=155), 60.5 (q, CH_3; J(C, H)=144), 85.9 (d, C-5; J(C, H)=170), 90.4 (d, C-4), 118.0 (s, C-7), 122.5 (d, C-1; J(C, H)=159), 124.2 (d, C-2; J(C, H)=158), 148.2 (d, C-8; J(C, H)=179), 211.8 (s, CO) IR (neat): 1955, 1965, 1980, 2040 mass spectrum: $[M-nCO]^+$ (n=0 to 3)	[66, 78, 79, 81]
b	CH_3O, H; ring positions 1–6; $Fe(CO)_3$	Va uncrystallizable oil (3:1 mixture with No. 35a) ^{1}H NMR ($CDCl_3$): 3.22 (m, H-3), 3.64 (s, OCH_3), ca. 3.7 (H-6, hidden), 5.1 to 6.0 (H-1, 2, 4, 5), 6.24 (s, H-8) ^{13}C NMR ($CDCl_3$): 57.9 (d, C-3; J(C, H)=155), 60.4 (q, CH_3; J(C, H)=144), 64.0 (t, C-6; J(C, H)=151), 85.3 (d, C-5; J(C, H)=170), 91.1 (d, C-4), 116.7 (s, C-7), 116.9 (d, C-1; J(C, H)=159), 127.5 (d, C-2; J(C, H)=158), 145.3 (d, C-8; J(C, H)=179), 211.8 (s, CO) IR (neat): 1955, 1965, 1980, 2040 mass spectrum: $[M-nCO]^+$ (n=0 to 3)	[66, 78, 79, 81]

References on pp. 177/9

Table 4 [continued]

No.	$RR'C_7H_6Fe(CO)_3$	method of preparation (yield in %), properties and remarks	Ref.
36	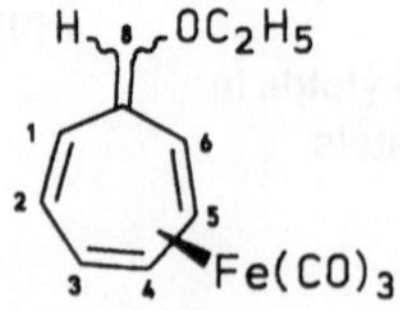 H, OC_2H_5; $Fe(CO)_3$	Va (50) ^{1}H NMR ($CDCl_3$): 1.37 (m, CH_3), 3.15 (m, H-3), 3.5 to 4.3 (m, H-6, CH_2), 5.0 to 6.0 (m, H-1,2,4,5), 6.18 (br, s, H-8)	[66]
*37	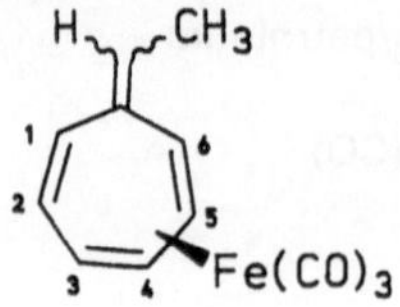 H, CH_3; $Fe(CO)_3$	Vb (19), VI (70) yellow oil (1:1 mixture of isomers), air-sensitive ^{1}H NMR (CS_2): 1.36 (d, CH_3; J=7), 1.47 (d, CH_3; J=7), 2.76 (m, H-3), 3.45 (d, H-6; J=8), 3.64 (d, H-6; J=8), 5.25 (m, H-1,2,4,5,8) ^{1}H NMR ($CDCl_3$): 1.56 (d, 1.5H; J=7), 1.67 (d, 1.5H; J=7), 3.02 (m, 1H), 3.71 (0.5H; J=8), 3.90 (0.5H; J=8), 5.40 (m, 4H), 5.74 (m, 1H; J=5) IR (n-heptane): 1979, 1988, 2050 (CO)	[26, 65]
38	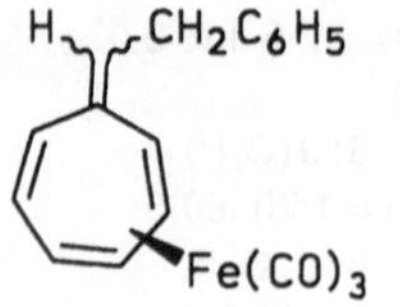 H, $CH_2C_6H_5$; $Fe(CO)_3$	Vc, VI yellow crystals (mixture of isomers) IR (hexane): three sharp bands (CO) mass spectrum: $[M]^+$	[23]
*39a	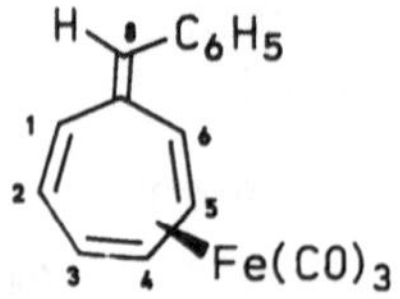 H, C_6H_5; $Fe(CO)_3$	Vc, VI yellow crystals, air-stable ^{1}H NMR ($C_6D_5CD_2H$): 2.81 (m, H-3), 4.46 (br, d, H-6; J(H-5,6)=6), 5.66 (m, H-1,2,4,5; J(H-2,8)=2), 6.53 (s, H-8), 7.35 (m, C_6H_5) IR (heptane): 1976, 1993, 2052 mass spectrum: $[M]^+$	[23, 26, 81]
b	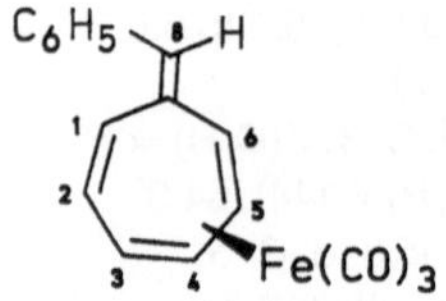 C_6H_5, H; $Fe(CO)_3$	Vc, VI yellow crystals, air-stable ^{1}H NMR ($C_6D_5CD_2H$): 2.81 (m, H-3), 3.72 (br, d, H-6; J(H-5,6)=7, J(H-4,6)= J(H-1,6)=J(H-3,6)≈2), 5.73 (ddd, H-2,4,5; J(H-2,3)=8, J(H-1,2)=11, J(H-2,8)=2), 6.19 (br, d, H-1; J(H-1,3)≈2), 6.65 (br, s, H-8), 7.35 (m, C_6H_5) IR (heptane): 1976, 1993, 2052 mass spectrum: $[M]^+$	[23, 26, 81]

References on pp. 177/9

Table 4 [continued]

No.	$RR'C_7H_6Fe(CO)_3$	method of preparation (yield in %), properties and remarks	Ref.
*40	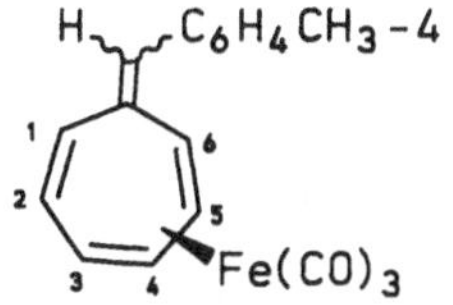 H, $C_6H_4CH_3$-4 ... $Fe(CO)_3$	VI (95) mixture of isomers m.p. 95°, subl.p. 100°/1 Torr, deep yellow crystals, air-stable ^{1}H NMR (CS_2): 2.12 (s, 1.65H, CH_3), 2.17 (s, 1.35H, CH_3), 2.88 (t, H-3; J = 8), 3.64 (d, 0.55H, H-6; J = 6), 4.09 (d, 0.45H, H-6; J = 6), 5.45 (m, H-1, 2, 4, 5), 6.24 (s, 0.45H, H-8), 6.48 (s, 0.55H, H-8), 6.92 (m, C_6H_4) IR (heptane): 1976, 1992, 2051	[26]

compounds of the type $R_2C{=}C_7H_6Fe(CO)_3$:

No.	$RR'C_7H_6Fe(CO)_3$	method of preparation (yield in %), properties and remarks	Ref.
*41	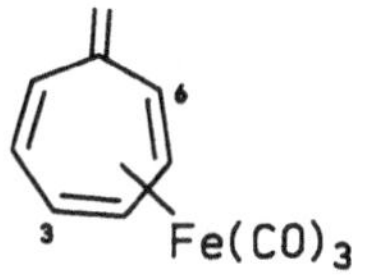 $Fe(CO)_3$	Vc (ca. 1), VI yellow liquid ^{1}H NMR (CS_2): 2.93 (t, 1H), 3.63 (d, 1H; J = 7), 4.89 (s, 1H), 5.19 (s, 1H), 5.44 (m, 4H) IR ($CHCl_3$): 1980, 2040 (CO) mass spectrum: $[M-nCO]^+$ (n = 1 to 3), $[M-3CO-C_2H_2]^+$, $[C_8H_8]^+$, $[C_6H_5]^+$, $[Fe]^+$	[24, 26]
42	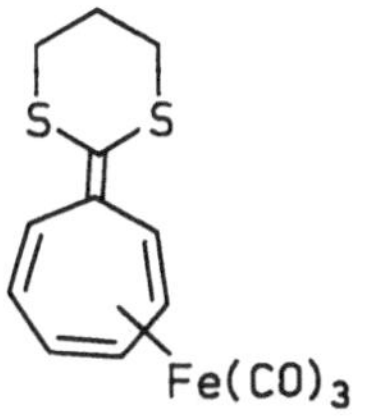 S S ... $Fe(CO)_3$	see No. 26, p. 170 red oil, air-sensitive ^{1}H NMR ($CDCl_3$): 3.13 (1H + 6H), 4.30 (split t, 1H; J = 7, J = 4), 5.13 to 5.83 (m, 4H) IR (cyclohexane): 1969, 1984, 2041 (CO)	[39]
*43	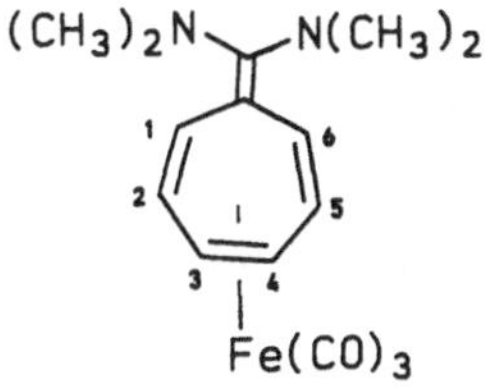 $(CH_3)_2N$, $N(CH_3)_2$ ... $Fe(CO)_3$	IIa (39) m.p. 84 to 86°, green-red crystals ^{1}H NMR (CD_2Cl_2): 2.80 (s, CH_3), 4.23 (br, s, 2H), 4.87 (br, s, 4H) ^{13}C NMR (CD_2Cl_2): 41.1 (q, CH_3); 74.5, 95.8, 100.1 (3 m, C-1 to 6), 93.3 (s, C-7), 161.0 (s, C-8), 215.5 (s, CO) IR (KBr): 1512, 1575, 1905, 1931, 1940, 1992 IR (n-hexane): 1952, 1965, 2027 (CO) UV (cyclohexane): λ_{max} (log ε) = 272 (4.1), 336 (4.15), 442 (4.07) nm mass spectrum: $[M-nCO]^+$ (n = 0 to 3)	[78]

References on pp. 177/9

Table 4 [continued]

No.	$RR'C_7H_6Fe(CO)_3$	method of preparation (yield in %), properties and remarks	Ref.
44	[structure: $(CH_3)_2C{=}C_7H_6Fe(CO)_3$; ring positions 1–6]	IVa, see No. 26, p. 170 distillable yellow oil, air-sensitive ^{1}H NMR: 1.75 (d, 2CH_3), 2.95 (m, H-2), 3.95 (d, H-1; J(1,2)=7.6), 5.10 to 5.75 (m, H-3 to 6) ^{1}H NMR (CS_2): 1.64 (s, CH_3), 1.74 (s, CH_3), 2.95 (t, H-3; J=7), 3.95 (d, H-6; J=8), 5.42 (m, H-1,2,4,5) IR (cyclohexane): 1975, 1985, 2055 IR (n-heptane): 1978, 1989, 2052 mass spectrum: $[M]^+$	[23, 26, 29]
*45	[structure: $(C_6H_5)_2C{=}C_7H_6Fe(CO)_3$; ring positions 1–6]	IVa (70), see No. 26, p. 170 m.p. 138 to 140° (from CH_2Cl_2/C_2H_5OH), distillable yellow oil or orange-red crystals ^{1}H NMR ($CDCl_3$): 3.02 (br, t, H-3; J=8), 4.05 (d, H-6; J=8), 5.34 (br, t, H-2; J=8), 5.6 (m, H-1,4,5), 7.1 (m, C_6H_5) IR: 1940, 2040 (CO) IR (thin film): 1978, 2025 (CO) UV (cyclohexane): $\lambda_{max}(\varepsilon)$=232 (18700), 263 (15900), 333 (12400), 394 (7400) nm mass spectrum: $[M]^+$	[23, 29, 61]
*46	[structure: cyclooctatetraenylidene–$C_7H_6Fe(CO)_3$; positions 1–12]	see No. 26, p. 170 reddish brown oil ^{1}H NMR ($CDCl_3$): 3.16 (dddd, H-3; J=7.5, J=7.5, J=1.5, J=1.1), 4.05 (ddd, H-6; J=8.0, J=1.5, J=1.1), 5.43 (ddd, H-5; J=8.0, J=5.0, J=1.5), 5.54 (ddd, H-4; J=7.5, J=5.0, J=1.1), 5.69 (dd, H-2; J=11.0, J=7.5), 5.82 (ddd, H-1; J=11.0, J=1.5, J=1.1), 5.97 (ddd, H-8 or H-11; J=12.0, J=3.5, J=3.5), 6.12 to 6.23 (m, H-7,9,10,12,8 or 11) ^{13}C NMR ($CDCl_3$): 57.1, 60.2, 86.1, 91.4, 119, 126.6, 127.7, 129.4, 129.7, 131.8, 132.2, 133.1, 210.7 IR (neat): 1960, 2035 UV (C_2H_5OH): $\lambda_{max}(\log\varepsilon)$=341 (4.05), 419 (4.07) nm mass spectrum: $[M-nCO]^+$ (n=0 to 3)	[82]

References on pp. 177/9

Table 4 [continued]

No.	$RR'C_7H_6Fe(CO)_3$	method of preparation (yield in %), properties and remarks	Ref.
47	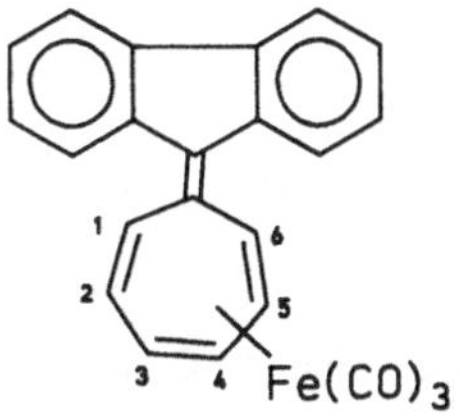 (9-fluorenylidene)cycloheptatriene, positions 1–6, $Fe(CO)_3$	see No. 26, p. 170 m.p. 165 to 166° (from $CHCl_3$/hexane), deep red crystals ^1H NMR ($CDCl_3$): 3.10 (br, t, H-3; J=8), 4.80 (br, d, H-6; J=8), 5.78 (m, H-4,5), 5.98 (dd, H-2; J=8, J=11), 6.47 (d, H-1; J=11); 7.2, 7.7 (m, 8H, aromatic) IR: 1970, 1988, 2045 UV (cyclohexane): $\lambda_{max}(\varepsilon)$ = 253 (21400), 275 (15800), 285 (13200), 366 (7600), 454 (14200) nm	[61]
*48 a	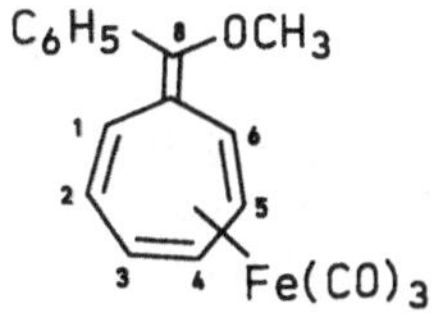 C_6H_5, OCH_3 (C-8); ring positions 1–6; $Fe(CO)_3$	Va data from a 4:1 mixture of Nos. 48a and b: ^1H NMR ($CDCl_3$): 3.10 (m, H-3), 3.43 (s, OCH_3), 4.66 (m, H-6); 5.2 to 5.8 (m, H-1,2,4,5) ^{13}C NMR ($CDCl_3$): 57.4 (C-6), 58.6 (C-3), 62.7 (CH_3), 86.9 (C-5), 91.1 (C-4), 119.0 (C-7), 122.0 (C-1), 126.5 (C-2); 128.2, 128.6, 130.2, 134.2 (C_6H_5); 129.9 (C-8), 209.1 (CO) IR (Nujol): 1965, 1970, 2040 mass spectrum: $[M-nCO]^+$ (n=0 to 3)	[66]
b	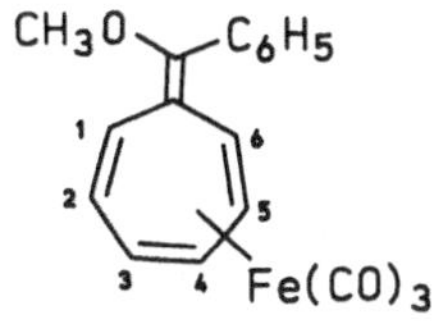 CH_3O, C_6H_5; ring positions 1–6; $Fe(CO)_3$	Va ^1H NMR ($CDCl_3$): 3.10 (m, H-3), 3.32 (s, OCH_3), 3.86 (br, d, H-6; J=8), 5.2 to 5.8 (m, H-1,2,4,5) IR (Nujol): 1965, 1970, 2040 mass spectrum: $[M-nCO]^+$ (n=0 to 3)	[66]
49	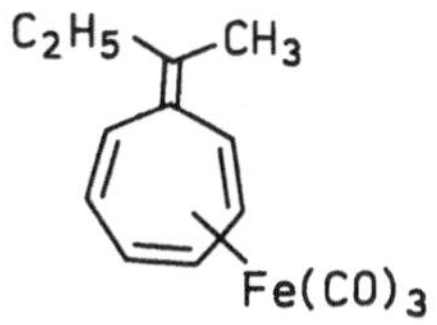 C_2H_5, CH_3; $Fe(CO)_3$	VI (90) distillable yellow oil, air-sensitive IR (n-heptane): 1978, 1987, 2050	[26]
*50	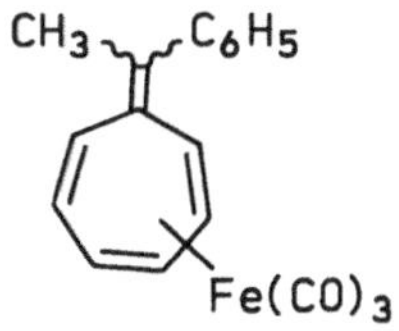 CH_3, C_6H_5; $Fe(CO)_3$	–	[59]

References on pp. 177/9

Table 4 [continued]

No.	$RR'C_7H_6Fe(CO)_3$	method of preparation (yield in %), properties and remarks	Ref.
*51	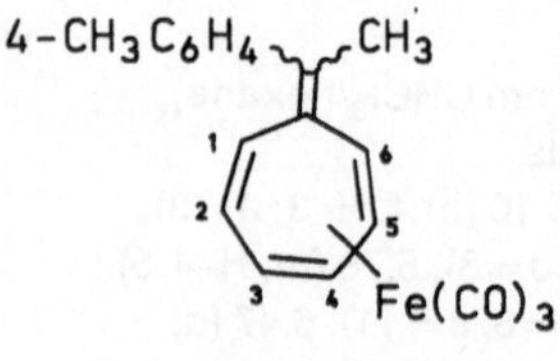 4-$CH_3C_6H_4$, CH_3 (structure: 1–6 numbered ring, Fe(CO)$_3$)	VI (90) m.p. 105°, subl. p. 100°/1 Torr, deep yellow crystals, air-stable ^{1}H NMR: 1.84 (s, CH_3), 1.97 (s, CH_3), 2.18 (s, CH_3-4), 2.22 (s, CH_3-4), 2.87 (t, H-3; J=7), 3.89 (d, H-6; J=7), 4.03 (d, H-6; J=7), 5.4 (m, H-1,2,4,5), 6.96 (m, C_6H_4) IR (n-heptane): 1974, 1992, 2050	[26]

Ionic complexes:

No.	$RR'C_7H_6Fe(CO)_3$	method of preparation (yield in %), properties and remarks	Ref.
52	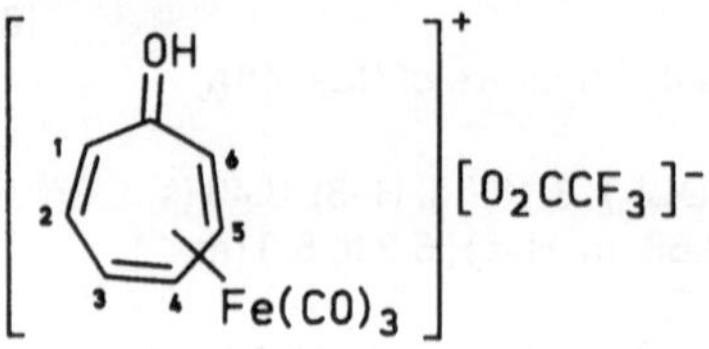 [OH, 1–6 numbered ring, Fe(CO)$_3$]$^+$ $[O_2CCF_3]^-$	see No. 26, p. 169 ^{1}H NMR (CF_3CO_2H/liquid SO_2/SO_2F_2/CD_2Cl_2 at −45°): 2.47 (H-3; J(H-2,3)=8.0), 3.04 (H-6; J(H-1,6)=2.5), 5.05 (H-1; J(H-1,3)=1.0), 6.6 to 6.9 (m, H-4,5; J(H-3,4)=7.2), 6.85 (H-2; J(H-1,2)=11.0) ^{13}C NMR (CF_3CO_2H/liquid SO_2/SO_2F_2/$CDCl_2$ at −60°): 52.0, 58.9 (2d, C-3,6; J=156); 95.0, 99.3 (2d, C-4,5; J=178, J=176); 158 (d, C-2; J=162), 121.2 (d, C-1; J=168); 202.5, 204.0 (CO); 211 (C-7), 214.5 (CO)	[51, 69]

supplements:

compounds of the type 1-R(7-R')$C_7H_6Fe(CO)_3$:

No.	$RR'C_7H_6Fe(CO)_3$	method of preparation (yield in %), properties and remarks	Ref.
53	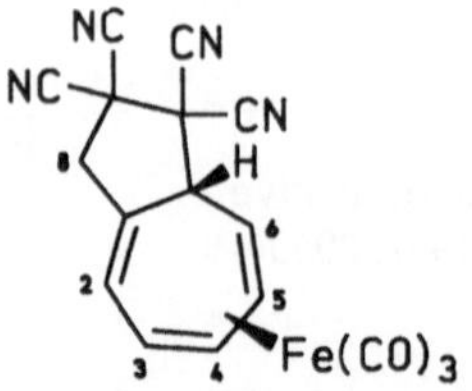 (structure: NC, CN, NC, CN, H, numbered 2–8, Fe(CO)$_3$)	see No. 41, p. 175 yellow crystals ^{1}H NMR (acetone-d_6): 3.0 to 4.2 (m, H-3,6,7,8), 5.87 (m, H-4,5), 6.09 (d, H-2; J=8) IR (CH_2Cl_2): 1994.6, 2060.8	[79]
54	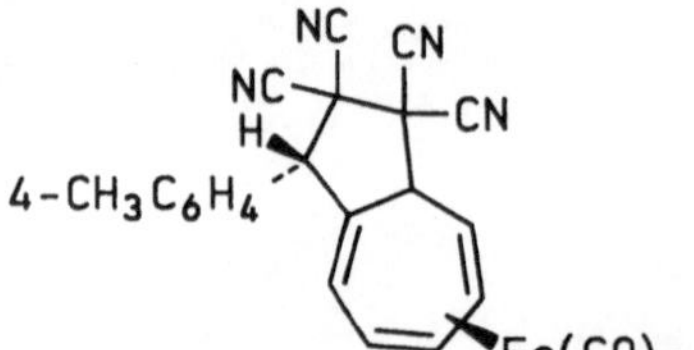 (structure: NC, CN, NC, CN, H, 4-$CH_3C_6H_4$, Fe(CO)$_3$)	see No. 40, p. 173 ^{1}H NMR almost identical with that of No. 9, p. 150 ^{13}C NMR also confirms the presence of only one structural form	[79]

References on pp. 177/9

Table 4 [continued]

No.	$RR'C_7H_6Fe(CO)_3$	method of preparation (yield in %), properties and remarks	Ref.
compound of the type 1-R(2-R')$C_7H_6Fe(CO)_3$:			
*55	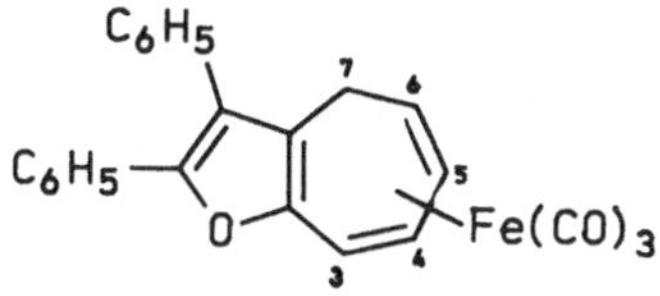 C_6H_5, C_6H_5, O, 3, 4, 5, 6, 7, Fe(CO)$_3$	m.p. 177 to 178°, orange-red crystals (from CH_2Cl_2/hexane) ^1H NMR ($CDCl_3$): 2.59 (br, d, H-7; J=4), 3.54 (m, H-6), 3.75 (m, H-3), 5.40 (m, 2H, H-4,5), 7.22 (m, C_6H_5) ^{13}C NMR ($CDCl_3$): 29.39 (C-7); 55.39, 65.89 (C-3,6); 91.64, 92.50 (C-4,5); 128.76, 130.41, 131.19, 132.60, 133.33 (aromatic C); 215.47 (CO) IR ($CHCl_3$): 1595 (furan C=C); 1975, 2045 (CO) mass spectrum: $[M+1]^+$, $[M+1-nCO]^+$ (n=0 to 3)	[89]
compounds of the type X=$C_7H_6Fe(CO)_3$ (X=NR):			
*56a	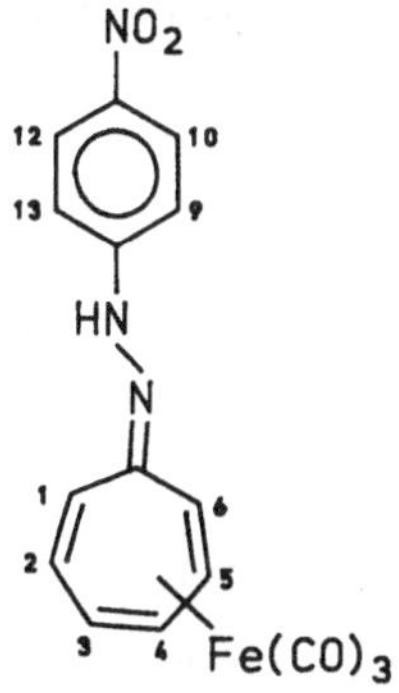 NO_2, 9, 10, 12, 13, HN, N, 1, 2, 3, 4, 5, 6, Fe(CO)$_3$	–	[86]
b	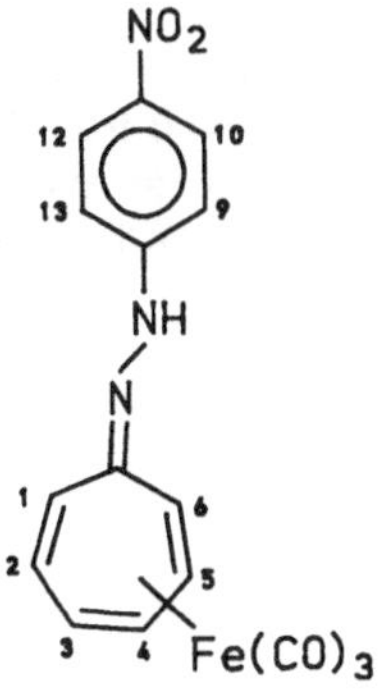 NO_2, 9, 10, 12, 13, NH, N, 1, 2, 3, 4, 5, 6, Fe(CO)$_3$	–	[86]

References on pp. 177/9

*Further information:

1-CH_3(7-HO)C_7H_6Fe(CO)$_3$ (Table **4**, No. **1**) is very sensitive and oxidizes readily to VIII. Reaction with HBF_4 in $O(COCH_3)_2$ gives $[CH_3C_7H_6Fe(CO)_3]BF_4$ [33].

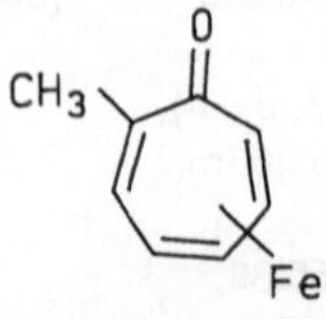

VIII 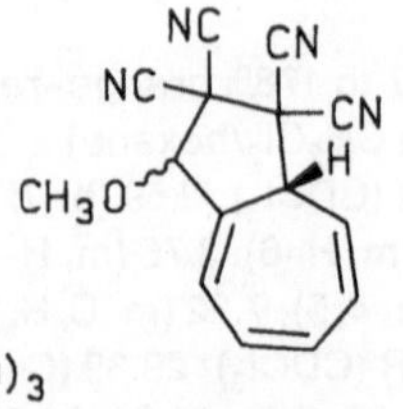 IX 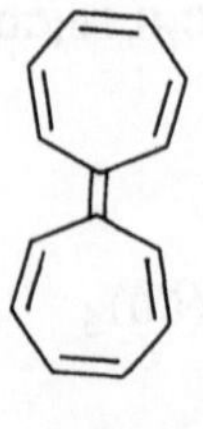 X 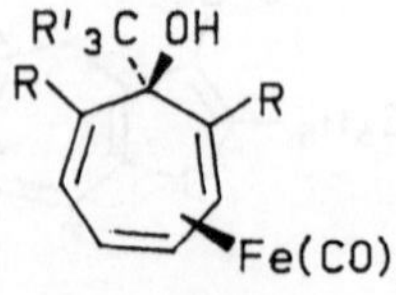 XI

1-(4-$CH_3C_6H_4$)(7-C_7H_7)C_7H_6Fe(CO)$_3$ (Table **4**, No. **5**). Nothing is said concerning the structure of the product obtained by Method Ib (p. 145). Thus, a structure like 7-(1′-(4-$CH_3C_6H_4$))-$C_7H_6C_7H_7Fe(CO)_3$ (see 1.4.1.4.1.2.4.2.4.1, Table 3, No. 60, p. 132) is also possible. The product is proposed as a fuel or antiknock additive and as a starting compound for metallic mirrors when decomposed >400 °C [8].

$C_{17}H_{15}N_3O_2$Fe(CO)$_3$ (Table **4**, No. **6**) in CH_2Cl_2 is absorbed on a silica gel column and left for 24 h. Elution with $CH_3CO_2C_2H_5$ affords water-soluble No. 3 in 85% yield (hydrolysis of No. 6) [66].

$C_{15}H_{10}N_4$OFe(CO)$_3$ (Table **4**, No. **7**). The presence of a 4:5 ratio of isomers No. 7a and b is confirmed by its ^{13}C NMR spectrum in acetone-d_6: δ=47.9, 48.6 (C-6); 50.3, 50.6 (C-3); 53.86, 55.1 (C-7); 57.96, 61.2 (OCH_3); 86.1, 87.9 (C-8); 90.8, 91.1 (C-5); 95.9 (C-4); 136.2, 138.5 (C-2) ppm [79]. Oxidative decomplexation of the 4:5 isomeric mixture of Nos. 7a and b with $[NH_4]_2[Ce(NO_3)_6]$ in CH_3CN at 0 °C gives a mixture of isomers IX, with the trans isomer as major product [66].

$C_{20}H_{12}N_4$Fe(CO)$_3$ (Table **4**, No. **9**). The coupling constant J(H-7,8)=2.5 Hz is probably due to the presence of a favorable W conformation in part of the molecule [59].

$C_{20}H_{12}N_4$Fe(CO)$_3$ (Table **4**, No. **12**). Attempted removal of the Fe(CO)$_3$ group by the reaction with $ON(CH_3)_3$ unexpectedly affords X in 62.5% yield [82].

7-HO(7-CH_3)C_7H_6Fe(CO)$_3$ (Table **4**, No. **13**). The deuterated complex XI (R=R′=D) is prepared from No. 26 (deuterated at C-1,6) and $LiCD_3$, complex XI (R=D, R′=H) from No. 26 (deuterated at C-1,6) and $LiCH_3$, and complex XI (R=H, R′=D) from No. 26 and $LiCD_3$ (cf. No. 26). Complex XI (R=D, R′=H) shows in the ^{1}H NMR spectrum in $CDCl_3$ chemical shifts at δ=1.35 (s, 3H), 1.9 (s, 1H), 2.96 (dt, 1H; J=7.5), 5.2 to 5.6 (m), and 5.75 (d, 1H;

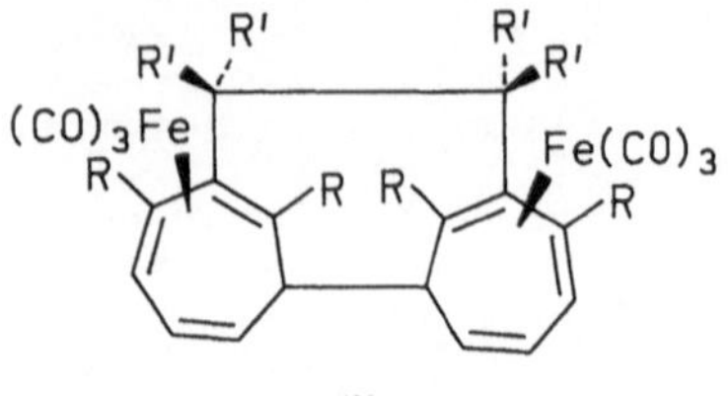

XII 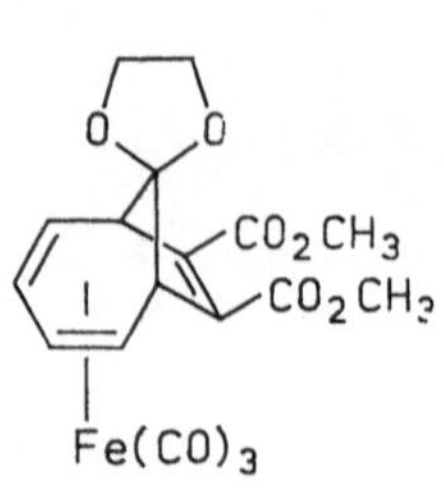 XIII 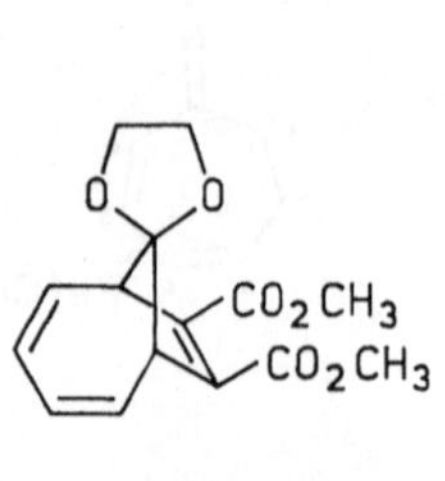 XIV

J = 8.5) ppm. Dehydration of No. 13 on silica gel gives XII (R = R′ = H). Similar dehydration of the deuterated complex XI with R = R′ = D, R = D, R′ = H or R = H, R′ = D gives the appropriate complexes XII [33]. Protonation of No. 13 with CF_3CO_2H [67] or HBF_4 in $O(COCH_3)_2$ [33] gives $[CH_3C_7H_6Fe(CO)_3]^+$ [33, 67]. Similar protonation of the deuterated complexes with HBF_4 in $O(COCH_3)_2$ followed by addition of $N(C_2H_5)_3$ gives the appropriate complexes XII with R = R′ = D; R = D, R′ = H or R = H, R′ = D [33].

7-CH_3O_2C(7-CH_3)$C_7H_6Fe(CO)_3$ (Table **4**, No. **14**). The major and minor isomers of No. 14, obtained by Method Ia (pp. 144/5), are identified as the endo- and exo-methyl isomers. Although it would be advantageous to know the stereochemistry at C-7, no simple way of obtaining that information is found [81]. Both isomers exhibit fluxional behavior, see 1.4.1.4.1.2.4.2.4.1, XXX, p. 143. Spin saturation transfer (SST) experiments on the major isomer of No. 14 in toluene-d_8 are performed by saturating H-6 and observing the change in signal intensity for H-1. For the minor isomer of No. 14, a nonaromatic solvent, octane-d_{18}, is chosen to provide adequate separation of the H-1 resonance from those of H-5 and H-4. In the SST experiments on this isomer, H-1 is saturated and the integral of the overlapping multiplets H-6 and H-3 measured by using the methyl resonance of the CO_2CH_3 groups as an integration standard. The results of these SST experiments are presented in the following table [81]:

t in °C	$\frac{M(0)}{M(\infty)}$	T_1 of H-1 (s)	$k \cdot 10^2$ (s^{-1})	$\Delta G^{\neq}$ (kcal/mol)
major isomer:				
85	1.20	(6.8)	2.94	23.5
89	1.33	6.8	4.85	23.4
98	1.46	5.8	7.93	23.7
108	1.79	(5.0)	15.8	23.8
minor isomer:				
77	1.17	(6.7)	15.8	23.1
86	1.27	(7.1)	2.48	23.4
92	1.58	7.4	3.86	23.3

The rate of 1,3 iron shift in No. 14 is given by $k = [(M(0)/M(\infty)) - 1]/T_{1(A)}$, where k is the rate constant for exchange in a two-site, equal population system, $A \rightleftharpoons B$. $T_{1(A)}$ is the spin-lattice relaxation time of the nucleus at site A, and M(0) is the normal equilibrium magnetization of the nucleus at site A after the nucleus is saturated at site B. Experimentally, $M(0)/M(\infty)$ is determined by comparing the signal areas for nucleus A with and without saturation of nucleus B. $T_{1(A)}$ can be conveniently measured by the fast inversion-recovery method [81]. The averaged values of $\Delta G^{\neq} = 23.6$ kcal/mol for the major isomer and $\Delta G^{\neq} = 23.3$ kcal/mol for the minor isomer [64, 81] are not consistent with a transition state for the 1,3 iron shift which resembles a $^4LFe(CO)_3$ species with ^4L = norcaradiene [81].

7-endo-HO(7-exo-R)$C_7H_6Fe(CO)_3$ (Table **4**, Nos. **15** to **17** with R = i-C_3H_7, cyclo-C_3H_5, C_6H_5). Protonation of these complexes with CF_3CO_2H in CD_2Cl_2 gives $[RC_7H_6Fe(CO)_3]^+$ [67].

7-CH_3O_2C(7-C_6H_5)$C_7H_6Fe(CO)_3$ (Table **4**, No. **18**) is of unknown stereochemistry but only one isomer is isolated by Methods Ia, IVa (pp. 144/5 and 147) [60, 81]. Compound No. 18 exhibits fluxional behavior, see 1.4.1.4.1.2.4.2.4.1, XXX, p. 143 [60, 64, 81]. Spin saturation transfer (SST) experiments on No. 18 have been performed, and the results are shown in the following table (see No. 14) [60, 81]:

References on pp. 177/9

t in °C	$\frac{M(0)}{M(\infty)}$	T_1 of H-1 (s)	$k \cdot 10^2$ (s^{-1})	ΔG^* (kcal/mol)
80	1.12	7.7	1.56	23.6
80	1.13	7.6	1.71	23.5
95	1.20	7.9	2.53	24.3
95	1.10	3.4	2.94	24.2
105	1.36	4.1	8.77	24.0

The averaged value $\Delta G^* = 23.9$ kcal/mol [60, 64, 81] is surprisingly slightly greater than that for $C_7H_8Fe(CO)_3$ and does not necessarily rule out a possible norcaradiene intermediate for this fluxional process [60], but a coordinatively unsaturated iron tricarbonyl intermediate is also discussed for this 1,3 iron shift [81].

$C_2H_4O_2C_7H_6Fe(CO)_3$ (Table **4**, No. **19**) is irradiated with $CH_3O_2CC{\equiv}CCO_2CH_3$ in tetrahydrofuran at 0 °C to give XIII (20% yield) and XIV (5% yield). Irradiation of No. 19 in tetrahydrofuran at −78 °C followed by addition of $CH_3O_2CC{\equiv}CCO_2CH_3$ in the dark gives only XIV at room temperature [34].

$C_{13}H_{16}ClONC_7H_6Fe(CO)_3$ (Table **4**, No. **20**) gives XV in 72% yield on oxidation with $ON(CH_3)_3$ in CH_3CN at room temperature for 24 h [75].

XV XVI XVII XVIII

$R(C_6H_5)_2CN_3C_7H_6Fe(CO)_3$ (Table **4**, Nos. **21** to **25** with $R = C_6H_4Cl$-4, C_6H_4Br-4, $C_6H_4OCH_3$-4, $C_6H_4CH_3$-4, $C_6H_4C_7H_7$-4) are oxidized by excess $ON(CH_3)_3$ in CH_3CN at room temperature within 12 h. Analysis shows the presence of XVI and XVII, small amounts of XVIII, and trace amounts of other nonidentified products. The following table gives the yield of products [71]:

compound No.	total yield of XVI and XVII (%)	ratio XVI:XVII	XVIII (%)
21 ($R = C_6H_4Cl$-4)	50	10:90	—
22 ($R = C_6H_4Br$-4)	69	10:90	—
23 ($R = C_6H_4OCH_3$-4)	58	70:30	8
24 ($R = C_6H_4CH_3$-4)	55	60:40	—
25 ($R = C_6H_4C_7H_7$-4)	61	45:55	—

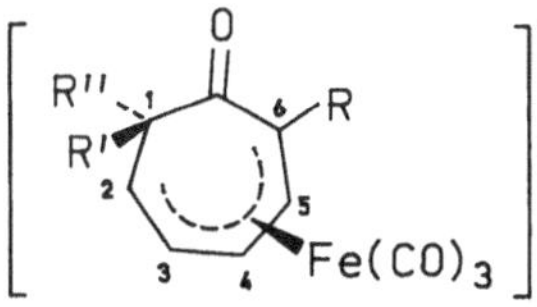

XIX 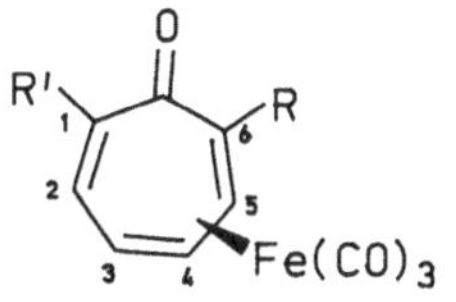 XX 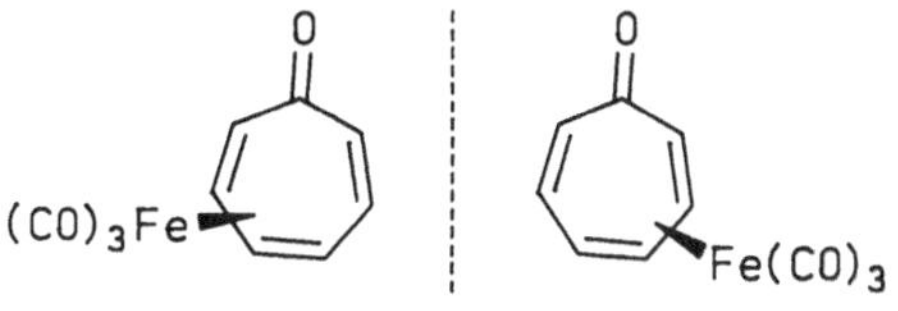 XXI

$C_7H_6OFe(CO)_3$ (Table **4**, No. **26**). Provided that precautions are taken to exclude air and prevent prolonged exposure to light, No. 26 can be gas-chromatographed at temperatures up to 150 °C with no apparent thermal decomposition. Teflon columns employing E-301 methyl silicone gum as the stationary phase give best results [25]. Compound No. 26 is also formed by refluxing I (p. 145) in C_6H_6 [82]. It is also obtained by the reaction of XIX (R=R′=R″=H, X=BF_4) with K_2CO_3 [27, 30] or with OCH_3^-/CH_3OH or LiC_6H_5 [42]. Analogous reaction of XIX (R=R′=H, R″=D, X=BF_4) with K_2CO_3 gives No. 26, too [27]. Analogous reaction of $[C_7H_3D_3OFe(CO)_3]^+$ (cation in XIX with R=R′=R″=D) with $N(C_2H_5)_3$ in CH_2Cl_2 followed by the usual work-up gives XX (R=R′=D) [33, 45] in 82% yield (isotope purity >91%) [27, 33]. Similar reaction of this cation with K_2CO_3 in CH_2Cl_2 followed by chromatography with ether/hexane (2:1) and sublimation at 46 °C/0.02 Torr gives XX (R=R′=D) in 25% yield (isotopic purity >92%) [27]. Treatment of a mixture of XIX with R=R′=D, R″=H and R=R′=R″=D; X=BF_4 (deuterium incorporation 68% at the 1-exo, 54% at the 1-endo, and 25% at the 6 positions) with $N(C_2H_5)_3$ in CH_2Cl_2 affords an 80% total yield of XX (R=H, R′=D, 41%, and R=D, R′=H, ca. 4% [54]) and XX (R=R′=D). Deuterium incorporation at the 6 and 1 positions in XX (R=R′=D) is found to be 25 and 52%, respectively [27, 45, 54], cf. [85]. The possible existence of No. 26 in two enantiomeric forms is first noted in [9]. Irradiation of No. 26 with left or right circularly polarized light at λ=380 to 500 nm induces stable optical activity in the complex by an asymmetric destruction process. Thus, irradiation of XXI in n-hexane solution for 6 h yields α_D(observed) = +0.012°±0.002°. At shorter irradiation times, a smaller α_D is observed. When the solution is irradiated by left circularly polarized light for ca. 6 h the observed α_D is −0.010°±0.002°. A specific rotation of 700° for the pure enantiomer is calculated [58].

The modification No. 26b is monomeric in solution [1], and diamagnetic with $\chi_{mol}^{293\,K}$ = −(89.05±2.5) × 10^{-6} cm^3/mol [9]. Its dipole moment in C_6H_6 at 25.0 °C is μ_D=4.30±0.05 D or $\mu_{15\%}$=4.25±0.05 D [9].

The ^1H NMR spectrum in $CDCl_3$ in the presence of Eu(dpm)$_3$ (dpm=2,2,6,6-tetramethylheptane-3,5-dionate) shows chemical shifts at δ=2.73 (H-3), 3.13 (H-6), 5.03 (H-2), 6.56 (H-1), and 6.73 (H-4,5) ppm [21] with maximum shift change observed for H-6 [21, 35]. The ^1H shifts relative to H-4 as 1.0 in benzene-d_6 in the presence of Eu(dpm)$_3$ are δ=1.00 (H-4), 1.18 (H-3), 1.26 (H-2), 1.46 (H-5), 4.40 (H-1), 4.63 (H-6) ppm and in the presence of Eu(fod)$_3$ (fod=1.1.1.2.2.3.3-heptafluoro-7,7-dimethyloctane-4,6-dionate) are δ=1.00 (H-4), 1.09 (H-3), 1.17 (H-2), 1.51 (H-5), 4.49 (H-1), 4.76 (H-6) ppm [35]. The ^1H NMR spectrum of the deuterated derivative XX (R=R′=D) in $CDCl_3$ shows chemical shifts at δ=2.72 (quint (pair of overlapping t), H-3; J(H-3,4)=5.0), 6.37 (d, H-4,5; J(H-4,5)=5.8), 6.56 (dd, H-2; J(H-2,3)=8.0) ppm [27, 33].

The ^1H NMR spectra in $CDCl_3$ in the range 30 to 120 °C [15] and in ClC_6H_5 in the range 40 to 120 °C [37] are not temperature dependent [15, 37], which rules out a fluxional process (cf. XXI) [15, 37, 58], cf. [81]. The relative shifts of the ^{13}C NMR spectra in benzene-d_6 with C-7 taken as δ=10.0 are shown in the following table [35]:

References on pp. 177/9

concentration of shift reagent	H-1	H-2	H-3	H-4	H-5	H-6	H-7
0.02 M Eu(dpm)$_3$	3.7	2.8	0.8	1.7	2.0	2.8	10.0
0.05 M Eu(dpm)$_3$	3.9	3.3	1.2	1.8	2.3	2.9	10.0
0.11 M Eu(dpm)$_3$	4.0	3.5	1.2	1.8	2.4	3.1	10.0

For values in the presence of La(dpm)$_3$ see [35]. The CO resonances broaden above 273 K and merge to a singlet [51] at 315 K with $\delta = 209.0$ ppm [46]. From linewidths at 253 K the rate constants for scrambling can be estimated as 30 s^{-1} [51]. The activation parameters for this scrambling are $\Delta H^{*} = 11.6 \pm 0.5$ kcal/mol, $\Delta S^{*} = -3.3 \pm 2$ e.u. [46], $\Delta G^{*}_{298} = 12.7$ kcal/mol [46, 52], and assuming log A = 13.2, $\Delta G^{*}_{298} = 12.5$ kcal/mol [51, 52]. Modification No. 26a crystallizes [6, 12, 17] in the space group $P2_1/c\text{-}C^5_{2h}$ with the unit cell parameters a = 6.52, b = 12.08, c = 12.90 Å and $\beta = 101.2°$; Z = 4 molecules per unit cell [12]. The measured density is 1.65 ± 0.02 g/cm^3 [9, 12] and the calculated 1.64 g/cm^3 [12]. The main bond distances and angles are shown in **Fig. 15** [12].

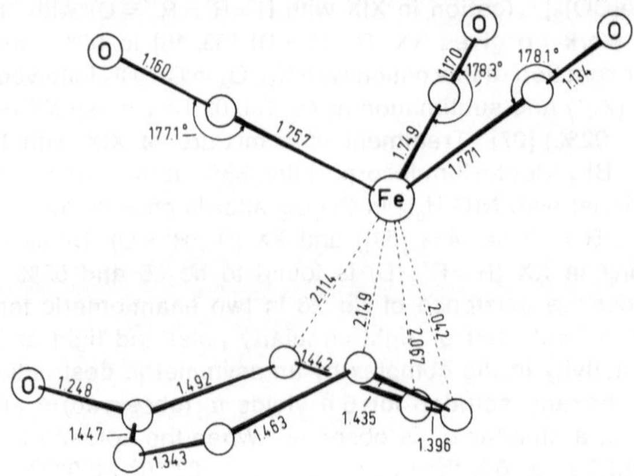

Fig. 15. Molecular structure of $C_7H_6OFe(CO)_3$ (No. 26a) [12].

The electronic structure of No. 26 is examined by simple molecular orbital theory [3], and it is found that experimental data on No. 26 can be satisfactorily explained by a localization energy approximation of the Hückel type [14]. For an application of topology and group theory to No. 26 see [16, 48].

XXII XXIII XXIV XXV

Compound No. 26 is soluble in organic solvents. It is stable in the crystalline state but is decomposed slowly in solution by air and light [1, 9]. No. 26 decomposes at 120 °C [15]. The significant ions in the negative ion mass spectrum are $[M - nCO]^{\dot{-}}$ (n = 0 to 3) and $[M - C_7H_6O - nCO]^{\dot{-}}$ (n = 0, 1) [63]. The significant ions in the proton transfer mass spectrum

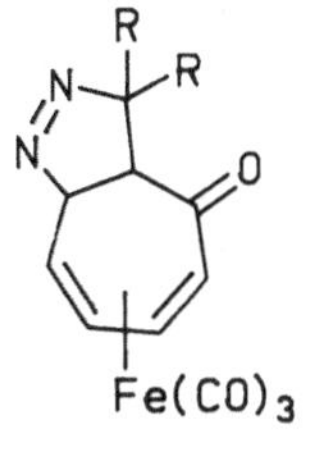

XXVI XXVII (a, b) XXVIII

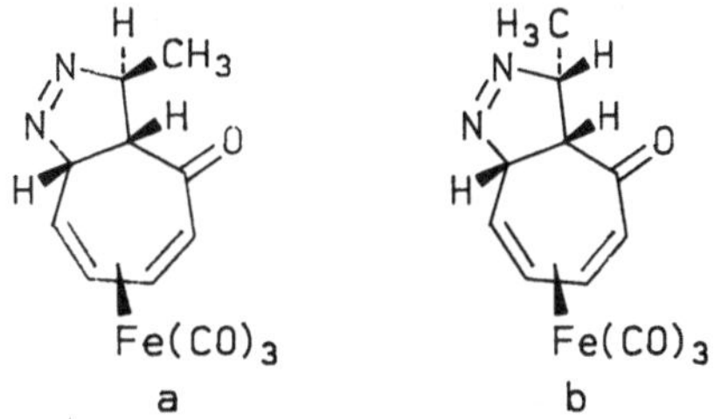

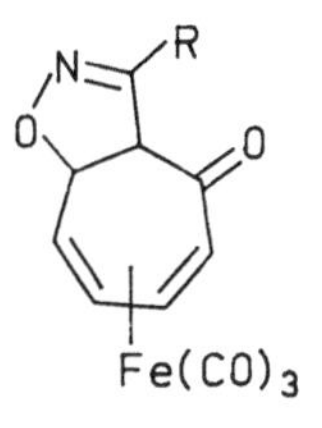

with H_2 are $[M+H-nCO]^+$ (n=0 to 3), $[M-nCO]^{+\cdot}$ (n=0 to 3), $[M-H-nCO]^+$ (n=0 to 3), $[C_7H_7O]^+$ [70]; with CH_4 [28, 70], the same ions, but with different percentage of the total ion current, are observed [70]. The proton transfer mass spectrum with i-C_4H_{10} shows $[M+H-nCO]^+$ (n=0, 1), $[M]^{+\cdot}$, $[M-H]^+$, and $[C_7H_7O]^+$ and, with NH_3, $[M+NH_4]^+$, $[M+H-nCO]^+$ (n=0, 1, 3), and $[M-H]^+$ [70]. No. 26 cannot be hydrogenated by an Adams catalyst in CH_3CO_2H at normal pressure, and hydrogenation with Zn/CH_3CO_2H gives scarcely soluble dimers and polymeric hydrocompounds (not identified [9]), but reaction with H_2 (150 atm) on a Pd catalyst (10% charcoal) in petroleum ether at 40 °C for 66 h gives XXII (20% yield) and XXIII with R=R′=R″=R‴=H (61% yield) [4, 9, 30]. Similar reaction in C_2H_5OH at 65 °C for 6 h gives XXIII with R=R′=R″=R‴=H (46%) and small amounts of isomeric mono- to tetranuclear compounds [9]. Similar reaction with D_2 (167 atm)/Pd (10% charcoal) in hexane at 40 °C for 66 h gives presumably dideuterated XXII and XXIII (R=R′=H, R″=R‴=D) [27]. Oxidation with $FeCl_3 \cdot 6H_2O$ in dioxane at 80 °C gives $FeCl_2 \cdot xH_2O$ and tropone (30%) [9]. Oxidation with H_2O_2/NaOH in CH_3OH at 0 °C for 60 min followed by 2 h at room temperature gives tropone in 95% yield [83]. Decomplexation with $ON(CH_3)_3$ in refluxing C_6H_6 gives tropone in 68% yield after 1 h [40] and in 71% yield after 12 h [36]. The protonation of No. 26 with H_2SO_4 is nonstereoselective and probably occurs from either the exo or endo side of the ring [27, 30]. Thus, reaction with conc. H_2SO_4 (98%) at 0 °C gives XIX (R=R′=R″=H) [19, 20, 69]. Dissolving No. 26 in CH_2Cl_2, followed by extraction with conc. H_2SO_4 and pouring the resulting solution after 15 min into an excess of Na_2CO_3 gives XXIV (R=R′=R″=R‴=H) [27]. Reaction of No. 26 with D_2SO_4 (99%) at 0 °C gives XIX (R′=R″=D, R=H) [19]. Further reaction for 2.5 h [19] or 12 h in CH_2Cl_2 [33] gives XIX (R=R′=R″=D) [19, 27, 33]. Reaction of No. 26 with D_2SO_4 [27] in CH_2Cl_2 [45] for 12 h [20, 27] or 5 h at 0 °C [45] followed by quenching with K_2CO_3/H_2O [20, 27] or K_2CO_3/CH_3OH [45] gives XXIV with R=R′=R″=D and R‴=H or CH_3 [27, 45]. Similar reaction with D_2SO_4 for 12 h and pouring the resulting solution into Na_2CO_3/CH_3OH at 0 °C gives XXIV (R=R′=R″=D, R‴=CH_3) [33]. No deuteration is observed with CF_3CO_2D [20, 69], but with CF_3CO_2D/D_2SO_4 (6:1) at room temperature (5 min), followed by dilution with CH_2Cl_2 and neutralization with K_2CO_3, the formation of XXIII (R=R′=R″=D, R‴=H) is observed [27]. This protonation is stereoselective and occurs from the exo side of the ring at a coordinated double bond (C-6) [27, 30]. On addition of No. 26 in CD_2Cl_2 to a mixture of CF_3CO_2H/liquid SO_2/SO_2F_2 (2:3:3) at −78 °C protonation occurs at the oxygen atom yielding a deep blood-red solution of No. 52 [47, 51, 69], cf. [27]. Warming of No. 52 to 0 °C gives previously observed (see above) XIX with R=R′=R″=H [51, 69] or XIX with R=R′=R″=D if CF_3CO_2D is used [47, 51]. Quenching a solution of No. 26 in $HOSO_2F$/liquid SO_2 in cold Na_2CO_3/CH_3OH gives XXIV (R=R′=R″=H, R‴=CH_3). Reaction of No. 26 with $DOSO_2F$/liquid SO_2 at −34 to −14 °C gives XIX (R=R′=H, R″=D) [19]. Dissolving No. 26 in $D_2SO_4/DSi(C_2H_5)_3$ at room temperature and dilution of the reaction mixture with CH_2Cl_2 after 5 min, followed by neutralization with K_2CO_3, gives XXIII (R=R′=R″=H, R‴=D). Similar reaction with $CF_3CO_2H/HSi(C_2H_5)_3$ gives XXIII (R=R′=R″=R‴=H), $CF_3CO_2D/DSi(C_2H_5)_3$ gives XXIII (R=R′=H, R″=R‴=D), and $DSi(C_2H_5)_3$ gives XXIII (R=R′=R″=R‴=D) [27], cf. [20, 30]. These reduc-

References on pp. 177/9

tions and that with H_2/Pd (see above) occur by a selective process involving cis addition of hydrogen from the exo side of the ring [27, 30]. Reaction with $P(C_6H_5)_3$ [2, 9] in C_6H_6 in a sealed tube at 100 °C for 5 h gives $(CO)_3Fe(P(C_6H_5)_3)_2$ [2, 9], tropone in 50 [2] or 61% yield [9], and $C_7H_6OFe(CO)_2P(C_6H_5)_3$ [9]. Warming with $H_2NNHC_6H_5$ in C_2H_5OH/CH_3CO_2H gives No. 33 in 15% yield. Similarly, $H_2NNHC_6H_3(NO_2)_2$-2,4 in C_2H_5OH/H_3PO_4 at room temperature gives No. 34 [9]. Similar reaction with $H_2NNHC_6H_4NO_2$-4 gives No. 56 [86]. Refluxing $Fe_2(CO)_9$ with No. 26 in ether gives I (see p. 145) in 25 to 30% yield [82]. Reaction of No. 26 with 1,4-$(HO)_2C_6H_4$ [5, 9], (ratio 2:1) in ether followed by filtration, concentration of the filtrate, and addition of small amounts of petroleum ether gives red crystals (81% yield) [9] of **2·$C_6H_7OFe(CO)_3$·HOC_6H_4OH-p**, m.p. 108 to 109.5 °C (dec.) [5, 9] which sublime in vacuum at 100 °C (IR spectrum (KBr): 1631 (C=O); 1992, 2066 (CO) cm^{-1}) [9]. Reaction with HSR (R=C_3H_7-i, C_6H_5)/$(C_2H_5)_2OBF_3$ in CH_3OH gives $RSC_7H_7Fe(CO)_3$, but similar reaction with $HSCH_2CH_2SH$ gives $(CO)_3FeC_7H_7SCH_2CH_2SC_7H_7Fe(CO)_3$ [68]. No. 26 reacts with $LiCH_3$ in ether at −78 °C for 30 [33] to 45 min [67]. The reaction mixture is hydrolyzed with H_2O and dilute HCl (1:4), the organic layer is washed with H_2O, dried ($MgSO_4$), and chromatographed on AR silica (cc-7). Elution with hexane gives small amounts of XII (R=H, R′=CH_3), and elution with C_6H_6 gives No. 13 in 75% yield. Similar reaction with $LiCD_3$ at −78 °C gives XI (R=H, R′=D), and analogous reaction of XX (R=R′=D) with $LiCH_3$ or $LiCD_3$ gives XI with R=D, R′=H, or R=R′=D [33]. The above reaction mixture can also be quenched with an H_2O/ether/tetrahydrofuran mixture. The resulting mixture is diluted with hexane and H_2O, and the organic layer dried (Na_2SO_4) and concentrated. The residue is diluted with hexane and filtered through glass wool. Cooling the hexane solution gives No. 13. Similarly, No. 17 is obtained as a solid and Nos. 15 and 16 are prepared analogously [67]. Compound No. 44 is prepared from No. 26 and i-C_3H_7MgBr in ether at −60 to −50 °C. After 30 min the reaction mixture is hydrolyzed with $[NH_4]Cl/H_2O$ and the organic layer is washed with H_2O, dried ($MgSO_4$), and evaporated. The resulting oil is placed on top of a silica gel column (made up with toluene), left for 15 h, and eluted with toluene to give No. 44 in 60% yield [23, 26]. Reaction of No. 26 with XXV gives No. 42 in 40% yield [39]. Reaction of No. 26 with a twofold excess of $CH_3COCl/AlCl_3$ in CH_2Cl_2 at room temperature gives a mixture of the tautomeric complexes XX (R=$COCH_3$, R′=H or R′=$COCH_3$, R=H) [56]. Electrophilic reactions with dienophiles give 1:1 addition products. Thus, reaction with R_2CN_2 with R=H [38, 44], D [44], CH_3 [38] in ether gives XXVI [38, 44]. Similar reaction of CH_3CHN_2 gives XXVIIa and b in the ratio 2.7:1 [38]. The reaction with XLVIII in C_6H_6 gives LVI (p. 175) [84]. The reaction with C_6H_5CNO and 2,4,6-$(CH_3)_3C_6H_2CNO$ is stereospecific, yielding XXVIII (R=C_6H_5, $C_6H_2(CH_3)_3$-2,4,6) [50]. Reaction of $O{=}C(CF_3)_2$ gives XXIX [54], and the reaction with excess $O{=}C{=}C(C_6H_5)_2$ in C_6H_6 at room temperature affords, after chromatography on alumina with hexane, No. 45 in 20% yield [61]. Similar reaction with XXX gives No. 47 [61], and with XXXI, generated in situ by the reaction of cycloheptatrienylcarbonyl chloride and $N(C_2H_5)_3$, No. 46 is formed in 72.7% yield [82]. These reactions involve a [2+2] cycloaddition followed by decarboxylation [82]. Reaction with $C_6H_5C{\equiv}N{-}NC_6H_5$, generated in situ from $C_6H_5CCl{=}NNHC_6H_5$ and $N(C_2H_5)_3$ gives a mixture of XXXII and its regioisomer. A competitive reaction between $C_6H_5CCl{=}NNHC_6H_5$, tropone, and No. 26 (ratio 0.5:3:1) in the presence of $N(C_2H_5)_3$ in C_6H_6 at room temperature in the dark gives XXXII and its regioisomer (67%), No. 26, free ligand ^4L in XXXII (25%), tropone, and traces of unidentified compounds [49]. Frontier orbital calculations predict that 1,3-addition is the preferred cycloaddition mode for $(NC)_2C{=}C(CN)_2$ to No. 26 [74]. The first reaction product from No. 26 and $(NC)_2C{=}C(CN)_2$ [22, 54, 72, 80] is XXXIII (R=H) [54, 72]. Similar reaction of XX (R=H, R′=D) gives XXXIII with R=D [54]. Recrystallization of XXXIII (R=H) [45, 54] or warming to 30 °C [54] gives XXXIV, which was originally [54] regarded as the reaction product [45]. The observed rate constants for this $(NC)_2C{=}C(CN)_2$ addition in CH_2Cl_2 are shown in the following table [72]:

t in °C	$[(NC)_2C=C(CN)_2]$ (mmol/L)	$k_{obs} \cdot 10^3$ (s^{-1})
15.6	10.3	0.04
	19.8	0.08
	30.0	0.12
	39.9	0.16
19.9	10.3	0.055
	20.7	0.11
	35.7	0.19
	40.0	0.21
25.3	10.1	0.07
	20.3	0.12
	30.1	0.19
	40.2	0.26
	45.1	0.29
29.4	10.3	0.08
	19.8	0.17
	30.0	0.25
	39.8	0.34

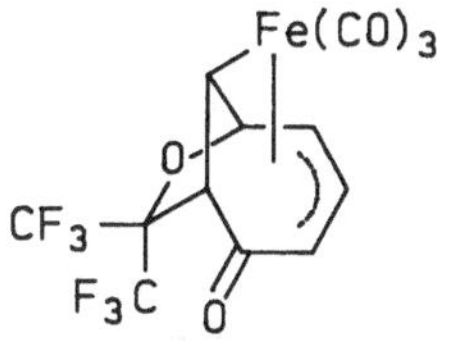

XXIX 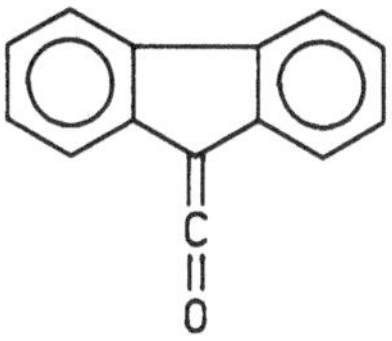 XXX 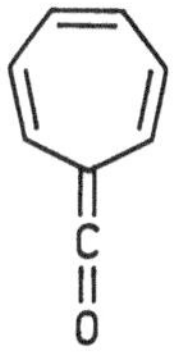 XXXI 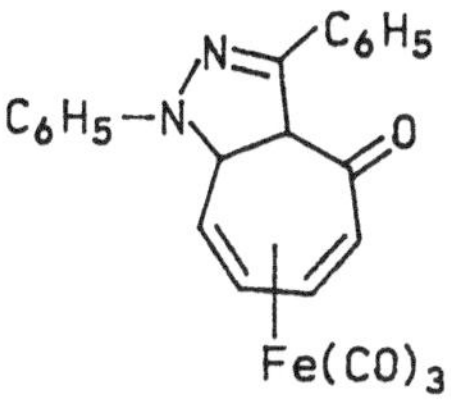 XXXII 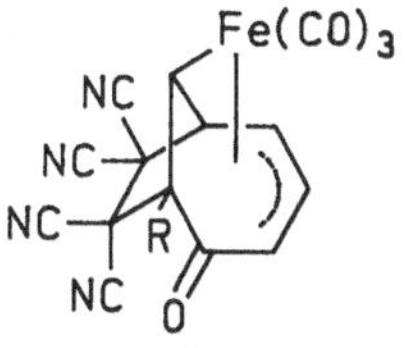 XXXIII

The second-order rate constant at 25.3 °C is $k_2 = (6.44 \pm 0.24) \times 10^{-3}$ $dm^3 \cdot mol^{-1} \cdot s^{-1}$. An electron-withdrawing effect of the C=O group is discussed. The thermodynamic activation parameters at 25 °C are $\Delta G^* = 85.38(\pm 0.08)$ kJ/mol, $\Delta H^* = 36.15(\pm 3.13)$ kJ/mol, and $\Delta S^* = -165.2(\pm 10.5)$ $J \cdot K^{-1} \cdot mol^{-1}$. These values are reasonable, and the negative entropy is as expected for an associative reaction [72]. Reaction of No. 26 with N-methyltriazolinedione gives XXXV [57, 62], and reaction with XXXVI gives XXXVII (X=O) [80]. Similarly reaction with XXXVIII gives XXXIX [80]. The reaction of No. 26 with monomeric cyclopentadiene in an autoclave at 70 to 80 °C gives XL, XLI, $(C_5H_5Fe(CO)_2)_2$, and XLII. The yield of XLI is 75% if $Fe_2(CO)_9$ is added to the reaction mixture [55]. Reaction of the dimer of XLIII with No. 26 yields a mixture of XLIV (14% yield) and XLV [50].

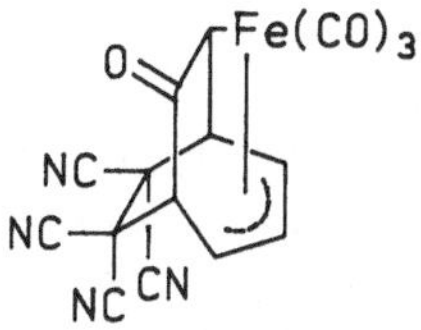

XXXIV 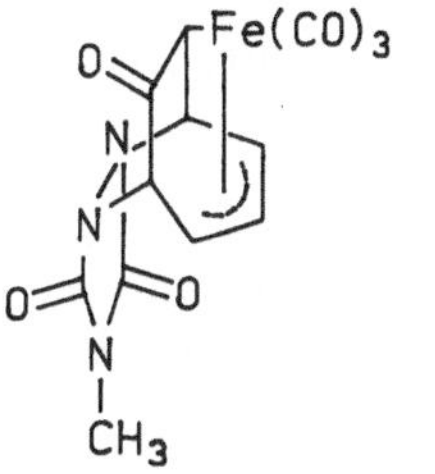 XXXV 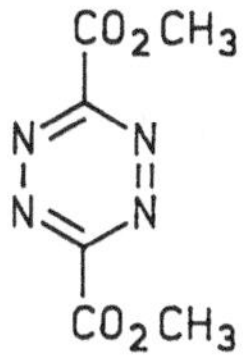 XXXVI 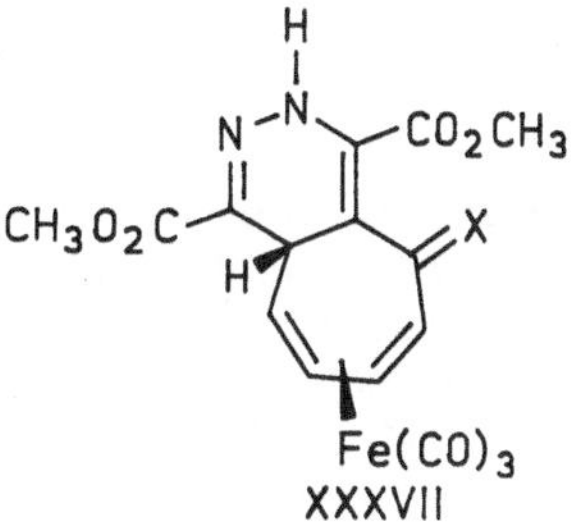 XXXVII

References on pp. 177/9

$HN{=}C_7H_6Fe(CO)_3$ (Table **4**, No. **27**). Reaction of $C_7H_8Fe(CO)_3$ (C_7H_8 = cycloheptatriene) with picryl azide gives a product for which an aziridine or an imine structure can be proposed from the ^{1}H NMR data. The X-ray analysis confirms the imine structure shown in Table 4, p. 155. Crystals of No. 27 are monoclinic, space group $P2_1/c-C^5_{2h}$ with the parameters a = 10.910, b = 14.329, c = 12.239 Å, β = 109.84°; Z = 4 molecules per unit cell [41]. Frontier orbital calculations predict that 1,3-addition is the preferred cycloaddition mode for $(NC)_2C{=}C(CN)_2$ [74].

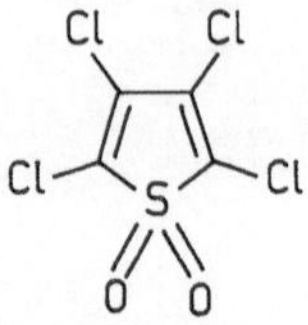

XXXVIII 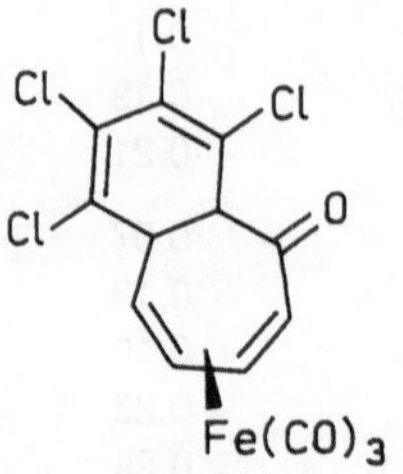 XXXIX 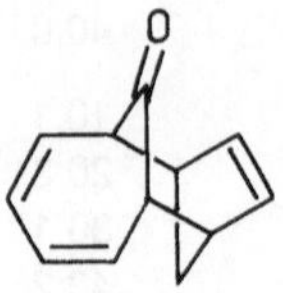 XL 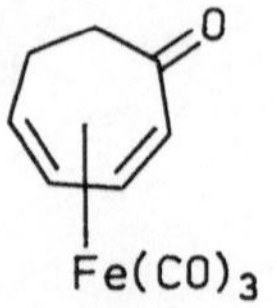 XLI

$RN{=}C_7H_6Fe(CO)_3$ (Table **4**, Nos. **28** to **32** with R = C_6H_4Cl-4, C_6H_4Br-4, $C_6H_4OCH_3$-4, $C_6H_4CH_3$-4, $C_6H_4C_7H_7$-4). No. 28 reacts with $Fe_2(CO)_9$ and $OC(C(CH_3)_2Br)_2$ in C_6H_6 at 45 °C for 8 h. The mixture is filtered, and the filtrate is evaporated to give crude No. 20 which is recrystallized (20% yield) [75]. Complexes No. 28 to 32 react with $C_6H_5CCl{=}NNHC_6H_5$ and $N(C_2H_5)_3$ (ratio 1:1:3) in CH_3CN at room temperature for 6 h. After usual work-up (see below) chromatography affords pure complexes No. 21 to 25 (>80% yield). The competition reaction of No. 31 with $C_6H_5C{\equiv}NNC_6H_5$ (generated in situ from $C_6H_5CCl{=}NNHC_6H_5$ and $N(C_2H_5)_3$) and II (R = $C_6H_4CH_3$-4, see p. 145) in the ratio 1.5:1.0:1.5 affords 64.5% of XLVI and XLVII (R = $C_6H_4CH_3$-4), and 30.5% of No. 24. The reaction mixture is left in the dark until $C_6H_5CCl{=}NNHC_6H_5$ disappears. The mixture is diluted with CH_2Cl_2, and the organic layer is washed with H_2O and dried. After evaporation the residue is chromatographed as described in Method II [71]. Reaction of No. 31 with XXXVI in CH_2Cl_2 gives XXXVII (X = $NC_6H_4CH_3$-4) [80].

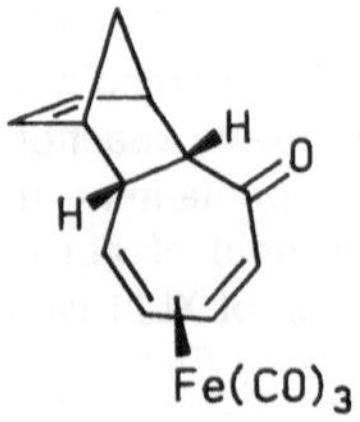

XLII 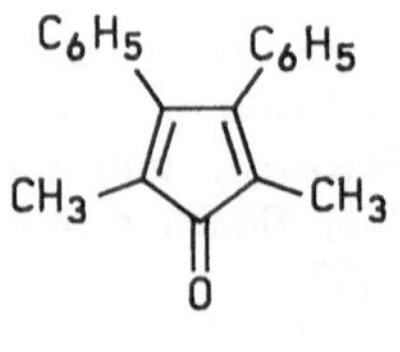 XLIII 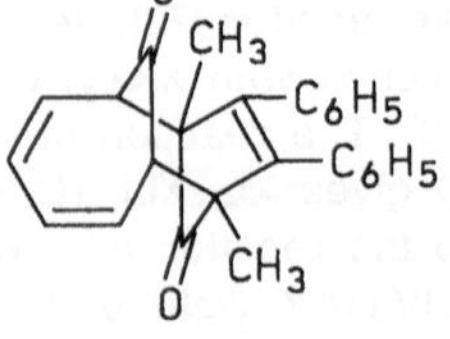 XLIV 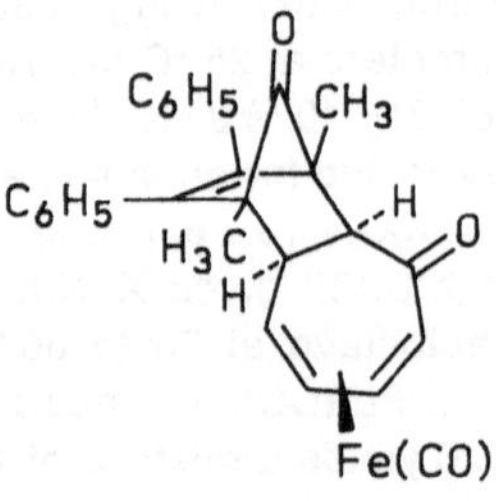 XLV

$CH_3OCH{=}C_7H_6Fe(CO)_3$ (Table **4**, No. **35**). The mixture of isomers No. 35a and b decomposes readily if not kept in solution under N_2. Treatment with XLVIII in CH_2Cl_2 at −78 °C for 1 h, followed by removal of the solvent and recrystallization of the residue gives No. 6 (25% yield) [66]. Treatment of the same 3:1 mixture of isomers No. 35a and b with freshly sublimed $(NC)_2C{=}C(CN)_2$ in CH_2Cl_2 at room temperature, followed by work-up as above, gives a 4:5 mixture (not separated) of isomers No. 7a and b in 62% yield [66, 79].

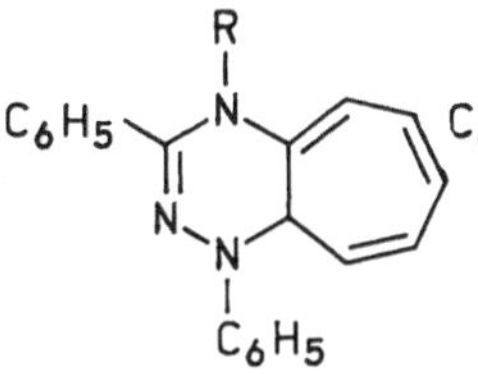

XLVI 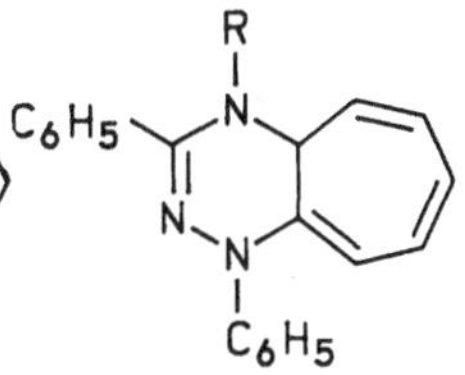 XLVII 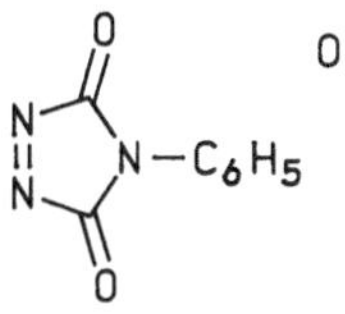 XLVIII 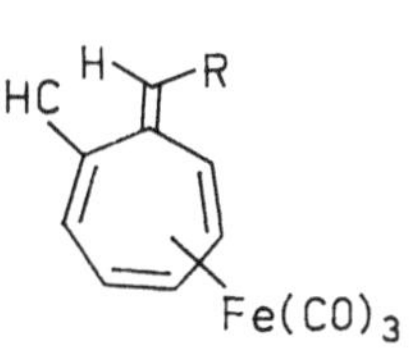 IL

RCH=$C_7H_6Fe(CO)_3$ (Table **4**, Nos. **37**, **39**, **40** with R=CH_3, C_6H_5, $C_6H_4CH_3$-4). The mixture of isomers No. 39 melts at 101 °C and sublimes at 100 °C/1 Torr. All three compounds are potentially fluxional but static on the ^{1}H NMR time scale. Attempts to detect any fluxional character were not successful up to 100 °C. The fact that pure IL (R=$C_6H_4CH_3$-4) is prepared in 70% yield from the 1:1 mixture of isomers No. 40 (see below) and that the recovered No. 40 (10%) is also a 1:1 mixture implies a fluxional nature for No. 40 [26]. Compound No. 39 exists in solution as a 4:3 mixture of two isomers No. 39a and b. Comparison of the chemical shifts of H-6 in these two isomers and those in No. 35 suggests that the major isomer No. 39b contains the C_6H_5 group trans to the bound diene unit [81]. In performing spin saturation transfer experiments (SST) on No. 39, it is necessary to saturate a signal from one isomer and monitor the decrease in intensity for a signal from the other isomer. Therefore, H-6 of the minor isomer, No. 39a, is saturated while the signal intensity for the peak due to H-1 of the major isomer, No. 39b, is measured. The results of the SST experiments are summarized in the following table (compare No. 14, p. 165):

t in °C	$\frac{M(0)}{M(\infty)}$	T_1 (s)	$k \cdot 10^2$ (s^{-1})	$\Delta G^{\ne}$ (kcal/mol)
37	1.24	3.6	6.67	19.8
47	1.72	4.1	17.6	19.8
53	2.18	(4.4)	26.8	19.9
60	2.90	(4.6)	41.3	20.0
63	3.80	(4.7)	59.6	20.0
67	4.15	4.8	65.6	20.2
73	7.2	(4.8)	129	20.1

These data apply to the conversion of No. 39a to b; however, since K_{eq} is close to 1, rates and activation energies for the conversion of No. 39b to a will be quite similar. The free (averaged) energy of activation of $\Delta G^{\ne}$=20 kcal/mol for No. 39 is more than 2 kcal/mol less than that for 1,3 shift in $C_7H_8Fe(CO)_3$ [64, 81]. This supports a mechanism in which the intermediate contains an $Fe(CO)_3$ group bound to C-3,4 [81]. The compounds No. 39 and 40 react with $(CH_3)_2NCHO/OPCl_3$ in ether to give IL (R=C_6H_5 or $C_6H_4CH_3$-4) [26]. The reaction of Nos. 37 [59, 65], 39 [59, 79], and 40 [79] with $(NC)_2C{=}C(CN)_2$ gives 1:1 adducts. Reaction of No. 37 with $(NC)_2C{=}C(CN)_2$ in CH_2Cl_2 at room temperature for 1 h followed by removal of solvent and recrystallization gives a mixture of isomers No. 8a and b (50% yield) [65]. Similar reaction of No. 39 gives No. 9 [59]. If No. 39 and $(NC)_2C{=}C(CN)_2$ are dissolved in CH_3NO_2, frozen immediately with liquid N_2, and the temperature is allowed to rise and the solvent removed rapidly in vacuum, La and b with R=H (15%) and No. 9 (60%) are obtained, separated by fractional crystallization from CH_2Cl_2/hexane. Similar reaction of No. 40 gives No. 54. The rate of $(NC)_2C{=}C(CN)_2$ addition to No. 39, forming La and b (R=H), is measured at 25 °C in CH_3NO_2 solution under pseudo-first-order conditions with $(NC)_2C{=}C(CN)_2$ in excess, and the results are given in the following table [79]:

References on pp. 177/9

$[(NC)_2C{=}C(CN)_2]$ (mmol/L)	$k_{obs} \cdot 10^2$ (s^{-1})
2.17	5.60
3.96	9.5
6.00	13.8
8.02	18.0
10.06	22.1
20.00	40.0
60.00	135.7

The second-order rate constant for the formation of La and b (R=H) is $k_2 = 22.4 \pm 0.45\ L \cdot mol^{-1} \cdot s^{-1}$ at 25 °C. It is possible to follow the isomerization of Lb (R=H) to No. 9 using the ν(CO) bands. The measured rate constants at 25 °C are $k_4 = (1.39 \pm 0.05) \times 10^{-4} \cdot s^{-1}$ in CH_2Cl_2, $k_4 = (2.49 \pm 0.02) \times 10^{-3} \cdot s^{-1}$ in CH_2Cl_2/CH_3NO_2 (1:1), and $k_4 > 5.5 \times 10^{-3} \cdot s^{-1}$ in CH_3NO_2. The latter is too fast to measure accurately. The addition of $(NC)_2C{=}C(CN)_2$ to No. 39 in CH_2Cl_2 at 25 °C gives directly $(1.36 \pm 0.05) \times 10^{-4} \cdot s^{-1}$ for k_4. La and b (R=H) isomerize at different rates, but give the same product No. 9. The solvent dependence of k_4 provides support for an ionic intermediate in the isomerization of Lb [79].

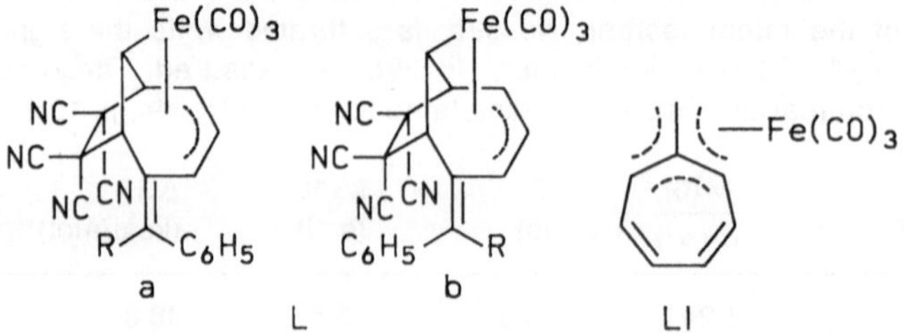

Fe(CO)3 Fe(CO)3 NC NC CN NC R C6H5 a NC NC CN NC C6H5 R b L —Fe(CO)3 LI

$H_2C{=}C_7H_6Fe(CO)_3$ (Table **4**, No. **41**) is also discussed as an intermediate in the reaction of $[CH_3C_7H_6Fe(CO)_3]BF_4$ with $N(C_2H_5)_3$; XII with R=R′=H (see 2.8.1.1 in "Organoiron Compounds" C5, 1981, Table 4, No. 42, p. 52) is the final product [32, 33]. Compound No. 41 appears to be less stable than its isomer LI [31]. For a detailed theoretical investigation of the minimum energy pathways for the haptotropic rearrangements in this compound see [87]. Compound No. 41 dimerizes slowly at room temperature to give XII (R=R′=H). Thermal decomposition at 100 °C for 10 min or heating in C_6H_6 at 78 °C for 20 h gives XII (R=R′=H) [24]. The quasimolecular ion is most abundant in the chemical ionization spectrum [24, 28]. For the methane chemical ionization spectrum see [28]. The compound No. 41 is not protonated by CH_3CO_2H [24], but protonation occurs with CF_3CO_2H [24, 30]. Treatment of the protonated species with K_2CO_3 in CH_2Cl_2 gives No. 41 in low yield [24]. Compound No. 41 suffers reduction to $C_8H_{10}Fe(CO)_3$ in the presence of $CF_3CO_2H/HSi(C_2H_5)_3$. Reaction

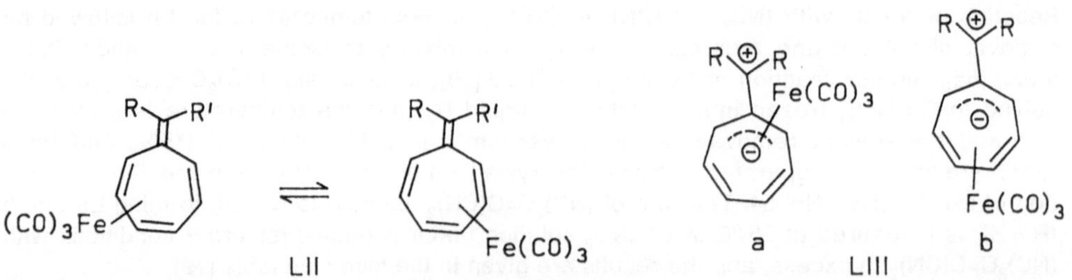

R R′ (CO)3Fe ⇌ R R′ Fe(CO)3 LII

R ⊕ R Fe(CO)3 ⊖ a R ⊕ R ⊖ Fe(CO)3 b LIII

References on pp. 177/9

of No. 41 with $CH_3O_2CC{\equiv}CCO_2CH_3$ gives a 1:1 adduct of the composition $C_{14}H_{14}O_4Fe(CO)_4$ [24, 30]. Frontier orbital calculations predict that 1,3-addition is the preferred cycloaddition mode for $(NC)_2C{=}C(CN)_2$ [74]. In fact, compound No. 41 reacts rapidly (0.5 h) with $(NC)_2C{=}C(CN)_2$ in CH_2Cl_2 at 25 °C to give No. 53 in 50% yield [79].

$[(CH_3)_2N]_2C{=}C_7H_6Fe(CO)_3$ (Table **4**, No. **43**) shows a temperature dependent ^1H NMR spectrum (for figures of the ^1H NMR spectra in CD_2Cl_2 at −49 and −71 °C see [78]); it is therefore fluxional as shown in LII (R=R′=$N(CH_3)_2$), but contribution of forms like LIII (R=$N(CH_3)_2$) cannot be excluded. Reaction with $(NC)_2C{=}C(CN)_2$ gives LIV (R=R′=$N(CH_3)_2$) [78].

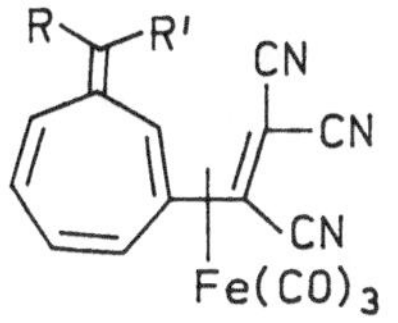

LIV 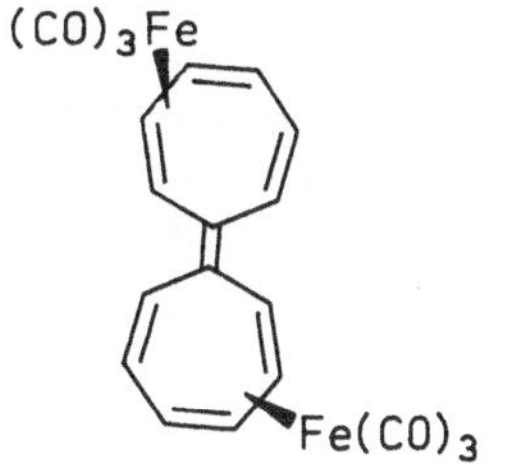 LV 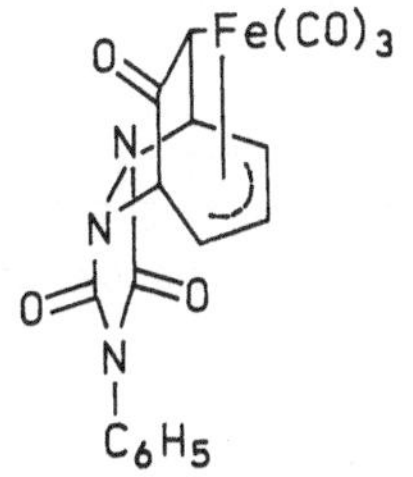 LVI

$(C_6H_5)_2C{=}C_7H_6Fe(CO)_3$ (Table **4**, No. **45**) reacts with $(NC)_2C{=}C(CN)_2$ in CH_2Cl_2 to give only L (a=b, R=C_6H_5). The observed rate constants in CH_2Cl_2 at 25 °C under pseudo first-order conditions, with $(NC)_2C{=}C(CN)_2$ in excess, are given in the following table [79]:

$[(NC)_2C{=}C(CN)_2]$ (mmol/L)	$k_{obs} \cdot 10^2$ (s^{-1})
1.93	1.09
4.06	2.59
5.86	3.77
8.13	5.18
10.05	6.20

The second-order rate constant for this addition is $k_2 = 6.31 \pm 0.18\ L \cdot mol^{-1} \cdot s^{-1}$ [79].

$C_7H_6{=}C_7H_6Fe(CO)_3$ (Table **4**, No. **46**) is air-sensitive and gradually polymerizes on standing at room temperature. It reacts with $Fe_2(CO)_9$ in boiling C_6H_6 to give LV in 16.6% yield. The reaction with $(NC)_2C{=}C(CN)_2$ affords No. 12 in 70.5% yield [82].

$CH_3O(C_6H_5)C{=}C_7H_6Fe(CO)_3$ (Table **4**, No. **48**). The 4:1 mixture of isomers No. 48a and b reacts with freshly sublimed $(NC)_2C{=}C(CN)_2$ in CH_2Cl_2 at room temperature for 1 h. Removal of solvent and recrystallization gives a mixture of isomers No. 10a and b in 26% yield [66].

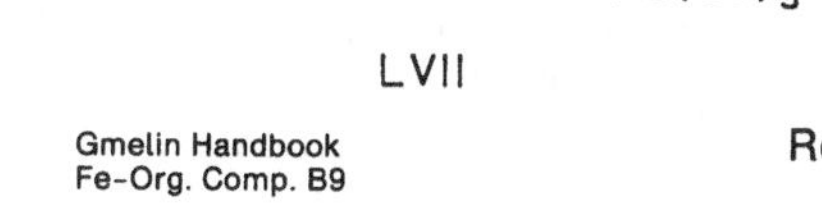

LVII 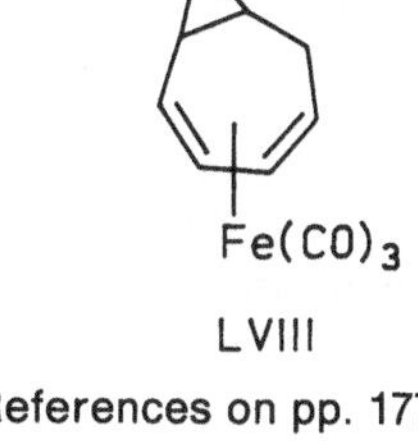 LVIII 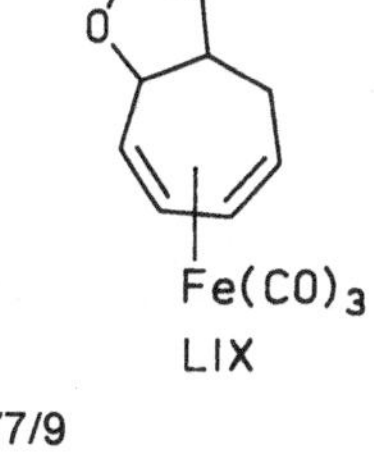 LIX 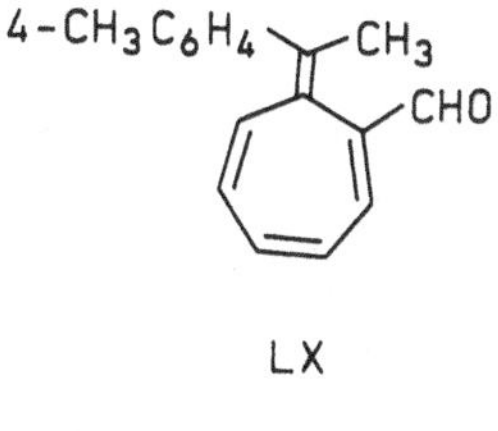 LX

References on pp. 177/9

$CH_3(C_6H_5)C{=}C_7H_6Fe(CO)_3$ (Table **4**, No. **50**) gives No. 11 in the reaction with $(NC)_2C{=}C(CN)_2$ [59].

$CH_3(4\text{-}CH_3\text{-}C_6H_4)C{=}C_7H_6Fe(CO)_3$ (Table **4**, No. **51**) gives LX (p-tolyl group trans to CHO) in the reaction with $POCl_3$ in $(CH_3)_2NCHO$ [26].

$(C_6H_5)_2C_2OC_7H_6Fe(CO)_3$ (Table **4**, No. **55**). $N_2C(C_6H_5)COC_6H_5$ is dissolved in $C_7H_8Fe(CO)_3$ (C_7H_8 = cycloheptatriene) and added dropwise with stirring over a period of 40 min to $C_7H_8Fe(CO)_3$ containing a copper catalyst (soluble Cu^{II} chelate of salicylaldehyde-aniline Schiff's base) at 65 °C. Heating is continued for 20 min, and the mixture is then chromatographed with hexane as solvent and eluent. Excess $C_7H_8Fe(CO)_3$ and LVII are eluted first, followed by No. 55 (9% yield). Compounds LVIII and LIX are discussed as intermediates in this reaction and are unstable under the reaction conditions. Compound No. 55 crystallizes in the monoclinic space group $P2_1/c\text{-}C_{2h}^5$ with the unit cell parameters a = 9.653(3), b = 18.243(4), c = 11.293(2) Å, β = 102.54(2)°; Z = 4 molecules per unit cell. D_{calc} = 1.451 g/cm³. The main bond distances and angles are shown in **Fig. 16** [89].

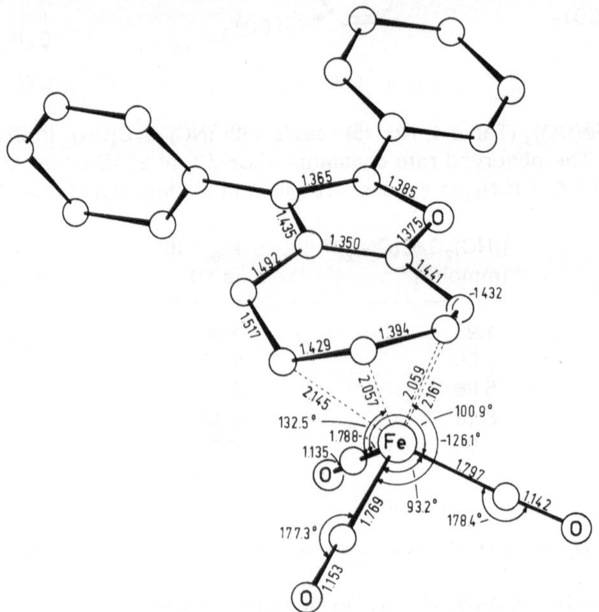

Fig. 16. Molecular structure of $(C_6H_5)_2C_2OC_7H_6Fe(CO)_3$ (No. 55) [89].

$4\text{-}O_2NC_6H_4NHN{=}C_7H_6Fe(CO)_3$ (Table **4**, No. **56**). To a solution of $C_7H_8Fe(CO)_3$ (C_7H_8 = cycloheptatriene) in acetone is added an acetone solution of $[N_2C_6H_4NO_2\text{-}4]BF_4$. After 5 min the solvent is removed and the residue is extracted with boiling $CHCl_3$. The solvent is removed, and the residue is dissolved in a minimum of acetone and chromatographed on alumina with hexane. Elution with acetone/ether (1:9) gives a red product contaminated with diacetone alcohol. Recrystallization from ether/hexane gives No. 56 (17% yield) as a red light-sensitive solid. Compound No. 56 can also be obtained by the reaction of No. 26 with $H_2NNHC_6H_4NO_2\text{-}4$ (see "Further information" for No. 26, p. 170). Compound No. 56 can also be prepared by column chromatography of freshly prepared $[1\text{-}(4\text{-}O_2NC_6H_4N_2)C_7H_8Fe(CO)_3]BF_4$ (C_7H_8 = hepta-2,4-dienyl) on Brockman alumina (activity II). ¹H NMR spectroscopy shows that No. 56 is a mixture of two isomers No. 56a and b. In acetone-d_6 the two isomers occur in approximately equal concentrations. H′ refers to the less abundant isomer in acetone-d_6. ^{1}H NMR (acetone-d_6): δ=3.16 (m, H-3 or H-3′), 3.35 (m, H-3 or H-3′), 4.03 (m, H-6,6′); 5.59 (m, 1H), 6.03 (m, 6H), 6.32 (m, 1H) (H-1,1′,2,2′,4,4′,5,5′); 7.27 (t, H-10,10′,12,12′), 8.13 (d, H-9,9′,13,13′), 9.84 (br, s, NH), 10.04 (br, s, NH) ppm. At −90 °C the signals at 8.13 and 7.27 ppm are replaced by doublets (J=9 to 10 Hz) at 8.33, 8.20, 7.65, and 7.55 and a triplet at 7.08 ppm. In benzene-d_6, however, the isomer ratio is 2:1 and this allows a partial assignment. ^{1}H NMR (benzene-d_6): δ=2.45 (m, H-3), 2.88 (m, H-6), 5.56 (m, H-1,2,4,5), 6.54 (d, H-9,13; J(H-9,10)=J(H-12,13)=9), 7.37 (s, NH), 8.00 (d, H-10,12) ppm for one isomer and δ=2.25 (m, H-3′), 3.75 (m, H-6′), 4.86 (m, H-1′,2′,4′,5′), 6.53 (d, H-9′,13′; J(H-9′,10′, H-12′,13′)=9), 7.08 (s, NH), 8.03 (d, H-10′,12′) ppm for the other. Similarly the ^{13}C NMR spectrum in acetone-d_6 gives δ=49.0, 57.8 (C-3,6); 89.0, 95.0 (C-4,5); 113.0 (C-10,12), 126.5 (C-9,13); 125.1, 133.1 (C-1,2); 149.0, 151.1 (C-8,11), 211 (CO) ppm for one isomer and δ=55.6, 65.5 (C-3′,6′); 89.2, 95.6 (C-4′,5′); 110.4, 140.8 (C-1′,2′); 112.7 (C-10′,12′), 126.5 (C-9′,13′); 148.1, 151.3 (C-8′,11′), 211 (CO) ppm for the other. Analogous ^{13}C NMR spectrum in benzene-d_6: δ=46.3, 57.8 (C-3,6); 87.3, 93.3 (C-4,5); 112.6 (C-10,12), 141.1 (C-7); 146.9, 149.3 (C-8,11) ppm and δ=53.9, 64.5 (C-3′,6′); 87.5, 93.9 (C-4′,5′); 112.1 (C-10′,12′); 108.0, 139.9 (C-1′,2′); 141.1 (C-7′); 146.4, 148.9 (C-8′,11′), 210.1 (CO) ppm. Both isomers show in the IR spectrum (n-hexane) ν(CO) bands at 1987, 1999, 2059 cm^{-1}. The complex No. 56 is slightly soluble in hexane but very soluble in polar solvents such as acetone to give deep orange solutions. Sample decomposition precluded a high temperature NMR study of the isomerization No. 56a ⇌ b. Oxidation with $ON(CH_3)_3 \cdot 2H_2O$ in acetone for 2 h liberates the organic ligand in 87% yield [86].

References:

[1] W. Hübel, E. Weiss (Chem. Ind. [London] **1959** 703). — [2] E.H. Braye, C. Hoogzand, W. Hübel, U. Krüerke, R. Merényi, E. Weiss (Intern. Conf. Coord. Chem., London 1959, Spec. Publ. No. 13, pp. 190/7). — [3] D.A. Brown (J. Inorg. Nucl. Chem. **13** [1960] 212/7). — [4] W. Hübel (unpublished results from [7]). — [5] E. Weiss, R. Merényi, W. Hübel (Chem. Ind. [London] **1960** 407/8).

[6] R.P. Dodge, V. Shomaker (unpublished results from [7]). — [7] D.L. Smith, L.F. Dahl (J. Am. Chem. Soc. **84** [1962] 1743/4). — [8] G. Wilkinson (Brit. 894852 [1962]). — [9] E. Weiss, W. Hübel, J. Nielsen, A. Gerondal, J. Vandervalle (Chem. Ber. **95** [1962] 1179/85). — [10] E. Weiss, W. Hübel, R. Merényi (Chem. Ber. **95** [1962] 1155/69).

[11] R.B. King (Inorg. Chem. **2** [1963] 807/10). — [12] R.P. Dodge (J. Am. Chem. Soc. **86** [1964] 5429/31). — [13] R.H. Herber, R.B. King, G.K. Wertheim (Inorg. Chem. **3** [1964] 101/7). — [14] B.J. Nicholson (J. Am. Chem. Soc. **88** [1966] 5156/65). — [15] R. Wenzl (Diss. Univ. Köln 1969, pp. 1/136).

[16] R.B. King (J. Am. Chem. Soc. **91** [1969] 7217/23). — [17] D.L. Smith (Diss. Univ. Wisconsin from [18, Ref. 37]). — [18] N.L. Clinton, C.P. Lillya (J. Am. Chem. Soc. **92** [1970] 3058/64). — [19] A. Eisenstadt, S. Winstein (Tetrahedron Letters **1971** 613/6). — [20] G.C. Farrant, G.T. Rodeheaver, D.F. Hunt (5th Intern. Conf. Organometal. Chem., Moscow 1971, Vol. 1, Abstr. No. 20).

[21] M.I. Foreman, D.G. Leppard (J. Organometal. Chem. **31** [1971] C31/C33). — [22] D.J. Ehntholt (Diss. Univ. New York 1971; Diss. Abstr. Intern. B **32** [1972] 4486). — [23] B.F.G. Johnson, J. Lewis, P. McArdle, G.L.P. Randall (Chem. Commun. **1971** 177/8). — [24] G.T. Rodeheaver, G.C. Farrant, D.F. Hunt (J. Organometal. Chem. **30** [1971] C22/C24). — [25] E.J. Forbes, M.K. Sultan, P.C. Uden (Anal. Letters **5** [1972] 927/30).

[26] B.F.G. Johnson, J. Lewis, P. McArdle, G.L.P. Randall (J. Chem. Soc. Dalton Trans. **1972** 2076/83). – [27] D.F. Hunt, G.C. Farrant, G.T. Rodeheaver (J. Organometal. Chem. **38** [1972] 349/65). – [28] D.F. Hunt, J.W. Russell, R.L. Torian (J. Organometal. Chem. **43** [1972] 175/83). – [29] J.A.S. Howell, B.F.G. Johnson, P.L. Josty, J. Lewis (J. Organometal. Chem. **39** [1972] 329/33). – [30] G.T. Rodeheaver (Diss. Univ. Virginia 1973; Diss. Abstr. Intern. B **34** [1973] 1433).

[31] R.C. Kerber, D.J. Ehntholt (J. Am. Chem. Soc. **95** [1973] 2927/34). – [32] A. Eisenstadt (unpublished results from [33]). – [33] A. Eisenstadt, J.M. Guss, R. Mason (J. Organometal. Chem. **80** [1974] 245/56). – [34] R.E. Davis, T.A. Dodds, T.H. Hseu, J.C. Wagnon, T. Devon, J. Tancrede, J.S. McKennis, R. Pettit (J. Am. Chem. Soc. **96** [1974] 7562/3). – [35] B.F.G. Johnson, J. Lewis, P. McArdle, J.R. Norton (J. Chem. Soc. Dalton Trans. **1974** 1253/8).

[36] Y. Shvo, E. Hazum (J. Chem. Soc. Chem. Commun. **1974** 336/7). – [37] A. Eisenstadt (J. Organometal. Chem. **97** [1975] 443/51). – [38] M. Franck-Neumann, D. Martina (Tetrahedron Letters **1975** 1759/62). – [39] R. Gompper, W. Reiser (Tetrahedron Letters **1976** 1263/4). – [40] Y. Shvo, E. Hazum (J. Chem. Soc. Chem. Commun. **1975** 829/30).

[41] V.N. Narasimhachari (Diss. Univ. Texas 1975; Diss. Abstr. Intern. B **36** [1975] 745). – [42] A. Eisenstadt (J. Organometal. Chem. **113** [1976] 147/56). – [43] H. Behrens, K. Geibel, R. Kellner, H. Knöchel, M. Moll, E. Sepp (Z. Naturforsch. **31b** [1976] 1021/2). – [44] B.F.G. Johnson, J. Lewis, D. Wege (J. Chem. Soc. Dalton Trans. **1976** 1874/80). – [45] Z. Goldschmidt, Y. Bakal (Tetrahedron Letters **1976** 1229/32).

[46] L. Kruczynski, J. Takats (Inorg. Chem. **15** [1976] 3140/7). – [47] C.P. Lewis (Diss. Univ. North Carolina 1976; Diss. Abstr. Intern. B **38** [1977] 692/3). – [48] R.B. King (Israel J. Chem. **15** [1976/77] 181/8). – [49] M. Bonadeo, C. de Micheli, R. Gandolfi (J. Chem. Soc. Perkin Trans. I **1977** 939/44). – [50] M. Bonadeo, R. Gandolfi, C. de Micheli (Gazz. Chim. Ital. **107** [1977] 577/8).

[51] M.S. Brookhart, C.P. Lewis, A. Eisenstadt (J. Organometal. Chem. **127** 1977] C14/C18). – [52] F.A. Cotton, B.E. Hanson (Israel. J. Chem. **15** [1977] 165/73). – [53] D. Cunningham, P. McArdle, H. Sherlock, B.F.G. Johnson, J. Lewis (J. Chem. Soc. Dalton Trans. **1977** 2340/4). – [54] M. Green, S.M. Heathcock, T.W. Turney, D.M.P. Mingos (J. Chem. Soc. Dalton Trans. **1977** 204/11). – [55] M. Franck-Neumann, D. Martina (Tetrahedron Letters **1977** 2293/6).

[56] M. Franck-Neumann, F. Brion, D. Martina (Tetrahedron Letters **1978** 5033/6). – [57] G.D. Andreetti, G. Bocelli, P. Sgarabotto (J. Organometal. Chem. **150** [1978] 85/92). – [58] S. Litman, A. Gedanken, Z. Goldschmidt, Y. Bakal (J. Chem. Soc. Chem. Commun. **1978** 983/4). – [59] P. McArdle (J. Organometal. Chem. **144** [1978] C31/C33). – [60] K.J. Karel, M. Brookhart (J. Am. Chem. Soc. **100** [1978] 1619/21).

[61] Z. Goldschmidt, S. Antebi (Tetrahedron Letters **1978** 1225/8). – [62] R. Gandolfi (unpublished results from [57]). – [63] M.R. Blake, J.L. Garnett, I.K. Gregor, S.B. Wild (J. Organometal. Chem. **178** [1979] C37/C42). – [64] K.J. Karel (Diss. Princeton Univ. 1978; Diss. Abstr. Intern. B **39** [1979] 4353). – [65] Z. Goldschmidt, Y. Bakal (J. Organometal. Chem. **168** [1979] 215/25).

[66] Z. Goldschmidt, Y. Bakal (J. Organometal. Chem. **179** [1979] 197/204). – [67] C.P. Lewis, W. Kitching. A. Eisenstadt, M. Brookhart (J. Am. Chem. Soc. **101** [1979] 4896/906). – [68] M. Cavazza, G. Morganti, F. Pietra (J. Org. Chem. **45** [1980] 2001/4). – [69] R.F. Childs, A. Varadarajan (J. Organometal. Chem. **184** [1980] C28/C32). – [70] M.R. Blake, J.L. Garnett, I.K. Gregor, D. Nelson (J. Organometal. Chem. **188** [1980] 203/10).

[71] R. Gandolfi, L. Toma (Tetrahedron **36** [1980] 935/41). – [72] S.K. Chopra, M.J. Hynes, P. McArdle (J. Chem. Soc. Dalton Trans. **1981** 586/9). – [73] K. Broadley, N.G. Connelly, R.M. Mills, M.W. Whiteley, P. Woodward (J. Chem. Soc. Chem. Commun. **1981** 19/20). – [74] S.K. Chopra, G. Moran, P. McArdle (J. Organometal. Chem. **214** [1981] C36/C38). – [75] T. Ishizu, K. Harano, M. Yasuda, K. Kanematsu (J. Org. Chem. **46** [1981] 3630/4).

[76] T. Jenny, W. von Philipsborn, J. Kronenbitter, A. Schwenk (J. Organometal. Chem. **205** [1981] 211/22). – [77] S. Zobl-Ruh, W. von Philipsborn (Helv. Chim. Acta **64** [1981] 2378/82). – [78] J. Daub, A. Hasenhündl, K.M. Rapp (Chem. Ber. **115** [1982] 2643/51). – [79] S.K. Chopra, M.J. Hynes, G. Moran, J. Simmie, P. McArdle (Inorg. Chim. Acta **63** [1982] 177/81). – [80] T. Ban, K. Nagai, Y. Miyamoto, K. Harano, M. Yasuda, K. Kanematsu (J. Org. Chem. **47** [1982] 110/6).

[81] K.J. Karel, T.A. Albright, M. Brookhart (Organometallics **1** [1982] 419/30). – [82] N. Morita, T. Asao (Chem. Letters **1982** 1575/8). – [83] M. Franck-Neumann, M.P. Heitz, D. Martina (Tetrahedron Letters **24** [1983] 1615/6). – [84] Z. Goldschmidt, Y. Bakal (Tetrahedron Letters **1977** 955/8). – [85] Z. Goldschmidt, S. Antebi (J. Organometal. Chem. **259** [1983] 119/25).

[86] N.G. Connelly, A.R. Lucy, J.B. Sheridan (J. Chem. Soc. Dalton Trans. **1983** 1465/8). – [87] T.A. Albright, P. Hofmann, C.P. Lillya, P.A. Dobosh (J. Am. Chem. Soc. **105** [1983] 3396/411). – [88] K. Broadley, N.G. Connelly, R.M. Mills, M.W. Whiteley, P. Woodward (J. Chem. Soc. Dalton Trans. **1984** 683/8). – [89] Z. Goldschmidt, S. Antebi, I. Goldberg (J. Organometal. Chem. **260** [1984] 105/13).

1.4.1.4.1.2.4.2.4.3 Other Substituted Cycloheptatrieneirontricarbonyl Complexes

Most of the complexes shown in Table 5, pp. 182/90, can be prepared by the following methods:

Method I: $Fe(CO)_5$ reacts with an organic substrate.

a. $Fe(CO)_5$ and 2,5,7-triphenylnorcaradiene are photolyzed in C_6H_6. After 10 h, it is necessary to filter the solution to remove $Fe_2(CO)_9$, but no evidence of an iron tricarbonyl complex is found. After an additional 14 h of photolysis, the solution is very dark. The volatiles are evaporated, and the residue is chromatographed by medium-pressure liquid chromatography with hexane to give two yellow bands and one very broad red band. The first yellow band contains unreacted 2,5,7-triphenylnorcaradiene, and the second yellow band gives No. 15. The red band is collected in two fractions. The first contains 1,4,7-triphenylcycloheptatriene, No. 3, and minor amounts of unidentified products. The second fraction contains predominantly No. 3. Photolysis of $Fe(CO)_5$ and 3,4-dimethyl-7-phenyl-7-carbomethoxynorcaradiene in C_6H_6 for 12 h followed by work-up as before but elution with hexane or hexane/C_6H_6 gives at least six separate bands; most of the ^1H NMR spectra are too complex to permit full characterization of the compounds present. The fourth band contains a small amount of an oil with an ^1H NMR spectrum suggesting the presence of the two isomers of No. 21 [32]. Photolysis of $Fe(CO)_5$ with a mixture of isomers I a and b ($R=CH_3$, $R'=H$), or I a, b, c ($R=H$, $R'=CH_3$) in C_6H_6 for 12 h followed by work-up as in Method I b gives Nos. 23 or 24, respectively [28]. Photolysis of $Fe(CO)_5$ with $^3LMo(CO)_2C_7H_7$ in C_6H_6 for 24 to 27 h [27] or with $C_5H_5Mo(CO)_2C_7H_7$ in ether for 4 d [19] followed by chromatography on neutral alumina (activity II for Nos. 28 to 31 [27], grade 0 for No. 32 [19]) gives the pure products. No. 32 elutes with CH_2Cl_2 [19].

References on pp. 196/7

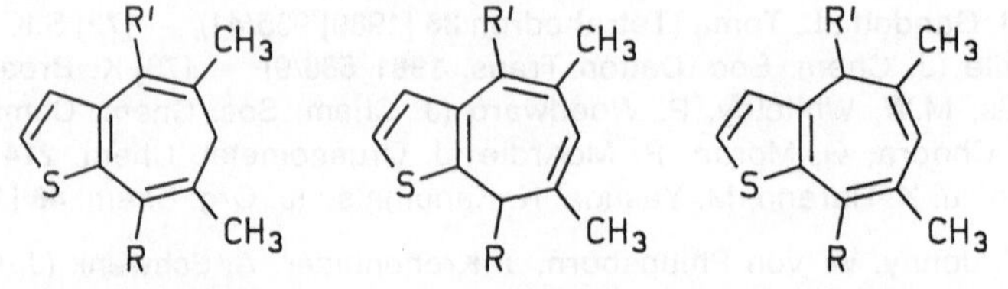

b. $Fe(CO)_5$ and a mixture of IIa and b are heated in a sealed tube at 110 to 120 °C for 4 d. After evaporation of the solvent and chromatography of the residue with pentane, pure No. 17 is obtained [28].

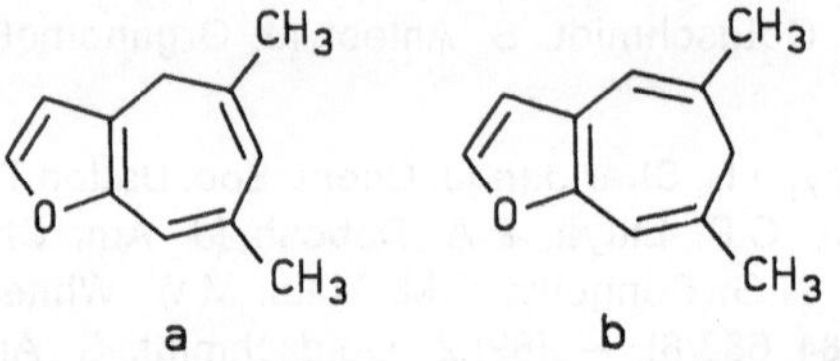

Method II: $Fe_2(CO)_9$ reacts with an organic substrate.

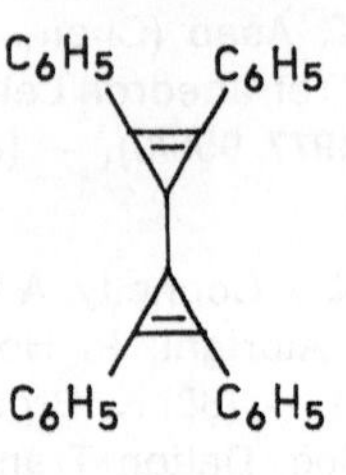

a. Reaction of $Fe_2(CO)_9$ with $C_5H_5Mo(CO)_2C_7H_7$ in n-heptane at 80 °C for 24 h followed by work-up as in Method Ia gives No. 32 [19].

b. $Fe_2(CO)_9$ reacts with the free ligand ^4L in C_6H_6 at 55 °C for 1.5 h. The reaction mixture is filtered, evaporated, and the residue is crystallized from hexane or alternatively chromatographed on neutral alumina (activity III) to give Nos. 4, 5, 7 in 60 to 80% yield [23, 24]. Refluxing $Fe_2(CO)_9$ and 5,6,7-trimethylfuro[b]tropilidene in C_6H_6 for 8 h followed by filtration, removal of the solvent, and chromatography of the residue on silica gel G 60 gives No. 22 [31]. Refluxing with a mixture of IIa and b for 24 h followed by work-up as in Method Ib gives No. 17. Similar reaction with a mixture of isomers I (R=R'=H) gives No. 18 [28].

c. $Fe_2(CO)_9$ and III react in toluene at room temperature for 20 h. The volatiles are removed and the residue is dissolved in toluene/hexane (1:3). Chromatography on silica gel with toluene/hexane (1:5 to 1:1) gives a green band ($Fe_3(CO)_{12}$), followed by a red-yellow to brown compound, and finally No. 27 as the main fraction [16].

d. $Fe_2(CO)_9$ reacts with $CH_3C{\equiv}C(CH_2)_2C{\equiv}CCH_3$ in tetrahydrofuran to give two new compounds in low yields which cannot be completely characterized. One is probably a cyclopentadienone dimer, $(C_8H_{10}O)_2$, and the other may be a cycloheptatrienone iron tricarbonyl complex, No. 33 [27].

Method III: $Fe_3(CO)_{12}$ is heated with an organic substrate in an inert solvent.

a. $Fe_3(CO)_{12}$ is refluxed with 2,5,7-triphenylnorcaradiene in C_6H_6 for 8 to 12 h. The solution is filtered through Celite and alumina and the solvent is removed. The residue is extracted with hexane, and the extracts are chromatographed (medium-pressure liquid chromatography) with hexane. The first two yellow bands correspond to 2,5,7-triphenylnorcaradiene and No. 15. A third yellow band, although the major product in this reaction (ca. 30% yield), cannot be completely characterized (^{1}H NMR ($CDCl_3$): δ=2.5 to 3.3 (m, 5H), 6.3 (s, 2H), 7.3 to 7.8 (m, $3C_6H_5$) ppm; IR (benzene-d_6): 1980, 2050 (CO) cm^{-1}) [32]. Refluxing $Fe_3(CO)_{12}$ and 2,3,5-triphenyltropone (^{4}L) in C_6H_6 for 5 h followed by filtration and chromatography on silica gel gives No. 25a and b [13].

b. $Fe_3(CO)_{12}$ and $C_6H_5C{\equiv}CH$ (ratio 0.75:0.1) are refluxed in petroleum ether (b.p. 60 to 70 °C)/C_6H_6 for 1.75 h. Chromatography on acidic Al_2O_3 gives $(C_6H_5C_2H)_5Fe(CO)_4$, the isomers No. 25a and b, $O{=}C_5H_2(C_6H_5)_2Fe(CO)_3$ (see 1.4.1.4.1.2.2.2.4 in "Organoiron Compounds" B7, 1981, p. 164), $(C_6H_5C_2H)_4$-$Fe(CO)_3$, $O{=}C_5(C_6H_5)_2H_2Fe(CO)_3$ (see 1.4.1.4.1.2.2.2.4 in "Organoiron Compounds" B7, 1981, p. 163), and $C_6H_3(C_6H_5)_3COFe_2(CO)_5$ (see 2.6.1.2.1 in "Organoiron Compounds" C5, 1981, p. 7) [1, 2, 6], cf. [3, 13]. Similar products are obtained at 60 °C after 3 h [4], at 75 °C after 0.5 h [1, 2, 6], or at 80 °C after a few minutes [4]. The volatiles are evaporated, and the products are dissolved in CS_2 and chromatographed as before. At 75 °C after 0.5 h $(C_6H_5C_2H)_5Fe(CO)_4$ is the predominant product, and no $C_6H_3(C_6H_5)_3COFe_2$-$(CO)_5$ is found [1, 2, 6]. Similarly, $Fe_3(CO)_{12}$ reacts with 4-$BrC_6H_4C{\equiv}CH$ in light petroleum ether at 65 °C for 3 h. The reaction mixture is filtered, and the residue is dissolved in CS_2. Both solutions are chromatographed on alumina with light petroleum ether to give $(4\text{-}BrC_6H_4C_2H)_2Fe_2(CO)_6$ (2%) and 1,2,4-$(4\text{-}BrC_6H_4)_3C_6H_3$. Elution with CS_2 gives $(4\text{-}BrC_6H_4C_2H)_3Fe_2(CO)_6$ (3%) and 1,2,4-$(4\text{-}BrC_6H_4)_3C_6H_3$. Elution with CS_2 or CS_2/C_6H_6 gives 1,3,5-$(4\text{-}BrC_6H_4)_3C_6H_3$, and further elution with C_6H_6 containing some ether gives $(4\text{-}BrC_6H_4C_2H)_5Fe(CO)_4$ (small amount), elution with C_6H_6/ether gives No. 26 (tautomer a) as a red oil which crystallizes upon addition of ether. Further elution with C_6H_6/ether gives a tris(p-bromophenyl)tropone, and elution with acetone or acetone/CH_3OH gives $O{=}C_5(C_6H_4Br\text{-}4)_2H_2Fe(CO)_3$ (see 1.4.1.4.1.2.2.2.4 in "Organoiron Compounds" B7, 1981, p. 164). Similar reaction in C_6H_6 at 60 °C for 4.5 h followed by chromatography of the CS_2 solution on acidic alumina with C_6H_6/ether (small amount) gives No. 26 (tautomer b) and $(4\text{-}BrC_6H_4C_2H)_5Fe(CO)_4$; pure tautomer b is obtained by additional chromatography with C_6H_6/ether (1:1) or pure ether. Further elution with C_6H_6/CH_3OH gives $O{=}C_5(C_6H_4Br\text{-}4)_2H_2Fe(CO)_3$ and $O{=}C_5H_2(C_6H_4Br\text{-}4)_2Fe(CO)_3$ (see 1.4.1.4.1.2.2.2.4 in "Organoiron Compounds" B7, 1981, p. 165 [13].

Method IV: ^{4}LFe(CO)$_3$ complexes are the starting compounds.

a. Complex IV reacts with a twofold excess of $CH_3COCl/AlCl_3$ in CH_2Cl_2 at room temperature to give a ca. 4:1 mixture of tautomers No. 6a and b in 80% yield. Tautomer 6b is obtained by cooling a CH_2Cl_2/ether solution of the mixture of tautomers to −78 °C, where it crystallizes first. This subsequently displaces the equilibrium completely in its favor [29].

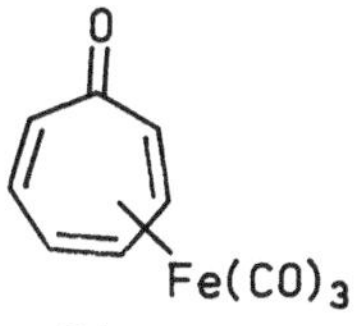

O
Fe(CO)$_3$
IV

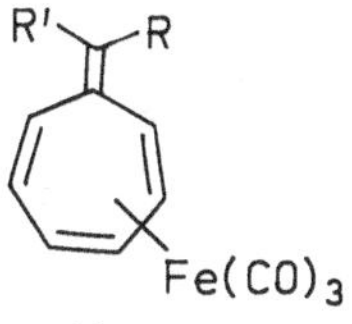

R′ R
Fe(CO)$_3$
V

References on pp. 196/7

b. Complexes V (R=H, R'=C_6H_5, $C_6H_4CH_3$-4) react with $OHCN(CH_3)_2/OPCl_3$ in ether at 0 °C for 72 h. The reaction mixture is poured into ice-water and left at room temperature for 15 h, followed by ether extraction, washing with H_2O, drying ($MgSO_4$), and chromatography on silica gel with $CH_3CO_2C_2H_5$/ toluene (1:10) to give Nos. 12 and 13 [21].

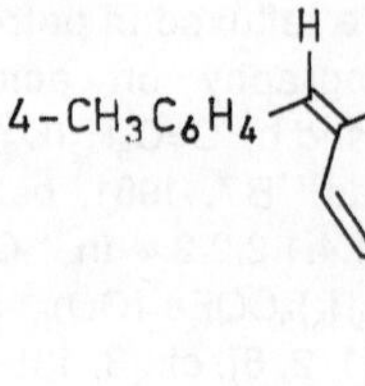

VI 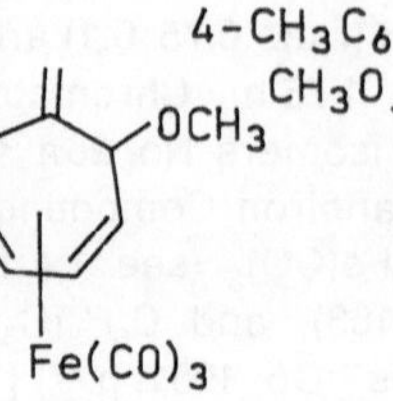 VII 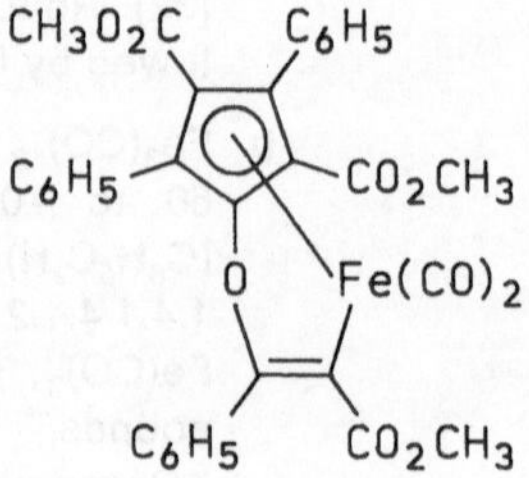 VIII

c. A mixture of the isomers VI and VII reacts with $(NC)_2C=C(CN)_2$ in C_6H_6 at 80 °C for 3 h. All volatiles are removed at room temperature, and the residue is chromatographed on silica gel with $CH_3CO_2C_2H_5$/toluene (1:5) to give Nos. 1 and 2 in 25% total yield [21].

A compound of the composition $(C_6H_5C_2CO_2CH_3)_3Fe(CO)_4$, initially thought to have a cycloheptatriene structure [7], has been found instead to be VIII [15, 17].

Table 5
Other Compounds of the Type $^4LFe(CO)_3$ with ^4L=Substituted Cycloheptatriene.
Further information on numbers preceded by an asterisk is given at the end of the table, pp. 190/6.
For abbreviations and dimensions see p. VIII.

No.	compound	method of preparation (yield in %), properties and remarks	Ref.
trisubstituted complexes:			
1	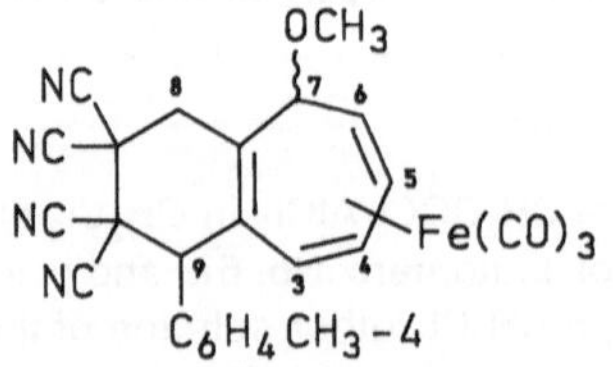	IVc red-brown oil ^{1}H NMR (CS_2): 2.33 (s, CH_3), 2.8 to 4.4 (m, H-3,6,7,8,9), 2.97 (s, OCH_3), 3.40 (s, OCH_3), 6.62 (m, H-4,5), 7.10 (m, C_6H_4) IR (liquid film): 1960, 2070 (CO); 2140 (CN) mass spectrum: $[M-nCO]^+$ (n=0 to 3), contaminated with toluene	[21]
2	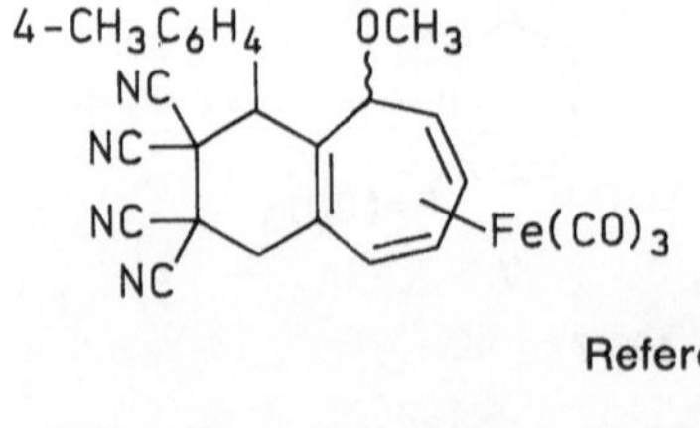	IVc	[21]

Table 5 [continued]

No.	compound	method of preparation (yield in %), properties and remarks	Ref.
3	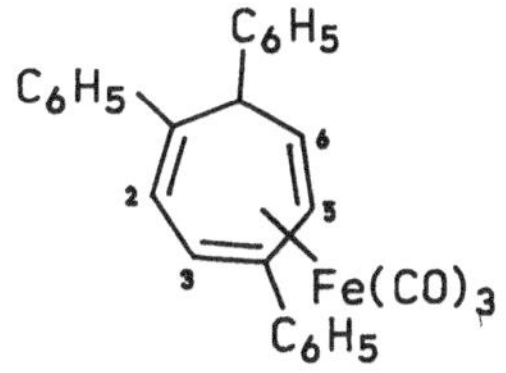	I a ^{1}H NMR (toluene-d_8): 3.48 (dd, H-6; J(H-5, 6) = 8, J(H-6, 7) = 5.5), 3.82 (dd, H-3; J(H-3, 5) = 2, J(H-2, 3) = 8.5), 4.28 (dd, H-7; J(H-2, 7) = 1), 5.30 (dd, H-5), 6.50 (dd, H-2), 7.0 to 7.6 (m, C_6H_5) IR ($CDCl_3$): 1990, 2050	[32]
*4	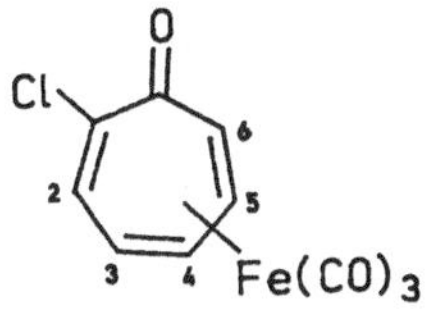	II b (60 to 80) m.p. 130 to 131°, red-orange solid ^{1}H NMR ($CDCl_3$): 2.61 (ddd, H-3), 3.22 (quint, H-6), 6.32 (m, H-4, 5), 6.73 (d, H-2; J(H-2, 3) = 10) mass spectrum: $[M]^+$	[24]
*5	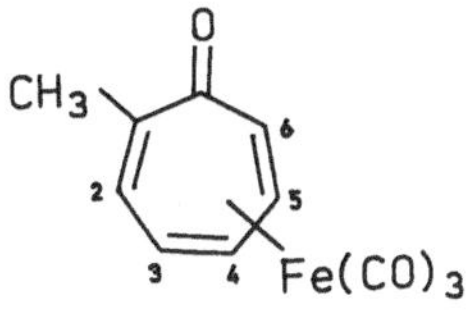	II b (45 or 60 to 80) m.p. 71 to 72°, red needles ^{1}H NMR ($CDCl_3$): 1.48 (d, CH_3; J(H-2, CH_3) = 1.5), 2.68 (br, t, H-3), 3.15 (m, H-6), 6.36 (m, H-2, 4, 5) IR (n-hexane): 1996, 2005, 2062 (CO) mass spectrum: $[M]^+$	[23, 24, 26]
*6a	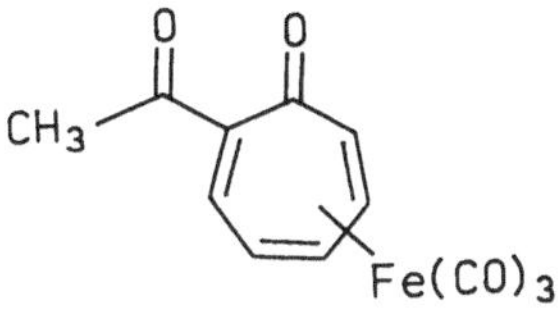	IV a m.p. 70°, purple crystals ^{1}H NMR ($CDCl_3$): 2.22 (s, CH_3), 2.74 (m, 1H), 3.21 (d, 1H; J = 8), 6.28 to 6.59 (m, 2H), 7.48 (d, 1H; J = 8) IR ($CHCl_3$): 1585, 1635, 1685, 2000 to 2015, 2080	[29]
b	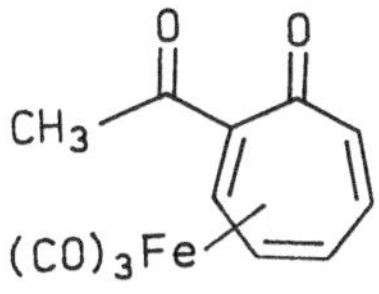	IV a m.p. 102°, red crystals ^{1}H NMR ($CDCl_3$): 2.34 (s, CH_3), 2.98 (m, 1H), 5.28 (d, 1H; J = 10), 6.38 (dd, 1H), 6.83 (dd, 1H), 6.95 (m, 1H) IR ($CHCl_3$): 1615, 1635, 1690, 2000 to 2015, 2080	[29]
*7	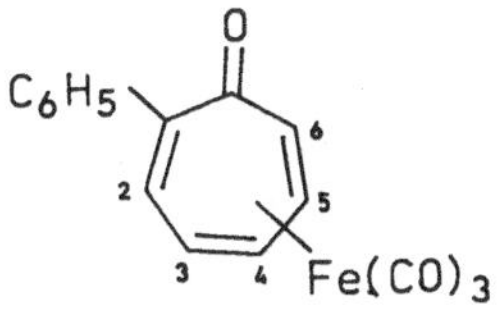	II b m.p. 146 to 148° (dec.), red-orange solid ^{1}H NMR ($CDCl_3$): 2.8 (br, t, H-3), 3.3 (m, H-6), 6.3 (m, H-4, 5), 6.62 (d, H-2), 7.15 (s, C_6H_5) mass spectrum: $[M]^+$	[24]

References on pp. 196/7

Table 5 [continued]

No.	compound	method of preparation (yield in %), properties and remarks	Ref.
*8	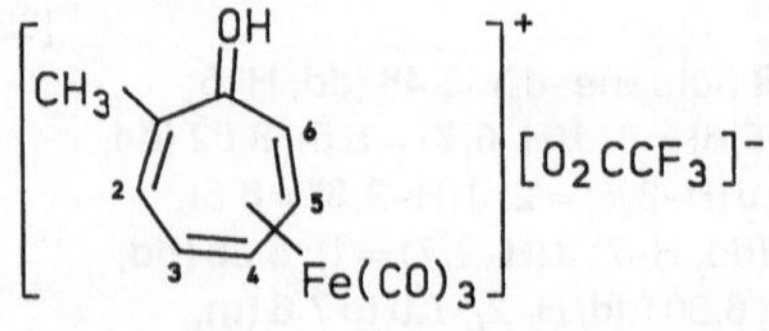 $[(2\text{-}CH_3C_7H_5\text{-}1\text{-}OH)Fe(CO)_3]^+[O_2CCF_3]^-$	see "Further information", pp. 190/1 blood-red ^{1}H NMR (CF_3CO_2H/liquid SO_2/SO_2F_2/CD_2Cl_2 at −40°): 1.50 (CH_3; J(H-2, CH_3) = 1.4), 2.77 (H-3), 3.14 (H-6), 6.70 to 6.94 (H-4, 5), 6.72 (H-2)	[26]
*9	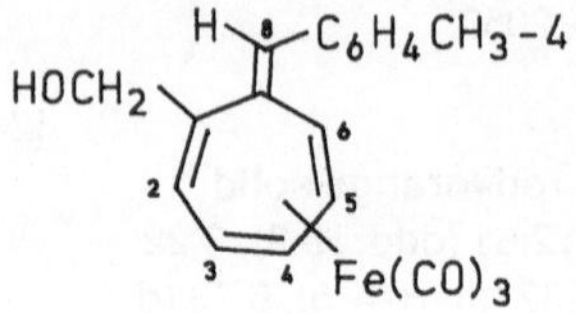 $HOCH_2$, H–C(8)–$C_6H_4CH_3$-4, Fe(CO)$_3$	see No. 13, p. 191 yellow gum ^{1}H NMR (CS_2): 2.26 (s, CH_3), 2.80 (d, H-3; J = 8), 2.90 (s, OH), 3.95 (m, CH_2), 4.20 (d, H-6; J = 8), 5.25 (m, H-5), 5.45 (m, H-4), 5.76 (d, H-2; J = 8), 6.66 (s, H-8), 7.06 (m, C_6H_4) IR (liquid film): 1960, 2045 (CO); 3350 (OH)	[21]
10	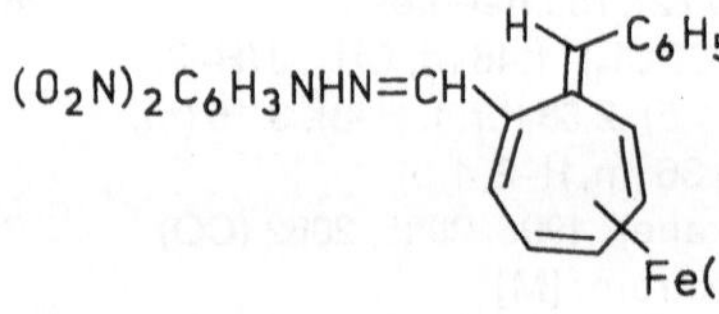 $(O_2N)_2C_6H_3NHN{=}CH$, H–C–C_6H_5, Fe(CO)$_3$	see No. 12, p. 191	[21]
11	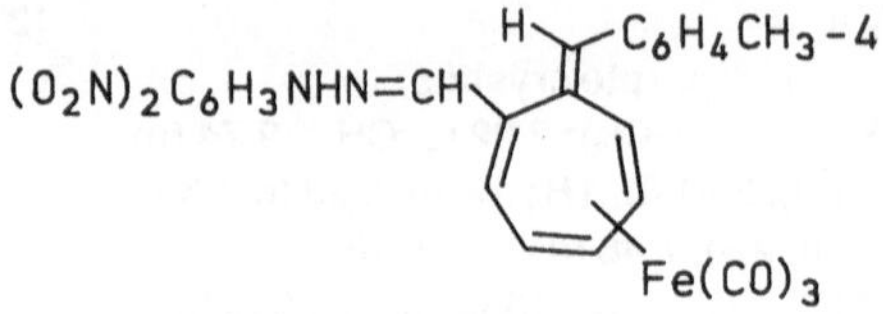 $(O_2N)_2C_6H_3NHN{=}CH$, H–C–$C_6H_4CH_3$-4, Fe(CO)$_3$	see No. 13, p. 191	[21]
*12	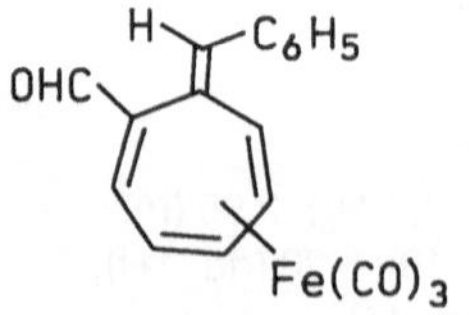 OHC, H–C–C_6H_5, Fe(CO)$_3$	IVb (70) m.p. 105°, orange crystalline solid IR (CS_2): 1686 (CHO); 1989, 2000, 2060 (CO)	[21]
*13	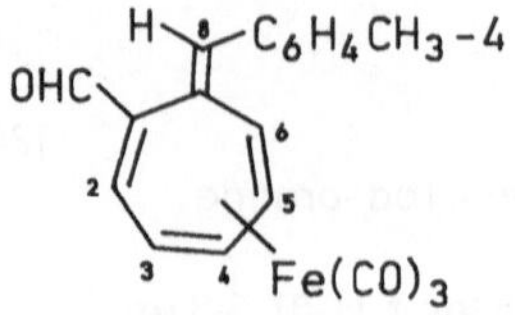 OHC, H–C(8)–$C_6H_4CH_3$-4, Fe(CO)$_3$	IVb (70) m.p. 97°, orange solid ^{1}H NMR: 2.30 (s, CH_3), 3.14 (t, H-3; J = 8), 4.45 (d, H-6; J = 7), 5.38 (t, H-5; J = 6), 5.68 (t, H-4; J = 7), 6.62 (d, H-2), 7.23 (m, C_6H_4), 8.20 (s, H-8), 9.24 (s, CHO) IR (CS_2): 1684 (CHO); 1987, 1999, 2059 (CO)	[21]

References on pp. 196/7

Table 5 [continued]

No.	compound	method of preparation (yield in %), properties and remarks	Ref.
*14	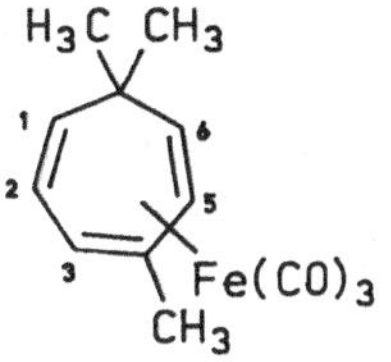	b.p. 150°/0.1 Torr, oil ^{1}H NMR ($CDCl_3$): 1.02 (s, CH_3-7), 1.15 (s, CH_3-7), 2.05 (s, CH_3-4), 2.85 (dd, H-3; J=2, J=8), 3.00 (br, d, obscured, H-6), 4.87 (dd, H-1; J=2, J=11), 5.10 (dd, H-5; J=2, J=7), 5.65 (dd, H-2; J=8, J=11) IR ($CHCl_3$): 1970, 2050 mass spectrum: $[M+1-nCO]^+$ (n=0 to 3)	[35]
*15	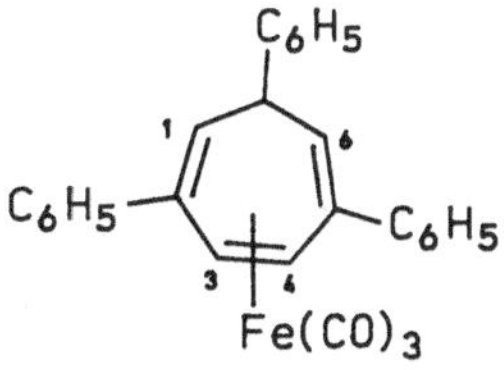	I a yellow oil ^{1}H NMR (toluene-d_8): 3.58 (t, H-7; J(H-1,7)=J(H-6,7)=2), 3.70 (dd, H-4; J(H-3,4)=8, J(H-4,5)=2), 4.20 (q, H-1; J(H-1,6)=J(H-1,3)=2), 5.48 (dd, H-3), 5.83 (q, H-6; J(H-4,6)=2), 7.20 to 7.75 (m, C_6H_5) IR ($CDCl_3$): 1995, 2055	[32]
*16	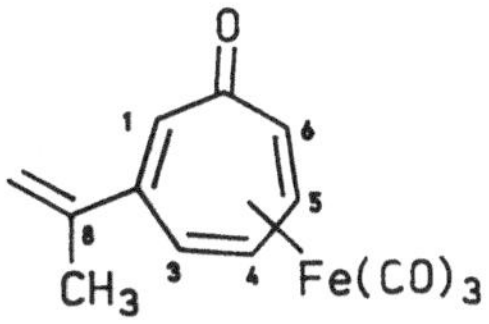	m.p. 101°, yellow-orange crystals ^{1}H NMR ($CDCl_3$): 1.87 (s, CH_3), 5.17 (s, CH_2), 5.19 (s, CH_2), 5.29 (s, H-1 or H-6) IR ($CHCl_3$): 1615 (C=O)	[29]

tetrasubstituted complexes (continued with No. 34):

No.	compound	method of preparation (yield in %), properties and remarks	Ref.
*17	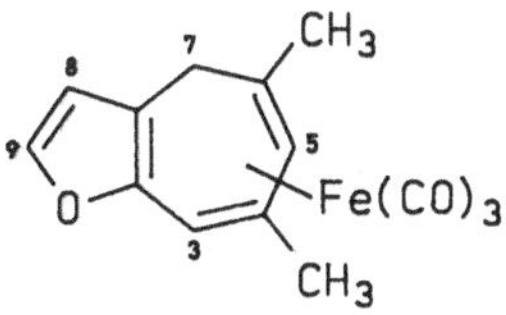	I b (7), II b (11) m.p. 84° (from pentane), yellow crystals ^{1}H NMR ($CDCl_3$): 1.81 (s, CH_3-6), 2.15 (s, CH_3-4), 2.71 (q, H-7), 3.40 (m, H-3), 5.20 (m, H-5), 6.00 (m, H-8), 7.05 (d, H-9)	[28]
*18	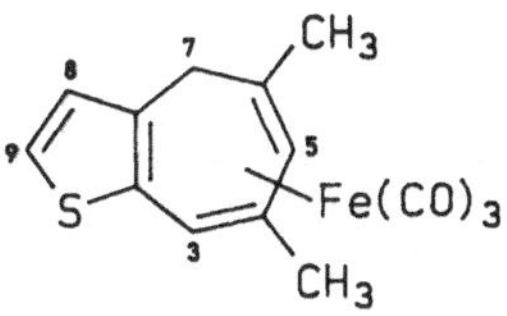	II b (10) m.p. 70° (from pentane), yellow crystals ^{1}H NMR ($CDCl_3$): 1.80 (s, CH_3-6), 2.15 (s, CH_3-4), 2.91 (q, H-7), 3.55 (m, H-3), 5.17 (m, H-5), 6.50 (d, H-8), 6.75 (d, H-9) mass spectrum: $[M-nCO]^{+\cdot}$ (n=0 to 3)	[28, 30]

References on pp. 196/7

Table 5 [continued]

No.	compound	method of preparation (yield in %), properties and remarks	Ref.
*19	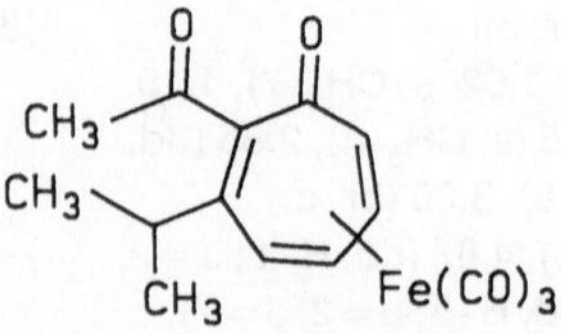	red-orange crystals ^{1}H NMR ($CDCl_3$): 0.98 (d, CH_3; J=7), 1.18 (d, CH_3; J=7), 2.18 (s, CH_3), 2.48 (7 lines, CH (isopropyl); J=7) IR ($CHCl_3$): 1610, 1700 (C=O)	[29]
*20	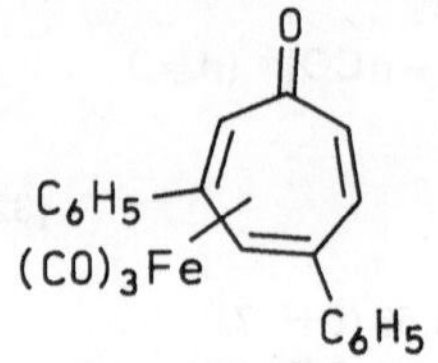	–	[18]
21	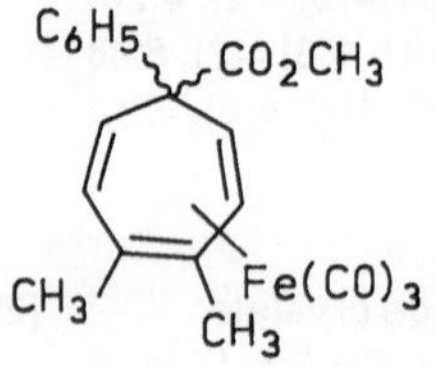	I a (small amount) yellow oil	[32]

pentasubstituted complexes (continued with No. 35):

No.	compound	method of preparation (yield in %), properties and remarks	Ref.
22	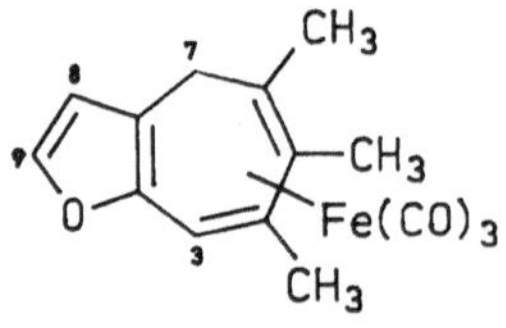	II b (22) m.p. 97° (from 1:1 ether/hexane), yellow crystals ^{1}H NMR ($CDCl_3$): 1.82 (s, CH_3-6), 1.88 (d, CH_3-4), 2.22 (s, CH_3-5), 2.72 (m, H-7), 3.40 (m, H-3), 6.18 (d, H-8), 7.05 (d, H-9) IR ($CHCl_3$): 1952, 1975, 2038 (CO) mass spectrum: $[M-nCO]^+$ (n=0 to 3) and further fragments given	[31]
23	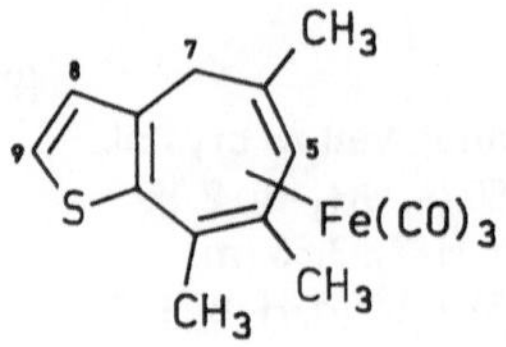	I a (12) m.p. 94° (from hexane), yellow-orange solid ^{1}H NMR ($CDCl_3$): 1.80 (s, CH_3-6), 2.05 (s, CH_3-3), 2.21 (s, CH_3-4), 2.98 (q, H-7), 5.10 (s, H-5), 6.62 (d, H-8), 6.95 (d, H-9) mass spectrum: $[M-nCO]^{+\cdot}$ (n=1 to 3) and further fragments given	[28, 30]

References on pp. 196/7

Table 5 [continued]

No.	compound	method of preparation (yield in %), properties and remarks	Ref.
24	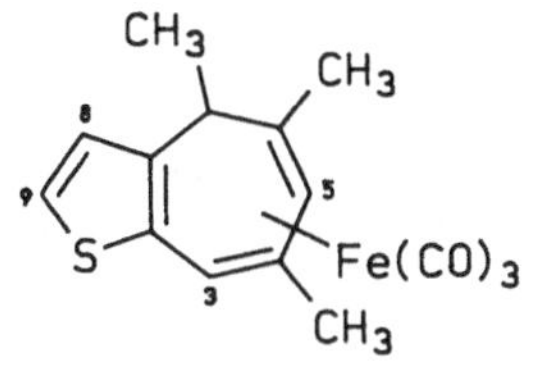	I a (13) m.p. 97° (from pentane), yellow solid ^{1}H NMR ($CDCl_3$): 1.37 (d, CH_3-7), 1.86 (s, CH_3-6), 2.13 (s, CH_3-4), 3.17 (q, H-7), 3.57 (m, H-3), 5.37 (m, H-5), 6.55 (d, H-8), 6.76 (d, H-9) mass spectrum: $[M-nCO]^{+\cdot}$ (n = 1 to 3) and further fragments given	[28, 30]
*25a	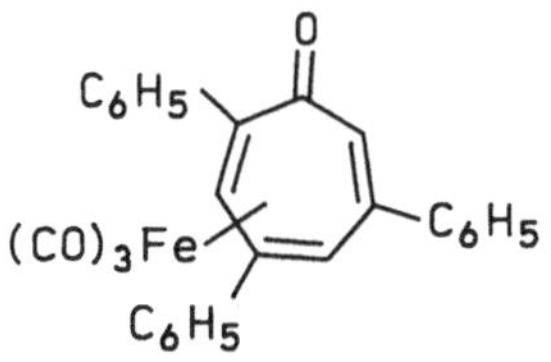	III a (17.5), III b m.p. 154 to 155° (dec.), m.p. 156 to 158° (dec.), orange rods, orange crystals (from C_6H_6/petroleum ether), orange-red crystals IR (KBr): 1623 (C=O); 2000, 2062 (CO)	[1 to 4, 6, 10, 13]
b	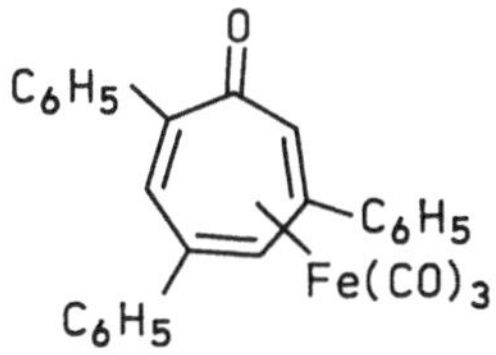	III a (33), III b m.p. 145 to 150° (dec.), dec.p. 148 to 150°, or m.p. 162 to 165° (dec.), dark red rods, red crystals (from C_6H_6/petroleum ether) IR (KBr): 1629 (C=O); 1996, 2062 (CO)	[1 to 4, 6, 10, 13]
*26	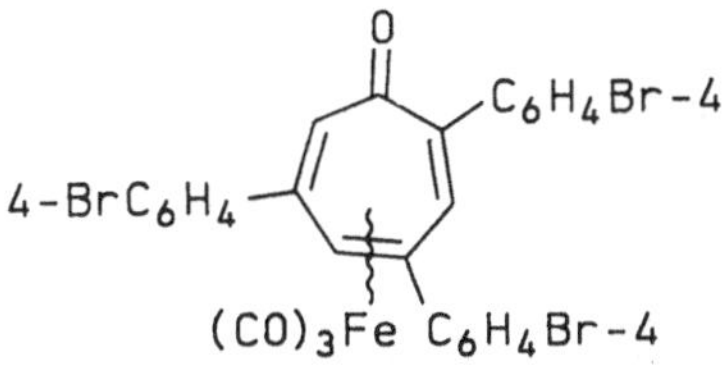	two tautomers, not assigned (compare No. 25, pp. 193/4)	[3, 13]
		tautomer a: III b (20) m.p. 199 to 205° (dec.), well-shaped dark red crystals (from CS_2/light petroleum ether or from C_6H_6/CH_3OH) IR (KBr): 1618 (C=O); 1992, 2000, 2058	[13]
		tautomer b: III b (20) m.p. 154 to 165° (dec.), thick red-brown crystals (from C_6H_6), m.p. 166 to 175° (dec.), thin yellow needles (from ether) IR (KBr): 1618 (C=O); 2008, 2066 (CO)	[13]

References on pp. 196/7

Table 5 [continued]

No.	compound	method of preparation (yield in %), properties and remarks	Ref.
hexasubstituted complex:			
*27	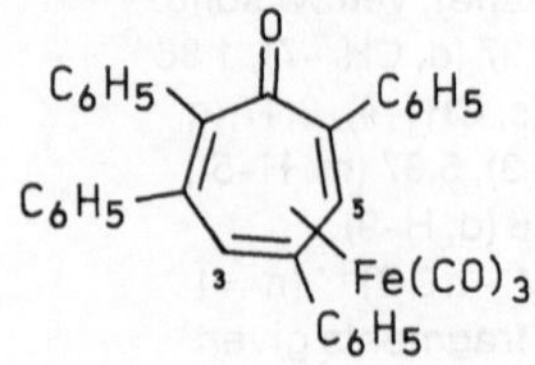 [structure: tetraphenyl-tropone $Fe(CO)_3$ complex; C_6H_5 substituents, O, positions 3, 5, $Fe(CO)_3$]	IIc (32) m.p. 187 to 188° (dec.), pale red prisms (from toluene/hexane) ^1H NMR: 3.56 ^{13}C NMR ($CDCl_3$): 55.6 (C-3), 94.8 (C-5), 109.3 (C-4), 126.1 to 130.8 (9 lines, C_6H_5), 131.8 (C-1 or 2), 136.8 (C_6H_5), 138.2 (C_6H_5), 142.7 (C_6H_5), 144.4 (C_6H_5), 156.3 (C-1 or 2), 195.1 (C-7), 204.3 (CO), 215.5 (CO) IR (hexane): 1618 (C=O); 2001, 2006, 2064 (CO)	[16]
Mo-complexes:			
*28	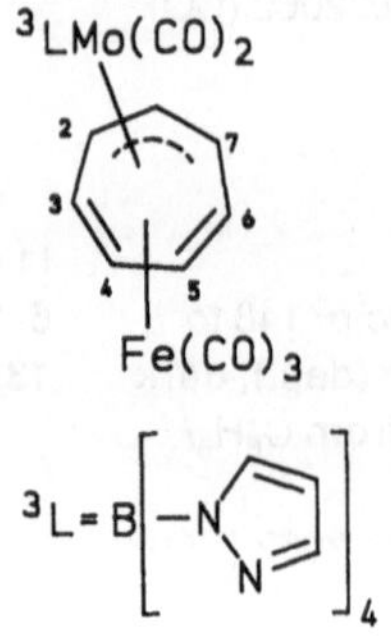 $^3LMo(CO)_2$ / $Fe(CO)_3$ on C_7H_7 ring (positions 2 to 7) $^3L = B[-N(pyrazolyl)]_4$	I a dec. 220° (from C_6H_6) ^1H NMR (CD_2Cl_2 at −56°): 4.05 (complex, H-3, 6), 4.71 (dt, H-1), 5.18 (t, H-2, 7), 5.80 (dd, H-4, 5) ^1H NMR (CD_2Cl_2 at 25°): 4.87 (br, C_7H_7); 6.17 (t), 6.35 (t), 6.55 (t), 7.25 (d), 7.86 (m), 8.23 (d), 8.69 (d) (pyrazolyl) ^1H NMR (toluene-d_8/cyclohexane at 83°): 4.55 (sharp s, C_7H_7) IR (CS_2/CCl_4): 610, 760, 803, 815, 850, 927, 1050, 1068, 1075, 1095, 1114, 1189, 1215, 1296, 1310, 1385, 1408, 1437, 2990, 3030, 3150 IR (cyclohexane): 1870, 1949, 1978, 1987, 2050 (CO)	[22]
*29	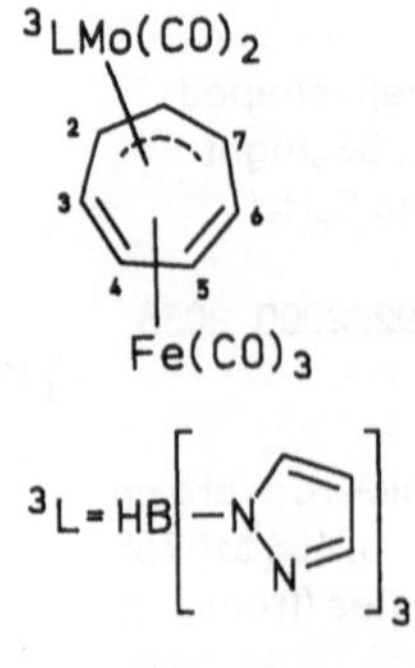 $^3LMo(CO)_2$ / $Fe(CO)_3$ on C_7H_7 ring (positions 2 to 7) $^3L = HB[-N(pyrazolyl)]_3$	I a m.p. 212° ^1H NMR (CD_2Cl_2 at −42°): 3.98 (complex, H-3, 6), 4.59 (dt, H-1), 5.13 (t, H-2, 7), 5.52 (dd, H-4, 5) ^1H NMR (CD_2Cl_2 at 25°): 4.80 (br, C_7H_7); 6.17 (t), 6.29 (t), 7.45 (d), 7.53 (d), 7.79 (d), 8.60 (d) (pyrazolyl) ^1H NMR (toluene-d_8 at 83°): 4.50 (sharp s, C_7H_7) IR (CCl_4/CS_2): 609, 723, 761, 790, 815, 990, 1055, 1075, 1130, 1210, 1225, 1309, 1356, 1407, 2989, 3040	[22]

References on pp. 196/7

Table 5 [continued]

No.	compound	method of preparation (yield in %), properties and remarks	Ref.
		IR (cyclohexane): 1874, 1950, 1980, 1990, 2050 (CO) IR (CS_2): 2470 (BH)	
*30	$^3LMo(CO)_2$ / $Fe(CO)_3$ (C_7H_7 ring, positions 1–7) $^3L = HB[\text{-N-pyrazolyl-3,4-}(CH_3)_2]_3$	Ia dec. 195° (from CH_2Cl_2/hexane) ^1H NMR (CD_2Cl_2 at −114°): 4.41 (complex, H-3,6), 4.84 (t, H-2,7), 5.47 (t, H-1), 6.09 (dd, H-4,5) ^1H NMR (CD_2Cl_2 at 25°): 2.52 (s), 2.79 (s), 3.03 (s), 3.41 (s) (CH_3); 5.24 (sharp s, C_7H_7); 6.21 (s), 6.30 (s) (pyrazolyl) IR (CCl_4/CS_2): 609, 630, 694, 783, 825, 987, 1043, 1070, 1080, 1210, 1380, 1420, 1440, 2927, 2960, 2985 IR (cyclohexane): 1842, 1925, 1970, 1979, 2040 (CO) IR (CS_2): 2520 (BH)	[22]
*31	$^3LMo(CO)_2$ / $Fe(CO)_3$ (C_7H_7 ring, positions 1–7) $^3L = H_2B[\text{-N-pyrazolyl-3,4-}(CH_3)_2]_2$	Ia	[22]
*32	$C_5H_5Mo(CO)_2$ / $Fe(CO)_3$ (C_7H_7 ring, positions 1–7)	Ia (55), IIa (ca. 5) m.p. 176° (dec.), red crystals (from CH_2Cl_2/n-hexane at −78° or from C_2H_5OH) ^1H NMR (CD_2Cl_2 or $CDCl_3$ at $< -50°$): 3.83 (complex, H-3,6), 4.50 (t, H-2,7), 4.98 (distorted t, H-1; J(H-1,2) ≈ 7.0), 5.21 (C_5H_5), 5.40 (dd, H-4,5) ^1H NMR ($CDCl_3$/toluene-d_8 (4:1) at $< -50°$): 3.77 (complex, H-3,6), 4.37 (t, H-2,7), 4.82 (distorted t, H-1; J(H-1,2) ≈ 7.0), 5.09 (C_5H_5), 5.37 (dd, H-4,5) ^1H NMR ($CDCl_3$ at 50°): 4.57 (s, C_7H_7), 5.13 (s, C_5H_5) IR (cyclohexane): 1880, 1950, 1975, 1980, 2040 (CO)	[19, 20, 22]

References on pp. 196/7

Table 5 [continued]

No.	compound	method of preparation (yield in %), properties and remarks	Ref.
compound with unknown structure:			
33	$(C_8H_{10})_3COFe(CO)_3$	IId	[27]
supplements:			
tetrasubstituted complex:			
34	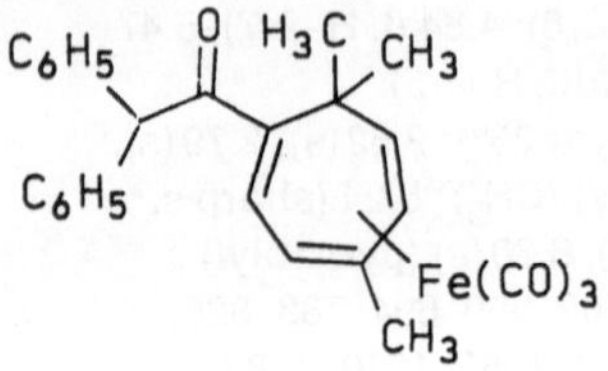	see No. 14, p. 191 m.p. 117 to 118°, crystals (from CH_3OH) ^1H NMR ($CDCl_3$): 1.13 (s, CH_3), 1.44 (s, CH_3), 2.04 (s, CH_3), 2.96 (m, 2H), 5.13 (br, d, 1H; J=8), 5.56 (s, 1H), 7.2 (m, $2C_6H_5$, CH vinyl) IR ($CHCl_3$): 1655, 1975, 2050	[35]
pentasubstituted complex:			
*35	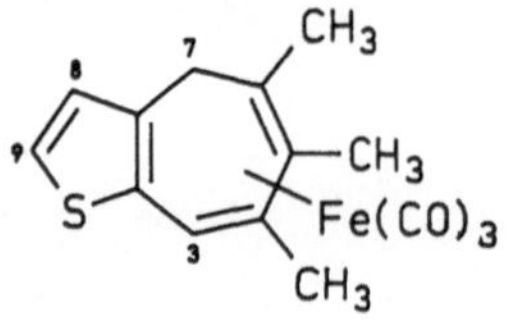	m.p. 92°, crystals (from ether/hexane 1:2) ^1H NMR: 1.90 (s, CH_3-6), 2.04 (s, CH_3-4), 2.08 (s, CH_3-5), 2.58 (q, H-7); 5.42 (s), 5.60 (s, H-3), 6.75 (d, H-8), 7.01 (d, H-9) IR: 1952, 1972, 2034	[33]

*Further information:

$1\text{-}RC_7H_5OFe(CO)_3$ (Table **5**, Nos. **4** to **7** with R=Cl, CH_3, CH_3CO, C_6H_5) and **$[1\text{-}CH_3C_7H_6O\text{-}Fe(CO)_3]O_2CCF_3$** (Table **5**, No. **8**). Both tautomers No. 6a and b are stable in the solid state, and no interconversion is observed after longer periods at 0 °C [29]. The complexes No. 4, 5, and 7 give IX (R=Cl, CH_3, C_6H_5) on protonation with 98% H_2SO_4 in CD_2Cl_2. The initial protonation occurs at the uncoordinated double bond at C-6 and is followed by rapid 1,2 shifts of the $Fe(CO)_3$ moiety. Protonation of No. 4 with $HOSO_2F$ at −78 °C occurs stereoselectively from the endo rather than exo side (in relation to Fe) of the uncoordinated double bond [24]. From the mixture of the interconverting (in solution) tautomers No. 6a and b, tautomer No. 6a is quantitatively obtained by dissolution of the mixture in $HOSO_2F$, followed by H_2O treatment, neutralization, and CH_2Cl_2 extraction [29]. Compound No. 5 undergoes

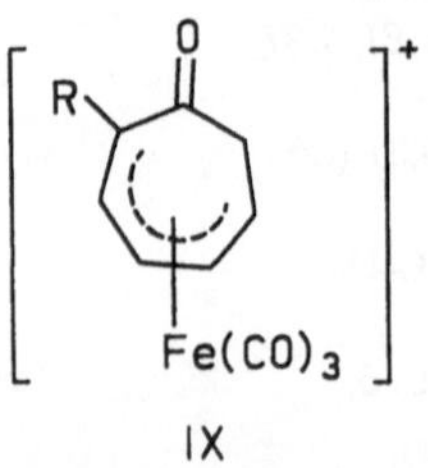

IX 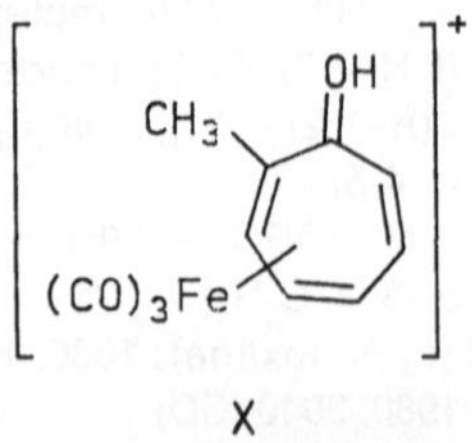 X 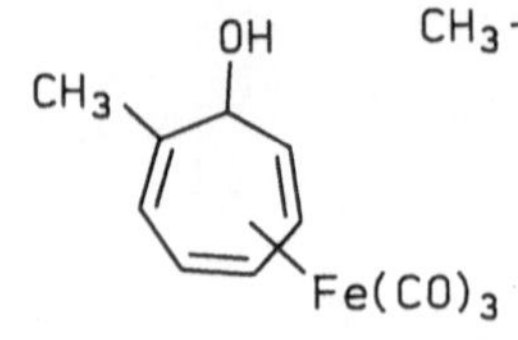 XI 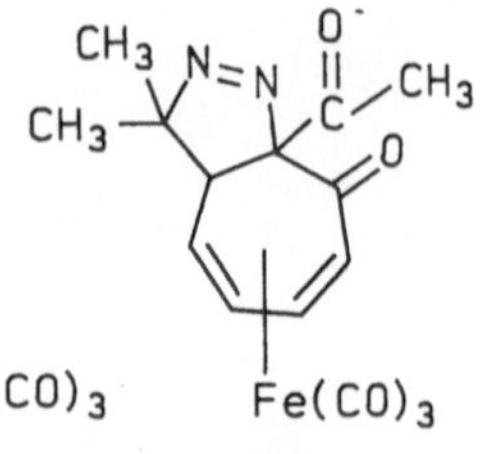 XII

References on pp. 196/7

kinetically controlled oxygen protonation with CF_3CO_2H in liquid SO_2/SO_2F_2 at −78 °C to give No. 8. On warming of No. 8 to 0 °C an orange solution of IX ($R=CH_3$) is obtained. It is of interest that during the tautomerization of No. 8 to IX ($R=CH_3$) no NMR evidence for the tautomer X is detected [25, 26]. Reduction of No. 5 with excess $LiAlH_4$ gives XI [23]. Reaction of tautomer No. 6a with $(CH_3)_2CN_2$ gives XII [29].

1-R(R'R''C=)C_7H_5Fe(CO)$_3$ (Table **5**, Nos. **9**, **12**, **13** with $R=CH_2OH$, $R'=H$, $R''=C_6H_4CH_3$-4; $R=CHO$, $R'=H$, $R''=C_6H_5$ or $C_6H_4CH_3$-4). Compound No. 9 is not fluxional at room temperature, and retains the geometry of No. 13. Compounds No. 12 and 13 are stereochemically rigid. The ^1H NMR spectra show that Fe and CHO are disposed as shown in Table 5, p. 184. Addition of the chemical shift reagent Eu(dpm)$_3$ (dpm = 2,2,6,6-tetramethyl-3,5-heptanedionate) to solutions of Nos. 12 and 13 demonstrate that the 4-$CH_3C_6H_4$ group is trans to CHO as shown in Table 5, p. 184. For the chemical shifts produced by No. 13 in the presence of Eu(dpm)$_3$ see [21]. Heating No. 13 in toluene at 100 °C for 20 h gives 90% No. 13 unchanged and 2% of V with $R=C_6H_4CH_3$-4, $R'=H$ (see p. 181). Compound No. 13 is reduced by $NaBH_4$ in C_2H_5OH at room temperature within 1 h. The reaction mixture is extracted with H_2O/ether, washed with H_2O, dried ($MgSO_4$), and chromatographed on alumina (activity V) with toluene to give No. 9. Protonation of No. 9 with HPF_6 in ether gives XIII. The compounds No. 12 and 13 readily form No. 10 and 11 with $H_2NNHC_6H_3(NO_2)_2$ [21].

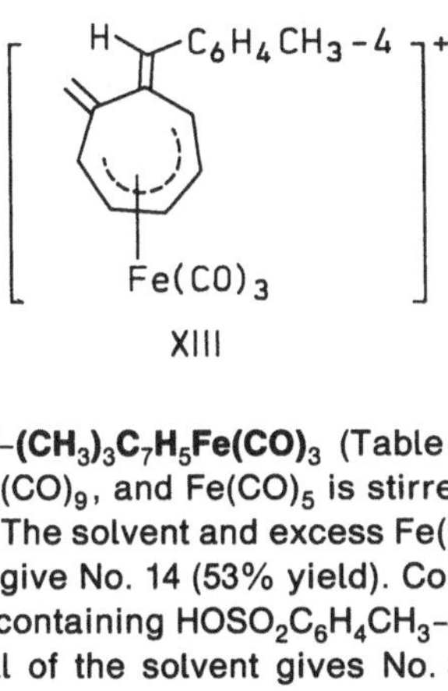

XIII 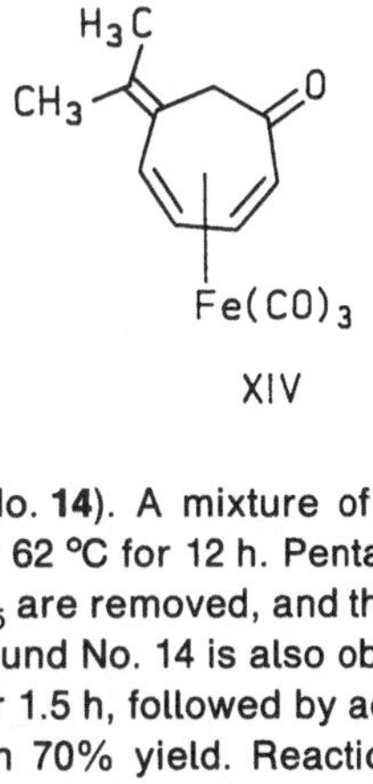 XIV 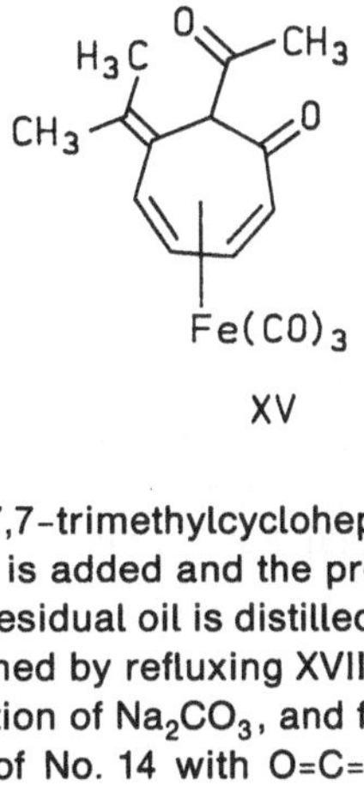 XV

4,7,7-$(CH_3)_3C_7H_5$Fe(CO)$_3$ (Table **5**, No. **14**). A mixture of 4,7,7-trimethylcycloheptatriene (^{4}L), $Fe_2(CO)_9$, and $Fe(CO)_5$ is stirred at 62 °C for 12 h. Pentane is added and the precipitate filtered. The solvent and excess $Fe(CO)_5$ are removed, and the residual oil is distilled (Kugelrohr) to give No. 14 (53% yield). Compound No. 14 is also obtained by refluxing XVIII, p. 194, in C_6H_6 containing $HOSO_2C_6H_4CH_3$-4 for 1.5 h, followed by addition of Na_2CO_3, and filtration. Removal of the solvent gives No. 34 in 70% yield. Reaction of No. 14 with $O=C=C(C_6H_5)_2$ in C_6H_6 at room temperature gives XIX, p. 194. However, refluxing No. 14 with $O=C=C(C_6H_5)_2$ in C_6H_6 for 12 h followed by removal of the solvent and chromatography on Kieselgel 60 with hexane/CH_2Cl_2 gives No. 34 in 12% yield. Reaction with $(NC)_2C=C(CN)_2$ in CH_2Cl_2 gives XX and XXI, p. 196 [35].

2,5,7-$(C_6H_5)_3C_7H_5$Fe(CO)$_3$ (Table **5**, No. **15**). In spin saturation transfer experiments, observed T_1's are short (1.5 to 3.0 s), and no more than a 10% decrease in the signal area for H-6 can be measured upon saturation of H-1, even at 105 °C. No reliable ΔG^{*} values can be calculated from this experiment [32].

2-CH_2=(CH_3)CC_7H_5OFe(CO)$_3$ (Table **5**, No. **16**) is prepared in 50% yield by dehydrogenation of XIV with an excess of MnO_2 in refluxing CH_2Cl_2 within 2 h. Decomplexation of No. 16 with $ON(CH_3)_3$ in boiling CH_2Cl_2 liberates the organic ligand ^{4}L (82% yield) [29].

$C_{11}H_{12}$OFe(CO)$_3$ (Table **5**, No. **17**) crystallizes in the monoclinic space group $P2_1/c$-C^5_{2h} with the unit cell parameters a = 12.42(2), b = 7.19(1), c = 29.31(3) Å, β = 93.00(5)°; Z = 4 molecules per unit cell or 2 molecules/asymmetric unit. The calculated density is 1.53 g/cm^3.

References on pp. 196/7

The main bond distances and angles for the two different molecules are shown in **Fig. 17**a and b.

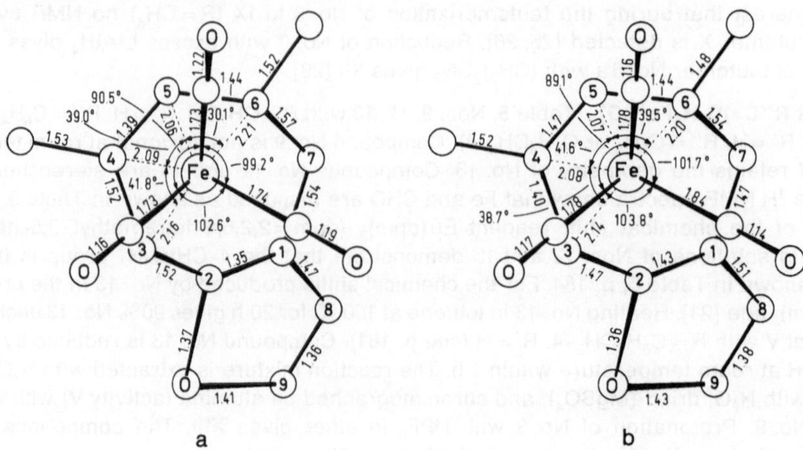

Fig. 17. Structure of the two different molecules in $C_{11}H_{12}OFe(CO)_3$ (No. 17) [28].

The atoms O, C(1), C(2), C(3), C(6), C(7), C(8), C(9) and C(3), C(4), C(5), C(6) both form a plane in both of the molecules; the angle between these two planes is 42°4′ in molecule a and 38°6′ in molecule b. Exposure of complex No. 17 to sunlight for 3 d liberates the organic ligand ^4L as a mixture of IIa and b (p. 180) in the ratio 93% and 7% [28].

$C_{11}H_{12}SFe(CO)_3$ (Table **5**, No. **18**) crystallizes in the triclinic space group $P\bar{1}$-C_i^1, S_2^1 with the unit cell parameters a = 11.00(2), b = 9.66(2), c = 6.77(1) Å, α = 88.05(5)°, β = 79.35(5)°, and γ = 79.40(5)°; Z = 2 molecules per unit cell. The calculated density is 1.51 g/cm³. The main bond distances and angles are shown in **Fig. 18**. The atoms O, C(1), C(2), C(3), C(6), C(7), C(8), C(9) and C(3), C(4), C(5), C(6) form a plane and the angle between these planes is 41.2° [28]. For the mass spectral fragmentation see [30].

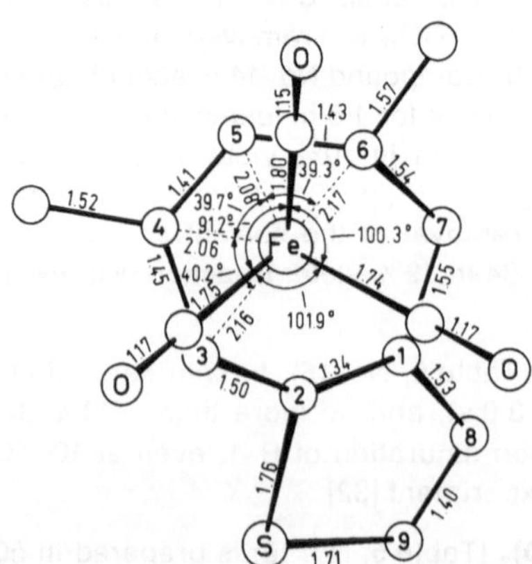

Fig. 18. Molecular structure of $C_{11}H_{12}SFe(CO)_3$ (No. 18) [28].

1-$CH_3CO(2$-$(CH_3)_2CH)C_7H_4Fe(CO)_3$ (Table **5**, No. **19**) is prepared by warming XII in CH_2Cl_2 at 30 °C (98% yield). Complex No. 19 is itself thermally labile and is almost quantitatively converted to XV by heating in C_6H_6 at 80 °C for 2 h [29].

2,4-$(C_6H_5)_2C_7H_4OFe(CO)_3$ (Table **5**, No. **20**) shows electrochemical reversible reduction to a brown radical anion, combined with further reduction at more negative potentials to a dianion. The first polarographic wave for $-E_{1/2}$ is 1.61 V and the second 1.92 V. The cathodic potential, $-E^{cath}_{p/2}$, at half-height is 1.6 and 2.0 V, and the anodic, $-E^{a}_{p/2}$, is 1.5 and 1.8 V. All these data are obtained from a three-electrode potentiostatic geometry, with 2×10^{-3} M substrate, and 1×10^{-1} M $[N(C_4H_9)_4]ClO_4$ in $(CH_2OCH_3)_2$. Reference: 1×10^{-3} M $AgClO_4/Ag$ and 1×10^{-1} M $[N(C_4H_9)_4]ClO_4$ in $(CH_2OCH_3)_2$ at 22 °C. ESR studies of the dianion, one-line spectrum, $\Delta H = 5$ Gauss, $g = 2.0247$, indicate that it is not in the triplet state [18].

2,4,6-$(C_6H_5)_3C_7H_3OFe(CO)_3$ (Table **5**, No. **25**). Refluxing $C_6H_3(C_6H_5)_3COFe_2(CO)_5$ (see 2.6.1.2.1 in "Organoiron Compounds" C5, 1981, p. 7) in C_6H_6 followed by chromatography on silica gel gives both tautomers No. 25a and b in 46% total yield. Longer reaction times decrease the yield, and at 55 °C no decomposition is observed [1, 2, 5, 6, 9, 13]. But passing a solution of $C_6H_3(C_6H_5)_3COFe_2(CO)_5$ through an alumina column, C_6H_6/ether and $CH_3CO_2C_2H_5$ elutes No. 25a and b (total 57%) which are separated by chromatography on silica gel [13]. Both tautomers are monomeric in solution [1, 2, 4, 6]. The UV spectra of tautomers, No. 25a and b resemble that of 2,4,6-triphenyltropone, see figure in [13]. The K-edge fine structure shows an absorption at ca. 2 to 5 eV characteristic for CO compounds [9]. The Debye-Scherrer diagrams are different for both tautomers [1, 2, 6]. An X-ray structure analysis should show that each of No. 25a and b exist in two optical isomers. If the space group of tautomer No. 25b is $P2_12_12_1$, the crystals used for X-ray consist of only one enantiomer. Sorting by hand gives no suitable single crystals [13]. Tautomer No. 25b crystallizes in the orthorhombic space group $P2_12_12_1-D_2^4$, V^4 with the unit cell parameters a = 19.53(4), b = 7.62(2), c = 15.03(3) Å; Z = 4 molecules per unit cell [7, 8]. The measured and calculated density is 1.41 g/cm^3. The main bond distances and angles are shown in **Fig. 19** [7]. The tropone ring is bent [7, 8, 12] with atoms C(4) and C(7) in two approximate planes (one plane comprises C(4), C(5), C(6), C(7), and the other plane contains the remaining ring atoms). The dihedral angle between the two planes is 139° [7].

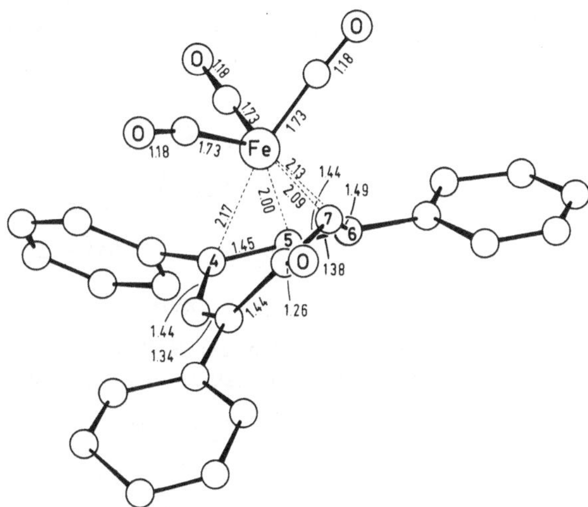

Fig. 19. Molecular structure of 2,4,6-$(C_6H_5)_3C_7H_3OFe(CO)_3$ (No. 25) [7].

Both tautomers are highly soluble in C_6H_6, acetone, and tetrahydrofuran, and slightly soluble in petroleum ether and C_2H_5OH [1, 2, 6], cf. [3]. The following table gives the data for the thermal tautomerization in C_6H_6 [13]:

References on pp. 196/7

tautomer No.	t in °C	time (h)	No. 25a (%)	No. 25b (%)
25a	25	138	80	8
b	25	138	6	91
25a	80	4	29	63
b	80	4	31	62

An equilibrium is observed at 80 °C (No. 25a:b = 1:2), No. 25b is favored by only about 0.5 kcal/mol [13, 14], cf. [34]. Above 160 °C decomposition occurs with exclusive formation of 1,3,5-$(C_6H_5)_3C_6H_3$ [1, 2, 5, 6]. No. 25b shows electrochemical reversible reduction to a brown-red radical anion, combined with further reduction at more negative potentials to a red dianion. The first polarographic wave for $-E_{1/2}$ is 1.62 V and the second 1.98 V. The cathodic potential, $-E^{cath}_{p/2}$, at half-height is 1.6 and 1.9 V, and the anodic potential, $-E^{a}_{p/2}$, is 1.5 and 1.8 V. For the experimental conditions see No. 20. An ESR study of the dianion gives a one-line spectrum with $\Delta H = 10.0$ Gauss, $g = 2.0251$, which indicates that it is not in the triplet state [18]. Both tautomers give oily products and XVI in the reaction with H_2 (150 atm) on Pd/C in C_6H_6 at 50 °C within 12 h. Reaction with CH_3CO_2H/Zn in $CH_3CO_2C_2H_5$ gives decomposition products. $LiAlH_4$ or $NaBH_4$ decompose No. 25b. Refluxing No. 25b in C_6H_6 for 2.25 h with $P(C_6H_5)_3$ in C_6H_6 in an open system gives No. 25a, by tautomerization, and XVII [13]. Reaction of No. 25a with $P(C_6H_5)_3$ in a sealed tube in C_6H_6 for 6 h gives $(CO)_3Fe(P(C_6H_5)_3)_2$ and 2,4,6-triphenyltropone (82%). Similar reaction of No. 25b for 9 h gives 2,4,6-triphenyltropone in 72% yield [13], cf. [5]. No reaction occurs with 2,4-$(O_2N)_2C_6H_3NHNH_2$ in C_2H_5OH/HCl or in the presence of H_3PO_4, even at elevated temperatures (5 h at 80 °C); No. 25b is partly recovered. $NH_2OH \cdot HCl$/NaOH at 80 °C fails to give the oxime with No. 25b, but it decomposes into oily products [13].

It is proposed that interesting substituted cyclic compounds can be prepared from complexes like No. 25. No. 25 may be used to stabilize long chain hydrocarbon polymers or oils, as a dielectric in electrical capacitors, as a fuel additive, and as a starting compound for metal plating [1, 6].

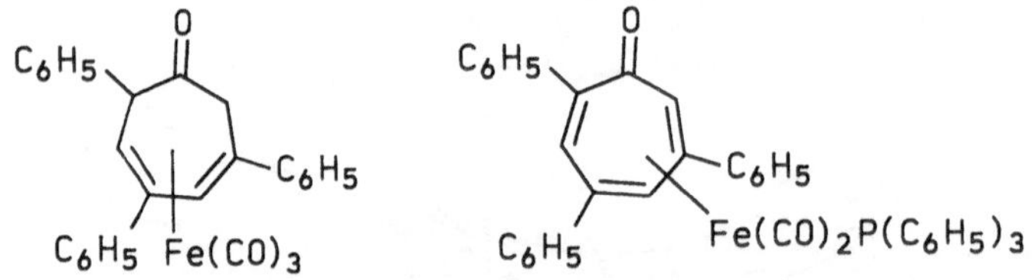

XVI XVII

2,4,6-$(4\text{-}BrC_6H_4)_3C_7H_3OFe(CO)_3$ (Table 5, No. **26**). Refluxing tautomer a in xylene for 3 h gives tris(p-bromophenyl)tropone. Reaction of tautomer a with 10% excess of $P(C_6H_5)_3$ in

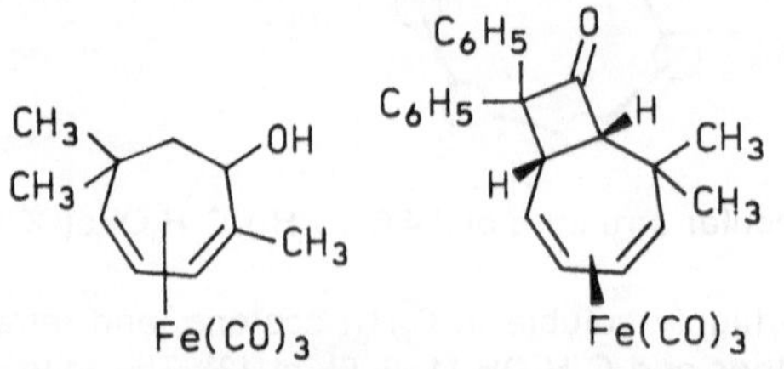

XVIII XIX

References on pp. 196/7

a sealed tube in C_6H_6 at 110 °C for 9 h gives tris(p-bromophenyl)tropone and 65% of $(CO)_3Fe(P(C_6H_5)_3)_2$ [13].

1,2,4,6-$(C_6H_5)_4C_7H_2OFe(CO)_3$ (Table **5**, No. **27**). Only the tautomer shown in Table 5, p. 188, is formed. It can exist as two enantiomers. The complex crystallizes in the monoclinic space group $P2_1ca\text{-}C^5_{2v}$ with the unit cell parameters a=727.5(4), b=1897.7(24), c=1939.9(21) ppm, β=92.62(6)°; Z=4 molecules per unit cell. The calculated density is 1.368 g/cm³. The main bond distances and angles are given in **Fig. 20**. The two planes in the organic ligand ⁴L with the atoms C(3), C(4), C(5), C(6) and C(1), C(2), C(3), C(6), C(7) form an angle of ca. 44°. The complex is scarcely soluble in hexane and ether, slightly soluble in toluene, and is soluble in $CHCl_3$ and acetone. Reaction of No. 27 with $[NH_4]_2[Ce(NO_3)_6]$ in acetone at −10 °C/room temperature liberates the organic ligand ⁴L (78% yield) [16].

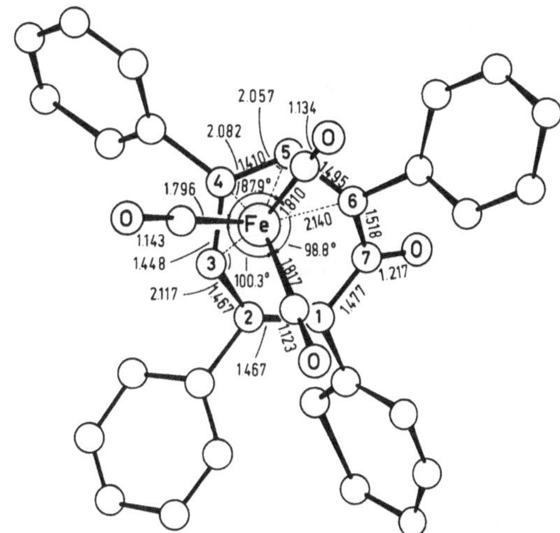

Fig. 20. Molecular structure of 1,2,4,6-$(C_6H_5)_4C_7H_2OFe(CO)_3$ (No. 27) [16].

$^3LMo(CO)_2C_7H_7Fe(CO)_3$ (Table **5**, Nos. **28** to **31** with $^3L=B(N_2C_3H_3)_4$, $HB(N_2C_3H_3)_3$, $HB(N_2C_3H(CH_3)_2\text{-}3,5)_3$, and $H_2B(N_2C_3H(CH_3)_2\text{-}3,5)_2$) are fluxional. Compound No. 30 shows a faster rate of fluxional behavior compared to Nos. 28, 29, 31. It is not clear why the C_7H_7 ring in No. 30 should encounter less of a barrier to "whizzing" as compared to the others. Possibly the bulky CH_3 group in No. 30 might cause the ring to be less securely bound and provide a sort of acceleration. The chemical shifts of the CH_3 groups are very sensitive to the solvent used [22].

$C_5H_5Mo(CO)_2C_7H_7Fe(CO)_3$ (Table **5**, No. **32**) is monomeric in $CHCl_3$ [19]. The molecule is fluxional [19, 20, 22] and gives an ¹H NMR spectrum consistent with the structure given in the Table 5, p. 189, only at temperatures below about −50 °C. Above this temperature the complex pattern of resonances due to the C_7H_7 ring collapses; and at room temperature and above, these seven protons give only a single resonance. The Arrhenius activation energy for the rearrangement in $CDCl_3$/toluene-d_8 (4:1) is 13±1 kcal/mol, $\Delta S^{*}=+2$ e.u. at 251 K and −2 e.u. at 273 K. The low-temperature spectrum collapses unsymmetrically, and detailed analysis, including comparison with computer simulated spectra, excludes all rearrangement pathways except 1,2 shifts or a mixture of 1,2 and 1,3 shifts [19]. The compound crystallizes at 25 °C in the orthorhombic space group $Pna2_1\text{-}C^9_{2v}$ with the unit

References on pp. 196/7

cell parameters a = 11.838(6), b = 6.773(3), c = 20.918(8) Å; Z = 4 molecules per unit cell. The observed density is 1.80(1) g/cm³ compared with the calculated 1.77 g/cm³. The main bond distances and angles are shown in **Fig. 21**. There are three planes in the C_7H_7 ring: plane I: C(1), C(2), C(7), plane II: C(3), C(4), C(5), C(6), and plane III: C(6), C(7), C(2), C(3). The angle between plane I and III is 150.9°, between plane II and III 144.3°, and between plane I and II 6.4° [20]. The compound is air-stable [19].

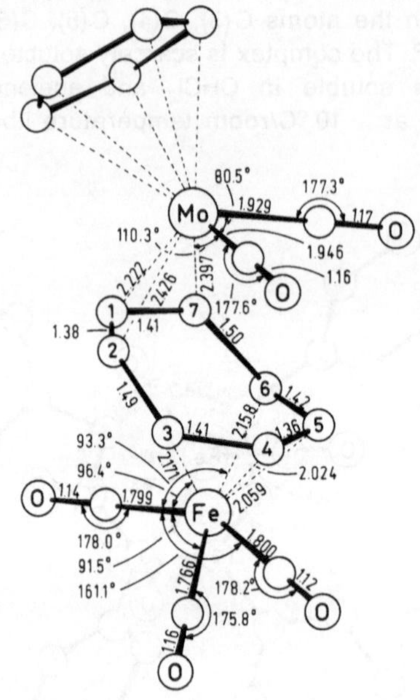

Fig. 21. Molecular structure of $C_5H_5Mo(CO)_2C_7H_7Fe(CO)_3$ (No. 32) [20].

XX XXI XXII

$C_{12}H_{14}SFe(CO)_3$ (Table **5**, No. **35**) is prepared from XXII (H in 3, 5 and 7 position) and $Fe_2(CO)_9$ in boiling C_6H_6. After 12 h, the solid formed is filtered, and the filtrate is evaporated. The residue is chromatographed on silica gel with pentane as eluent to give No. 35 in 15% yield [33].

References:

[1] K.W. Hübel, Society European Research Associates S.A. (Belg. 567743 [1958]; C.A. **1960** 1541). – [2] K.W. Hübel, Society European Research Associates S.A. (Ger. Offen. 1169932 [1958/64]). – [3] W. Hübel, E.H. Braye, A. Clauss, E. Weiss, U. Krüerke, D.A. Brown, G.S.D. King, C. Hoogzand (J. Inorg. Nucl. Chem. **9** [1959] 204/10). – [4] W. Hübel, E.H.

Braye, I. Caplier, R. Vannieuwenhoven (J. Inorg. Nucl. Chem. **10** [1959] 250/68). – [5] E.H. Braye, C. Hoogzand, W. Hübel, U. Krüerke, R. Merényi, E. Weiss (Advan. Chem. Coord. Compounds **1959** 190/7).

[6] W. Hübel, Union Carbide Corp. (Brit. 885514 [1961]; C.A. **57** [1962] 868). – [7] D.L. Smith, L.F. Dahl (J. Am. Chem. Soc. **84** [1962] 1743/4). – [8] D.L. Smith (Diss. Univ. Wisconsin 1962; Diss. Abstr. **23** [1962] 1948). – [9] A. Schneider (Z. Physik. Chem. [Frankfurt] **31** [1962] 249/73). – [10] E.H. Braye, W. Hübel (unpublished results from [11]).

[11] E. Weiss, W. Hübel, J. Nielsen, A. Gerondal, J. Vandervalle (Chem. Ber. **95** [1962] 1179/85). – [12] P. Dodge, V. Schomaker (unpublished results 1961 from [13]). – [13] E.H. Braye, W. Hübel (J. Organometal. Chem. **3** [1965] 25/37). – [14] H.W. Whitlock, Y.N. Chuah (J. Am. Chem. Soc. **87** [1965] 3605/8). – [15] R.J. Doedens (Diss. Univ. Wisconsin 1966; Diss. Abstr. **26** [1966] 5059).

[16] J. Klimes, E. Weiss (Chem. Ber. **115** [1982] 2175/80). – [17] L.F. Dahl, R.J. Doedens, W. Hübel, J. Nielsen (J. Am. Chem. Soc. **88** [1966] 446/52). – [18] R.E. Dessy, R.L. Pohl (J. Am. Chem. Soc. **90** [1968] 1995/2001). – [19] F.A. Cotton, C.R. Reich (J. Am. Chem. Soc. **91** [1969] 847/53). – [20] F.A. Cotton, B.G. De Boer, M.D. LaPrade (Proc. 23rd Intern. Congr. Pure Appl. Chem., Boston 1971, Vol. 6, pp. 1/30).

[21] B.F.G. Johnson, J. Lewis, P. McArdle, G.L.P. Randall (J. Chem. Soc. Dalton Trans. **1972** 2076/83). – [22] J.L. Calderon, A. Shaver, F.A. Cotton (J. Organometal. Chem. **57** [1973] 121/6). – [23] A. Eisenstadt, J.M. Guss, R. Mason (J. Organometal. Chem. **80** [1974] 245/56). – [24] A. Eisenstadt (J. Organometal. Chem. **97** [1975] 443/51). – [25] C.P. Lewis (Diss. Univ. North Carolina **1976**; Diss. Abstr. Intern. B **38** [1977] 692/3).

[26] M.S. Brookhart, C.P. Lewis, A. Eisenstadt (J. Organometal. Chem. **127** [1977] C14/C18). – [27] R.S. Dickson, C. Mok, G. Connor (Australian J. Chem. **30** [1977] 2143/51). – [28] M. El Borai, R. Guilard, P. Fournari, Y. Dusausoy, J. Protas (J. Organometal. Chem. **148** [1978] 285/301). – [29] M. Franck-Neumann, F. Brion, D. Martina (Tetrahedron Letters **1978** 5033/6). – [30] M. El Borai, M.F. Abdel-Megeed (Phosphorus Sulfur **9** [1980] 165/73).

[31] M. El Borai (Transition Metal Chem. [Weinheim] **7** [1982] 213/4). – [32] K.J. Karel, T.A. Albright, M. Brookhart (Organometallics **1** [1982] 419/30). – [33] M. El Borai, A. Akelah, M.A. Hassen (Egypt. J. Chem. **24** [1981] 91/6). – [34] N.G. Connelly, A.R. Lucy, J.B. Sheridan (J. Chem. Soc. Dalton Trans. **1983** 1465/8). – [35] Z. Goldschmidt, S. Antebi (J. Organometal. Chem. **259** [1983] 119/25).

1.4.1.4.1.2.4.2.5 ^{4}L is a Cyclohepta-1,4-diene

This type of compound is only stable if a bridgehead atom between the two double bonds exists (compare structures in Table 6, pp. 200/4) to prevent isomerization of the double bonds. Most of the compounds described in this section can be prepared by the following methods:

Method I: $Fe(CO)_5$ is the starting material, which reacts photochemically or thermally with an organic substrate.

a. $Fe(CO)_5$ and I are irradiated in pentane for 3 h. The mixture is filtered, and the solvent is removed to give a brown residue, from which No. 21 (35%) is recovered by distillation at 80 to 100 °C/10 Torr. The remaining residue is further distilled at 100 to 120 °C/0.008 Torr to afford a mixture containing II (major) and No. 21 (minor). The mixture is dissolved in a minimum volume of C_2H_5OH. Upon cooling to −20 °C, II precipitates. The filtrate is evaporated and the residue is distilled carefully at 118 to 120 °C/0.008 Torr to give II first and then No. 21 (11%). Similar irradiation of $Fe(CO)_5$ and the free ligand ^4L (see Formula III) in pentane for 5 h, followed by work-up as before, distillation of the residue at 100 to 110 °C/8 Torr to recover III (21%), and subsequent distillation of the remaining residue at 110 to 120 °C/0.006 Torr gives No. 21 [7].

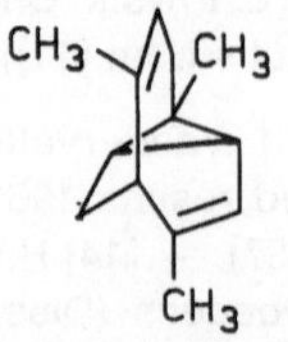

I 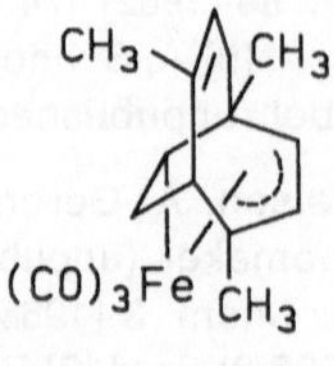 II 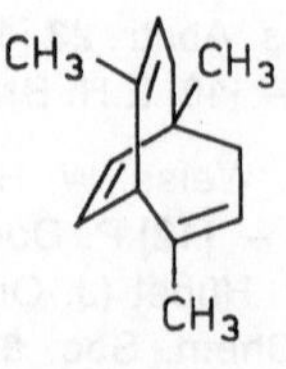 III

b. $Fe(CO)_5$ reacts with the free organic ligand ^4L in methylcyclohexane to give No. 1 [1].

c. Complex IV rearranges to No. 16 in refluxing heptane whether or not $Fe(CO)_5$ is present. The isomerization proceeds more rapidly in the presence of $Fe(CO)_5$, but none the less proceeds at an appreciable rate in its absence. Refluxing $Fe(CO)_5$ and V in heptane for 10 h gives No. 16 as the only isolable material besides starting complex V [13].

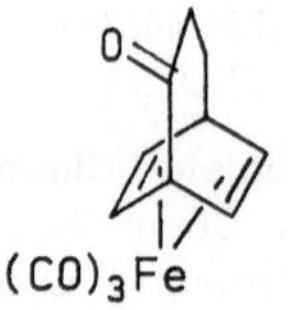

IV 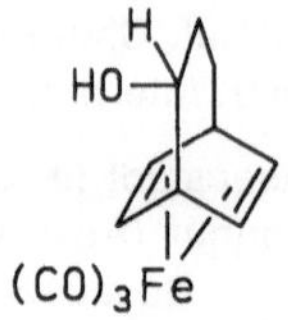 V  VI 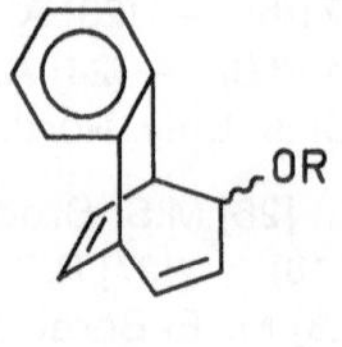 VII

Method II: $Fe_2(CO)_9$ reacts with the free organic ligand ^4L in an inert solvent.

a. $Fe_2(CO)_9$ and ^4L (see Formula VI) react in hexane at 55 °C for 2 h. The solvent is removed, the residue is dissolved in hexane, and the major amount of $Fe_3(CO)_{12}$ is crystallized by cooling to −78 °C. The mother liquor is chromatographed on neutral silica gel with petroleum ether to give No. 22 together with small amounts of $Fe_3(CO)_{12}$. The second pale yellow band contains uncomplexed and unidentified byproducts [9].

b. Treatment of $Fe_2(CO)_9$ with each of the isomeric free ligand ^4L (see Formula VII with $R=CH_3$) in hexane/C_6H_6 gives isomeric No. 18, purified by basic alumina chromatography. This reaction gives better yields of No. 18 than treatment with VII (R=H) to give No. 17. The latter reaction gives rise to considerable amounts of No. 20 [3]. Similarly, $Fe_2(CO)_9$ and an epimeric mixture of VIII (exo and endo) reacts in hexane/C_6H_6 to give No. 2 [4], and reaction with IX in hexane/C_6H_6 gives IV, No. 14, and V. No. 14 is separated by chromatography [2].

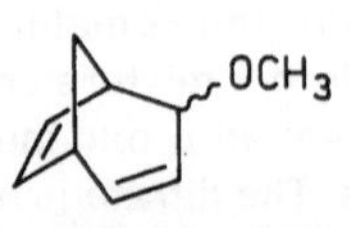

VIII 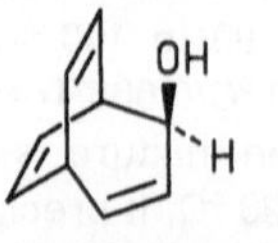 IX  X 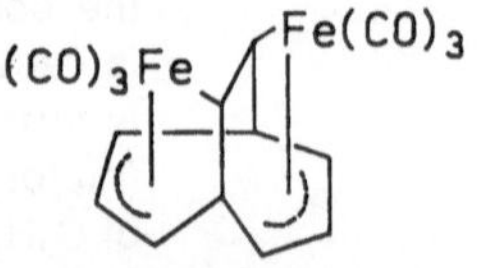 XI

References on p. 207

c. $Fe_2(CO)_9$ and bullvalene (Formula X) react in C_6H_6 at 80 to 85 °C under CO pressure (100 atm) for 24 h. Petroleum ether is added to the mixture which is cooled to −30 °C. XI crystallizes within 24 h. The mother liquor is chromatographed on silica gel with C_6H_6 to give XII and XI. Further elution with ether gives No. 23 [8]. Similarly, $Fe_2(CO)_9$ and XIII are stirred in C_6H_6 at 47 °C for 2 h. The reaction mixture is filtered and evaporated. Separation of the residue on basic alumina gives, in order of elution, No. 15 and XIV [12]. Treatment with XV in C_6H_6 gives No. 20 [3].

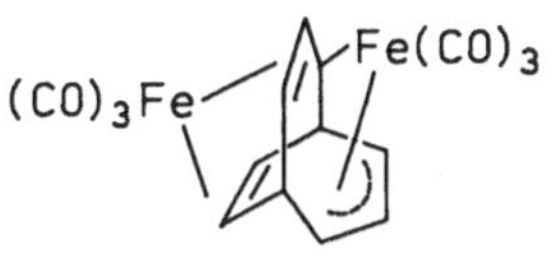

$(CO)_3Fe$ $Fe(CO)_3$ OH CH_3 $(CO)_3Fe$ OH CH_3 O

XII 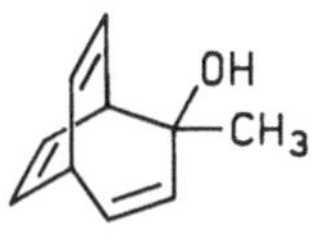 XIII 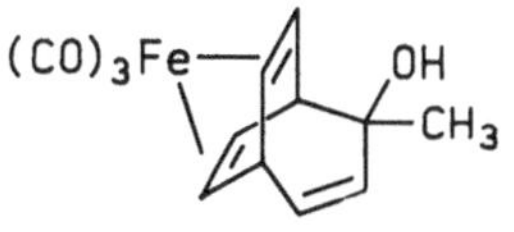 XIV 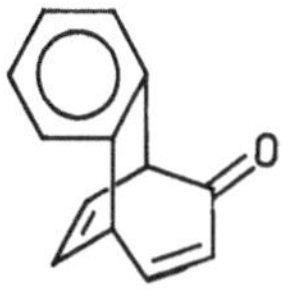 XV

d. $Fe_2(CO)_9$ and XVI react in tetrahydrofuran at 45 °C for 2 h. The mixture is filtered, evaporated, and the residue is chromatographed on 5% basic alumina to give No. 19. The yield drops down with C_6H_6 as solvent [12].

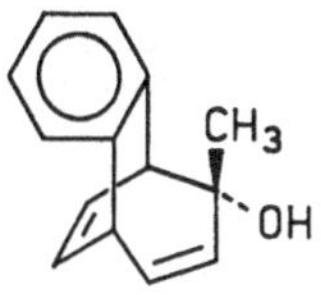

O CH_3 CH_3 NC OH $(CO)_3Fe$

XVI 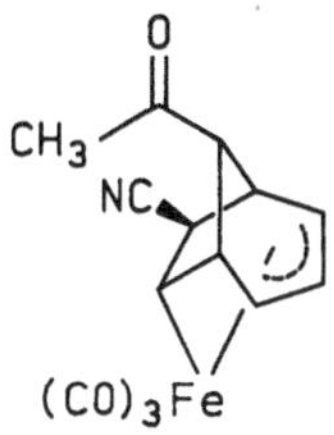 XVII

Method III: $^5LFe(CO)_3$ complexes (see Formula XX, p. 205) react with Lewis bases in inert solvents.

a. XX ($R=COCH_3$, $X=PF_6$, p. 205) reacts with $NaOCH_3/CH_3OH$ at room temperature for 10 min. The reaction mixture is poured into H_2O, extracted with ether, dried ($MgSO_4$), and chromatographed on silica gel with toluene to give a very small amount of $C_8H_8Fe(CO)_3$. Further elution with 10% $CH_3CO_2C_2H_5$/toluene gives No. 8 [14, 15]. Similarly, XX ($R=COCH_3$, $X=PF_6$, p. 205) is added in small amounts to a saturated aqueous solution of NaCN with vigorous stirring. After 30 min, work-up as before but elution with 20% $CH_3CO_2C_2H_5$/toluene gives XVII from the slower running band, and the faster running band gives another isomer, probably No. 12 (ratio XVII: No. 12 = 14:1) [14].

b. XX ($R=COCH_3$, $X=PF_6$, p. 205) reacts with $LiSC_3H_7$-i in CH_3CN at 0 °C [6, 14]. After 5 min, the reaction mixture is slowly hydrolyzed with ice-cold 1% HCl solution, ether is added, and the aqueous layer is washed with H_2O and dried (K_2CO_3). The solvent is removed, and the residue is chromatographed on alumina (grade 1) eluting first with C_6H_6 to remove the leading pale red band, then with C_6H_6/1% $CH_3CO_2C_2H_5$ to collect No. 9 which is further purified by distillation. The complexes No. 10 and 11 are prepared in a similar manner. No satisfactory analysis for No. 11 is reported [14]. Similar reaction of XX ($R=COC_6H_5$, $X=PF_6$, p. 205) with $LiSC_3H_7$-i [6, 14] in CH_3CN at 0 °C followed by work-up as before but with C_6H_6 as eluent gives No. 13 [14].

References on p. 207

Table 6
Compounds of the Type $^4LFe(CO)_3$ with ^4L = Substituted Cyclohepta-1,4-diene.
Further information on numbers preceded by an asterisk is given at the end of the table, pp. 204/7.
For abbreviations and dimensions see p. VIII.

No.	compound	method of preparation (yield in %), properties and remarks	Ref.
*1	(CO)$_3$Fe	Ib m.p. 30 to 31° (from CH_3OH), pale yellow plates IR: 1957, 2032 (CO) UV (C_2H_5OH): $\lambda_{max}(\varepsilon)$ = 215 (22400) nm mass spectrum: $[M]^+$	[1]
*2	OCH_3 (CO)$_3$Fe	IIb	[4]
*3	OH, CH_3, 9, 8, 1, 2, 7, 5, 6, 3, 4 (CO)$_3$Fe	see No. 9, p. 204 ^{1}H NMR (CS_2): 0.99 (d, CH_3; J=9), 1.30 (m, H-8), 1.74 (br, m, OH), 1.92 (m, H-2endo), 2.04 (m, H-2exo), 2.5 to 3.4 (m, H-1, 3 to 7), 3.64 (m, H-9) IR (Nujol, mixture with No. 4): 1958, 1970, 2033 (CO) IR (cyclohexane, mixture with No. 4): 3400 (OH)	[14]
*4	OH, CH_3, 9, 8, $SCH(CH_3)_2$, 1, 7, 5, 6, 3, 4 (CO)$_3$Fe	see No. 9, p. 204 ^{1}H NMR (CS_2): 1.03 (d, CH_3-9; J=9), 1.10 (dd, 2CH_3; J=9, J=4), 1.86 (d, H-8), 2.0 (br, m, OH), 2.5 to 3.4 (m, H-1, 3 to 7, SCH), 3.64 (br, m, H-2,9) IR (Nujol, mixture with No. 3): 1958, 1970, 2033 (CO) IR (cyclohexane, mixture with No. 3): 3400 (OH)	[14]
*5	OH, C_6H_5 (CO)$_3$Fe	see No. 13, p. 206	[14]
*6	OH, C_6H_5, $SCH(CH_3)_2$ (CO)$_3$Fe	see No. 13, p. 206	[14]

Table 6 [continued]

No.	compound	method of preparation (yield in %), properties and remarks	Ref.
*7	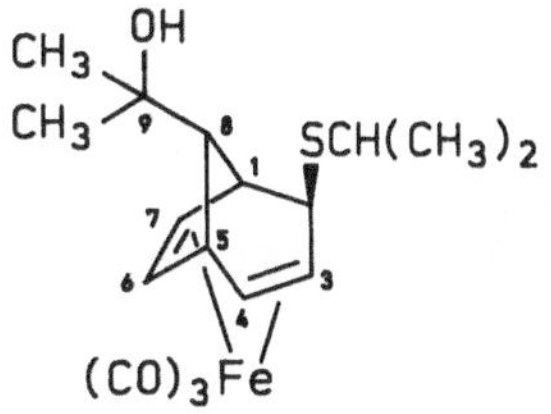	see No. 9, pp. 204/5 distillable yellow oil ^{1}H NMR (CS_2): 1.06 (s, OH), 1.14 (s, 2CH_3-9), 1.28 (dd, 2CH_3; J=4, J=9), 1.82 (s, H-8), 2.30 (t, H-1; J=4), 2.85 to 3.20 (H-3, 4, 5 or 6, 7, SCH), 3.34 (t, H-5 or 6; J=4), 3.74 (dd, H-2; J=2, J=4) IR (thin film): 3400 (OH) IR (cyclohexane): 1959, 1970, 2033 (CO) mass spectrum: $[M-nCO]^+$ (n=0 to 3)	[14]
*8	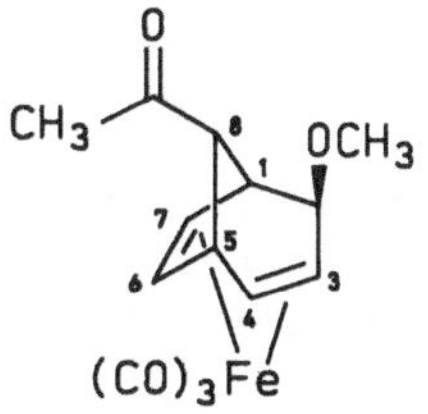	III a (low yield) b.p. 120°/0.05 Torr, yellow oil ^{1}H NMR (CS_2): 2.55 (m, H-1); 2.7, 3.1, 3.2 (m, H-3 to 7); 2.8 (s, H-8), 3.3 (s, OCH_3), 3.8 (m, H-2) IR (CCl_4): 1070, 1090 (C-O); 1703 (C=O); 1963, 1978, 2043 (CO) mass spectrum: $[M-nCO]^+$ (n=0 to 3)	[14, 15]
*9	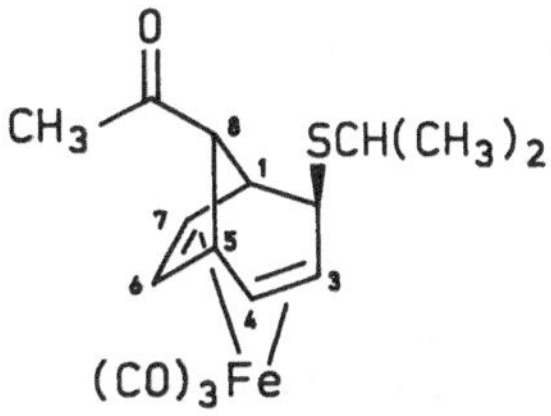	III b (50) b.p. 45 to 50°/0.001 to 0.05 Torr, yellow oil ^{1}H NMR (CS_2): 1.25 (dd, 2CH_3; J=3.5, J=6), 1.96 (s, CH_3CO), 2.54 (t, H-1; J=5), 2.76 (s, H-8), 2.92 (m, H-3, 4, 7, SCH), 3.31 (m, H-5, 6), 3.74 (dd, H-2; J=2, J=5) IR (cyclohexane): 1715 (C=O); 1962, 1973, 2035 (CO) mass spectrum: $[M-nCO]^+$ (n=0, 2, 3)	[6, 14]
10	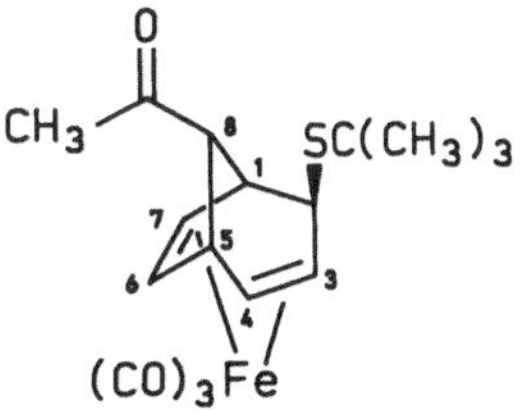	III b ^{1}H NMR (CS_2): 1.34 (s, C_4H_9-t), 1.99 (s, $COCH_3$), 2.54 (t, H-1; J=5), 2.63 (s, H-8), 3.0 (m, H-3, 4, 7), 3.32 (m, H-5, 6), 3.74 (dd, H-2; J=2, J=5) IR (cyclohexane): 1718 (C=O); 1961, 1971, 2035 (CO) mass spectrum: $[M-nCO]^+$ (n=0, 2, 3)	[14]
11	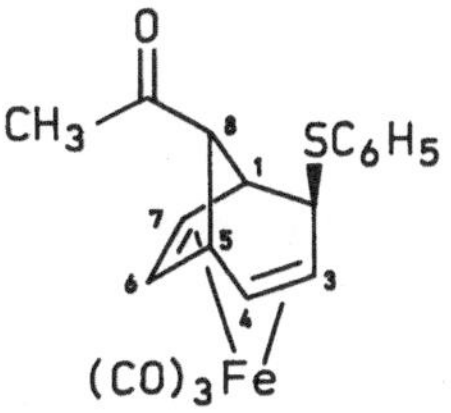	III b ^{1}H NMR (CS_2): 1.98 (s, CH_3), 2.62 (t, H-1; J=5), 3.0 (m, H-3, 4, 7, 8), 3.40 (m, H-5, 6), 4.22 (dd, H-3; J=2, J=5), 7.28 (m, C_6H_5) IR (cyclohexane): 1715 (C=O); 1965, 1977, 2037 (CO) mass spectrum: $[M-nCO]^+$ (n=0 to 3)	[14]

Compound structures (drawn formulas, labels as printed):
*7: $(CH_3)_2C(OH)$ at C-9/C-8, $SCH(CH_3)_2$ at C-1, $(CO)_3Fe$
*8: $CH_3C(O)$ at C-8, OCH_3, $(CO)_3Fe$
*9: $CH_3C(O)$ at C-8, $SCH(CH_3)_2$, $(CO)_3Fe$
10: $CH_3C(O)$ at C-8, $SC(CH_3)_3$, $(CO)_3Fe$
11: $CH_3C(O)$ at C-8, SC_6H_5, $(CO)_3Fe$

References on p. 207

Table 6 [continued]

No.	compound	method of preparation (yield in %), properties and remarks	Ref.
12	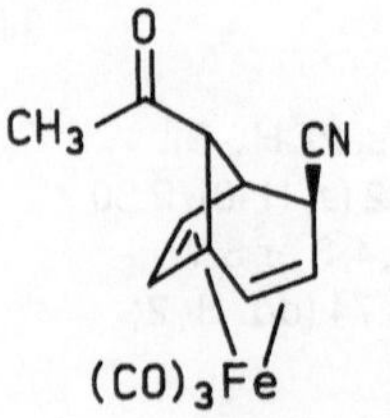 O, CH_3, CN, $(CO)_3Fe$	IIIa (9) yellow crystals ^{1}H NMR ($CDCl_3$): 2.15 (br, s, 3H), 2.8 (br, s, 3H), 3.15 (m, 2H), 3.55 (m, 3H) IR (cyclohexane): 1720 (C=O); 1976, 1985, 2046 (CO)	[14]
*13	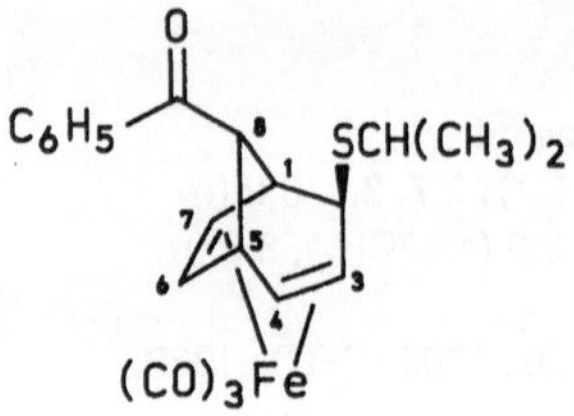 O, C_6H_5, $SCH(CH_3)_2$, $(CO)_3Fe$	IIIb (80) m.p. 65 to 65.5° (from pentane), yellow crystals ^{1}H NMR (CS_2): 1.36 (dd, 2CH_3; J=3.5, J=6), 2.60 (t, H-1; J=5), 2.90 to 3.18 (m, H-3,4,7, SCH), 3.43 (t, H-5; J=5), 3.57 (t, H-6; J=5), 3.66 (H-8), 3.86 (dd, H-2; J=2, J=5), 7.36 (m, H′-3,4,5 in C_6H_5), 7.99 (dd, H′-2,6 in C_6H_5; J=7, J=2) IR (cyclohexane): 1685 (C=O); 1962, 1972, 2035 (CO) mass spectrum: $[M-nCO]^+$ (n=0 to 3)	[6, 10, 14]
*14	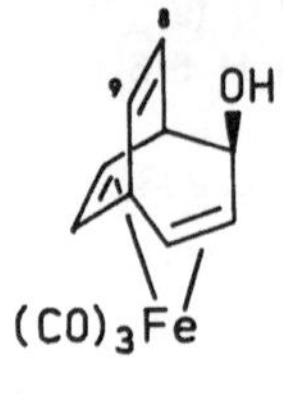 OH, $(CO)_3Fe$	IIb (10) m.p. 56 to 57° ^{1}H NMR: 6.32 (6 line m, H-8, 9) IR: 1972, 2038 (CO); 3595 (OH, indicating that this is the exo isomer as shown)	[2]
15	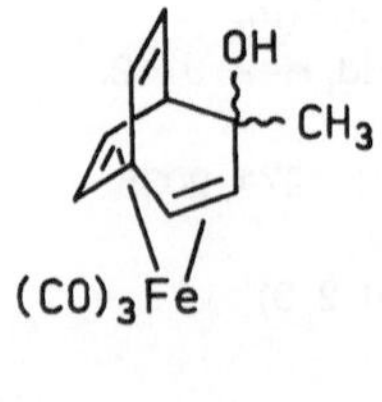 OH, CH_3, $(CO)_3Fe$	IIc	[12]
*16	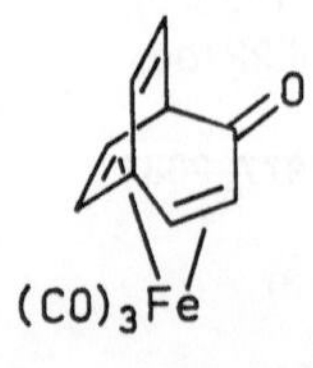 O, $(CO)_3Fe$	Ic (70) m.p. 180 to 181° ^{1}H NMR ($CDCl_3$): 2.46 (br, d, 1H), 3.60 (complex, 5H), 6.35 (br, q, 2H) IR ($CHCl_3$): 1660, 1990, 2060	[13]

References on p. 207

Table 6 [continued]

No.	compound	method of preparation (yield in %), properties and remarks	Ref.
*17	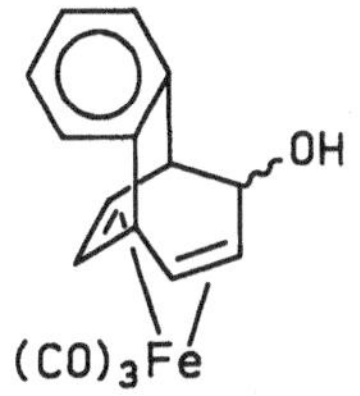	IIb	[3]
*18	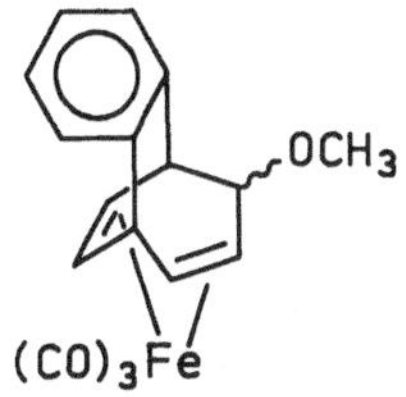	IIb isomer a: oil IR (hexane): 1950, 1965, 2025 (CO) isomer b: m.p. 78 to 79° IR (hexane): 1960, 1987, 2040 (CO)	[3]
*19	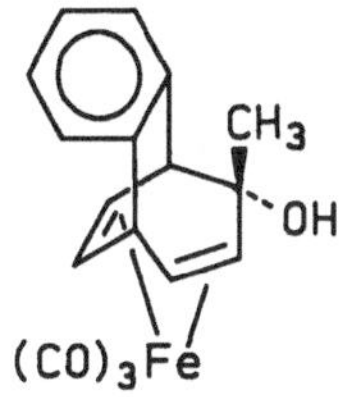	IId (18 or 62) m.p. 82 to 84°, yellow complex ^{1}H NMR ($CDCl_3$): 0.85 (s, CH_3), 1.25 (s, OH), 2.9 to 3.7 (m, 4H), 3.8 to 4.2 (m, 2H) IR (hexane): 1975, 2030 (CO) mass spectrum: $[M-nCO]^+$ (n = 1 to 3)	[12]
*20	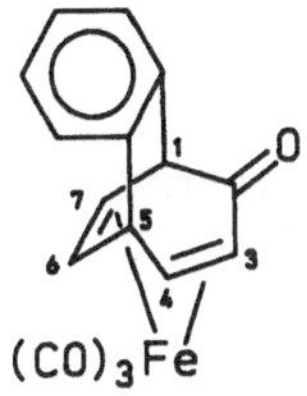	IIc (quantitative) m.p. 171 to 175° (dec.), crystals (from hexane) ^{1}H NMR ($CDCl_3$): 2.51 (dd, H-3; J(H-3,4) = 8.0, J(H-1,3) = 1.5), 3.6 to 4.3 (m, H-1,4,6,7), 7.19 (A_2B_2 pattern, aromatic H) IR (hexane): 1660 (C=O); 1975, 1990, 2050 (CO)	[3]
*21	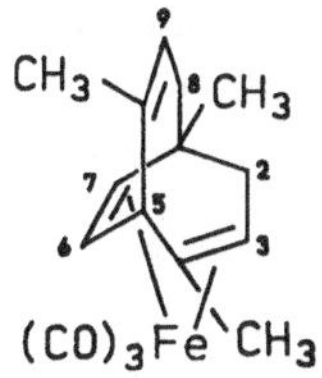	Ia (46) b.p. 118 to 120°/0.008 Torr, yellow oil ^{1}H NMR ($CDCl_3$): 1.31 (s, CH_3-1), 1.82 (br, s, CH_3-10), 1.96 (s, CH_3-4), 2.07 to 2.36 (m, 4H), 2.57 (d, H-5; J(H-5,6) = 8.0), 2.59 to 2.74 (m, H-3), 3.11 (t, H-6; J(H-6,7) = 8.0), 3.53 (d, H-7), 5.44 to 5.60 (m, H-9) IR (film): 1022, 1288, 1372, 1385, 1445; 1955, 2025 (CO); 2880, 2910, 2030, 2960 UV (hexane): $\lambda_{max}(\varepsilon)$ = 216 (21040), 228 (sh, 2200) nm mass spectrum: $[M-nCO]^+$ (n = 0 to 3)	[7]

References on p. 207

Table 6 [continued]

No.	compound	method of preparation (yield in %), properties and remarks	Ref.
*22	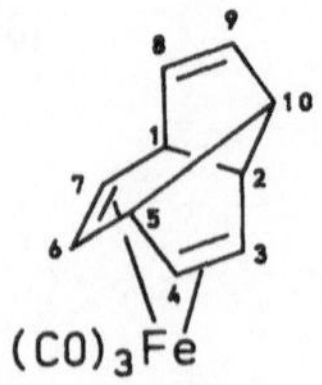 (structure; (CO)$_3$Fe)	IIa (78) yellow oil ^{13}C NMR ($CDCl_3$): 34.2 (2C); 35.0, 55.1, 55.2, 57.7, 61.9, 65.3 (C-1 to 7, 10); 122.1, 139.9 (C-8,9); 216.0 (CO) IR (hexane): 1951, 1965, 2035.3 (CO) mass spectrum: $[M-nCO]^+$ (n=0 to 3) and further fragments given	[9]
*23	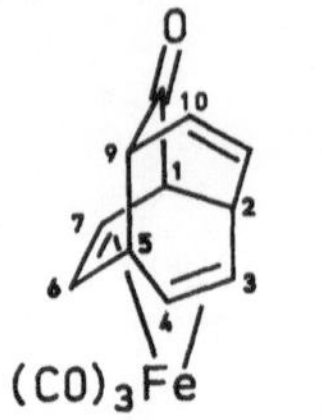 (structure; (CO)$_3$Fe)	IIc (80) m.p. 100 to 102° (from ether at −60°), yellow mushroom-shaped crystals ^{1}H NMR (CS_2): 2.35 (H-1; J(H-1,7)=8, J(H-1,2)=7), ca. 3 (H-9; J(H-5,9)= J(H-9,10)=7), 5.52 (H-10; J(H-10,11)=8, J(H-2,10)=1.5), 6.13 (H-11; J(H-2,11)=6) ^{13}C NMR (benzene-d_6): 23.3, 30.3, 32.0, 45.2; 47.2 (2C), 57.3, 67.1; 121.5 (C=C); 137.4 (C=C); 206.3 (C=O); 214.8 (CO) IR (KBr): 600, 620, 630, 700, 1355; 1720 (C=O); 2960, 3040 IR (hexane): 1966, 1977, 2039 (CO) mass spectrum: $[M-nCO]^+$ (n=0 to 4), $[M-4CO-C_2H_2]^+$	[5, 8]

*Further information:

$C_8H_{10}Fe(CO)_3$ (Table **6**, No. **1**). Hydride abstraction with $[C(C_6H_5)_3]BF_4$ in CH_2Cl_2 gives $[C_8H_9Fe(CO)_3]BF_4$ (see Formula XVIII) [1].

2-R(8-R′)$C_8H_8Fe(CO)_3$ (Table **6**, Nos. **2** to **9** with R=OCH_3, R′=H; R=H or SC_3H_7-i, R′=$CH(CH_3)OH$; R=H, R′=$CH(C_6H_5)OH$; R=SC_3H_7-i, R′=$CH(C_6H_5)OH$ or $C(CH_3)_2OH$; R=OCH_3 or SC_3H_7-i, R′=$COCH_3$). Compound No. 9 is reduced by dropwise addition of $NaBH_4$ to a solution of No. 9 in C_2H_5OH at −15 °C over a period of 15 min. The reaction mixture is stirred at −15 °C for 40 min, then acetone and H_2O (1:10) are added, followed by ether and sufficient H_2O to achieve a good phase separation. The organic layer is dried (K_2CO_3), concentrated, and chromatographed on alumina (grade 1). The first red band elutes with C_6H_6 containing a mixture of unidentified $Fe(CO)_3$ complexes, and the second yellow band contains starting material No. 9 (46%). Another yellow band elutes in 25% $CH_3CO_2C_2H_5/C_6H_6$ which contains a mixture of Nos. 3 and 4 which cannot be separated further by chromatography. In an alternative procedure, t-C_4H_9OH in ether is added at −78 °C to $LiAlH_4$ in ether. The mixture is stirred at −78 °C for 30 min, then No. 9 in ether is added dropwise and the reaction is stirred for another 3 h at −78 °C. Addition of 1% aqueous HCl solution and work-up as above gives Nos. 4 and 3 as a 2:1 mixture. A variety of conditions are examined for the reduction, but formation of No. 3 cannot be suppressed. Best results are obtained with $Li[HAl(OC_4H_9\text{-}t)_3]$ in ether at low temperature. Compound No. 9 reacts with $LiCH_3$ in ether at 0 °C for 4 h. Hydrolysis with 1% aqueous HCl gives a yellow organic layer, which is washed with H_2O once, and dried (K_2CO_3). Chromatography on alumina (grade 1) with 1% $CH_3CO_2C_2H_5/C_6H_6$ elutes starting material. Elution with 35% $CH_3CO_2C_2H_5/C_6H_6$ gives No. 7 (37% yield) which is further purified by distillation. Nucleophilic attack on No. 8 with CH_3MgI proves unsatisfactory, resulting in either decomposition or recovery of the starting material [14]. Complex No. 8 was originally formulated as XIX [15] but is now identified as No. 8 (see Table 6, p. 201) [14].

To assign individual proton resonances, spin-decoupling experiments are performed in the presence of $Eu(tmhd)_3$ (tmhd = 2,2,6,6-tetramethylheptane-3,5-dionate) for No. 9. The chemical shifts are δ = 1.68 (d, $2CH_3$), 3.61 (m, H-3,9), 3.83 (t, H-4), 4.27 (t, H-7), 4.70 (dd, H-2), 4.83 (t, H-6), 5.24 (br, t, H-1), 5.40 (s, $COCH_3$), 5.87 (s, H-8), 6.16 (t, H-5) ppm [14].

Oxidation of No. 8 with $[NH_4]_2[Ce(NO_3)_6]$ in C_2H_5OH liberates the organic ligand ^4L [15].

Protonation of No. 1 with HBF_4 in ether gives XX (R = H, X = BF_4). Protonation of Nos. 4, 6 (as a mixture with 3 and 5, respectively), and 7 with 60% aqueous HBF_6 in ether at 0 °C gives XX (R = $CH(CH_3)OH$, $CH(C_6H_5)OH$, $C(CH_3)_2OH$; X = PF_6); the compounds No. 3 and 5 are not protonated thus providing a convenient method of purification [14].

$[Fe(CO)_3]^+ [BF_4]^-$ XVIII — XIX — $[(CO)_3Fe]^+ X^-$ XX — $[(CO)_3Fe]^+ [BF_4]^-$ XXI

2-$(CH_3)_2CHS(8-C_6H_5CO)C_8H_8Fe(CO)_3$ (Table **6**, No. **13**) shows in the ^1H NMR spectrum in the presence of $Eu(fod)_3$ (fod = 6,6,7,7,8,8,8-heptafluoro-2,2-dimethyloctane-3,5-dionate) chemical shifts at δ = 1.75 (d, $2CH_3$), 3.70 (m, H-3,4,9), 4.26 (t, H-7), 4.81 (d, H-2); 5.23, 5.31 (overlapping t) (H-1,6); 6.64 (s, H-8), 6.76 (br, t, H-5), 7.34 (m, H′-3,4,5 in C_6H_5), 10.4

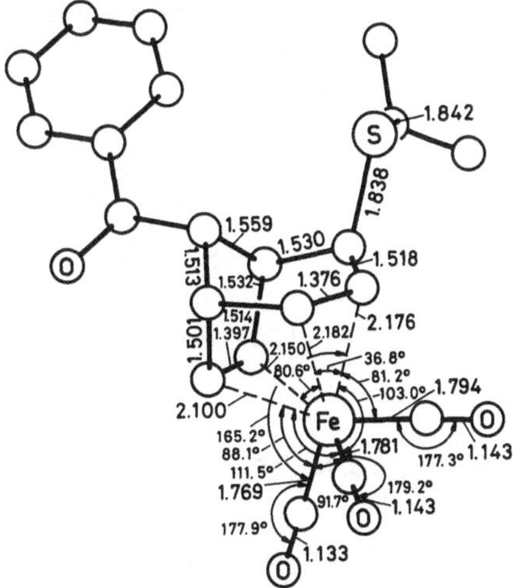

Fig. 22. Molecular structure of 2-$(CH_3)_2CHS(8-C_6H_5CO)C_8H_8Fe(CO)_3$ (No. 13) [10].

References on p. 207

(m, H'-2,6 in C_6H_5) ppm [14]. The compound crystallizes in the triclinic space group $P\bar{1}-C_i^1$, S_2^1 with the unit cell parameters a=9.437(6), b=10.249(5), c=11.545(6) Å, α=84.67(5)°, β=71.44(4)°, γ=67.54(5)°; Z=2 molecules per unit cell. The calculated density is 1.441 g/cm³. The main bond distances and angles are shown in **Fig. 22**, p. 205 [10].

Oxidation with $[NH_4]_2[Ce(NO_3)_6]$ in C_2H_5OH at 0 °C liberates the organic ligand ⁴L (98% yield). Reduction as for No. 9 gives a mixture of Nos. 5 and 6 [14]. Protonation with 60% aqueous HPF_6 gives XX ($R=COC_6H_5$) [6].

2-HOC$_9$H$_9$Fe(CO)$_3$ (Table **6**, No. **14**) reacts with HBF_4 in $O(COCH_3)_2$; addition of ether yields XXI [2].

C$_9$H$_8$OFe(CO)$_3$ (Table **6**, No. **16**) crystallizes in the triclinic space group $P1-C_i^1$ with the unit cell parameters a=6.809(3), b=13.199(6), c=7.090(3) Å, α=110.26(5)°, β=114.30(5)°, γ=94.31(5)°; Z=2 molecules per unit cell. The main bond distances and angles are shown in **Fig. 23** [13].

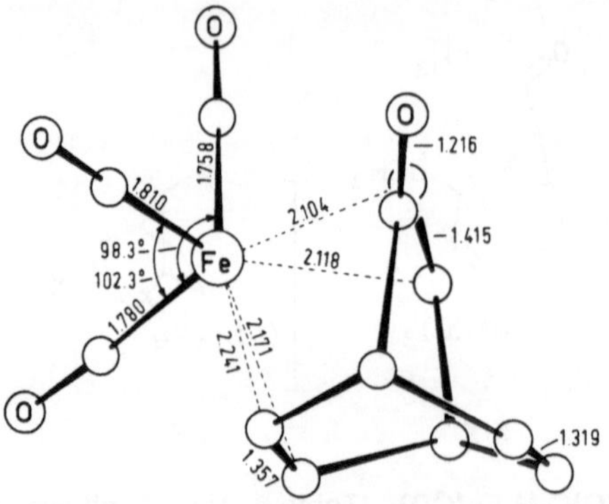

Fig. 23. Molecular structure of $C_9H_8OFe(CO)_3$ (No. 16) [13].

2-ROC$_{13}$H$_{11}$Fe(CO)$_3$ (Table **6**, Nos. **17** and **18** with R=H, CH_3). Attempts to make stereochemical assignments of the epimeric derivatives No. 17 and 18 by ¹H NMR techniques (solvent dependency, shift reagents like Eu complexes) are unsuccessful. Treatment of either isomer a or b of No. 18 with HBF_4 in $O(COCH_3)_2$, followed by ether addition, leads to XXII (R=H) [3].

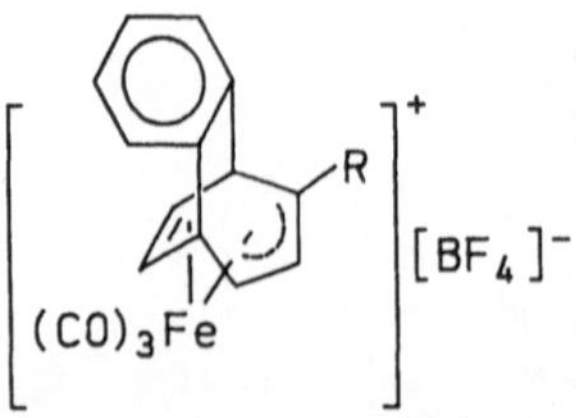

XXII

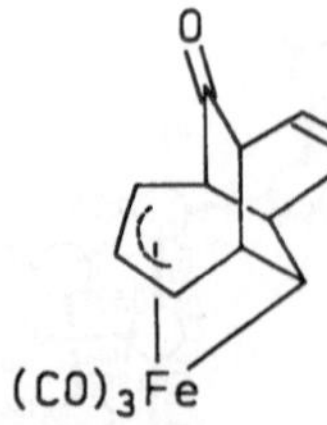

XXIII

XXIV

2-HO(2-CH$_3$)C$_{13}$H$_{10}$Fe(CO)$_3$ (Table **6**, No. **19**) gives XXII ($R=CH_3$) in 80% yield in the reaction with HBF_4 in $O(COCH_3)_2$ at 0 °C [12].

C$_{13}$H$_{10}$OFe(CO)$_3$ (Table **6**, No. **20**). The proton H-5 shows a triplet in the ¹H NMR spectrum. Compound No. 20 does not react with $LiAlH_4$ in ether at −78 °C. Reaction with $NaBH_4$ in C_2H_5OH or with $LiAlH_4$ in ether at 0 °C gives VII (R=H, p. 198) [3].

References on p. 207

1,4,10-$(CH_3)_3C_{10}H_9Fe(CO)_3$ (Table **6**, No. **21**) in hexane is added to silica gel (mesh 70 to 230), and the suspension is stirred vigorously for 24 h at room temperature. Work-up gives the free organic ligand ^{4}L (see Formula III, p. 198) [7].

$C_{10}H_{10}Fe(CO)_3$ (Table **6**, No. **22**) is also obtained by treating $(CO)_3Fe(C_{10}H_{10})Fe(CO)_3$ (see 2.5.1.2 in "Organoiron Compounds" C3, 1980, p. 96) in C_6H_6 in a sealed tube at 130 °C for 15 h (quantitative yield) [9, 11].

$C_{11}H_{10}OFe(CO)_3$ (Table **6**, No. **23**) is also prepared by carbonylation of XII (p. 199) with CO (100 atm) at 75 °C [5] or 80 to 85 °C [8] within 24 h [5, 8]. On addition of $Pr(fod)_3$ (fod= 1,1,1,2,2,3,3-heptafluoro-7,7-dimethyl-4,6-octanedionate) the values of H-1 and H-9 shift to higher field [8]. Treatment of No. 23 in C_6H_6 in a sealed tube at 100 °C for 3 d gives XXIII [5, 8]. Oxidation with $FeCl_3 \cdot 6H_2O$ in ether at 0 °C liberates the organic ligand ^{4}L [5, 8] in 80% yield [8]. Reaction with CO in C_6H_6, traces of acids, or air gives XXIV. Thus, with CO (100 atm) in C_6H_6 in the presence of traces of $(C_2H_5)_2OBF_3$, in an autoclave at 100 °C, XXIV is obtained in 99% yield after 2 d [8].

References:

[1] T. Margulis, L. Schiff, M. Rosenblum (J. Am. Chem. Soc. **87** [1965] 3269/70). — [2] A. Eisenstadt, S. Winstein (Tetrahedron Letters **1970** 4603/6). — [3] A. Eisenstadt (J. Organometal. Chem. **38** [1972] C32/C34). — [4] A. Eisenstadt (J. Organometal. Chem. **60** [1973] 335/42). — [5] R. Aumann (Angew. Chem. **88** [1976] 375).

[6] A.D. Charles, P. Diversi, B.F.G. Johnson, K.D. Karlin, J. Lewis, A.V. Rivera, G.M. Sheldrick (J. Organometal. Chem. **128** [1977] C31/C34). — [7] K. Hayakawa, H. Schmid (Helv. Chim. Acta **60** [1977] 1942/60). — [8] R. Aumann (Chem. Ber. **110** [1977] 1432/41). — [9] R. Aumann, H. Averbeck, C. Krüger (J. Organometal. Chem. **160** [1978] 241/53). — [10] V. Rivera, G.M. Sheldrick (Acta Cryst. B **34** [1978] 3374/6).

[11] R. Aumann, H. Averbeck, H. Wörmann (9th Intern. Conf. Organometal. Chem., Dijon 1979, Abstr. No. D33). — [12] S. Abramson, A. Eisenstadt (Tetrahedron Letters **36** [1980] 105/15). — [13] J.C. Barborack, S.L. Watson, A.T. McPhail, R.W. Miller (J. Organometal. Chem. **185** [1980] C29/C33). — [14] A.D. Charles, P. Diversi, B.F.G. Johnson, J. Lewis (J. Chem. Soc. Dalton Trans. **1981** 1906/17). — [15] B.F.G. Johnson, J. Lewis, G.L.P. Randall (J. Chem. Soc. A **1971** 422/9).

1.4.1.4.1.2.5 ^{4}L is an Eight-Membered Ring System

General References:

B.E. Mann, Dynamic NMR Spectroscopy in Inorganic and Organometallic Chemistry, Ann. Rept. NMR Spectrosc. **12** [1982] 263/86.

J.C. Green, Gas Phase Photoelectron Spectra of d- and f-Block Organometallic Compounds, Struct. Bonding [Berlin] **43** [1981] 37/112.

R. Hoffmann, Theoretical Organometallic Chemistry, Science **211** [1981] 995/1002.

H. Behrens, Four Decades of Metal Carbonyl Chemistry in Liquid Ammonia: Aspects and Prospects, Advan. Organometal. Chem. **18** [1980] 1/53.

G.I. Fray, R.G. Saxton, The Chemistry of Cyclooctatetraene and Its Derivatives, Cambridge Univ. Press, London 1978, pp. 1/492, 79/81.

L.A. Paquette, The Renaissance in Cyclooctatetraene Chemistry, Tetrahedron **31** [1975] 2855/83.

B.E. Mann, ^{13}C NMR Chemical Shifts and Coupling Constants of Organometallic Compounds, Advan. Organometal. Chem. **12** [1974] 135/213.

L.A. Fedorov, Stereochemical Non-Rigidity (Internal Rotation and Metallotropy) of Organic Derivatives of Transition Metals Investigated by Nuclear Magnetic Resonance Spectroscopy, Usp. Khim. **42** [1973] 1481/514; Russ. Chem. Rev. **42** [1973] 678/95.

1.4.1.4.1.2.5.1 ⁴L is a Cyclooctadiene

All possible examples of the tautomers $^{4}LFe(CO)_3$ with ^{4}L = cycloocta-1,3-diene, cycloocta-1,4-diene, and cycloocta-1,5-diene have been prepared as shown in Table 7, pp. 210/4.

It appears that the tendency of iron carbonyls to cause the migration of double bonds to produce the isomeric conjugated dienes is so great that the $^{4}LFe(CO)_3$ complexes of nonconjugated dienes (cf. Table 7) can only be prepared when a diene is used in which simple migration cannot occur [8] (remark: but cf. the existence of Nos. 7, 9, 10, 16 to 18, obtained by Methods I to III).

Most of the compounds described in Table 7 can be prepared by one of the following methods:

Method I: $Fe(CO)_5$ reacts photochemically or thermally with cyclooctapolyene in an inert solvent.

a. $Fe(CO)_5$ and the free organic ligand ^{4}L are irradiated in C_6H_6 at 15 to 20 °C [2, 16, 19, 33] for 19.5 to 48 h [16, 19, 33]. Thus, irradiation of $Fe(CO)_5$ and cycloocta-1,3-diene [16, 19, 33] for 20 h [19] or 48 h [33] followed by chromatography on silica gel gives No. 1 [33]. This compound is also obtained if the solvent of the reaction mixture is stripped off at room temperature at ca. 0.01 Torr, and the dark brown slush is dissolved in pentane. Filtration of the dark brown solid followed by cooling with dry ice/acetone gives No. 1 [19]. Similar irradiation of $Fe(CO)_5$ and cycloocta-1,5-diene with sunlight for 12 h [2] or extended irradiation with UV light [16] for 19.5 h followed by removal of the volatiles as above, dissolving the very thick dark oil in pentane, and filtration from dark brown pyrophoric material gives No. 7 on cooling as above [19]. Similar irradiation of $Fe(CO)_5$ and cyclooctatetraene for 44 h gives Nos. 1 and 7, separated by chromatography on silica gel with pentane, and $C_8H_{10}Fe(CO)_4$ (C_8H_{10} = cycloocta-1,3,5-triene) [33].

b. Some of the compounds are obtained by refluxing $Fe(CO)_5$ and the free organic ligand ^{4}L in an inert solvent. Thus, refluxing $Fe(CO)_5$ with cycloocta-1,5-diene [7], in the presence of a trace of hydroquinone in n-decane at 130 °C for 8 h followed by filtration and distillation [10] gives No. 7. Refluxing 3-methylcyclohexa-1,5-diene for 2 h followed by filtration, evaporation of the filtrate, dissolving the residue in low-boiling petroleum ether, and chromatography on alumina gives No. 9 [10]. Refluxing $Fe(CO)_5$ with 3-ethylcycloocta-1,5-diene in $(CH_2OCH_3)_2$ for 6 h or with 3-ethyl-7-octylcycloocta-1,5-diene in n-nonane for 14 h or in $O(C_2H_4OCH_3)_2$ for 6 h followed by filtration, evaporation, and sublimation gives Nos. 10 or 16, respectively [10]. Refluxing the free organic ligand ^{4}L of No. 18 with $Fe(CO)_5$ in hexane/C_6H_6 gives No. 18 and I [23].

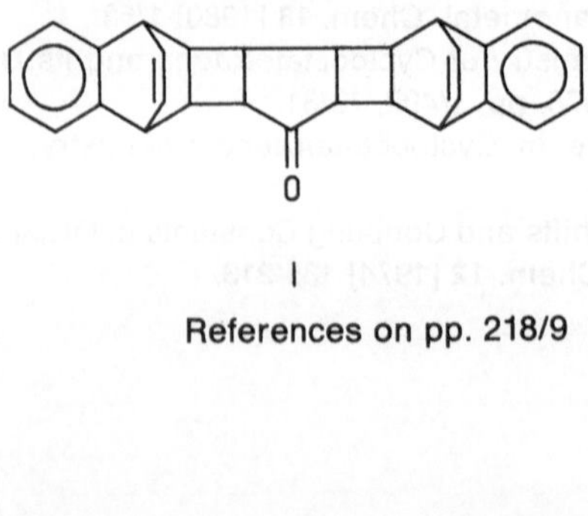

O

I

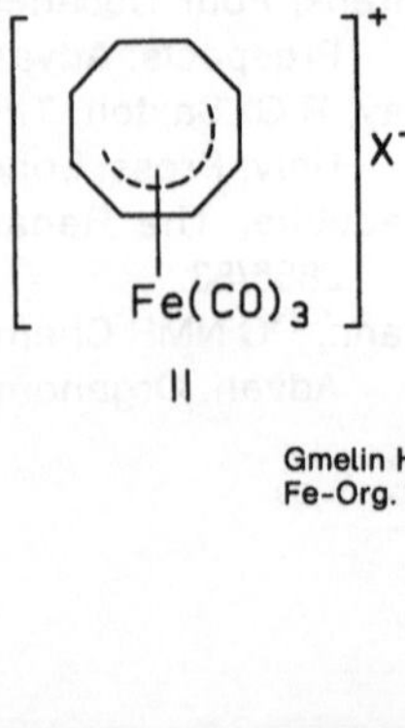

X^-

Fe(CO)$_3$

II

References on pp. 218/9

c. Heating $Fe(CO)_5$ and cycloocta-1,5-diene in a sealed tube at 130 °C for 17 h [7], or in an autoclave in an N_2 atmosphere (7 atm) at 220 °C for 2 h [3], or in the presence of a trace of hydroquinone under N_2 (17 atm) at 222 °C for 2 h [10] gives No. 7. The volatiles are removed [3, 7, 10], and the residue is distilled at 60 °C/high vacuum [7]. Chromatography on neutral [7] Al_2O_3 (activity II [7]) [7, 10] with C_6H_6 [7] or petroleum ether (b.p. 30 to 60 °C) [10] gives pure No. 7.

Method II: $Fe_2(CO)_9$ reacts thermally or photochemically with the free organic ligand ^4L in an inert solvent.

a. Irradiation of $Fe_2(CO)_9$ and cycloocta-1,5-diene (1,5-C_8H_{12}) gives 1,5-$C_8H_{12}(Fe(CO)_4)_2$ (main product) and No. 7 [16].

b. Treating $Fe_2(CO)_9$ and the free organic ligand ^4L of No. 17 in C_6H_6 at 60 °C gives No. 17 [11]. Similar treatment of ^4L from No. 18 in hexane/C_6H_6 at 55 °C gives No. 18 [23]. Refluxing $Fe_2(CO)_9$ with 2-ethylcycloocta-1,3-diene in tetrahydrofuran for 24 h or with 3-methylcycloocta-1,5-diene for 2 h gives Nos. 2 and 9. The reaction mixture is filtered, evaporated, and the residue is dissolved in low-boiling petroleum ether and chromatographed on alumina to give the pure products No. 2 and 9 [10].

c. $Fe_2(CO)_9$ reacts with excess cycloocta-1,5-diene in a sealed tube in ether at 130 to 150 °C for 50 h. The reaction mixture is filtered on Kieselgur. Removal of the volatiles gives No. 7 [58].

Method III: $Fe_3(CO)_{12}$ reacts with a cycloocta-1,5-diene in boiling C_6H_6 for 2 to 12 h followed by filtration, evaporation of the filtrate, and sublimation of the residue [1, 5, 6, 10]. The compound obtained by this method may be No. 1 in view of the tendency for iron carbonyl to catalyze the rearrangement of cycloocta-1,5-diene to cycloocta-1,3-diene. The reactions of this product (No. 1 or No. 7) with PR_3 (R = OCH_3, CH_3) cannot clarify this question [15], cf. 1.4.1.3.1.1.2.1 in "Organoiron Compounds" B6, 1981, Table 4, Nos. 59 and 60, p. 78. Similar reaction of $Fe_3(CO)_{12}$ with 3-methylcycloocta-1,5-diene for 2 h followed by work-up as described in Method IIb gives No. 9 [10].

Method IV: $[^5LFe(CO)_3]X$ complexes (see Formulas II, III) react with Lewis bases.

a. Reduction of II (X = BF_4) with $NaBH_4$ in H_2O deposits a black gummy like material which is extracted with ether, and the extracts are chromatographed on alumina to give No. 1, $Fe_3(CO)_{12}$ and an unidentified compound of the composition $C_{17}H_{24}OFe(CO)_3$ [33]. Similar reduction with 10% aqueous $NaBH_4$ covered with ether at 0 °C for 30 min followed by ether extraction, washing the extracts with H_2O, drying ($MgSO_4$), and chromatography on silica gel with C_6H_6 [34] gives No. 1 in 10% yield [24, 34] and V (R = H) [24]. Reaction of II with $NaOCH_3/CH_3OH$ gives IV and No. 3a, and the intermediate IV smoothly forms No. 3b [36, 37]. Addition of $P(C_6H_5)_3$ to a solution of II gives No. 4 [37].

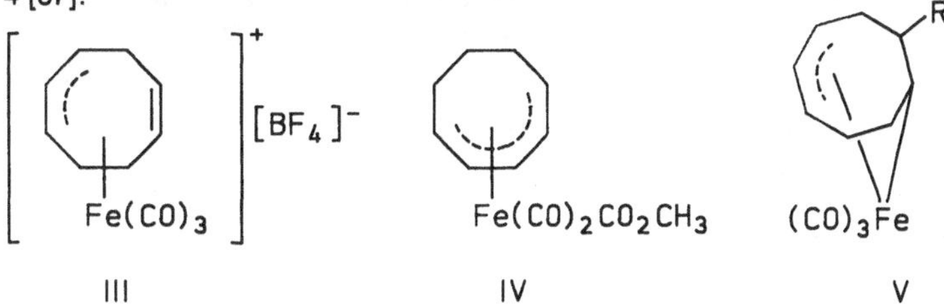

References on pp. 218/9

b. Reduction of III with $NaBH_4$ in H_2O at 0 °C for 5 min followed by ether extraction and separation on alumina gives No. 7 and V (R=H) in the ratio 1:9 [33]. Reaction of III with NaCN or $NaCH(COCH_3)_2$ in H_2O gives mixtures of V (R=CN) and No. 8 in the ratio 70:30 or of V ($R=CH(COCH_3)_2$) and No. 12 (ratio 1:1). Similar reaction of III with RMgX ($R=CH_3$, X=I; $R=CH_2{=}CHCH_2$, X=Br) in ether gives a mixture of complexes V ($R=CH_3$ or $CH_2CH{=}CH_2$) and Nos. 9 and 11, respectively. Reaction of III with $NaCH(CO_2C_2H_5)_2$ in tetrahydrofuran gives V ($R=CH(CO_2C_2H_5)_2$) and No. 13 (ratio 60:40), and similar reaction with $NaC(CO_2C_2H_5)_2C_6H_5$ gives V ($R=C(CO_2C_2H_5)_2C_6H_5$) and No. 14 (ratio 60:40) [33].

Table 7
Complexes of the Type $^4LFe(CO)_3$ where ^4L is a Cyclooctadiene.
Further information on numbers preceded by an asterisk is given at the end of the table, pp. 214/8.
For abbreviations and dimensions see p. VIII.

No.	compound	method of preparation (yield in %), properties and remarks	Ref.
^4L is a cycloocta-1,3-diene:			
*1	(cycloocta-1,3-diene, positions 1–8)$Fe(CO)_3$	I a (25), III (10), IV a (10) m.p. 36°, m.p. 36.5 to 37° (from pentane), subl.p. 35°/0.01 Torr, yellow crystals or needles ^{1}H NMR (benzene-d_6): 1.2 (m, H-6,7), 2.0 (m, H-5,8), 3.8 (m, H-1,4), 4.7 (m, H-2,3) ^{1}H NMR (C_6H_6): 1.12, 1.82 (m, H-5 to 8), 2.99 (m, H-1,4), 4.77 (m, H-2,3) ^{13}C NMR: 25.2, 28.9 (C-5 to 8); 63.2, 91.8 (C-1 to 4); 212.2 (CO) ^{13}C NMR (toluene-d_8 at 195 K): 25.6, 29.2 (C-5 to 8); 63.2 (C-1,4), 91.8 (C-2,3), 210.6 (basal CO), 216.9 (apical CO) ^{57}Fe NMR (C_6H_6): 169.4±4 (downfield from $Fe(CO)_5$) ^{57}Fe-γ (−196°): δ=0.12 (relative Fe), Δ=1.36 IR (KBr): 1965, 2070 (CO) IR (cyclohexane): 1971, 1974, 2043 (CO) IR (heptane or decalin): 1980, 1984, 2053 (CO) UV (pentane): λ_{max}=324 nm mass spectrum: $[M-nCO]^+$ (n=0 to 3)	[1, 5, 6, 8, 15, 16, 19, 27, 32 to 34, 42 to 44, 48]
*2	(C_2H_5-cyclooctadiene)$Fe(CO)_3$	IIb	[10]

References on pp. 218/9

Table 7 [continued]

No.	compound	method of preparation (yield in %), properties and remarks	Ref.
*3a	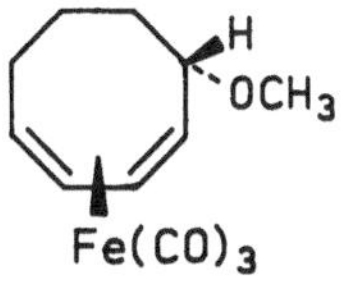 H, OCH_3, $Fe(CO)_3$	IVa mass spectrum: $[M-nCO]^+$ (n=0 to 3)	[36, 37]
b	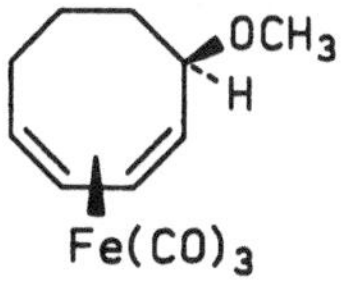 OCH_3, H, $Fe(CO)_3$	IVa mass spectrum: $[M-nCO]^+$ (n=0 to 3)	[36, 37]
4	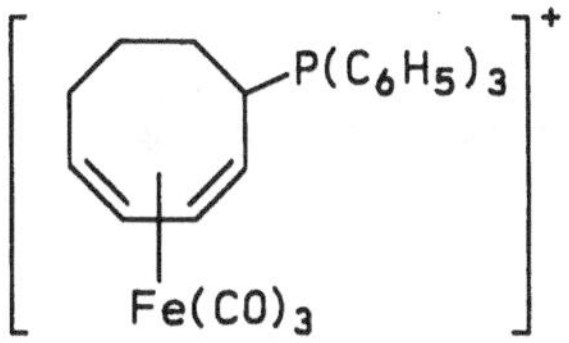 $[\ldots P(C_6H_5)_3 \ldots Fe(CO)_3]^+$	IVa IR (CH_2Cl_2): 1980, 2062 (CO)	[37]
*5	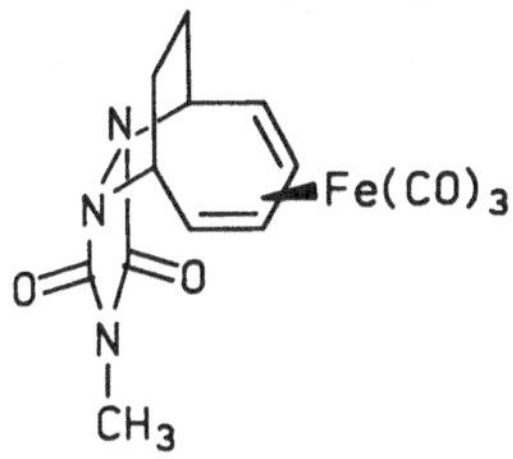 N, N, O, O, N, CH_3, $Fe(CO)_3$	–	[53]
^{4}L is a cycloocta-1,4-diene:			
*6	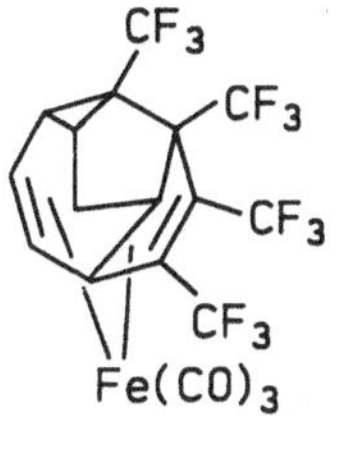 CF_3, CF_3, CF_3, CF_3, $Fe(CO)_3$	crystals	[40, 54]
^{4}L is a cycloocta-1,5-diene:			
*7	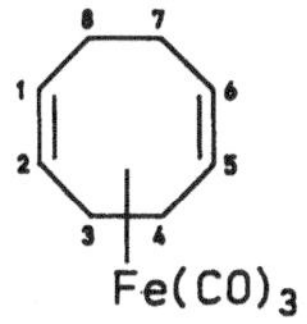 1–8, $Fe(CO)_3$	I a (47), I b (4.4), I c (3.2), II a, II c, III (10), IV a, IV b (10) m.p. 60 to 62°, m.p. 61 to 63°, m.p. 76°, m.p. 90 to 90.5° (from pentane), subl.p. 35 to 60°/0.1 Torr, b.p. 47°/2 Torr, b.p. 50°/2 Torr, b.p. 56°/ca. 10^{-3} Torr, b.p. 90°/5 Torr, orange liquid or oil which forms a yellow solid just below ambient temperature, stable yellow, orange, or orange-yellow crystals	[1 to 8, 10, 12 to 17, 19, 21, 24, 25, 27, 33, 58]

References on pp. 218/9

Table 7 [continued]

No.	compound	method of preparation (yield in %), properties and remarks	Ref.
		^{1}H NMR (benzene-d_6): 2.0 (m, 4H), 3.38 (m, 8H) ^{1}H NMR (C_6H_6): 1.2 to 2.0 (m, H-3,4,7,8), 3.80 (m, H-1,2,5,6) ^{13}C NMR (−130°): 33.8 (C-3,4,7,8), 83.7 (C-1,2,5,6), 218.8 (s, CO) ^{57}Fe NMR (C_6H_6): 380.4 ± 1.5 (downfield from $Fe(CO)_5$) ^{57}Fe-γ (294 ± 1 K): δ = 0.23 (relative $Na_2[Fe(CN)_5NO] \cdot 2H_2O$), Δ = 1.83 ^{57}Fe-γ (78 K): δ = 0.228 (relative $Na_2[Fe(CN)_5NO] \cdot 2H_2O$), Δ = 1.83 IR: 697, 709, 730, 754, 791, 818, 866, 871, 929, 954, 995, 1031, 1043, 1089, 1142, 1177, 1186, 1235, 1269, 1326, 1383, 1443, 1463, 1692, 1980, 2049, 2805, 2941, 2985 IR (liquid): 1965, 2049 (CO) IR (cyclohexane): 1976, 2018 (CO) IR (cyclohexane): 1967, 1975, 2043 (CO) IR (cyclohexane): 1951, 1961, 2027 (CO) IR (CS_2): 1977, 2010, 2080 (CO) UV (pentane): λ_{max} = 434 nm mass spectrum: $[M - nCO]^+$ (n = 0 to 3), $[C_4H_6Fe]^+$, $[C_8H_{12}]^+$	
8	NC Fe(CO)$_3$	IVb	[33]
*9	CH_3 Fe(CO)$_3$	Ib, IIb, III, IVb (small amount)	[10, 33]
*10	C_2H_5 Fe(CO)$_3$	Ib	[10]
11	CH_2=CHCH$_2$ Fe(CO)$_3$ (ring positions 1–8)	IVb (23) viscous orange oil ^{1}H NMR (C_6H_6): 0.9 to 2.8 (m, H-3,4,7,8, CH_2), 3.31 (m, H-1,2,5,6), 4.93 (m, =CH_2), 5.59 (m, -CH=) IR (cyclohexane): 1948, 1962, 2036 (CO)	[24, 33]

References on pp. 218/9

Table 7 [continued]

No.	compound	method of preparation (yield in %), properties and remarks	Ref.
12	$(CH_3OC)_2HC$ $Fe(CO)_3$	IVb ^{1}H NMR ($CDCl_3$): 1.2 to 3.5 (m, 8 ring-H), 2.19 (s, CH_3), 2.23 (s, CH_3), 3.84 (m, 3 ring-H), 4.28 (m, CH) IR (cyclohexane): 1952, 1972, 2030 (CO)	[24, 33]
13	$(C_2H_5O_2C)_2HC$ $Fe(CO)_3$	IVb IR (cyclohexane): 1952, 1968, 2029 (CO)	[24, 33]
14	$C_6H_5(C_2H_5O_2C)_2C$ $Fe(CO)_3$	IVb IR (cyclohexane): 1951, 1970, 2027 (CO)	[33]
*15	O O $Fe(CO)_3$	^{57}Fe-γ (78 K): $\delta = 0.33 \pm 0.03$, $\Delta = 1.25 \pm 0.03$ ^{57}Fe-γ (methyltetrahydrofuran, 78 K): $\delta = 0.33$, $\Delta = 1.17$	[14]
*16	C_8H_{17} C_2H_5 $Fe(CO)_3$	Ib	[10]
17	$Fe(CO)_3$ O	IIb (13)	[11]
18	$Fe(CO)_3$	Ib (small quantity), IIb (60) dec.p. 124 to 126°, yellow-orange crystals ^{1}H NMR ($CDCl_3$): 2.84 (m, 2H), 4.00 (m, 2H), 4.16 (m, 2H), 4.51 (t, 2H), 7.01 (m, aryl, 4H) IR (hexane): 1963, 1975, 2035 (CO) mass spectrum: $[M-nCO]^+$ (n = 0 to 3), $[M-3CO-nHC_2H]^+$ (n = 1, 2), $[C_{14}H_{14}]^+$, $[C_{10}H_8]^+$	[23]

References on pp. 218/9

Table 7 [continued]

No.	compound	method of preparation (yield in %), properties and remarks	Ref.
*19	CH_3, O, CH_3, CF_3, CH_3, CF_3, O, CH_3, $Fe(CO)_3$	dark yellow crystals (from hexane/CH_2Cl_2 (1:1)) ^1H NMR ($CDCl_3$): 1.1 (s, 2CH_3), 1.9 (s, CH_3), 2.0 (s, CH_3) ^{19}F NMR ($CDCl_3$): 51.2 (q, CF_3; J(F,F)=10.0), 53.9 (q, CF_3) IR (hexane): 1728, 1749 (C=O); 2013, 2035, 2083 (CO) mass spectrum: $[M-nCO]^+$ (n=0 to 3), $[C_{14}H_{12}F_6O_2]^+$, $[C_{14}H_{12}F_4]^+$	[51]
compound of unknown constitution:			
*20	OH, $Fe(CO)_3$	–	[20]

*Further information:

1,3-$C_8H_{12}Fe(CO)_3$ (Table **7**, No. **1**) is also obtained by irradiation of 1,5-$C_8H_{12}Fe(CO)_3$ (No. 7) in pentane [19] or C_6H_6 [33] at 20 °C for 18 h. Filtration and cooling of the filtrate gives No. 1 in 59% yield [19]. Irradiation of No. 6 in the presence of cyclopenta-1,5-diene in n-pentane at −45 °C for 118 h gives Nos. 1 and 7 [28]. The compound No. 1 seems to be thermally less stable than No. 7 [16], but refluxing No. 7 in hexane gives No. 1 (half live ca. 1 h) [33]. These reactions show that $Fe(CO)_3$ prefers to be coordinated to conjugated rather than unconjugated dienes [38]. Successful gas chromatographic elution of No. 1 at temperatures up to 150 °C with no apparent thermal decomposition is achieved provided that precautions are taken to exclude air and prevent prolonged exposure to light. Teflon columns employing E-301 methyl silicone gum as the stationary phase give best results [26].

Compound No. 1 is monomeric in C_6H_6 (osmometry) [19]. The ^1H NMR pattern is similar to that of 1,3-$C_8H_{12}Fe(PR_3)_3$ (R=OCH_3, OC_2H_5, OC_3H_7-i) (see 1.4.1.1.1 in "Organoiron Compounds" B6, 1981, Table 1, Nos. 32 to 34, p. 12) and is different from No. 7 [43]. The coalescence temperature for the intramolecular site exchange of the CO groups is 215 K, and the activation parameters determined from the temperature dependent ^{13}C NMR spectrum are E_a=40.7 kJ/mol, $\Delta H^{\neq}$=38.9±2 kJ/mol, $\Delta S^{\neq}$=−10±5 J·mol^{-1}·K^{-1}, and $\Delta G^{\neq}_{298}$=41.8 ±3 kJ/mol [9], comparable with $\Delta H^{\neq}$=42.7±2.0 kJ/mol, $\Delta S^{\neq}$=−3±8 J·mol^{-1}·K^{-1}, $\Delta G^{\neq}_{298}$=43.5±2.0 kJ/mol in [42], cf. [47]. For a description of this compound as a resonance hybrid of valence-bond structures see [56, 57]. The He(I) photo electron spectrum gives the ionization energies 7.45, 8.27 (Fe); 8.87 (1a_2), 10.44 (1b_1), and 10.87 eV [18].

Oxidation with $[NH_4]_2[Ce(NO_3)_6]$ in C_2H_5OH at 0 °C liberates the organic ligand ^4L in 52% [19] or 99% yield [16]. Oxidation with $ON(CH_3)_3$ in boiling C_6H_6 gives the free organic ligand ^4L in 95% yield within 1 h [59]. Reaction with CO (85 atm) in cyclohexane at 25 °C for 50 h gives $Fe(CO)_5$ and cycloocta-1,3-diene (^4L) [33].

References on pp. 218/9

The compound No. 1 reacts smoothly with excess $P(C_6H_5)_3$ in n-heptane (Ar atmosphere) to form trans-$(CO)_3Fe(P(C_6H_5)_3)_2$. Compared with this second-order process (rate constant k_2), the first-order reaction to give 1,3-$C_8H_{12}Fe(CO)_2P(C_6H_5)_3$ (rate constant k_1) is negligible:

t in °C	$k_1 \cdot 10^5$ (s^{-1})	$k_2 \cdot 10^3$ ($L \cdot mol^{-1} \cdot s^{-1}$)
50.0	0.15 ± 0.22	1.05 ± 0.3
60.0	0.72 ± 0.32	3.15 ± 0.08
70.3	1.74 ± 0.60	7.79 ± 0.16
79.9	-0.44 ± 0.69	18.34 ± 0.03

For the second-order process, the activation parameters are $\Delta H^{\neq} = 20.76 \pm 0.22$ kcal/mol and $\Delta S^{\neq} = -8.0 \pm 0.6$ $cal \cdot mol^{-1} \cdot K^{-1}$. The relatively high activation enthalpy for a reaction of this type might result from a low degree of bond making in the transition state. However, the observed activation parameters can be composite values containing terms relating to dissociation of one double bond from the Fe atom, if the overall reaction involves this process as a preequilibrium rather than direct binuclear attack on No. 1. Indeed the relative positive value of $\Delta S^{\neq}$ tends to support this suggestion. When Ar is replaced by CO (1 atm) the observed rate constants remain unaltered as shown in the following table, indicating that loss of CO from No. 1 is not involved in the reaction mechanism:

t in °C	$[P(C_6H_5)_3]$ (mol/L)	$k_{obs} \cdot 10^4$ (s^{-1})	$k_{calc} \cdot 10^4$ (s^{-1})
70.3	0.0164	1.40	1.45
70.3	0.0870	6.90	6.85

However, $(CO)_4FeP(C_6H_5)_3$ and trans-$(CO)_3Fe(P(C_6H_5)_3)_2$ are formed under these conditions, more $(CO)_4FeP(C_6H_5)_3$ being formed at lower $P(C_6H_5)_3$ concentrations. The experimental conditions used were such that $k_1 \ll k_2$ $[P(C_6H_5)_3]$. Nonetheless, an upper limit of $k_1 \lesssim 1 \times 10^{-5}$ s^{-1} at 60 °C can be estimated [32]. Reaction with $NaN(Si(CH_3)_3)_2$ in C_6H_6 in a sealed tube gives a mixture of compounds, because not only 1,3-$C_8H_{12}Fe(CO)_2CN$ (see 1.4.1.3.1.2 in "Organoiron Compounds" B6, 1981, Table 5, No. 20, p. 97) is formed but also deprotonation occurs [52]. Reaction with $[C(C_6H_5)_3]BF_4$ in CH_2Cl_2 for 1 h gives $[C_8H_{11}Fe(CO)_3]BF_4$ (see Formula II, p. 208, with $X = BF_4$) [24, 33]. No reaction occurs if No. 1 and cycloocta-1,3-diene are irradiated in pentane at −5 °C for 100 h or without a solvent at −20 °C for 90 h [28]. Irradiation with $CF_2{=}CF_2$ in hexane gives a 3:1 mixture of VI (R=F) and VII [41], and similar irradiation with $CF_3CF{=}CF_2$ gives VI ($R = CF_3$) [39]. For a theoretical exploration of the reaction path of these alkene additions see [55].

VI

VII

VIII

IX

References on pp. 218/9

2-$C_2H_5C_8H_{11}Fe(CO)_3$ (Table **7**, No. **2**) is proposed as an antiknock agent, as a fuel additive, as an additive for lubricant compositions, and as a starting material for metal plating [10].

5-$CH_3OC_8H_{11}Fe(CO)_3$ (Table **7**, No. **3**). Isomer No. 3b(?) shows IR bands at 1080 (C–O); 1970, 2040 (CO) cm^{-1}. On evaporation, CH_3OH is lost to give VIII [37].

$CH_3N(CON)_2C_8H_{10}Fe(CO)_3$ (Table **7**, No. **5**) is prepared by treatment of IX with H_2 in the presence of PtO_2 (quantitative yield). Oxidation with $[NH_4]_2[Ce(NO_3)_6]$ liberates the organic ligand ^4L in 99% yield [53].

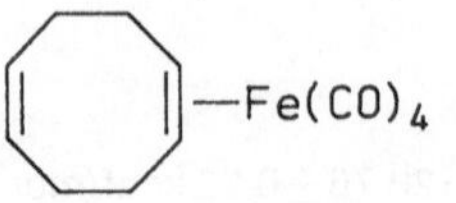

X 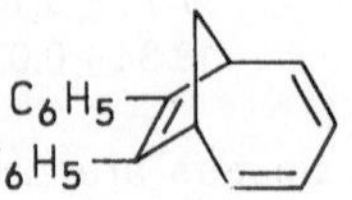 XI 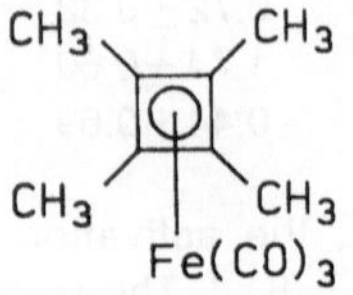 XII 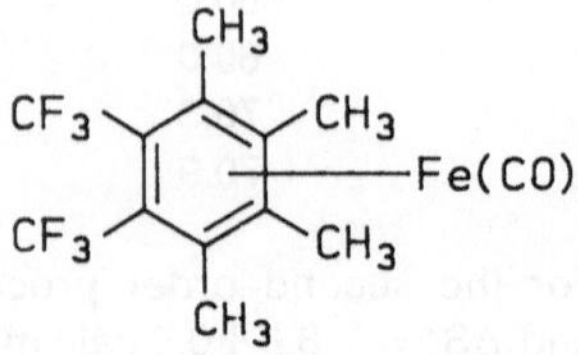 XIII

$(CF_3)_4C_{11}H_8Fe(CO)_3$ (Table **7**, No. **6**) is prepared by irradiation of $C_7H_8Fe(CO)_3$ (C_7H_8= cycloheptatriene) with $CF_3C{\equiv}CCF_3$ in hexane. Reaction with $P(OCH_2)_3CCH_3$ gives $(CF_3)_4C_{11}H_8Fe(CO)_2P(OCH_2)_3CH_3$ (see 1.4.1.3.1.1.2.1 in "Organoiron Compounds" B6, 1981, Table 4, No. 63, p. 79) [40, 54].

1,5-$C_8H_{12}Fe(CO)_3$ (Table **7**, No. **7**) is also obtained by irradiation of X [20]. Reaction of $(1,5\text{-}C_8H_{12})_2Fe$ with CO in methylcyclohexane at −130 °C gives traces of $Fe(CO)_5$ and No. 7; no isomerization is observed in this reaction [44, 45]. Successful purification by gas chromatographic elution of No. 7 at temperatures up to 150 °C, with no apparent thermal decomposition, is achieved provided that precautions are taken to exclude air and prevent prolonged exposure to light. Teflon columns employing E-301 methyl silicone gum as stationary phase give best results [26].

Osmometric [19] and cryoscopic molecular weight determinations [7] in C_6H_6 [7, 19] show No. 7 to be monomeric in solution. A 1:1 mixture of No. 7 and $1,5\text{-}C_8H_{12}(Fe(CO)_4)_2$ ($1,5\text{-}C_8H_{12}$=cycloocta-1,5-diene) obtained by Method IIa (p. 209) melts at 67.5 to 68.5 °C [16]. The diamagnetic [7] compound has a refractive index of $n_D^{20}=1.5765$ [2], and in C_6H_6 a dipole moment of $\mu_{25°}=3.09\pm0.04$ D [7, 21]. The ^{13}C NMR spectrum is essentially unchanged down to −130 °C, indicating that the coalescence temperature for the CO scrambling is below −130 °C [30, 42, 47]. For the UV spectrum in 95% aqueous C_2H_5OH see the figure in [2].

Compound No. 7 is fairly stable to air [2, 7] and can be stored under N_2 at −78 °C [6]. It seems to be thermally more stable than No. 1 [16], decomposes slightly at >140 °C, and rapid decomposition is observed at >190 °C [7]. Irradiation in pentane for 18 h [19] or in C_6H_6 [33] at room temperature gives No. 1 [19, 33] in 59% yield [19]. This isomerization reaction is an example of ligand isomerization by formal shift of the double bonds [19], cf. [38]. The compound does not rearrange in solution to No. 1 on standing [12], but refluxing No. 7 in hexane gives No. 1 (half live ca. 1 h) [33]. Oxidation with $[NH_4]_2[Ce(NO_3)_6]$ in C_2H_5OH [16, 19] at 0 °C [19] gives cycloocta-1,5-diene (99.9%) and cycloocta-1,3-diene (0.1%) in a total yield of 30% [19]. Reaction of No. 7 with D_2 in C_6H_6 at room temperature for 30 h gives cycloocta-1,3- and -1,5-diene and cyclooctene, for the isotopic distribution see [8]. Mass spectrometric analysis of the No. 1 formed in this reaction shows that D is incorporated in the ring in No. 1, if at all for the amount is <1% [19]. NaOH in CH_3OH/H_2O decomposes No. 7 [7].

References on pp. 218/9

Reaction with PR_3 (R=OCH_3, CH_3) at room temperature gives $(CO)_2Fe(P(OCH_3)_3)_3$ and trans-$(CO)_3Fe(P(CH_3)_3)_2$, respectively. Similar reaction at −15 °C gives $C_8H_{12}Fe(CO)_2$-$P(OCH_3)_3$ and trans-$(CO)_3Fe(P(CH_3)_3)_2$ together with $C_8H_{12}Fe(CO)_2P(CH_3)_3$. The positions of the C=C double bonds in ^{4}L are not clear, possibly isomerization occurs (see 1.4.1.3.1.1.2.1 in "Organoiron Compounds" B6, 1981, Table 4, Nos. 59, 60, p. 78) [15]. $P(C_6H_5)_3$ was first considered not to react with No. 7 in refluxing heptane within 17 h [7], but refluxing of No. 7 in C_6H_6 for 8 h gives $(CO)_3Fe(P(C_6H_5)_3)_2$ [16, 19, 24] in 58% yield and cycloocta-1,5- and cycloocta-1,3-diene in the ratio 1:4 [19]. The kinetic data indicate that the reaction is independent of $P(C_6H_5)_3$ concentration and the kinetic parameters are comparable with those obtained for $Ni(CO)_4$ with phosphines (see 1.1.5.4.1.3.6 in "Nickel-Organische Verbindungen" 1, 1975, pp. 238/9) [24]. Compound No. 7 does not give CO insertion products with $AlCl_3$ in C_6H_6 at room temperature within 16 h [35]. Reaction with CO (85 atm) in cyclohexane at 25 °C for 50 h gives $Fe(CO)_5$ and cycloocta-1,5-diene [33]. Reaction with $NaN(Si(CH_3)_3)_2$ gives a mixture of compounds because not only formation of Na[1,5-$C_8H_{12}Fe(CO)_2CN$], but also deprotonation occurs [52]. Reaction with $[C(C_6H_5)_3]BF_4$ gives III (p. 209) [33] erroneously formulated first as II (p. 208) [24], and reaction with $BrC(CH_3)_3$ at 60 °C gives cycloocta-1,3-, -1,4-, and -1,5-diene, isobutene, and CO [19]. Irradiation of No. 7 with cycloocta-1,5-diene in n-pentane at −45° for 118 h gives No. 1. No reaction is observed without a solvent [28]. Heating No. 7 together with cycloheptatriene and $C_6H_5C{\equiv}CC_6H_5$ gives XI in low yield [29].

Compound No. 7 is proposed as an antiknock fuel additive [3, 10], additive for lubricant compositions [10], and as a starting material for metal plating [3, 10].

3-$RC_8H_{11}Fe(CO)_3$ (Table **7**, Nos. **9** and **10** with R=CH_3, C_2H_5), **$C_2H_2O_2C_8H_{10}Fe(CO)_3$** (Table **7**, No. **15**), and **3-C_2H_5(7-C_8H_{17})$C_8H_{10}Fe(CO)_3$** (Table **7**, No. **16**). The compounds No. 9, 10, 16 are proposed as antiknock agents, as fuel additives, as additives for lubricant compositions, and as starting materials for metal plating [10]. The ^{1}H NMR spectrum for No. 15, recorded over a temperature range of −60 to +120 °C, does not show any appreciable change in appearance [14].

$(CF_3)_2(CH_3)_4C_8O_2Fe(CO)_3$ (Table **7**, No. **19**) is prepared by irradiation of $(CH_3)_4C_4Fe(CO)_3$ (see Formula XII, p. 216) and $CF_3C{\equiv}CCF_3$ in a Carius tube in hexane [49, 51] at −196 °C for 4 d [51]. The irradiation is repeated for 4 times followed by filtration and removal of

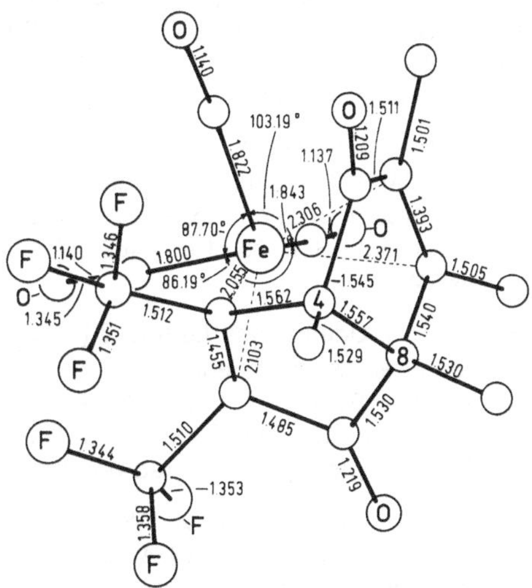

Fig. 24. Molecular structure of $(CF_3)_2(CH_3)_4C_8O_2Fe(CO)_3$ (No. 19) [51].

References on pp. 218/9

the solvent. Chromatography on Florisil with hexane gives $(CF_3)_2(CH_3)_4C_6Fe(CO)_3$ (see Formula XIII). Further elution with hexane/CH_2Cl_2 (1:1) gives No. 19 in 2.4% yield. The compound No. 19 crystallizes in the monoclinic space group $P2_1/n-C^5_{2h}$ with the unit cell parameters a=9.031(7), b=17.865(16), c=11.519(10) Å, β=105.49(6)°; Z=4 molecules per unit cell. The measured density (flotation) is 1.71 g/cm³, calculated 1.728 g/cm³. The main bond distances and angles are shown in **Fig. 24**, p. 217. The bicyclic dienone is folded by ca. 97.9° along C(4)-C(8). Individually, the two five-membered rings are not planar, with the C atom opposite the alkene function acting as the apex of an envelope conformation. Both atoms are folded towards the Fe; C(8) by ca. 28.5° and C(4) by ca. 31.2° [51].

$HOC_8H_{11}Fe(CO)_3$ (Table **7**, No. **20**) is mentioned as the reaction product of $[C_8H_{11}Fe(CO)_3]^+$ (C_8H_{11} = cycloocta-2,4-dienyl) with OH^- [20].

References:

[1] T.A. Manuel, F.G.A. Stone (Chem. Ind. [London] **1960** 231/2). – [2] A. Nakamura, N. Hagihara (Mem. Inst. Sci. Ind. Res. Osaka Univ. **17** [1960] 187/90; C.A. **1961** 6457). – [3] K.G. Ihrman, T.H. Coffield, Ethyl Corp. (U.S. 3093671 [1960/63]). – [4] A. Nakamura, N. Hagihara (unpublished results from [6]). – [5] T.A. Manuel (Diss. Harvard Univ. 1961; Diss. Abstr. **22** [1961] 1003/4).

[6] R.B. King, T.A. Manuel, F.G.A. Stone (J. Inorg. Nucl. Chem. **16** [1961] 233/9). – [7] W. Fröhlich (Diss. München Univ. 1961). – [8] R.B. King (J. Am. Chem. Soc. **84** [1962] 4705/8). – [9] P. Bischofberger, H.-J. Hansen (Helv. Chim. Acta **65** [1982] 721/5). – [10] K.G. Ihrman, T.H. Coffield, Ethyl Corp. (U.S. 3164621 [1963/65]).

[11] W.S. Trahanovsky, B.W. Surber, M.C. Wilkes, M.M. Preckel (J. Am. Chem. Soc. **104** [1982] 6779/81). – [12] F.A. Cotton, A. Davison, J.W. Faller (J. Am. Chem. Soc. **88** [1966] 4507/9). – [13] R. Grubbs, R. Breslow, R. Herber, S.J. Lippard (J. Am. Chem. Soc. **89** [1967] 6864/70). – [14] R.H. Herber (Symp. Faraday Soc. No. 1 [1967] 86/96). – [15] R. Mathieu, R. Poilblanc (Compt. Rend. C **264** [1967] 1035/6).

[16] E. Koerner von Gustorf, J.C. Hogan (Tetrahedron Letters **1968** 3191/4). – [17] R.H. Herber (Chem. Anal. [N.Y.] **26** [1969] 315/40). – [18] J.C. Green, P. Powell, J. van Tilborg (J. Chem. Soc. Dalton Trans. **1976** 1974/6). – [19] J.C. Hogan (Diss. Boston College 1969; Diss. Abstr. Intern. B **30** [1969/70] 4012). – [20] P. McArdle, H. Sherlock (Proc. 16th Intern. Conf. Coord. Chem., Dublin 1974, Abstr. No. R 89).

[21] R.D. Fischer (Diss. München Univ. 1961). – [22] Y. Shvo, E. Hazum (J. Chem. Soc. Chem. Commun. **1974** 336/7). – [23] Y. Menachem, A. Eisenstadt (J. Organometal. Chem. **33** [1971] C29/C32). – [24] F.A. Cotton, A.J. Deeming, P.L. Josty, S.S. Ullah, A.J.P. Domingos, B.F.G. Johnson, J. Lewis (J. Am. Chem. Soc. **93** [1971] 4624/6). – [25] R.H. Herber (Introd. Mössbauer Spectrosc. **1971** 138/54).

[26] E.J. Forbes, M.K. Sultan, P.C. Uden (Anal. Letters **5** [1972] 927/30). – [27] T. Jenny, W. von Philipsborn, J. Kronenbitter, A. Schwenk (J. Organometal. Chem. **205** [1981] 211/22). – [28] J. Buchkremer (Diss. Bochum Univ. 1973). – [29] R.E. Davis, T.A. Dodds, T.H. Hseu, J.C. Wagnon, T. Devon, J. Tancrede, J.S. McKennis, R. Pettit (J. Am. Chem. Soc. **96** [1974] 7562/3). – [30] L. Kruczynski (unpublished results from [31]).

[31] L. Kruczynski, J. Takats (J. Am. Chem. Soc. **96** [1974] 932/4). – [32] B.F.G. Johnson, J. Lewis, M.V. Twigg (J. Chem. Soc. Dalton Trans. **1974** 2546/9). – [33] A.J. Deeming, S.S. Ullah, A.J.P. Domingos, B.F.G. Johnson, J. Lewis (J. Chem. Soc. Dalton Trans. **1974** 2093/104). – [34] R. Edwards, J.A.S. Howell, B.F.G. Johnson, J. Lewis (J. Chem. Soc. Dalton

Trans. **1974** 2105/12). – [35] B.F.G. Johnson, J. Lewis, D.J. Thompson, B. Heil (J. Chem. Soc. Dalton Trans. **1975** 567/71).

[36] G. Schiavon, C. Paradisi, C. Boanini (Inorg. Chim. Acta **14** [1975] L5/L6). – [37] G. Schiavon, C. Paradisi (J. Organometal. Chem. **210** [1981] 247/51). – [38] M. Elian, R. Hoffmann (Inorg. Chem. **14** [1975] 1058/76). – [39] M. Green, B. Lewis, J.J. Daly, F. Sanz (J. Chem. Soc. Dalton Trans. **1975** 1118/27). – [40] M. Bottrill, R. Goddard, M. Green, R.P. Hughes, M.K. Lloyd, B. Lewis, P. Woodward (J. Chem. Soc. Chem. Commun. **1975** 253/5).

[41] A. Bond, B. Lewis, M. Green (J. Chem. Soc. Dalton Trans. **1975** 1109/18). – [42] L. Kruczynski, J. Takats (Inorg. Chem. **15** [1976] 3140/7). – [43] A.D. English, J.P. Jesson, C.A. Tolman (Inorg. Chem. **15** [1976] 1730/2). – [44] R.A. Cable, M. Green, R.E. Mackenzie, P.L. Timms, T.W. Turney (J. Chem. Soc. Chem. Commun. **1976** 270/1). – [45] P.L. Timms, T.W. Turney (unpublished results from [46]).

[46] P.L. Timms, T.W. Turney (Advan. Organometal. Chem. **15** [1977] 53/112). – [47] F.A. Cotton, B.E. Hanson (Israel J. Chem. **15** [1977] 165/73). – [48] D. Cunningham, P. McArdle, H. Sherlock, B.F.G. Johnson, J. Lewis (J. Chem. Soc. Dalton Trans. **1977** 2340/4). – [49] M. Bottrill, M. Green, A.J. Welch (unpublished results from [50]). – [50] M. Green (Ann. N.Y. Acad. Sci. **295** [1977] 160/83).

[51] A. Bond, M. Bottrill, M. Green, A.J. Welch (J. Chem. Soc. Dalton Trans. **1977** 2372/81). – [52] H. Behrens, M. Moll, W. Popp, P. Würstl (Z. Naturforsch. **32b** [1977] 1227/9). – [53] H. Olsen, J.P. Snyder (J. Am. Chem. Soc. **100** [1978] 285/7). – [54] R. Goddard, P. Woodward (J. Chem. Soc. Dalton Trans. **1979** 711/4). – [55] A. Stockis, R. Hoffmann (J. Am. Chem. Soc. **102** [1980] 2952/62).

[56] W.C. Herndon (J. Am. Chem. Soc. **102** [1980] 1538/41). – [57] W.C. Herndon (Israel J. Chem. **20** [1980] 276/80). – [58] S.S. Ullah, S.E. Kabir, A.K.M.A. Matin, A.K.F. Rahman (J. Bangladesh Acad. Sci. **5** [1981] 75/83).

1.4.1.4.1.2.5.2 ^{4}L is a Cyclooctatriene

The compounds of the type $^4LFe(CO)_3$ where ^4L = cyclooctatriene are listed in Table 8, pp. 222/5. The compounds with ^4L = cycloocta-1,3,5-triene (Nos. 1 to 7) can be prepared by the following methods:

Method I: $Fe(CO)_5$ reacts photochemically or thermally with a cyclooctatriene.

a. Excess $Fe(CO)_5$ and a mixture of cycloocta-1,3,5-triene/cycloocta-1,3,6-triene are irradiated for 16 h. Distillation or chromatography on alumina gives No. 1 and $C_8H_{10}Fe_2(CO)_6$ (see 2.6.1.2.4 in "Organoiron Compounds" C5, 1981, Table 3, No. 1, p. 28) [5]. Similar irradiation of excess $Fe(CO)_5$ with cycloocta-1,3,5-triene in C_6H_6 at room temperature for 3 h followed by chromatography on neutral alumina (grade III) gives No. 1 and small amounts of $C_8H_{10}Fe_2(CO)_6$. Rechromatography of the first band gives pure No. 1 [41]. Analogous irradiation of $Fe(CO)_5$ with cycloocta-2,4,6-trienone in C_6H_6 for 12 h followed by filtration and chromatography on basic alumina (activity II) with C_6H_6 gives No. 5 [65]. In a similar reaction $Fe(CO)_5$ is irradiated with I (X=CH_2) in C_6H_6 at 40 °C for 4 d. The volatiles are evaporated and the residue is chromatographed on basic alumina (activity II). Elution with hexane gives minor amounts of $C_9H_{10}Fe(CO)_3$ (C_9H_{10} = cyclonona-1,3,5,7-tetraene) and II, but chiefly No. 7; a second fraction from this elution contains predominantly $C_9H_{10}Fe(CO)_3$ but also II. Further elution with hexane/3% C_6H_6 gives $C_9H_{10}Fe_2(CO)_6$ (see 2.6.1.2.4 in "Organoiron Compounds" C5, 1981, Table 3,

 References on pp. 229/30

No. 6, p. 28). Further purification requires repeated fractional chromatography [38], cf. [56].

b. Prolonged heating of $Fe(CO)_5$ with cyclododeca-1,5,9-triene gives No. 6 [16].

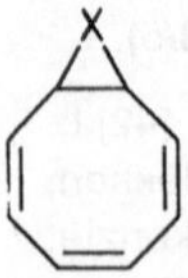

I 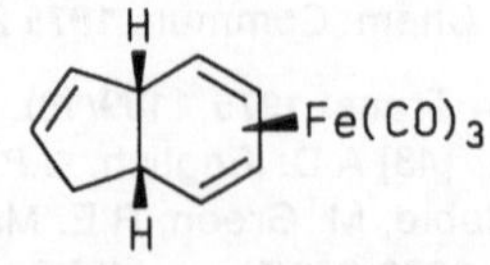 II 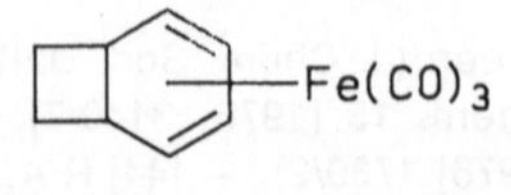 III 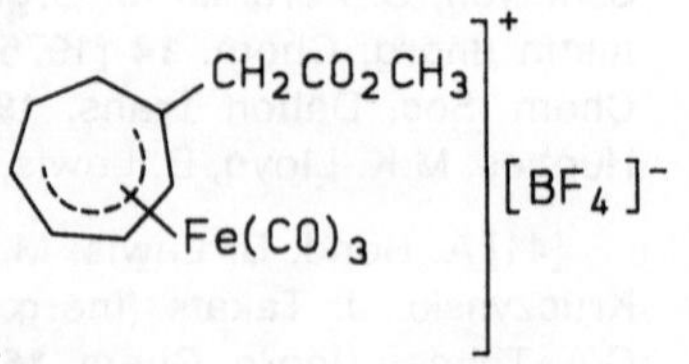 IV

Method II: $Fe_2(CO)_9$ reacts thermally with a cyclooctatriene in CH_3OH or in an inert solvent (C_6H_6, petroleum ether, ether etc.).

a. $Fe_2(CO)_9$ and cycloocta-1,3,6-triene are refluxed in petroleum ether (b.p. 130 °C) for 12 h. The reaction mixture is filtered, evaporated, and the residue is distilled at ca. 100 to 124 °C/5 Torr to give No. 1. Chromatography of this residue on neutral Al_2O_3 (activity II) with cyclohexane gives No. 1 and $C_8H_{10}Fe_2(CO)_6$ (see 2.6.1.2.4 in "Organoiron Compounds" C5, 1981, Table 3, No. 1, p. 28) [1 to 3, 5], cf. [19, 27]. Refluxing $Fe_2(CO)_9$ and I (X=O) in ether for 60 h followed by filtration, evaporation, extraction of the residue with petroleum ether, and chromatography on neutral alumina (activity III) with petroleum ether gives $C_8H_8Fe(CO)_3$, trans-$C_8H_8Fe_2(CO)_6$ (see 2.8.1.1 in "Organoiron Compounds" C5, 1981, Table 4, No. 25, p. 49) and C_8H_8O-$Fe_2(CO)_6$ (see 2.6.1.2.3 in "Organoiron Compounds" C5, 1981, Table 2, No. 15, p. 18). Elution with petroleum ether/C_6H_6 (1:1) gives $C_8H_8Fe_2(CO)_5$ (see 2.8.2.2 in "Organoiron Compounds" C5, 1981, Table 7, No. 1, p. 114) and 1-OHCC$_7$H$_7$-Fe(CO)$_3$. Further elution with petroleum ether/C_6H_6 (1:1) gives No. 5, C_8H_8O-$Fe_2(CO)_6$, and a large amount of polymer [31]. In a similar reaction $Fe_2(CO)_9$ and I (X=CH_2) in ether, C_6H_6, or petroleum ether at room temperature for 4 d followed by chromatography on alumina (activity II) with petroleum ether gives II. Rechromatography gives $C_9H_{10}Fe(CO)_4$ (C_9H_{10}=bicyclo[5.2.0]nona-2,5,8-triene), No. 7, two isomers of $C_9H_{10}Fe(CO)_3$ (C_9H_{10}=cyclonona-1,3,5,7-tetraene), and $C_9H_{10}Fe_2(CO)_6$ (see 2.6.1.2.4 in "Organoiron Compounds" C5, 1981, Table 3, No. 6, p. 28). No. 7 is contaminated by the two isomers $C_9H_{10}Fe(CO)_3$ [63]. Similarly, $Fe_2(CO)_9$ and I (X=CH_2) react in ether at 25 °C for 40 h. The reaction mixture is filtered, evaporated, and concentrated. Storage of the residual dark viscous oil at 0 °C for several days produces a crystalline mass which is separated to give $C_9H_{10}Fe_2(CO)_6$. The remaining oil is chromatographed at neutral alumina (activity I) with hexane to give a mixture of $C_9H_{10}Fe(CO)_3$ and II, and $C_9H_{10}Fe_2(CO)_6$ [38]. Later fractions contain $C_9H_{10}Fe(CO)_3$ and No. 7 [38, 63]. Gas-liquid-chromatography for No. 7 is unsuccessful since No. 7 decomposes in the injection port and elutes as a free ligand [38], cf. [56].

b. $Fe_2(CO)_9$ reacts with cycloocta-1,3,5-triene in CH_3OH at room temperature for 2 d. Removal of the volatiles, dissolution of the residue in hexane, and filtration followed by chromatography on neutral Al_2O_3 (activity III) with hexane and cooling to −30 °C gives $C_8H_{10}Fe_2(CO)_7$ (see 2.6.1.1 in "Organoiron

Compounds" C5, 1981, p. 2). Cooling to −80 °C gives No. 1, contaminated with III. No. 1 is purified by repeated recrystallization from hexane at −80 °C. Analogous reaction with I ($X=CH_2$) followed by work-up as before but chromatography on Al_2O_3 gives a first band containing $C_9H_{10}Fe(CO)_3$ (C_9H_{10}= cyclonona-1,3,5,7-tetraene) and No. 7. The second band contains pure $C_9H_{10}Fe_2(CO)_6$ (see 2.6.1.2.4 in "Organoiron Compounds" C5, 1981, Table 3, No. 6, p. 28). Further chromatography of the first band at silica gel (activity III) with hexane gives pure No. 7 [40].

Method III: $Fe_3(CO)_{12}$ reacts thermally with a cyclooctatriene.

a. $Fe_3(CO)_{12}$ and cycloocta-1,3,5-triene are refluxed in C_6H_6 for 17 h [4] or 20 h [14]. The reaction mixture is filtered, the filtrate is evaporated, and the residue is chromatographed on alumina with pentane to give III, No. 1, $C_8H_{10}Fe_2(CO)_6$ (see 2.6.1.2.4 in "Organoiron Compounds" C5, 1981, Table 3, No. 1, p. 28) [4, 14, 15] and hydrocarbons [14]. Heating $Fe_3(CO)_{12}$ and cyclooctatrienone in C_6H_6 at 90 °C for 4 h [13] or refluxing the components in C_6H_6 for 7 h [14] followed by filtration, evaporation, and extraction of the residue with ether gives No. 5 [13]. Filtration by gravity, evaporation, and chromatography of the residue on alumina with C_6H_6 gives No. 5 and $C_8H_8OFe_2(CO)_6$ (see 2.6.1.2.4 in "Organoiron Compounds" C5, 1981, Table 3, No. 2, p. 28) [14].

b. $Fe_3(CO)_{12}$ and a mixture of cycloocta-1,3,5-triene/cycloocta-1,3,6-triene/bicyclo[4.2.0]octa-2,4-diene (obtained by reduction of cyclooctatetraene with Zn dust and NaOH in C_2H_5OH) are refluxed in light petroleum ether (b.p. 100 to 120 °C) for 35 min. The reaction mixture is filtered and chromatographed on alumina (grade III) with light petroleum ether (b.p. 40 to 60 °C) to give $Fe(CO)_5$ and No. 1. If either the stated time or the temperature of the reaction is exceeded, the main product is III [15], cf. [6].

Method IV: Compounds of the type $[^5LFe(CO)_3]X$ where ^5L is a cyclic dienyl ligand (see Formulas IV to VI) react with Lewis bases.

$[BF_4]^-$ Fe(CO)$_3$ V; R, X^-, Fe(CO)$_3$ VI; Fe(CO)$_3$ VII; NC, Fe(CO)$_3$, NC, R, R VIII

a. Treating IV or VI ($R=CO_2CH_3$) with H_2O gives No. 2 [24], cf. [28].

b. Reduction of V with $NaBH_4$ in tetrahydrofuran at 25 °C gives VII and a small amount of $C_8H_8Fe(CO)_3$ separated by chromatography. Under different conditions a variety of compounds including III and No. 1 are obtained [12].

c. Treatment of VI ($R=H$, $X=PF_6$) with $N(C_2H_5)_3$ gives No. 1 [34], see "Chemical Behavior" in "Further information" for No. 1, p. 226. Treatment of VI ($R=H$, $X=PF_6$) with KCN gives No. 1 and $C_8H_{11}Fe(CO)_2COCN$ (C_8H_{11}=cycloocta-2,4-dienyl), separated by chromatography. When this reaction is carried on for a longer time the yield of No. 1 increases [62], cf. [28].

The compounds of the type $^4LFe(CO)_3$ where ^4L is a cycloocta-1,3,6-triene are prepared by 1,4 addition reactions to $C_8H_8Fe(CO)_3$ (see "Further information", pp. 228/9). The product of the analogous reaction of $C_8H_8Fe(CO)_3$ with $(NC)_2C{=}C(CN)_2$ was considered as a 1,2- [23, 25], 1,3- [32, 35], or 1,4-adduct [7 to 9, 11, 21] or it was mentioned as "1:1 adduct", e.g. [20]. An X-ray structure determination showed the product to be VIII (R = CN) [37]. Similarly, the adduct of $C_8H_8Fe(CO)_3$ and $(NC)_2C{=}C(CF_3)_2$ originally formulated as a 1,2-adduct [23, 25] is reformulated as VIII ($R = CF_3$) [35]. The claim that only 1,3-adducts are formed in the reaction of $RC_8H_7Fe(CO)_3$ (R = Br, CH_3, C_6H_5) with $(NC)_2C{=}C(CN)_2$ [39] seems uncertain, since work on $RC_8H_7Fe(CO)_3$ ($R = CH_3$, C_6H_5) showed that only the most insoluble isomer seems to have been characterized in each instance [50].

Table 8
Complexes of the Type $^4LFe(CO)_3$ where ^4L is a Cyclooctatriene.
Further information for all compounds is given at the end of the table, pp. 225/9.
For abbreviations and dimensions see p. VIII.

No.	compound	method of preparation (yield in %), properties and remarks	Ref.
1	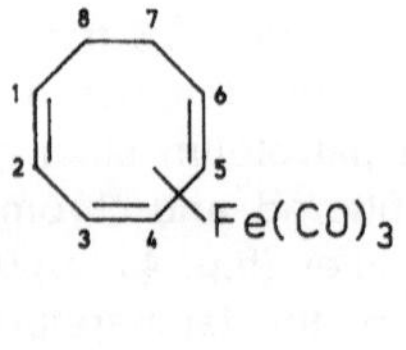 1, 2, 3, 4, 5, 6, 7, 8 $Fe(CO)_3$	I a (10 to 56), II a (22.0 to 27), II b, III a (5), III b (20), IV b, IV c m.p. −10° (from light petroleum ether at −78°), m.p. ca. 8°, m.p. 20° (from hexane at −80°), m.p. 24°, sublimes, b.p. 90°/5 Torr, dec. ca. 230°, yellow solid at low temperatures, or viscous orange, yellow, yellow-orange or golden-yellow oil at room temperature ^{1}H NMR (CS_2): 1.5 to 3.1 (complex, H-7,8), 3.14 (t, H-3), 3.49 (asym. t, H-6), 5 to 6 (complex, H-1,2,4,5) ^{13}C NMR (benzene-d_6): 22.3, 24.9 (C-7,8); 59.4, 60.3 (C-3,6); 88.5, 88.9 (C-4,5); 122.1, 131.9 (C-1,2); 211.5 (CO) ^{13}C NMR (CD_2Cl_2): 22.0, 24.7 (C-7,8); 59.0, 60.5 (C-3,6); 89.0, 89.3 (C-4,5); 122.3, 132.0 (C-1,2), 212.2 (CO) IR (liquid film): 701, 719, 771, 805, 862, 912, 957, 999, 1043, 1168, 1193, 1212, 1314, 1335, 1370, 1403, 1412, 1445, 1460; 1645 (uncoordinated C=C); 2941, 3049 IR (liquid film): 704, 722, 776, 809, 828, 847, 863, 890, 914, 934, 956, 962, 999, 1044, 1088, 1133, 1168, 1194, 1215, 1225, 1256, 1310, 1330, 1362, 1385, 1428, 1442, 1636, 1673, 1945, 1980, 2065, 2846, 2941, 2970, 3045 IR (n-hexane): 1984, 1996, 2052 (CO) IR (CH_2Cl_2): 1645 (uncoordinated C=C); 1970, 2040 (CO) mass spectrum: $[M - nCO]^+$ (n = 0 to 3)	[1 to 6, 12, 14, 15, 19, 27, 28, 34, 40, 41, 52, 62]

Table 8 [continued]

No.	compound	method of preparation (yield in %), properties and remarks	Ref.
2	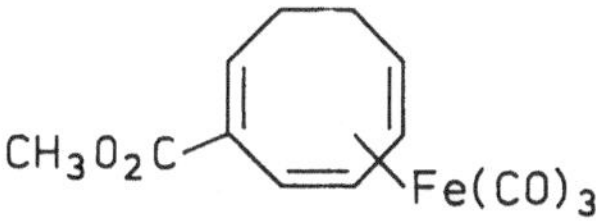 CH_3O_2C–C_8H_9 $Fe(CO)_3$	IVa	[24, 28]
3	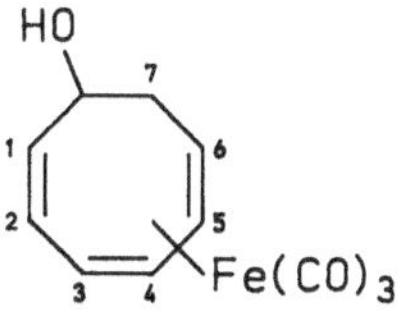 HO–C_8H_9 $Fe(CO)_3$	unstable yellow oil ^{1}H NMR (acetone-d_6): 2.0 (m, H-7), 3.30 (m, H-3, 6), 3.68 (m, H-8), 4.08 (d, OH; J = 6), 5.40 (m, H-4, 5), 5.92 (m, H-1, 2) mass spectrum: $[M]^+$	[51]
4	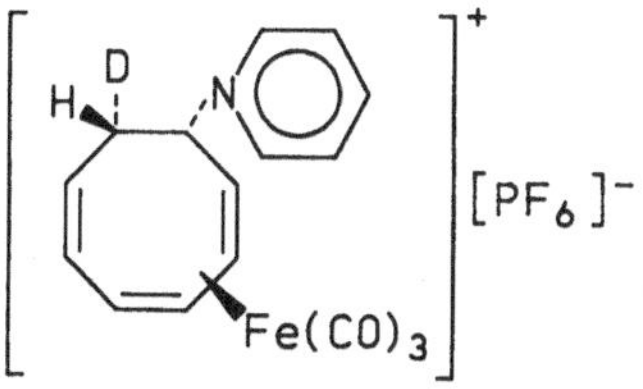 $[C_8H_8D(NC_5H_5)Fe(CO)_3]^+[PF_6]^-$	pale yellow crystals	[48]
5	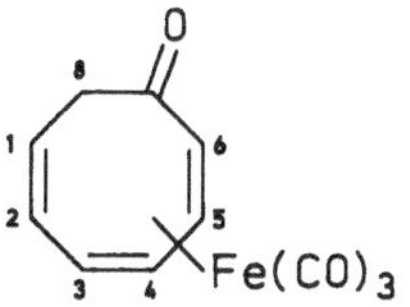 O=C_8H_8 $Fe(CO)_3$	I a (36), II a (0.9), III a (13 or 50) m.p. 98 to 100° (from ether/pentane), m.p. 103° (from CH_3OH), m.p. 103 to 104° (from light petroleum ether), m.p. 104 to 105° (dec.), dec.p. 94° (from ether), subl.p. 70 to 90°/ 0.25 Torr, orange solid or yellow needles or crystals with a sweet odor ^{1}H NMR (benzene-d_6): 2.20 (H-8endo), 2.57 (H-8exo), 2.79 (H-6), 2.80 (H-3), 4.72 (H-4), 5.31 (H-1), 5.45 (H-5), 5.73 (H-2) ^{1}H NMR ($CDCl_3$): 2.30 (ddd, H-8endo; J = 10.0, J = 2.0), 2.85 (m, H-8exo), 2.97 (d, H-6; J = 6.5), 3.37 (dd, H-3; J = 7.5), 5.42 to 5.75 (m, H-1, 4), 6.10 (ddd, H-2; J = 9.0), 6.25 (dd, H-5) ^{1}H NMR (CS_2): 2.29 (H-8endo; J(H-1, 8endo) = 5.6, J(H-2, 8endo) = 0.8, J(H-3, 8endo) = 1.1, J(H-6, 8endo) = 2.0), 2.86 (H-8exo; J(H-1, 8exo) = 9.7, J(H-2, 8exo) = 2.5, J(H-8endo, 8exo) = 11.3), 2.96 (H-6; J(H-4, 6) = 1.1), 3.39 (H-3; J(H-3, 4) = 8.5), 5.52 (H-1; J(H-1, 2) = 10.5), 5.56 (H-4; J(H-4, 5) = 5.8), 6.10 (H-2; J(H-2, 3) = 8.1), 6.27 (H-5; J(H-5, 6) = 6.9) ^{13}C NMR ($CDCl_3$): 42.4 (C-8), 53.5 (C-3 or 6), 60.1 (C-3 or 6), 90.9 (C-4 or 5), 92.1 (C-4 or 5), 120.9 (C-1 or 2), 135.5 (C-1 or 2), 208.2 (CO), 210.7 (C-7)	[13, 14, 18, 31, 50, 53, 56, 65]

References on pp. 229/30

Table 8 [continued]

No.	compound	method of preparation (yield in %), properties and remarks	Ref.
		^{57}Fe-γ (78 K): δ = 0.022 (referred to Cr (Co^{57}) source), Δ = 0.141 IR (KBr or Nujol): 686, 715, 803, 850, 915, 960, 975, 1105, 1165, 1245, 1265, 1282, 1355, 1405, 1428, 1433, 1460; 1635, 1650 (C=C and C=O); 1971, 1983, 2054 (CO); 2960 (CH) IR (cyclohexane): 1673, 1985, 2006, 2063	
6	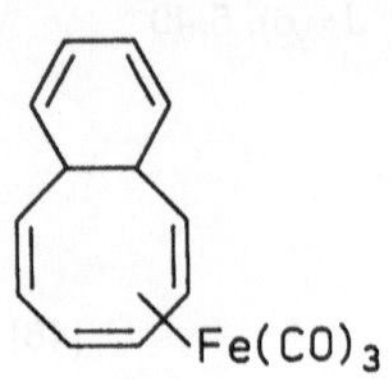 $Fe(CO)_3$	Ib	[16]
7	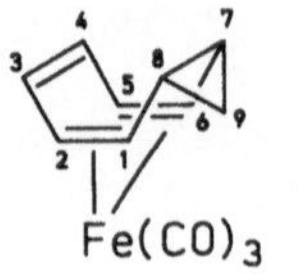 $Fe(CO)_3$	I (15), IIa (15), IIb yellow oil ^{1}H NMR (CS_2): 0.9 to 1.80 (m, H-7,8,9; J(H-6,7) ~ 0), 3.57 (d, H-1,6; J(H-1,2) = 9), 4.08 (d, H-2,5), 5.48 (br, s, H-3,4; J(H-2,3) ~ 0) ^{13}C NMR (benzene-d_6): 12.3, 21.2 (C-7 to 9); 77.5, 89.3 (C-1,2,5,6), 131.8 (C-3,4), 215.8 (CO) IR (n-hexane): 1965, 1985, 2040 (CO) mass spectrum: $[M]^+$	[38, 40, 56, 63]
8	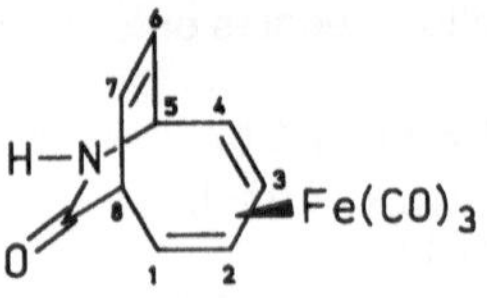 H—N, O, $Fe(CO)_3$	m.p. 143.5 to 144.5°, yellow crystals ^{1}H NMR: 3.3, 4.0 (H-5,8); 3.62, 5.55 (H-1 to 4); 6.38 (H-6,7), 6.9 (NH) IR: 1675 (C=O), 1990, 2050 (CO)	[37]
9	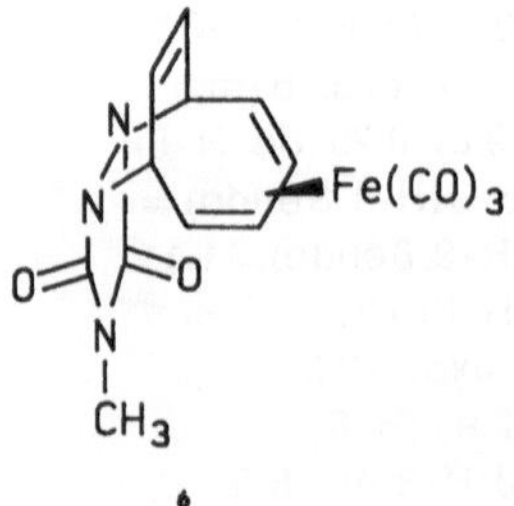 $Fe(CO)_3$, N—CH_3	m.p. 185° (dec.)	[57]
10	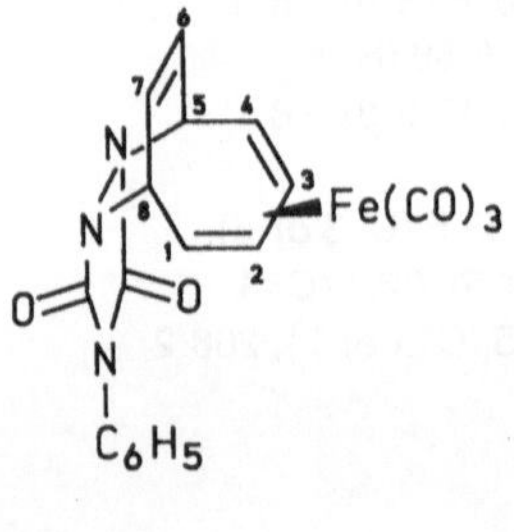 $Fe(CO)_3$, N—C_6H_5	m.p. 155°, m.p. 156 to 157°, pale yellow crystalline solid ^{1}H NMR ($CDCl_3$): 3.94 (H-1,4; J(H-1,2) = 8.10, J(H-1,3) = 1.4), 5.20 (H-5,8; J(H-1,8) = 9.2, J(H=5,6) = 7.60), 5.43 (H-2,3; J(H-2,3) = 4.13, J(H-1,4) = 1.40), 6.33 (H-6,7; J(H-6,7) = 8.88, J(H-5,7) = 0.90), 7.36 (C_6H_5) IR (cyclohexane): 1993, 1999, 2961 (CO)	[25, 57]

References on pp. 229/30

Table 8 [continued]

No.	compound	method of preparation (yield in %), properties and remarks	Ref.
11		m.p. 180° (dec., from CH_2Cl_2) ^{1}H NMR (acetone-d_6): 3.74 (s, 3H), 3.64 to 4.72 (m, 4H), 5.5 (d, 1H), 5.86 to 6.25 (m, 2H)	[50]
supplement:			
12		–	[64]

*Further information:

$\mathbf{C_8H_{10}Fe(CO)_3}$ (Table **8**, No. **1**) is also obtained by reduction of $C_8H_8Fe(CO)_3$ with a slight excess of K in tetrahydrofuran at room temperature yielding $[C_8H_8Fe(CO)_3]^-$ and $[C_8H_8Fe(CO)_3]^{2-}$. Hydrolysis of these anions followed by ether extraction, drying the organic layer ($MgSO_4$), evaporation, and chromatography of the residue on alumina with n-hexane gives $C_8H_{10}Fe(CO)_3$ (No. 1) in 30% yield and $C_8H_8Fe(CO)_3$ (ratio 1:1) [54], cf. $C_8H_8Fe(CO)_3$, p. 243. Electrochemical reduction of $C_8H_8Fe(CO)_3$ in the presence of a proton donor ($[HN(CH_3)_3]Br$) at -1.3 V gives an orange solution which, treated as above, gives No. 1 in 100% yield. This illustrates the advantage of the electrochemical preparation of No. 1 in a series where the starting materials are costly [54]. Reduction of $C_8H_8Fe(CO)_3$ with H_2 (1 atm)/Raney Ni in C_2H_5OH at room temperature for 1.5 h followed by filtration, evaporation, and chromatography of the residue on alumina with petroleum ether (b.p. 30 to 40 °C) gives No. 1, too [66]. Heating IX [43] above 40 °C in C_6H_6 [55] or removal of CH_3OH from X gives No. 1, too. Thermal isomerization of XI (with H instead of D) in octane at 120.6±0.1 °C gives III (p. 220) and No. 1 [47]. Similar heating of XI in C_6H_6 gives XII and XIII; XIV and XV are discussed as intermediates. Further isomerization of intermediate XV in C_6H_6 at 60 to 70 °C gives also XII [55]. The product obtained by Method III (p. 221) was first formulated as XVI [1 to 3] or III (p. 220) [1, 4], but no XVI can be isolated by Method IIIb (p. 221) [15]. Therefore, the product is formulated as No. 1 [15], cf. [1, 4, 14].

IX X XI XII

References on pp. 229/30

The diamagnetic [1, 2] compound No. 1 is monomeric in cyclohexane (cryoscopy) [1, 2] or C_6H_6 (ebullioscopy) [15], and its dipole moment is $\mu_{25°}=2.37\pm0.03$ D in cyclohexane [1, 2]. Refractive index: $n_D^{20}=1.6580$ [5]. The ^1H NMR spectrum in CS_2 is exceedingly complex [4]. The ^{13}C NMR spectrum at 203 K in CD_2Cl_2 shows chemical shifts of the ^{13}CO at $\delta=209.5$ and 210.8 (basal CO); 215.3 (apical CO) ppm [52]; thus, the sharp peak at 212.2 ppm at room temperature indicates the scrambling of the CO groups [42, 45, 52]. The activation parameters are $\Delta H^{*}=11.3\pm0.5$ kcal/mol, $\Delta S^{*}=0.3\pm2$ cal · K^{-1} · mol^{-1} [52], $\Delta G^{*}_{298}=11.2$ ±0.5 kcal/mol (same value assuming $\Delta S^{*}=0$) [42, 52].

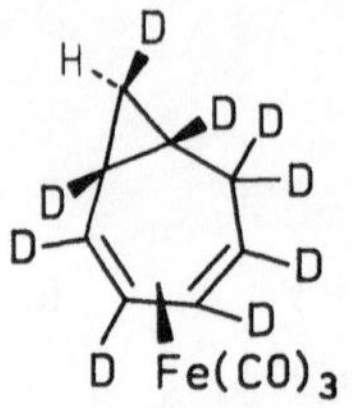

XIII 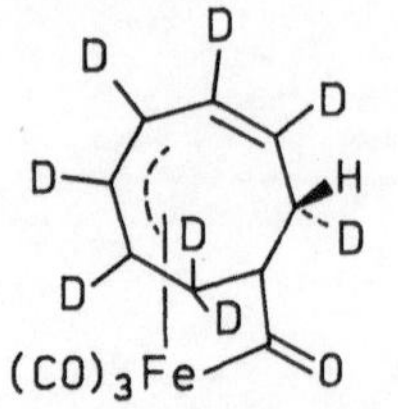 XIV 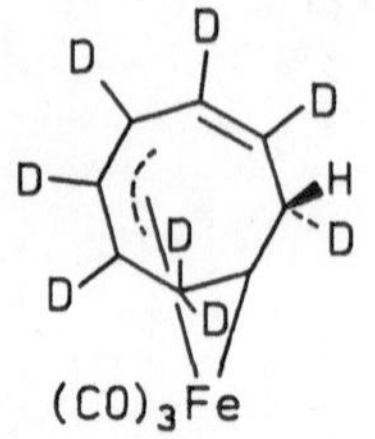 XV 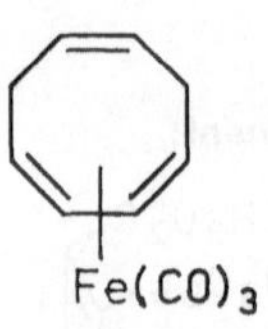 XVI

The compound is soluble in organic solvents [1 to 3] but not miscible with H_2O [1, 2]. It is said to be air-stable [1 to 3] but seems to be air-sensitive [15] or easily oxidized by air [4], or unstable to a considerable extent in air [5]. In solution, there exists an equilibrium between No. 1 and III (p. 221) [33, 36, 41, 47, 56]. No. 1 is shown to undergo electrophilic ring closure to III at 102 °C in hexane, heptane, or octane. The first-order rate constant for this process is determined to be $k_1=7\times10^{-5}\cdot s^{-1}$ corresponding to $\Delta G^{*}=29.3$ kcal/mol. Less than 1% of No. 1 remains at equilibrium thus, K_{eq} must be greater than 100 for No. 1$\rightleftharpoons$III [33, 36, 41, 47, 56] and for the reverse reaction $k_{-1}\leq7\times10^{-7}\cdot s^{-1}$, $\Delta G^{*}\geq32.7$ kcal/mol [41, 47]. Similar thermal isomerization of XII gives XVII [55]. Irradiation of No. 1 in tetrahydrofuran at −78 °C, followed by addition of $C_6H_5C{\equiv}CC_6H_5$ and warming to room temperature, gives XVIII [44]. Compound No. 1 decomposes when hydrogenation is tried [4]. Reaction with $NaBH_4$ occurs with great loss of the reactant and the resulting unstable pale yellow product decomposes on chromatography [40]. Protonation with 40% aqueous HBF_4 in ice-cold propionic anhydride gives $[C_8H_{11}Fe(CO)_3]BF_4$ (see Formula VI, p. 221, with $X=BF_4$) [15]. Similar protonation occurs with H_2SO_4 or CF_3CO_2H [15, 34]. A mixture of No. 1 and III (p. 221), always produced under the reaction conditions of Methods I to IV (pp. 219/21), with HPF_6 in ether precipitates VI ($R=H$, $X=PF_6$, p. 221), and III remains in solution. This behavior provides a useful method of separation of the two isomers as No. 1 can be regenerated by Method IVc (p. 221) [34]. Refluxing No. 1 with $Fe(CO)_5$ in ethylcyclohexane for 17 h gives III (p. 221) [4]. No ligand exchange is observed with benzylideneacetone in C_6H_6 at 60 °C within 3 h 20 min [59, 60] or with a fivefold excess of C_8H_8 at 25 °C within 24 h [41]. The compound No. 1 is proposed as an antiknock agent for gasoline, as an additive for fuels, for metal plating, as a drying agent, and as a chemical intermediate for the production of metal-containing polymer materials [66].

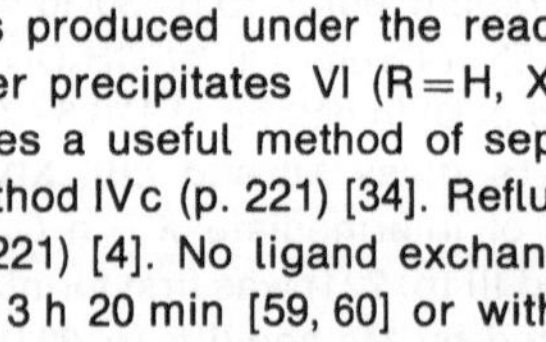

XVII 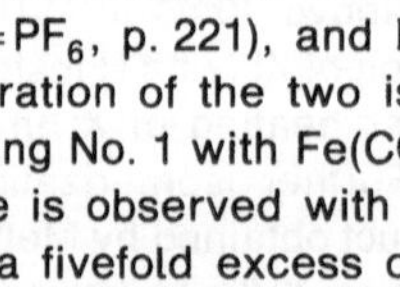 XVIII 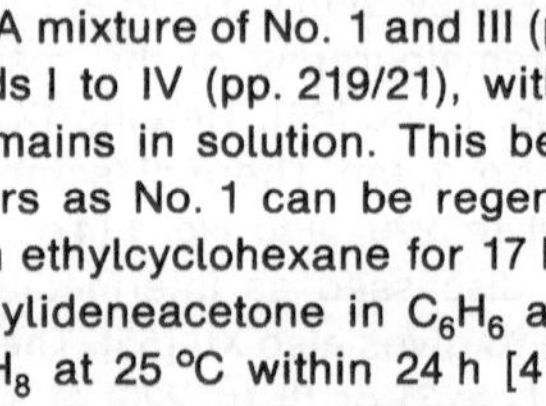 XIX

2-$CH_3O_2CC_8H_9Fe(CO)_3$ (Table **8**, No. **2**) gives the cation of VI (p. 221) on treatment with conc. H_2SO_4 [24].

8-$HOC_8H_9Fe(CO)_3$ (Table **8**, No. **3**) is prepared from $C_8H_8Fe(CO)_3$ and $(C_2H_5)_2OBF_3$ in CH_3CO_2H at room temperature within 3 h. The reaction mixture is poured into H_2O, extracted with CH_2Cl_2, and the organic layer is chromatographed on silica gel to give the product, probably No. 3. Protonation with HPF_6 gives XIX, and reaction with pyridine in $O(COCH_3)_2$ gives $C_8H_8Fe(CO)_3$ [51].

[7-$C_5H_5NC_8H_8DFe(CO)_3$]PF_6 (Table **8**, No. **4**) is prepared from XX in acetone at 25 °C within 30 min and can be precipitated with ether in quantitative yield. Compound No. 4 in solution slowly gives at 25 °C $C_8H_8Fe(CO)_3$ and $[DNC_5H_5]PF_6$ (NC_5H_5 = pyridine) [48].

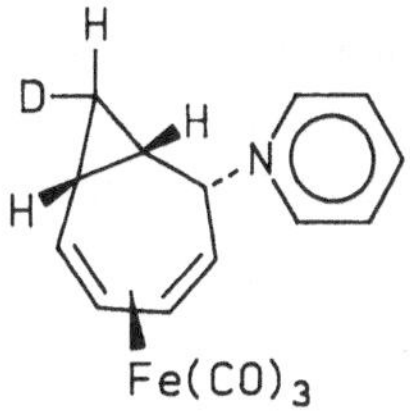

XX 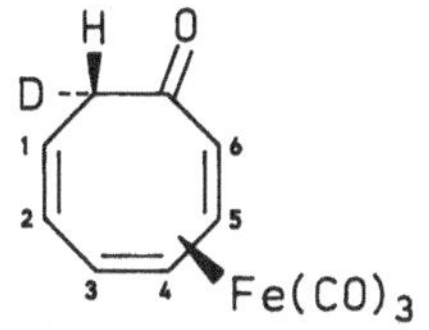 XXI 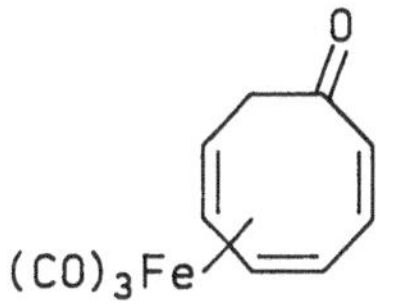 XXII 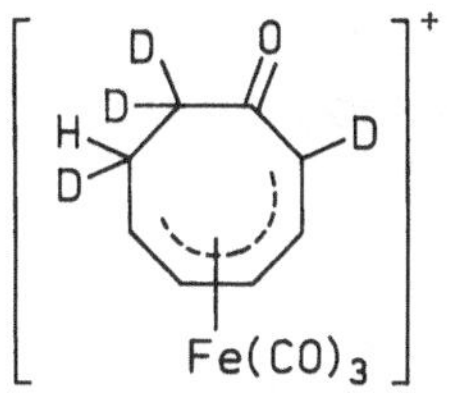 XXIII

$C_8H_8OFe(CO)_3$ (Table **8**, No. **5**) is also prepared by refluxing $Fe_2(CO)_9$ with methoxy-cyclooctatetraene in ether for 3 h. The reaction mixture is filtered, evaporated, and the residue is chromatographed on silica gel. Elution with petroleum ether gives $CH_3OC_8H_7$-$Fe(CO)_3$. Elution with petroleum ether/ether (5:1) gives No. 5 (9.7% yield), and further elution with ether gives $C_8H_8OFe_2(CO)_6$ (see 2.6.1.2.4 in "Organoiron Compounds" C5, 1981, Table 3, No. 2, p. 28) [50]. No. 5 and $C_8H_8OFe_2(CO)_6$ are identical with the appropriate products obtained by Method IIa (p. 220) [33]. Compound No. 5 is also prepared by the reaction of $CH_3OC_8H_7Fe(CO)_3$ in tetrahydrofuran with H_2SO_4/H_2O (1:1) at room temperature for 10 min. The reaction mixture is diluted with H_2O. Extraction with light petroleum ether gives No. 5 in 90% yield. Analogous reaction of $CH_3OC_8H_7Fe(CO)_3$ with D_2SO_4/D_2O (1:1) gives XXI [53]. The compound is monomeric in C_6H_6 (osmometry) [14], and thermodynamically more stable than its isomer XXII [53]. The ^1H NMR spectrum of the deuterated derivative XXI in $CDCl_3$ shows chemical shifts at δ=2.32 (d, H-8; J=9.5), 2.96 (d, H-6; J=6.5), 3.36 (dd, H-3; J=7.5), 5.42 to 5.70 (m, H-1,4), 6.10 (dd, H-2; J=9.0), 6.24 (dd, H-5) ppm [53]. The ^{13}C NMR spectrum in $CDCl_3$ at −46 °C shows ^{13}CO chemical shifts at δ=204.3, 207.1, 212.9 (CO); 213.5 (C-7) ppm; thus, the sharp peak at room temperature indicates CO scrambling. Using the width of the ^{13}C line (ca. 15 Hz at half-height) at 25 °C, the ΔG^{*} for scrambling can be roughly estimated as 15 kcal/mol. Those for simple diene complexes fall in the range of 7 to 12 kcal/mol. It is not clear what the structural features are which are responsible for this increase, although ring size or electronic effects of the ring carbonyl function may play a role [56]. The deuterated derivative XXI shows bands in the IR spectrum at 1670, 1985, 2005, and 2061 cm^{-1} [53]. It is not clear if the compound No. 5 is air-stable [14] or air-sensitive [13]. Polarographic, cyclic triangular, and potential electrolytic studies in 4×10^{-3} M solutions in $(CH_2OCH_3)_2$ with $[N(C_4H_9)_4]ClO_4$ as supporting electrolyte (10^{-3} M $AgClO_4/Ag$ as reference electrode) give $-E^1_{1/2}$=1.7 and $-E^2_{1/2}$=2.2 V, the cathodic potential is $-E_{p,c}$=1.9 V and the anodic $-E_{p,a}$=1.5 V. The electrochemical reduction to **[C_8H_8O-$Fe(CO)_3$]$^{\dot{-}}$** which shows in the ESR spectrum g=2.0403 and a hyperfine coupling of ΔH= 12.8 Gauss (signal gradually decreases with time) is reversible [22]. Dissolving compound No. 5 in D_2SO_4 gives XXIII. For a possible way in which this deuterium exchange can occur see [61]. Addition of $(NC)_2C{=}C(CN)_2$ in C_6H_6 gives XXIV [58].

References on pp. 229/30

XXIV XXV XXVI

$C_4H_4C_8H_8Fe(CO)_3$ (Table **8**, No. **6**) may alternatively have structure XXV [16].

$CH_2C_8H_8Fe(CO)_3$ (Table **8**, No. **7**). Mixtures of $C_9H_{10}Fe(CO)_4$, No. 7, and of the two isomers $C_9H_{10}Fe(CO)_3$ (C_9H_{10}=cyclononatetraene) undergo slow rearrangement to II (p. 220) [63].

$NHCOC_8H_8Fe(CO)_3$ (Table **8**, No. **8**). A solution of $C_8H_8Fe(CO)_3$ reacts readily in dry CH_2Cl_2 at 0 °C with $ONCSO_2Cl$ to give, after dechlorosulfonylation with C_6H_5SH and pyridine in cold acetone, No. 8 in 80% yield. The compound crystallizes in the presumed triclinic space group $P\bar{1}$-C_2^1, S_2^1 with the unit cell parameters a=14.317(7), b=14.428(6), c=6.701(5) Å, α=95.64(3)°, β=91.62(2)°, γ=114.89(2)°; Z=2 molecules per asymmetric unit. The measured density is 1.62 g/cm³ and the calculated one 1.60 g/cm³. Both independent molecules have the conformation shown in **Fig. 25** and in Table 8, p. 224. An H_2O molecule of crystallization is between the two independent molecules and destroys the potential subcell symmetry. Oxidation with $[NH_4]_2[Ce(NO_3)_6]$ in 95% C_2H_5OH liberates the organic ligand ^4L [37].

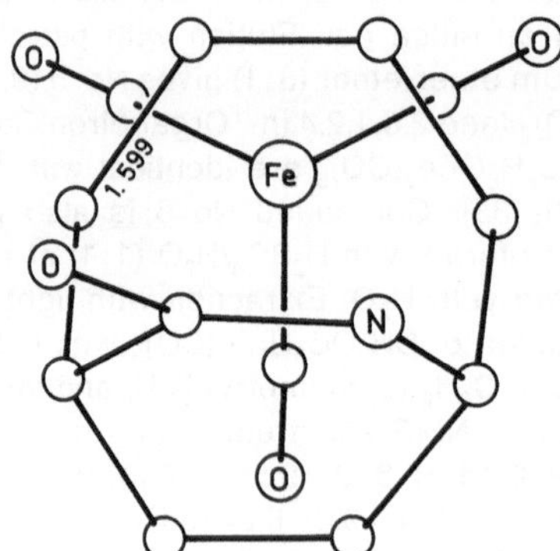

Fig. 25. Molecular structure of $NHCOC_8H_8Fe(CO)_3$ (No. 8) [37].

$CH_3N(CO)_2N_2C_8H_8Fe(CO)_3$ (Table **8**, No. **9**) is prepared by the reaction of $C_8H_8Fe(CO)_3$ with 4-methyl-2H,4H-1,2,4-triazole-3,5-dione in CH_2Cl_2 at 25 °C for 30 min to give No. 9 and XXVI ($R=CH_3$). Chromatography on silica gel with $CH_3CO_2C_2H_5$ gives pure No. 9 in

XXVII XXVIII

34% yield and XXVI (R = CH_3) in 16% yield [57]. Reaction with H_2/PtO_2 gives the dihydroderivative XXVII [30].

$C_6H_5N(CO)_2N_2C_8H_8Fe(CO)_3$ (Table **8**, No. **10**) is prepared from $C_8H_8Fe(CO)_3$ and 4-phenyl-2H,4H-1,2,4-triazole-3,5-dione in CH_2Cl_2 at 25 °C for 30 min. Chromatography on silica gel with $CH_3CO_2C_2H_5$ gives No. 10 in 33% yield and XXVI (R = C_6H_5) in 14% yield [57]. Similar reaction, but for 10 h and chromatography on Florisil with CH_2Cl_2, gives No. 10 in 10% yield. The compound is monomeric in $CHCl_3$ [25]. Refluxing No. 10 with an eightfold excess of $ON(CH_3)_3$ in C_6H_6 for 3 h liberates the organic ligand ^4L in 77% yield [49].

$CH_3O[(NC)_2C]_2C_8H_7Fe(CO)_3$ (Table **8**, No. **11**). $CH_3OC_8H_7Fe(CO)_3$ reacts with $(NC)_2C{=}C(CN)_2$ in CH_2Cl_2 at room temperature for 1 h. The volatiles are removed and the residual pale yellow solid is slurried with ether, filtered, and dried to give XXVIII. Removal of the solvent from the filtrate gives a pale yellow gum, containing a number of products (five visible methoxy signals). Attempted purification at that time proved unsuccessful. However, when purification is attempted at a later (ca. 1 month) date, it proves possible to isolate No. 11. Oxidation of No. 11 with $[NH_4]_2[Ce(NO_3)_6]$ in CH_3OH at room temperature for 36 h liberates the organic ligand ^4L in 77% yield [50].

$(CH_3O_2C)_2C_{10}H_{10}Fe(CO)_3$ (Table **8**, No. **12**) is prepared by irradiating $C_8H_8Fe(CO)_3$ and $CH_3O_2CC{\equiv}CCO_2CH_3$ in tetrahydrofuran at 0 °C [64].

References:

[1] C. Palm (Diss. München T.H. 1959). – [2] E.O. Fischer, C. Palm, H.P. Fritz (Chem. Ber. **92** [1959] 2645/57). – [3] E.O. Fischer, Badische Anilin- und Sodafabrik A.-G. (Ger. 1154472 [1959/63]). – [4] T.A. Manuel, F.G.A. Stone (J. Am. Chem. Soc. **82** [1960] 6240/2). – [5] A. Nakamura, N. Hagihara (Nippon Kagaku Zasshi **82** [1961] 1389/92).

[6] T.A. Manuel (Diss. Harvard Univ. 1961; Diss. Abstr. **22** [1961] 1003/4). – [7] G.N. Schrauzer, S. Eichler (Angew. Chem. **74** [1962] 585). – [8] R.H. Herber (TID-16124 [1962]; N.S.A. **16** [1962] No. 27037). – [9] G.K. Wertheim, R.H. Herber (unpublished results from [10]). – [10] G.K. Wertheim, R.H. Herber (J. Am. Chem. Soc. **84** [1962] 2274/5).

[11] A. Davison, W. McFarlane, L. Pratt, G. Wilkinson (J. Chem. Soc. **1962** 4821/9). – [12] A. Davison, W. McFarlane, G. Wilkinson (Chem. Ind. [London] **1962** 820/1). – [13] G.N. Schrauzer, H. Thyret (Chem. Ber. **96** [1963] 1755/61). – [14] R.B. King (Inorg. Chem. **2** [1963] 807/10). – [15] W. McFarlane, L. Pratt, G. Wilkinson (J. Chem. Soc. **1963** 2162/6).

[16] T.A. Manuel, E. Tornquist, R. Kochhar, R. Pettit (unpublished results from [17]). – [17] G.F. Emerson, J.E. Mahler, R. Kochhar, R. Pettit (J. Org. Chem. **29** [1964] 3620/4). – [18] R.H. Herber, R.B. King, G.K. Wertheim (Inorg. Chem. **3** [1964] 101/7). – [19] G.F. Emerson, J.E. Mahler, R. Pettit, R. Collins (J. Am. Chem. Soc. **86** [1964] 3590/1). – [20] R.B. King (Organometal. Syn. **1** [1965] 126/7).

[21] W. McFarlane, G. Wilkinson (Inorg. Syn. **8** [1966] 184/5). – [22] R.E. Dessy, R.B. King, M. Waldrop (J. Am. Chem. Soc. **88** [1966] 5112/7). – [23] M. Green, D.C. Wood (Chem. Commun. **1967** 1062). – [24] J. Lewis, A.W. Parkins (Chem. Commun. **1968** 1194/5). – [25] M. Green, D.C. Wood (J. Chem. Soc. A **1969** 1172/5).

[26] F.A. Cotton, T.J. Marks (J. Organometal. Chem. **19** [1969] 237/40). – [27] F.A. Cotton, W.T. Edwards (J. Am. Chem. Soc. **91** [1969] 843/7). – [28] J. Lewis, M. Schiavon (unpublished results from [29]). – [29] L.A.P. Kane-Maguire (J. Chem. Soc. A **1971** 1602/6). – [30] H. Olsen, J.P. Snyder (J. Am. Chem. Soc. **100** [1978] 285/7).

[31] H. Maltz, G. Deganello (J. Organometal. Chem. **27** [1971] 383/7). – [32] D.J. Ehntholt (Diss. State Univ. New York 1971; Diss. Abstr. Intern. B **32** [1972] 4486). – [33] K.J. Karel, M. Brookhart (J. Am. Chem. Soc. **100** [1978] 1619/21). – [34] J. Evans, B.F.G. Johnson, J. Lewis (J. Chem. Soc. Dalton Trans. **1972** 2668/75). – [35] D.J. Ehntholt, R.C. Kerber (J. Organometal. Chem. **38** [1972] 139/45).

[36] M.S. Brookhart, N.M. Lippman, B.F. Lewis (Abstr. 163rd Natl. Meeting Am. Chem. Soc., Boston, Mass., 1972, ORGN-16). – [37] L.A. Paquette, S.V. Ley, M.J. Broadhurst, D. Truesdell, J. Fayos, J. Clardy (Tetrahedron Letters **1973** 2943/6). – [38] E.J. Reardon, M. Brookhart (J. Am. Chem. Soc. **95** [1973] 4311/6). – [39] M. Green, S. Heathcock, D.C. Wood (J. Chem. Soc. Dalton Trans. **1973** 1564/9). – [40] A. Salzer, W. von Philipsborn (J. Organometal. Chem. **161** [1978] 39/49).

[41] M. Brookhart, N.M. Lippman, E.J. Reardon (J. Organometal. Chem. **54** [1973] 247/53). – [42] F.A. Cotton, B.E. Hanson (Israel J. Chem. **15** [1977] 165/73). – [43] R. Aumann (Angew. Chem. **85** [1973] 628/9). – [44] R. Pettit (Organotransition-Metal Chem. 1st Proc. Japan.-Am. Semin., Honolulu 1974 [1975], pp. 157/67; C.A. **85** [1976] No. 176360). – [45] L. Kruczynski, L.S. Man (unpublished results from [46]).

[46] L. Kruczynski, J. Takats (J. Am. Chem. Soc. **96** [1974] 932/4). – [47] M. Brookhart, R.E. Dedmond, B.F. Lewis (J. Organometal. Chem. **72** [1974] 239/45). – [48] R. Aumann (J. Organometal. Chem. **78** [1974] C31/C34). – [49] Y. Shvo, E. Hazum (J. Chem. Soc. Chem. Commun. **1974** 336/7). – [50] L.A. Paquette, S.V. Ley, S. Maiorana, D.F. Schneider, M.J. Broadhurst, R.A. Boggs (J. Am. Chem. Soc. **97** [1975] 4658/67).

[51] B.F.G. Johnson, J. Lewis, D.J. Thompson, B. Heil (J. Chem. Soc. Dalton Trans. **1975** 567/71). – [52] L. Kruczynski, J. Takats (Inorg. Chem. **15** [1976] 3140/7). – [53] B.F.G. Johnson, J. Lewis, D. Wege (J. Chem. Soc. Dalton Trans. **1976** 1874/80). – [54] N. El Murr, M. Riveccié, E. Laviron (Tetrahedron Letters **1976** 3339/40). – [55] R. Aumann (Chem. Ber. **109** [1976] 168/73).

[56] M.S. Brookhart, G.W. Koszalka, G.O. Nelson, G. Scholes, R.A. Watson (J. Am. Chem. Soc. **98** [1976] 8155/61). – [57] H. Olsen (Acta Chem. Scand. B **31** [1977] 635/6). – [58] Z. Goldschmidt, Y. Bakal (Tetrahedron Letters **1977** 955/8). – [59] C.R. Graham (Diss. North Carolina Univ. 1976; Diss. Abstr. Intern. B **38** [1977] 688/9). – [60] C.R. Graham, G. Scholes, M. Brookhart (J. Am. Chem. Soc. **99** [1977] 1180/8).

[61] R.F. Childs, A. Varadarajan (J. Organometal. Chem. **184** [1980] C28/C32). – [62] G. Schiavon, C. Paradisi (J. Organometal. Chem. **210** [1981] 247/51). – [63] G. Deganello, H. Maltz, J. Kozarich (J. Organometal. Chem. **60** [1973] 323/33). – [64] R.E. Davis, T.A. Dodds, T.H. Hseu, J.C. Wagnon, T. Devon, U. Tancrede, J.S. McKennis, R. Pettit (J. Am. Chem. Soc. **96** [1974] 7562/3). – [65] G.O. Nelson (Diss. North Carolina Univ. 1977; Diss. Abstr. Intern. B **39** [1978] 237/8).

[66] K.G. Ihrmann, T.H. Coffield, Ethyl Corp. (U.S. 3077489 [1960/63]).

1.4.1.4.1.2.5.3 ^{4}L is a Cyclooctatetraene

1.4.1.4.1.2.5.3.1 $C_8H_8Fe(CO)_3$ and Its $AlBr_3$ Adduct

1.4.1.4.1.2.5.3.1.1 Preparation and Formation of $C_8H_8Fe(CO)_3$

Preparation

From $Fe(CO)_5$. The compound $C_8H_8Fe(CO)_3$ can be prepared by photochemical [1, 3, 7 to 9, 11, 12, 16] or thermal [2, 4, 8, 9, 15, 22, 36] reaction of $Fe(CO)_5$ with cyclooctatetraene (C_8H_8). Photochemical reaction of $Fe(CO)_5$ with a slight excess of C_8H_8 in hexane [1] or

C_6H_6 [1, 8, 11] for 24 h [1, 8] gives $C_8H_8Fe(CO)_3$ in 72% yield [1] and $C_8H_8Fe_2(CO)_6$ (see 2.8.1.1 in "Organoiron Compounds" C5, 1981, Table 4, No. 25, p. 49) [1, 11] if short irradiation times are used [11]. Irradiation of $Fe(CO)_5$ with a fivefold molar excess of C_8H_8 in C_6H_6 or hexane for 24 h gives $C_8H_8Fe(CO)_3$ (58% yield) and $C_8H_8Fe_2(CO)_6$ (31% yield) [1]. Irradiation of an equimolar mixture of $Fe(CO)_5$ and C_8H_8 without a solvent for 8 h gives $C_8H_8Fe(CO)_3$ [3, 7, 12, 16] in 10% yield [3, 7]. Similar irradiation of excess $Fe(CO)_5$ with C_8H_8 without a solvent gives $C_8H_8Fe(CO)_3$ and $C_8H_8Fe_2(CO)_6$ [3]. Refluxing excess $Fe(CO)_5$ with C_8H_8 in ethylcyclohexane [2, 4, 9, 15, 35, 36] (if this solvent is not available, other unsaturated, aliphatic or alicyclic hydrocarbons of similar boiling point such as 2,2,5-trimethylhexane, n-octane, toluene, or xylene may be used [8, 15, 22]) for 24 h followed by removal of solvent gives a mixture of compounds. Sublimation of the residue gives $C_8H_8Fe(CO)_3$ (60% yield [2, 4, 9, 15] or 87% yield [36]), and chromatography of the residue with petroleum ether gives $C_8H_8Fe_2(CO)_6$. Further elution with CH_2Cl_2 gives $C_8H_8Fe_2(CO)_5$ (see 2.8.2.2 in "Organoiron Compounds" C5, 1981, Table 7, No. 1, p. 114) [2, 4, 9, 15, 36]. Refluxing in the presence of a trace of hydroquinone for 12.5 h followed by standing overnight and filtration gives $C_8H_8Fe(CO)_3$, and reduction in volume of the mother liquor gives $C_{16}H_{16}Fe(CO)_3$ (compare 1.4.1.4.1.2.3.2.1.1.4 in "Organoiron Compounds" B8, 1985, Table 12, No. 56, p. 277) [8]. In absence of solvent no $C_8H_8Fe(CO)_3$ is isolated [15]; but reaction of C_8H_8 with $Fe(CO)_5$ at 120 °C for 18 h gives $C_8H_8Fe(CO)_3$ [3].

From $Fe_2(CO)_9$ or $Fe_3(CO)_{12}$. Much higher yields are obtained by the thermal reaction of $Fe_2(CO)_9$ or $Fe_3(CO)_{12}$ with C_8H_8 than by the photochemical reaction of $Fe(CO)_5$ [7]. Thus, reaction of $Fe_2(CO)_9$ with C_8H_8 [14, 18, 20, 28, 30, 32] in boiling hexane for a few minutes [20], in C_6H_6 at 65 °C [28], or in boiling ether for 4 h [30] gives two isomers of $C_8H_8Fe_2(CO)_6$ (see 2.8.1.1 in "Organoiron Compounds" C5, 1981, Table 4, Nos. 24 and 25, p. 49) [14, 18, 28, 32] and $C_8H_8Fe_2(CO)_5$ (see 2.8.2.2 in "Organoiron Compounds" C5, 1981, Table 7, No. 1, p. 114) [18, 28, 32] in small amounts [28] and $C_8H_8Fe(CO)_3$ [14, 18, 20, 28, 30, 32] in good yield [7, 20], 86% [28]. $C_8H_8Fe(CO)_3$ is purified by chromatography [30]. Reaction of $Fe_3(CO)_{12}$ with C_8H_8 [5, 13, 14, 17, 26] possibly gives the same products as with $Fe_2(CO)_9$ [14]. Refluxing $Fe_3(CO)_{12}$ with C_8H_8 in light petroleum ether (b.p. 120 to 130 °C) for ca. 30 min [17] or 1 h [13] followed by filtration, evaporation, and dissolving the residue in hot C_2H_5OH [17] gives $C_8H_8Fe(CO)_3$ in 56 to 88 [17] or 90% yield [13] together with small amounts of $C_8H_8Fe_2(CO)_6$ [13]. Treatment of $Fe(CO)_5$ with NaOH in CH_3OH (formation of $Fe_3(CO)_{12}$) followed by oxidation with H_2O_2 in the presence of C_8H_8 gives $C_8H_8Fe(CO)_3$ [41].

Formation

From Mononuclear Organoiron Compounds. Small amounts of $C_8H_8Fe(CO)_3$ together with $[N(P(C_6H_5)_3)_2][Fe(NO)_2I_2]$ are obtained by the reaction of $[N(P(C_6H_5)_3)_2][(CO)_3FeNO]$ with CH_3I and C_8H_8 in CH_2Cl_2 [29]. $^3LFe(CO)_3$ ($^3L = C_6H_5CH{=}CHCOCH_3$, see 1.3.4.3 in "Eisen-Organische Verbindungen" B5, 1978, Table 28, No. 39, p. 132) reacts with C_8H_8 [38, 39] in C_6H_6 at 57 °C for 48 h [39] to give $C_8H_8Fe(CO)_3$ in 74% yield [38, 39]. For a competitive reaction with cyclohexadiene see [39], and for the kinetics of this reaction see [40]. $C_8H_8Fe(CO)_3$ is also formed in the reaction of I with $O(COCH_3)_2$ in pyridine [33]. $C_8H_8Fe(CO)_3$ is slowly formed together with $[DNC_5H_5]PF_6$ by stirring a solution of II at 25 °C [25]. Reaction of III with C_8H_8 in boiling petroleum ether for 30 min followed by evaporation and chromatography with petroleum ether or C_6H_6 gives $C_8H_8Fe(CO)_3$ in 64% yield [31]. Neutralization of IV gives $(C_6H_5)_3CC_8H_7Fe(CO)_3$, $C_8H_8Fe(CO)_3$, $(C_6H_5)_3CC_8H_7$, and some unidentified substances [34]. Treating V with H_2O followed by extraction with C_6H_6 gives $C_8H_8Fe(CO)_3$ [10]. Similar treatment of V with a slight excess of aqueous KOH gives $C_8H_8Fe(CO)_3$ and decomposition products [13]. $C_8H_8Fe(CO)_3$ is also formed in the reaction of cyclo-$C_7H_8Fe(CO)_3$ with C_8H_8 [4, 5] in refluxing ethylcyclohexane after 4.5 h. The reaction mixture is evaporated, and the red-

References on pp. 233/4

black residue is partially dissolved in boiling $CHCl_3$ with charcoal. Filtration followed by sublimation or chromatography gives pure $C_8H_8Fe(CO)_3$ as the sole product [4]. Reaction of $(C_8H_8)_2Fe$ with CO (1 atm) in C_6H_6 at 20 °C for 24 h gives $C_8H_8Fe(CO)_3$ (>50% yield), C_8H_8, and small amounts of iron [19].

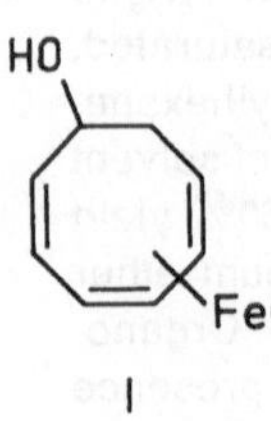

HO Fe(CO)$_3$ I; $[\ ]^+[PF_6]^-$ (H, D, N, Fe(CO)$_3$) II; C_6H_5, O, $(CO)_3Fe$ III; $[\ ]^+$ $C(C_6H_5)_3$, Fe(CO)$_3$ IV

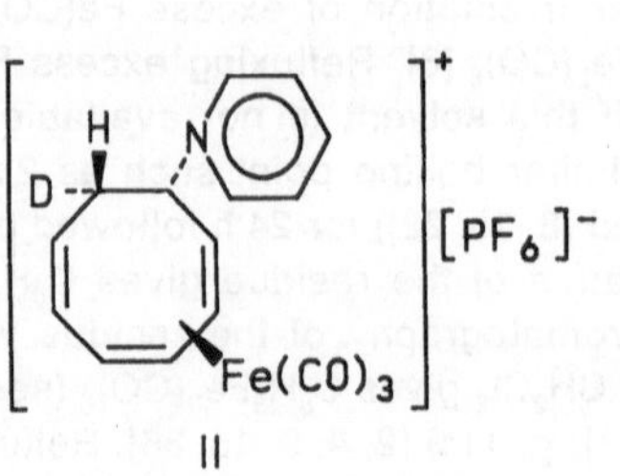

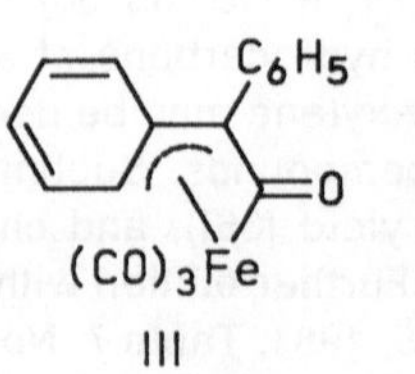

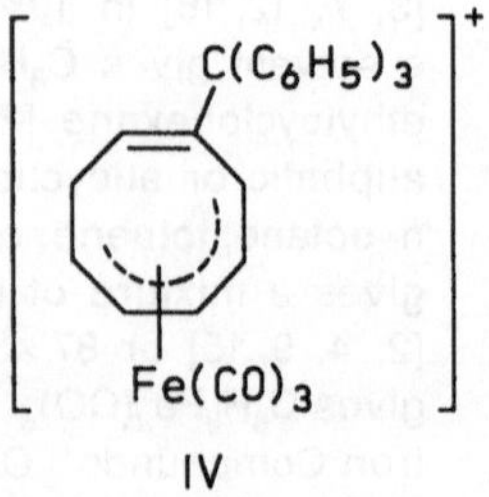

From Dinuclear Organoiron Compounds. Heating VI ($R=CH_2OSO_2C_6H_5$) and $Fe_2(CO)_9$ in hexane at 40 °C for 3 h followed by filtration through infusorial earth, evaporation of the filtrate, and chromatography of the residue on neutral alumina with hexane gives styrene (^2L), $^2LFe(CO)_4$, and $C_8H_8Fe(CO)_3$. Similar reaction between $Fe_2(CO)_9$ and VI ($R=CH_2OH$) in ether at room temperature for 17 h, followed by refluxing for 4 h and stirring at room temperature for another 4 d, work-up as before but distillation at reduced pressure gives VII, $C_8H_8Fe(CO)_3$, and a $^4LFe(CO)_3$ (^4L = hydroxymethylcycloheptatriene) [24]. Refluxing $Fe_2(CO)_9$ with VIII in ether for 60 h followed by filtration, evaporation, extraction of the residue with petroleum ether, and chromatography on neutral alumina (activity III) gives a mixture of compounds. Elution with petroleum ether gives $C_8H_8Fe(CO)_3$ (1.6% yield), trans-C_8H_8-$Fe_2(CO)_6$, and IX. Further elution with petroleum ether/C_6H_6 (1:1 to 1:2) gives $C_8H_8Fe_2(CO)_5$, X, XI, and XII (see 2.6.1.2.4 in "Organoiron Compounds" C5, 1981, Table 3, No. 2, p. 28). Cyclooctatetraene iron carbonyls which are formed in the reaction of VIII with $Fe_2(CO)_9$ under various conditions (petroleum ether, ether, CH_2Cl_2, hexane, C_6H_6, octane; reaction temperature: 25 to 125 °C) are probably produced via prior deoxygenation by $Fe_2(CO)_9$ followed by complex formation. The small amounts of aldehyde and ketone complexes, X to XII, formed in non-aromatic solvents might result from reaction of VIII or IX with traces of acid in the $Fe_2(CO)_9$. However, under conditions where no IX is isolated (refluxing in C_6H_6 or toluene), the yield of X greatly increases, and X becomes the principal product [23]. Refluxing IX (see 2.6.1.2.3 in "Organoiron Compounds" C5, 1981, Table 2, No. 15, p. 18) in toluene or C_6H_6 for 3 h followed by evaporation, extraction of the residue with petroleum ether, and chromatography on neutral alumina (activity III) with petroleum ether gives C_8H_8-$Fe(CO)_3$ (25% yield). Further elution with petroleum ether/C_6H_6 (1:1) gives $C_8H_8Fe_2(CO)_5$, trans-$C_8H_8Fe_2(CO)_6$, and X. Refluxing IX in n-heptane for 15 h followed by work-up as before gives $C_8H_8Fe(CO)_3$ (23.5%), trans-$C_8H_8Fe_2(CO)_6$, $C_8H_8Fe_2(CO)_5$, $(C_8H_8O)_2Fe_2(CO)_4$, and $(C_8H_8O)_2Fe(CO)_2$, but no traces of X.

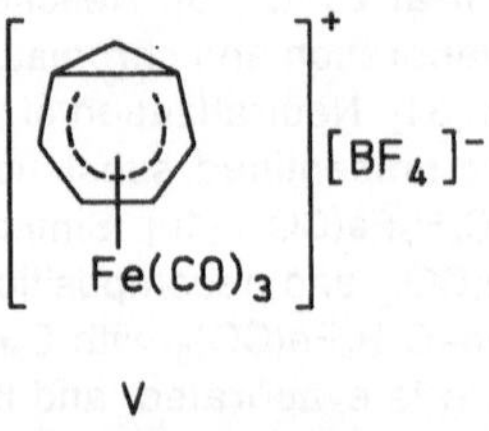

$[\ ]^+[BF_4]^-$ Fe(CO)$_3$ V; H, R VI; —Fe(CO)$_3$ VII; O VIII

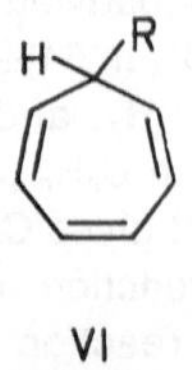

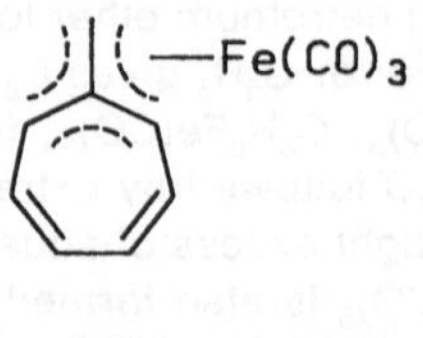

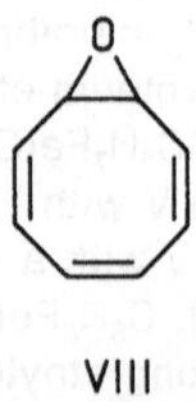

References on pp. 233/4

From Trinuclear Organoiron Compounds. Refluxing VIII with $Fe_3(CO)_{12}$ in C_6H_6 gives X and small amounts of $C_8H_8Fe(CO)_3$ together with other cyclooctatetraene and cyclooctatrienone complexes [23].

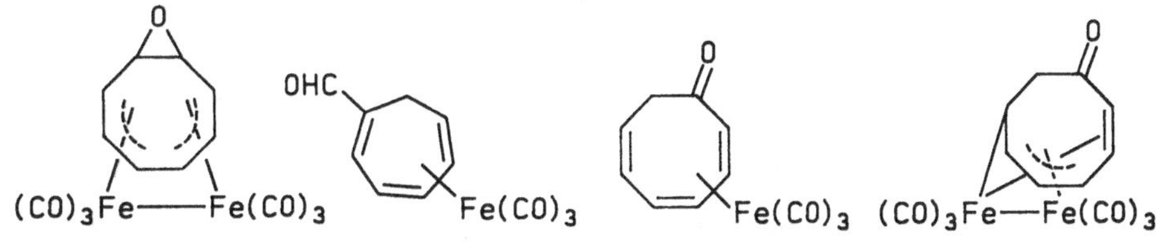

IX X XI XII

Formation of Deuterated Derivatives

Refluxing C_8D_8 and $Fe_2(CO)_9$ in C_6H_6 gives $C_8D_8Fe(CO)_3$ [37]. For the preparation of other deuterated derivatives see [21].

References:

[1] M.D. Rausch, G.N. Schrauzer (Chem. Ind. [London] **1959** 957/8). – [2] T.A. Manuel, F.G.A. Stone (Proc. Chem. Soc. **1959** 90). – [3] A. Nakamura, N. Hagihara (Bull. Chem. Soc. Japan **32** [1959] 880/1). – [4] T.A. Manuel, F.G.A. Stone (J. Am. Chem. Soc. **82** [1960] 366/72). – [5] R. Burton, G. Wilkinson (unpublished results from [6]).

[6] F.A. Cotton (J. Chem. Soc. **1960** 400/6). – [7] A. Nakamura, N. Hagihara (Mem. Inst. Sci. Ind. Res. Osaka Univ. **17** [1960] 187/90; C.A. **1961** 6457). – [8] K.G. Ihrman, T.H. Coffield, Ethyl Corp. (U.S. 3077489 [1960/63]). – [9] T.A. Manuel (Diss. Harvard Univ. 1961; Diss. Abstr. **22** [1961] 1003/4). – [10] G.N. Schrauzer (J. Am. Chem. Soc. **83** [1961] 2966/7).

[11] G.N. Schrauzer, S. Eichler (Angew. Chem. **74** [1962] 585). – [12] H.P. Fritz, H. Keller (Chem. Ber. **95** [1962] 158/73). – [13] A. Davison, W. McFarlane, L. Pratt, G. Wilkinson (J. Chem. Soc. **1962** 4821/9). – [14] G.F. Emerson, J.E. Mahler, R. Pettit, R. Collins (J. Am. Chem. Soc. **86** [1964] 3590/1). – [15] R.B. King (Organometal. Syn. **1** [1965] 126/7).

[16] R.T. Bailey, E.R. Lippincott, D. Steele (J. Am. Chem. Soc. **87** [1965] 5346/50). – [17] W. McFarlane, G. Wilkinson (Inorg. Syn. **8** [1966] 184/5). – [18] E.B. Fleischer, A.L. Stone, R.B.K. Dewar, J.D. Wright, C.E. Keller, R. Pettit (J. Am. Chem. Soc. **88** [1966] 3158). – [19] A. Carbonaro, A. Greco, G. Dall'Asta (Tetrahedron Letters **1967** 2037/40). – [20] F.A.L. Anet (J. Am. Chem. Soc. **89** [1967] 2491/2).

[21] L.A. Bock (Diss. Univ. California 1969; Diss. Abstr. Intern. B **30** [1970] 4967). – [22] R.B. King (Appl. Spectrosc. **23** [1969] 536/46). – [23] H. Maltz, G. Deganello (J. Organometal. Chem. **27** [1971] 383/7). – [24] R.C. Kerber, D.J. Ehntholt (J. Am. Chem. Soc. **95** [1973] 2927/34). – [25] R. Aumann (J. Organometal. Chem. **78** [1974] C31/C34).

[26] G. Wilkinson, R. Burton (unpublished results from [27]). – [27] F.A. Cotton (J. Organometal. Chem. **100** [1975] 29/41). – [28] R. Aumann, J. Knecht (Chem. Ber. **109** [1976] 174/9). – [29] N.G. Connelly, C. Gardner (J. Chem. Soc. Dalton Trans. **1976** 1525/7). – [30] G.A. Olah, S.H. Yu, G. Liang (J. Org. Chem. **41** [1976] 2383/6).

[31] J. Weichmann (Dipl.-Arbeit Univ. Regensburg 1981, pp. 1/111). – [32] C.E. Keller, G.F. Emerson, R. Pettit (J. Am. Chem. Soc. **87** [1965] 1388/90). – [33] B.F.G. Johnson, J. Lewis, D.J. Thompson, B. Heil (J. Chem. Soc. Dalton Trans. **1975** 567/71). – [34] B.F.G. Johnson, J. Lewis, J.W. Quail (J. Chem. Soc. Dalton Trans. **1975** 1252/7). – [35] R.B. King (Organometal. Syn. **1** [1965] 127/8).

[36] R. Grubbs, R. Breslow, R. Herber, S.J. Lippard (J. Am. Chem. Soc. **89** [1967] 6864/70). – [37] R. Aumann (Chem. Ber. **109** [1976] 168/73). – [38] G.O. Nelson (Diss. Univ. North Carolina 1977; Diss. Abstr. Intern. B **39** [1978] 237/8). – [39] M. Brookhart, G.O. Nelson (J. Organometal. Chem. **164** [1979] 193/202). – [40] P.M. Burkinshaw, D.T. Nixon, J.A.S. Howell (J. Chem. Soc. Dalton Trans. **1980** 999/1004).

[41] M. Franck-Neumann, M.P. Heitz, D. Martina (Tetrahedron Letters **24** [1983] 1615/6).

1.4.1.4.1.2.5.3.1.2 Physical Properties of $C_8H_8Fe(CO)_3$

Color and Thermodynamic Properties. The compound $C_8H_8Fe(CO)_3$ forms red [9, 10, 34, 103], intense red (from petroleum ether/ether) [97], beautiful [2] dark red (from light petroleum ether [2]) [31], red-brown (from hot C_2H_5OH [20, 38]) [38], reddish brown plate-like (by sublimation) [22] crystals [1, 2, 9, 10, 20, 22, 31, 34, 83, 97], long [3, 48] red (from hexane [3] or pentane [79]) [3], long [102] red-black [48, 102] needles [3, 48, 102] (from petroleum ether (b.p. 30 to 60 °C [102])) or platelets [48] or brown plates (from isooctane) [10], which melt at 82 to 84 °C, 82 to 86 °C, 86 to 88 °C [70], 90 to 92 °C [102], 92 °C [1], 93 °C [79], 93 to 94 °C [10], 93 to 95 °C [34], 94 °C [2, 31, 38], or 94 to 95 °C [3], and readily [9] sublime at 30 °C/2 Torr [10], 50 °C/0.01 Torr [20, 38], 50 °C/2 Torr [10], 60 to 80 °C/0.1 Torr [9, 34], 70 °C/0.01 Torr [79], 75 °C/0.1 Torr [65], 75 °C/0.2 Torr [63], 80 °C/5 Torr [2], or 80 to 90 °C/0.5 Torr [102] and decomposes at ca. 155 [1] or 160 °C [9]. $C_8H_8Fe(CO)_3$ is monomeric in C_6H_6 (by cryoscopy) [9] and its dipole moment in C_6H_6 is $\mu_{298} = 1.79$ [77]. Microcalorimetric measurements at temperatures 460 to 513 K of the thermal decomposition and iodination give a standard enthalpy of formation $\Delta H_f^o(c) = -237 \pm 12$ kJ/mol, $\Delta H_f^o(g) = -150 \pm 13$ kJ/mol and $\Delta H_{dec}^{298} = 203 \pm 11$ kJ/mol corresponding to $\Delta H_f^o(c)$. The enthalpy of disruption at 298 K corresponding to $C_8H_8Fe(CO)_3(g) \rightarrow Fe(g) + C_8H_8(g) + 3CO(g)$ is $\Delta H_{disrupt}^{298} = 532 \pm 14$ kJ/mol. Separate measurements by the vacuum-sublimation technique give $\Delta H_{subl}^{298} = 87 \pm 4$ kJ/mol. The bond enthalpy contribution for C_8H_8-Fe is evaluated as 180 ± 15 kJ/mol presuming that the bond enthalpy contribution to $\Delta H_{disrupt}$ from Fe-CO remains the same as in $Fe(CO)_5$ where $\bar{D}(Fe\text{-}CO) = 117.5 \pm 2$ kJ/mol [91].

NMR Spectra. Much work has been done on the ¹H NMR spectrum of diamagnetic [1] $C_8H_8Fe(CO)_3$ [1, 3 to 6, 9, 12, 13, 22, 25 to 27, 29, 32, 34, 36, 39, 41 to 43, 46, 47, 49, 62, 64, 67 to 71, 76, 78 to 80, 93, 102, 103]. The ¹H NMR spectrum at room temperature in different solvents shows a single resonance [1, 3, 6, 9, 25, 26, 34, 70, 76, 79, 93, 102, 103] which implies that no C atoms in the C_8H_8 moiety are preferentially bonded to the Fe atom, but that all 8 C atoms are equivalently bonded [1]. The chemical shifts (in ppm) are as

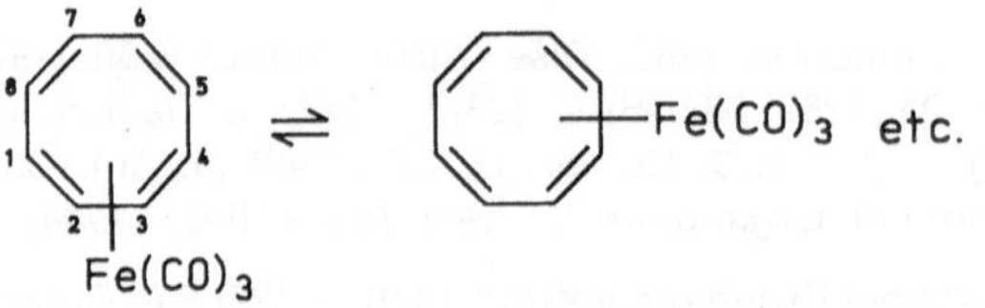

$\rightleftharpoons$ Fe(CO)₃ etc.

I 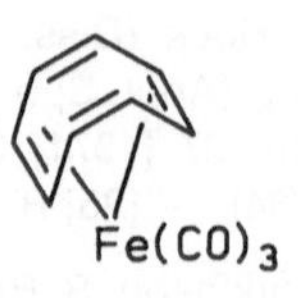 II

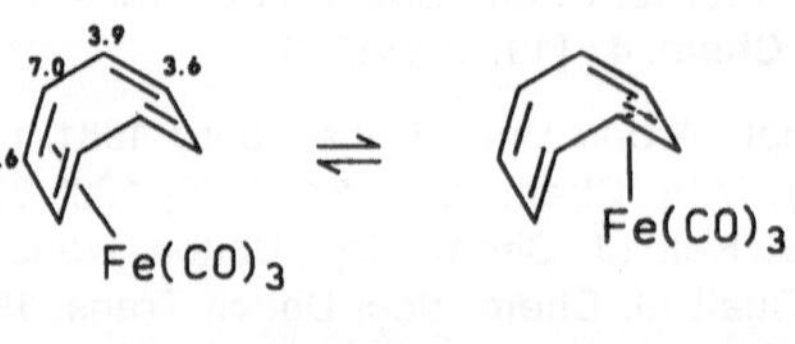

III

References on pp. 239/41

follows: δ=5.14 [67] or 5.2 [25, 27] in C_6H_6, 5.2 in benzene-d_6 [93, 99] and 4.86 in C_6H_6 in the presence of $(CH_3C_5H_4)_3Nd$ ($(CH_3C_5H_4)_3Nd$:$C_8H_8Fe(CO)_3$=1:2) [67], 5.39 in $CHCl_3$ [102], 5.15 in CCl_4 [25, 26], 5.12 [70], 5.18 [34, 79], or 5.27 [25, 16] in CS_2, 4.73 in acetone, 4.95 in CH_3CN, and 4.44 in n-$C_3H_7NH_2$ [25, 26]. For similar shifts in cyclo-C_6H_{12} see [1, 3, 9] and in H_2O or toluene-d_8 see [3]. The examination of the temperature dependence of the ^1H NMR spectrum [12, 32, 36, 41, 42, 46, 47, 49, 62, 64, 68, 69, 71, 78, 80, 102] as proposed in [12, 13] gave no conclusive results concerning the structure of $C_8H_8Fe(CO)_3$ in solution or the shift mechanism [80]. The ^1H NMR spectrum in Freon at −155 °C shows δ=4.55 (H-1,4), 4.78 (H-2,3), 6.19 (H-6,7), 6.45 (H-5,8) ppm [102].

In the case of dynamic $C_8H_8Fe(CO)_3$, an average value of δ=6.1 ppm is required for the four olefinic protons of the two double bonds not coordinated to Fe in order that the total average be equal to δ=5.2 ppm (C_6H_6) as observed, assuming that the other four protons have the same chemical shift as those in $C_7H_8Fe(CO)_3$ [29]. Investigation shows that the sharp singlet remains essentially a single sharp line down to very low temperatures (ca. −100 °C) [22, 32, 36, 39, 41]. However, by using still lower temperatures, the single line broadens rapidly below −100 °C and separates into two bands below −118 °C. At −130 °C the spectrum consists of two bands [41] of equal intensity at δ=4.6 and 6.4 ppm [36] or δ=4.56 and 5.94 ppm [42] separated by about 80 Hz [41]. This behavior is observed in a number of solvents (CH_3OH, $CH_3CHO/CDCl_3$ [41], $CHCl_2F/CF_3Br$) [36, 41]. In the Freon mixture at −150 °C the two bands are found to be unsymmetric and independent of temperature in the range of −140 to −150 °C. From these changes the rate of rearrangement is calculated to be $k \sim 200 \cdot s^{-1}$ at −120 °C and $\Delta G^* = 7.2$ kcal/mol [41]. There exist different explanations for this spectrum [41]. The complex may have an equivalent valence-tautomeric structure I, if the structure in solution is the same as in the solid [22, 32]; or it may be tube-bonded as in II [16, 25, 36, 47]. However, a 1,3 diene bonded model III is favored in [41]. A planar symmetrical model I, as proposed in [4, 5, 17], or a structure as shown in II (see, e.g. [25, 36]) is ruled out [2, 7, 8, 18, 41]. No bicyclic model is possible [41]. Furthermore, due to the lack of absorption around δ=3.0 ppm, at least two types of degenerate valence tautomers are involved in the rearrangement of $C_8H_8Fe(CO)_3$ at room temperature. This implies that the observed spectrum does not correspond to the rigid 1,3 diene bonded model discussed in [41] and assumes that the protons have the values indicated in III [42]. All these models are said to be incorrect and structure IV as a 1,3 bonded model similar to that in [41, 49] is favored on the basis of the chemical shifts at δ=3,0 (H-4), 5.4 (H-3), 6.1 (H-5), 6.4 (H-6) ppm in the ^1H NMR spectrum [47]. Ring whizzing might occur by 1,2 shifts [53] involving concomitant migration of double bond character (π-electron density) within the ring [49]. Pulse ^1H NMR studies on $C_8H_8Fe(CO)_3$ reveal that $C_8H_8Fe(CO)_3$ is fluxional, possessing considerable motional freedom in the solid state, which can be related to its bonding and reorientation in solution. The high resolution NMR spectrum shows a single ^1H resonance at room temperature, while at 128 K it shows a multiplet structure. The "wide line" ^1H NMR spectrum in the solid state is very markedly temperature dependent. At room temperature it has a line width of 1.9 G and a second moment of 1.0 G^2. On lowering the temperature to 250 K, a broad transition spectrum is observed; further lowering of the temperature results in formation of two bands with values of 9.4 G and 7.9 G^2, respectively. Calculation suggests that at low temperature the molecule is rigid in the lattice. Since X-ray diffraction (see below) results show no disorder in the ring C atoms, any motion must involve simultaneous distortion and reorientation of the ring, the ring moves, the C atoms change positions, and the molecule takes up the same position in the lattice as it had before rotation, see, e.g. [62, 64, 71]. The activation energy for these processes is E_a=9.4 kcal/mol [78]. For a line shape analysis of polycrystalline $C_8H_8Fe(CO)_3$ see also [68, 69]. The difficulties in the interpretation of the $C_8H_8Fe(CO)_3$ case arises from

References on pp. 239/41

unjustified attempts to treat the spectrum of this molecule observed at lowest temperatures as though it were the limiting spectrum [46, 80]. Thus, nothing in the earlier literature proves the mechanism. The ^{1}H NMR spectra obtained, e.g., in [41, 47] is an intermediate-exchange spectrum. The data correspond to a mean residence time (m.r.t.) of ca. 0.007 s, whereas a m.r.t. 30 times longer than this is required to give the slow-exchange spectrum. The true slow-exchange spectrum consists of four well-separated signals, not two broad humps [46, 80]. It is true that ^{1}H NMR spectra provide evidence as to the solution structures of $C_8H_8Fe(CO)_3$, e.g. [47], but they do not provide any evidence whatever concerning the mechanism of rearrangement (compare ^{13}C NMR spectra, below). There are so many unknown details of the structure of the transition state for 1,2 shifts [53] that an analysis of this question is certain to be very speculative [80].

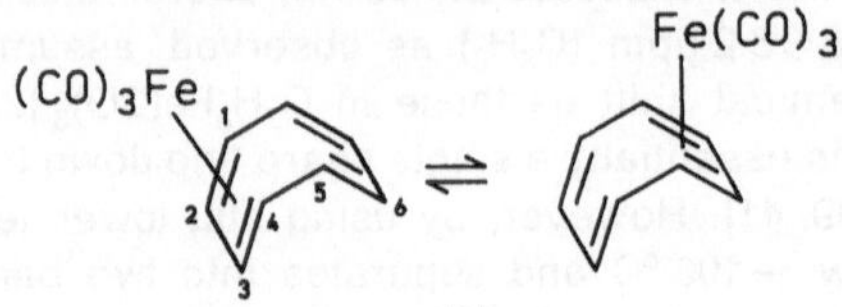

(CO)$_3$Fe ⇌ Fe(CO)$_3$

IV

The ^{13}C NMR spectrum of $C_8H_8Fe(CO)_3$ is temperature dependent [66, 80]. At −134 °C in $CHCl_2F/CF_2Cl_2$ (2:1) [66], or in $CHCl_2F$ with either 15% CD_2Cl_2 or acetone-d_6, or 10% acetone and 2% $Si(CH_3)_4$, the chemical shifts are at δ=73.7(s), 92.5(s) (C-1 to 4); 122.6(s), 128.8(s) (C-5 to 8); 211.7 (s, CO); 214.0 (s, 2CO) ppm (for assignment see Formula I) [66, 88]. At higher temperatures (−90 to +22 °C) in CH_2Cl_2/10% acetone-d_6/5% $Si(CH_3)_4$ [80] or in $CHCl_2F/CF_2Cl_2$ (2:1) [66] all ring signals coalesce to a singlet and, likewise the CO signals become a singlet [66, 80] with δ=104.4 (s, C-1 to 8), 213.2 (s, CO) ppm, respectively [66]. A similar spectrum is observed in liquid SO_2 [82]. The initial rates of broadening of the ring signals by lowering the temperature are unequal. Two of them broaden more quickly than the other two. This demonstrates that the process by which the ring C atoms achieve time-average equivalence is one in which two environments afford greater residence time than the other two. A line shape analysis for the CO resonances is considerably less accurate than for the ring resonances since the chemical shift differences involved are much smaller and the temperature range over which the entire process of transformation from a two line spectrum to a one line spectrum occurs is very small. The activation parameters for the two processes (CO scrambling and ring whizzing) are shown in the following table:

	ring whizzing	CO scrambling	Ref.
E_a (kcal/mol)	8.1(2)	8.3(9)	[80, 83]
log A	13.5(2)	13.3(9)	[80]
ΔH^{*} (kcal/mol)	7.8(2)	8.0(9)	[80]
ΔS^{*} (cal · K^{-1} · mol^{-1})	3.0(9)	3(6)	[80]
$\Delta G^{*}_{25°}$ (kcal/mol)	6.8(3)	7(2)	[80, 83]

Unfortunately, the values are so similar that it is impossible to say whether the two processes are truly simultaneous and, therefore, possibly interdependent or not. Thus, for $C_8H_8Fe(CO)_3$, any random exchange process, a process with a symmetrical intermediate or transition state, and also the 1,3 shift processes are all rigorously excluded. The possibility of 1,5 shifts can also be dismissed since this would lead to a high temperature spectrum consisting of two lines rather than one. It can be shown that the mechanism of rearrangement is a series of omnidirected 1,2 shifts of the $Fe(CO)_3$ group relative to the ring [80]. Cotton

References on pp. 239/41

et al. [80] disagree with the previous assertions of Rigatti et al. [66] that line shape analysis "leads to significantly lower rate constants" for the carbonyls and that "the carbonyl exchange process is independent of the ring-atom exchange process" [66]. There are so many unknown details of the structure of the transition state for 1,2 shifts that an analysis of this question is sure to be very speculative [80], see ^{1}H NMR spectrum above.

The ^{57}Fe NMR spectrum (0.67 M solution in C_6H_6) shows one chemical shift at $\delta = 300.5 \pm 1$ ppm relative to $Fe(CO)_5$ [58].

^{57}Fe Mössbauer Spectra. The ^{57}Fe Mössbauer spectrum has been measured [23, 24, 30, 48, 49, 54, 57, 60, 61, 72, 101, 102]. The results are shown in the following table (isomer shift δ (referred to $Na_2[Fe(CN)_5NO] \cdot 2H_2O$) and quadrupole splitting Δ in mm/s):

T in K	solvent	δ	Δ	Ref.
78 to 80	—	0.31±0.020	1.24 ±0.03	[48, 54, 57, 61, 72, 102]
	—	0.301	1.26	[101]
	—	0.375	1.19	[60]
	n-C_8H_{18}	0.36±0.02	1.23 ±0.02	[48, 54, 61, 102]
	methyltetrahydrofuran	0.36±0.06	1.121±0.03	[48, 57, 61, 102]
	$O_2NC_6H_5$	0.34±0.020	1.16 ±0.03	[48, 54, 57, 61]
	$O_2NC_6H_5$	0.36	1.21	[102]
	16% C_2H_5OH/42% i-C_3H_7OH/42% ether	0.34±0.020	1.16 ±0.03	[48, 54, 57, 61, 102]
	polymethacrylate	0.375	1.26	[57]
84	n-$C_4H_9C_6H_5$	0.359	1.178	[72]
298	neat	0.255	1.21	[60]

The frozen glassy matrix technique has been used in a study of the lattice dynamics of $C_8H_8Fe(CO)_3$. This technique is also suited for determining possible conformational changes on dissolution of crystalline solids in solvents which set to a glassy matrix at cryoscopic temperatures. A model consistent with the quasi-octahedral configuration of the Fe atom with respect to three CO groups and three of the eight C-C bonds is discussed [23]. The ring acts as a 1,3 diene [48, 49, 54]. The calculated Mössbauer spectrum for a temperature of 145±5 K gives 100±4 cm^{-1} for the C-Fe-C angle bending mode [54] compared with 101 cm^{-1} from the IR spectrum (see below) [31].

IR and Raman Spectra. The IR spectrum in the ν(CO) region [1, 2, 5, 8 to 10, 26, 28, 34 to 36, 43, 67, 73, 79, 90, 92, 100, 102, 104, 105] is measured in different solvents as shown in the following table:

solvent	ν(CO) (cm^{-1})	others (cm^{-1})	Ref.
KBr	1961, 2050	699, 716, 720	[2, 8]
hexane	1984, 1996, 2051	—	[79]
cyclohexane	1976, 1986, 2043	—	[26, 28, 100]
CCl_4	1972, 1990, 2050	1100, 1550, 2180	[55, 73, 102]
CH_2Cl_2	1978, 2050	—	[90, 92]
CS_2	1992, 2058	1416, 3012	[9, 34, 79, 104]

References on pp. 239/41

The IR spectrum in Nujol/KBr in the region from 200 to 2100 cm^{-1} [25 to 28, 31] shows bands at 262, 275, 329, 356, 388, 400, 417, 458, 498, 538, 563, 597, 610, 627, 698, 708, 715, 741, 765, 778, 806, 845, 868, 898, 918, ca. 945, 987, 1125, 1150, 1172, 1244, 1256, 1299, 1313, 1418, 1947, 1982, 1998, 2062 cm^{-1} [31]. Solutions in CS_2, CCl_4, and C_2Cl_4 show bands at 174, 224, 263, 330, 362, 389, 404, 461, 485, 498, 537, 561, 597, 607, 628, 672, 698, 707, 743, 765, 777, 804, 846, 864, 898, 922, 955, 984, 995, ca. 1040, 1075, ca. 1090, 1122, 1177, 1235, 1252, 1298, 1314, 1383, 1401, 1419, 1431, 1460, 1490, 1562, 1708, 1750, 1768, 1845, ca. 1895, 1942, 1976, 1993, 2061, 2879, ca. 2925, 3022, 3040, and 3075 cm^{-1} [31]. The Raman spectrum in C_6H_6 and CS_2 solution shows bands at 100, 137, 176, (238), 262p, 279, 330p, 391, 456, 498, 538, 563, 607, 643, 690p, 716, (730), 766, 782, 827, 849, 875, 889, 924, 951, 984, 1024, 1042, 1057, (1080), 1129, 1147, 1180, 1230p, 1255, 1303, 1312, (1406), 1431p, 1460, 1563p, (1639) cm^{-1} [31]. For a descriptive assignment of these frequencies see [31]. The symmetry force constant for the Fe-C bond is ca. 3.6 mdyn/Å [28]. These spectra indicate that the structure in solution is different from that in the solid state and a model with a 1,3- [31] or 1,5-structure is discussed [36]. The spectrum below 1600 cm^{-1} is too complex to provide any conclusive distinction between these structures [36] but agrees with IV, see p. 236 [47]. The IR spectrum is similar to that of $C_8H_8Ru(CO)_3$ (cf. ^1H NMR spectrum above and X-ray structure below) [45]. Fourier transform IR spectroscopic evidence, including ^{13}CO labelling and energy-factored CO force-field fitting is presented to show that photolysis of III (p. 234) in a CH_4 matrix at 12 K leads to the formation of a IV species via the $^2LFe(CO)_3$ intermediate ($^2L=C_8H_8$). The results are related to the fluxional behavior of $C_8H_8Fe(CO)_3$ in solution. For the observed and calculated band positions of terminal CO stretching bands for ^{13}CO enriched III and IV see [105].

UV Spectra. The UV spectra show the following bands in different solvents [2, 3, 8, 9, 25, 27, 102]. Neat: $\lambda_{max}=330$ nm, in 95% C_2H_5OH: $\lambda_{max}(\log\varepsilon)=300$ (3.99) nm [2] or 333 nm [102], in cyclohexane: $\lambda_{max}(\varepsilon)=304$ (11500) nm [9] or 333 (15000) nm [102], in $CHCl_3$: $\lambda_{max}(\log\varepsilon)=303$ (4.3), 430 (2.14) nm [25, 27], in mull: $\lambda_{max}=333$ nm [102].

Photoelectron Spectra. The He(I) photoelectron spectrum shows bands at 7.84, 8.74, 10.61, and 11.63 eV [98]. The first band is assigned to $1a_1$ ($3d_{z^2}$), 1e ($3d_{x^2-y^2}$), 1e ($3d_{xy}$) [94, 98], and to the ionization of π^* of the complexed diene moiety [94] and $3d_{xy}$ [94, 98]. Band two corresponds to $1a_2(\pi)$ and $1b_1(\pi)$ and band three to $1a_2(\pi)$. The fourth band at 11.63 is due to ejection of electrons from σ levels of C_8H_8 [94], cf. [98].

Structure. The compound crystallizes in the space groups Pnam-D_{2h}^{16}, V_h^{16} with the unit cell parameters a=6.544±0.008, b=13.455±0.023, and c=11.506±0.021 Å; Z=4 molecules per unit cell [12, 22]. The measured density is 1.57 g/cm³, calculated 1.60 g/cm³ [22]. The main bond distances and angles are shown in **Fig. 26** [22]. The structure shows that six

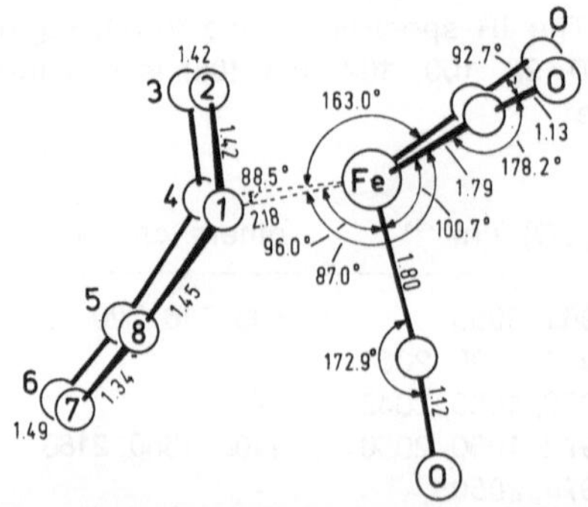

Fig. 26. Molecular structure of $C_8H_8Fe(CO)_3$ [22].

of the eight atoms of the ring are very nearly coplanar and that six of the eight bond angles of the ring are within about 3° of the 135° angle of a completely regular and planar ring [12, 22]. A comparison of overlap integrals, which are 0.25 between C-4,5 or C-1,8 in $C_8H_8Fe(CO)_3$ and 0.04 between corresponding atoms in $C_8H_8(Fe(CO)_3)_2$, indicates that the butadiene halves of the C_8H_8 ring interact more strongly in $C_8H_8Fe(CO)_3$ than in $C_8H_8(Fe(CO)_3)_2$ [12, 22]. The structure shows that the $Fe(CO)_3$ moiety is attached to a "butadiene" part of the C_8H_8 ring in a structural unit which is now seen to be quite general [11, 12, 14, 15, 21, 22].

It is clear from ^1H NMR results [32, 45] that $C_8H_8Fe(CO)_3$ and $C_8H_8Ru(CO)_3$ (see 3.22.2.1.1.1 in "Ruthenium" Erg.-Bd., 1970, pp. 524/5) must have the same or very similar structures in solution [51]. Both compounds are isostructural [45, 56], and the structure of solid $C_8H_8Ru(CO)_3$ is nearly isomorphous with that of $C_8H_8Fe(CO)_3$, suggesting that it is safe to regard the Ru compound as a close structural and dynamical analogy to the Fe compound. Therefore, the assumption made in [46], that the two molecules have analogous structural and dynamical properties in solution, is a reasonable one [56].

Theoretical Aspects of the Molecule. A theoretical discussion of the structure and the stability of $C_8H_8Fe(CO)_3$ in terms of molecular orbitals is given in [4], and the structure is discussed in terms of simple molecular orbital theory [5, 75] and ligand-field theory [7]. For a topological model see [59, 81]. A topological, Hückel-type orbital model is used to discuss the bonding in $C_8H_8Fe(CO)_3$ [85 to 87]. Justification derives from experimental studies and perturbation theory arguments [33, 40, 74, 84]. The sites of lowest localization energy are considered to be preferred sites of reaction [37]. However, the concept of "three-dimensional" aromaticity [89] is in agreement with the idea of describing $C_8H_8Fe(CO)_3$ as a resonance hybrid of valence bond structures [95, 96]. For an INDO calculation on $C_8H_8Fe(CO)_3$ and interpretation of the photoelectron spectrum see [94]. The correct name for this compound is proposed as (1,2,3,4-tetrahaptocyclooctatetraene)tricarbonyliron or (1-4-η-C_8H_8)-$Fe(CO)_3$ [52].

References:

[1] T.A. Manuel, F.G.A. Stone (Proc. Chem. Soc. **1959** 90). – [2] A. Nakamura, N. Hagihara (Bull. Chem. Soc. Japan **32** [1959] 880/1). – [3] M.D. Rausch, G.N. Schrauzer (Chem. Ind. [London] **1959** 957/8). – [4] D.A. Brown (J. Inorg. Nucl. Chem. **10** [1959] 39/48). – [5] F.A. Cotton (J. Chem. Soc. **1960** 400/6).

[6] R. Burton, G. Wilkinson (unpublished results from [5]). – [7] L.E. Orgel (An Introduction to Transition-Metal Chemistry Ligand-Field Theory, Methuen, London 1960, pp. 1/186, 157/77). – [8] A. Nakamura, N. Hagihara (Mem. Inst. Sci. Ind. Res. Osaka Univ. **17** [1960] 187/90). – [9] T.A. Manuel, F.G.A. Stone (J. Am. Chem. Soc. **82** [1960] 366/72). – [10] K.G. Ihrman, T.H. Coffield, Ethyl Corp. (U.S. 3077489 [1960/63]).

[11] L.F. Dahl, D.L. Smith (unpublished results from [12]). – [12] B. Dickens, W.N. Lipscomb (J. Am. Chem. Soc. **83** [1961] 4862/3). – [13] B. Dickens, W.N. Lipscomb (J. Am. Chem. Soc. **83** [1961] 489/90). – [14] R.E. Dodge, V. Schomaker (unpublished results from [12]). – [15] R.B. Woodward (unpublished results from [12]).

[16] G.N. Schrauzer, H. Thyret (Z. Naturforsch. **16b** [1961] 353/6). – [17] T.A. Manuel (Diss. Harvard Univ. 1961; Diss. Abstr. **22** [1961] 1003/4). – [18] H.P. Fritz (unpublished results from [19]). – [19] G.N. Schrauzer (J. Am. Chem. Soc. **83** [1961] 2966/7). – [20] A. Davison, W. McFarlane, L. Pratt, G. Wilkinson (J. Chem. Soc. **1962** 4821/9).

[21] B. Dickens (Diss. Univ. Minnesota 1962; Diss. Abstr. **23** [1962] 1526). – [22] B. Dickens, W.N. Lipscomb (J. Chem. Phys. **37** [1962] 2084/93). – [23] G. Wertheim, R.H. Herber

(J. Am. Chem. Soc. **84** [1962] 2274/5). – [24] R.H. Herber (TID-16124 [1962]; N.S.A. **16** [1962] No. 27037). – [25] H. Keller (Diss. München Univ. 1962, pp. 1/97).

[26] H.P. Fritz, H. Keller (Chem. Ber. **95** [1962] 820). – [27] H.P. Fritz, H. Keller (Chem. Ber. **95** [1962] 158/73). – [28] H.P. Fritz, E.F. Paulus (Z. Naturforsch. **18b** [1963] 435/8). – [29] J.E. Mahler, D.A.K. Jones, R. Pettit (J. Am. Chem. Soc. **86** [1964] 3589/90). – [30] R.H. Herber, R.B. King, G.K. Wertheim (Inorg. Chem. **3** [1964] 101/7).

[31] R.T. Bailey, E.R. Lippincott, D. Steele (J. Am. Chem. Soc. **87** [1965] 5346/50). – [32] C.E. Keller, G.F. Emerson, R. Pettit (J. Am. Chem. Soc. **87** [1965] 1388/90). – [33] S.F.A. Kettle (Inorg. Chem. **4** [1965] 1661/3). – [34] R.B. King (Organometal. Syn. **1** [1965] 126/7). – [35] F.A. Cotton (Proc. Robert A. Welch Found. Conf. Chem. Res. **11** [1966] 213/4 from [75]).

[36] F.A. Cotton, A. Davison, J.W. Faller (J. Am. Chem. Soc. **88** [1966] 4507/9). – [37] B.J. Nicholson (J. Am. Chem. Soc. **88** [1966] 5156/65). – [38] W. McFarlane, G. Wilkinson (Inorg. Syn. **8** [1966] 184/5). – [39] C.E. Keller (Diss. Univ. Texas 1966; Diss. Abstr. B **27** [1967] 2651). – [40] S.F.A. Kettle (J. Chem. Soc. A **1966** 420/2).

[41] C.G. Kreiter, A. Maasböl, F.A.L. Anet, H.D. Kaesz, S. Winstein (J. Am. Chem. Soc. **88** [1966] 3444/5). – [42] C.E. Keller, B.A. Shoulders, R. Pettit (J. Am. Chem. Soc. **88** [1966] 4760/1). – [43] J. Faller (Diss. Massachusetts Inst. Technol. 1966 from [44]). – [44] E. Fluck (in: V. Goldanskii, R.H. Herber, Chemical Applications of Mössbauer Spectroscopy: ^{57}Fe: Metal Organic Compounds, Academic, New York 1964, pp. 268/313, 306). – [45] F.A. Cotton, A. Davison, A. Musco (J. Am. Chem. Soc. **89** [1967] 6796/7).

[46] W.K. Bratton, F.A. Cotton, A. Davison, A. Musco, J.W. Faller (Proc. Natl. Acad. Sci. U.S. **58** [1967] 1324/8). – [47] F.A.L. Anet, H.D. Kaesz, A. Maasböl, S. Winstein (J. Am. Chem. Soc. **89** [1967] 2489/91). – [48] R.H. Herber (Symp. Faraday Soc. No. 1 [1967] 86/96). – [49] R. Grubbs, R. Breslow, R. Herber, S.J. Lippard (unpublished results from [50]). – [50] S.J. Lippard (Trans. N.Y. Acad. Sci. **29** [1967] 917/9).

[51] F.A. Cotton, W.T. Edwards (J. Am. Chem. Soc. **90** [1968] 5412/7). – [52] F.A. Cotton (J. Am. Chem. Soc. **90** [1968] 6230/2). – [53] R.H. Grubbs (Diss. Columbia Univ. 1968; Diss. Abstr. Intern. B **30** [1969] 560). – [54] R.H. Herber (Chem. Anal. [N.Y.] **26** [1969] 315/40). – [55] F. Farraone, F. Zingales, P. Uguagliatti, U. Belluco (Inorg. Chem. **7** [1968] 2362/4).

[56] F.A. Cotton, R. Eiss (J. Am. Chem. Soc. **91** [1969] 6593/7). – [57] S. Chandra, R.H. Herber (Mössbauer Eff. Methodol. **5** [1969] 45/64; C.A. **73** [1970] No. 82375). – [58] T. Jenny, W. von Philipsborn, J. Kronenbitter, A. Schwenk (J. Organometal. Chem. **205** [1981] 211/22). – [59] R.B. King (J. Am. Chem. Soc. **91** [1969] 7217/23). – [60] R.B. King, L.M. Epstein, E.W. Gowling (J. Inorg. Nucl. Chem. **32** [1970] 441/5).

[61] R.H. Herber (Introd. Mössbauer Spectrosc. **1971** 138/54). – [62] A.J. Campbell, C.A. Fyfe, E. Maslowsky (Chem. Commun. **1971** 1032). – [63] D.J. Ehntholt, R.C. Kerber (J. Organometal. Chem. **38** [1972] 139/45). – [64] A.J. Campbell, C.A. Fyfe, E. Maslowsky (J. Am. Chem. Soc. **94** [1972] 2690/2). – [65] M. Brookhart, E.R. Davis, D.L. Harris (J. Am. Chem. Soc. **94** [1972] 7853/8).

[66] G. Rigatti, G. Boccalon, A. Ceccon, G. Giacometti (J. Chem. Soc. Chem. Commun. **1972** 1165/6). – [67] A.E. Crease, P. Legzdins (J. Chem. Soc. Dalton Trans. **1973** 1501/7). – [68] V.O. Zavel'skii, E.I. Fedin (Zh. Strukt. Khim. **14** [1973] 376/7; J. Struct. Chem. [USSR] **14** [1973] 340/1). – [69] V.O. Zavel'skii, E.I. Fedin (Zh. Strukt. Khim. **14** [1973] 375/6; J. Struct. Chem. [USSR] **14** [1973] 338/9). – [70] R.C. Kerber, D.J. Ehntholt (J. Am. Chem. Soc. **95** [1973] 2927/34).

[71] A.J. Campbell (Diss. Guelph Univ., Can., 1974; Diss. Abstr. Intern. B **35** [1974] 2058). – [72] R.H. Herber, R.B. King, M.N. Ackermann (J. Am. Chem. Soc. **96** [1974] 5437/41). – [73] B.F.G. Johnson, J. Lewis, M.V. Twigg (J. Chem. Soc. Dalton Trans. **1974** 2546/9). – [74] M. Elian, R. Hoffmann (Inorg. Chem. **14** [1975] 1058/76). – [75] F.A. Cotton (J. Organometal. Chem. **100** [1975] 29/41).

[76] G. Wilkinson, R. Burton (unpublished results from [75]). – [77] H. Lumbroso, D.M. Bertin (J. Organometal. Chem. **108** [1976] 111/23). – [78] A.J. Campbell, C.E. Cottrell, C.A. Fyfe, K.R. Jeffrey (Inorg. Chem. **15** [1976] 1321/5). – [79] R. Aumann, J. Knecht (Chem. Ber. **109** [1976] 174/9). – [80] F.A. Cotton, D.L. Hunter (J. Am. Chem. Soc. **98** [1976] 1413/7).

[81] R.B. King (Israel J. Chem. **15** [1976/77] 181/8). – [82] D. Cunningham, P. McArdle, H. Sherlock, B.F.G. Johnson, J. Lewis (J. Chem. Soc. Dalton Trans. **1977** 2340/4). – [83] F.A. Cotton, B.E. Hanson (Israel J. Chem. **15** [1977] 165/73). – [84] T.A. Albright, P. Hofmann, R. Hoffmann (J. Am. Chem. Soc. **99** [1977] 7546/57). – [85] D.M.P. Mingos (J. Chem. Soc. Dalton Trans. **1977** 31/7).

[86] D.M.P. Mingos (J. Chem. Soc. Dalton Trans. **1977** 26/30). – [87] D.M.P. Mingos (J. Chem. Soc. Dalton Trans. **1977** 20/5). – [88] G.A. Olah, G. Liang, S. Yu (J. Org. Chem. **42** [1977] 4262/3). – [89] J. Aihara (Bull. Chem. Soc. Japan **51** [1978] 1541/2). – [90] K.D. Karlin, B.F.G. Johnson, J. Lewis (J. Organometal. Chem. **160** [1978] C21/C23).

[91] J.A. Connor, C.P. Demain, H.A. Skinner, M.T. Zafarani-Moattar (J. Organometal. Chem. **170** [1979] 117/30). – [92] B.F.G. Johnson, K.D. Karlin, J. Lewis (J. Organometal. Chem. **174** [1979] C29/C31). – [93] H. Behrens, M. Moll, W. Popp, H.-J. Seibold, E. Sepp, P. Würstl (J. Organometal. Chem. **192** [1980] 389/98). – [94] M.C. Böhm, R. Gleiter (Z. Naturforsch. **35b** [1980] 1028/30). – [95] W.C. Herndon (J. Am. Chem. Soc. **102** [1980] 1538/41).

[96] W.C. Herndon (Israel J. Chem. **20** [1980] 276/80). – [97] J. Weichmann (Dipl.-Arbeit Univ. Regensburg 1981, pp. 1/111). – [98] J.C. Green, P. Powell, J. van Tilborg (J. Chem. Soc. Dalton Trans. **1976** 1974/6). – [99] H. Behrens, M. Moll (Z. Anorg. Allgem. Chem. **416** [1975] 193/202). – [100] R.D. Fischer (Diss. München Univ. 1961).

[101] R.L. Collins, R. Pettit (J. Am. Chem. Soc. **85** [1963] 2332/3). – [102] R. Grubbs, R. Breslow, R. Herber, S.J. Lippard (J. Am. Chem. Soc. **89** [1967] 6864/70). – [103] M. Brookhart, G.O. Nelson (J. Organometal. Chem. **164** [1979] 193/202). – [104] A.C. Barefoot, E.W. Corcoran, R.P. Hughes, D.M. Lemal, W.D. Saunders, B.B. Laird, R.E. Davis (J. Am. Chem. Soc. **103** [1981] 970/2). – [105] R.B. Hitam, R. Narayanaswamy, A.J. Rest (J. Chem. Soc. Dalton Trans. **1983** 1351/6).

1.4.1.4.1.2.5.3.1.3 Chemical Behavior and Uses of $C_8H_8Fe(CO)_3$

Solubility and Thermal Decomposition. $C_8H_8Fe(CO)_3$ forms quite [3] air-stable crystals [1 to 3, 22, 23, 27], decomposing slowly on standing in air [15] and is stable to moisture [1]. In the liquid state $C_8H_8Fe(CO)_3$ undergoes decomposition in air at about 165 °C and at >200 °C under N_2 in a sealed tube [3]. $C_8H_8Fe(CO)_3$ is soluble in most organic solvents [1 to 3, 8, 27] including pyridine [2] and glacial acetic acid [2, 8, 10] to give dark red-brown [22] or red [23] solutions. The solutions are quite stable to air at room temperature [3, 10, 22, 23] being 80% recovered after 20 h UV irradiation in a stream of O_2 [8]. It is insoluble in H_2O [8, 27], dilute acids or bases [8] but dissolves in conc. H_2SO_4 and can be recovered unchanged on neutralization [2]; but decomposition occurs when these solutions are heated to 110 °C [2].

Mass Spectra. The mass spectra show the parent ion [34, 40, 42, 59, 60, 105], and the most common fragmentation is loss of the CO groups, followed by loss of C_2H_2 groups

 References on pp. 250/3

from the Fe hydrocarbon ion [40]. The following ions are found: $[C_8H_8Fe(CO)_n]^+$ (n=0 to 3), $[C_nH_nFe]^+$ (n=2 to 6), $[C_2HFe]^+$, $[Fe(CO)_n]^+$ (n=0 to 2), $[C_nH_n]^+$ (n=3 to 8), $[C_nH_{n-1}Fe]^+$ (n=4, 6), $[C_nH_{n-2}Fe]^+$ (n=4, 5), $[Fe]^+$ [40]. Similar ions are cited in [34, 42, 59]. Electron impact studies [40, 88, 104] show $[C_8H_8Fe(CO)_n]^{\dot{+}}$ (n=0 to 3), $[C_6H_6Fe]^{\dot{+}}$, $[C_8H_8]^{\dot{+}}$ and $[Fe]^{\dot{+}}$ as the main ions [104]. For the results of field desorption mass spectrometry of $C_8H_8Fe(CO)_3$ see [105]. The chemical ionization spectra [54, 94] with H_2 shows $[M+H-nCO]^+$, $[M-H-nCO]^+$, $[M-nCO]^{\dot{+}}$ (n=0 to 3], $[C_8H_8+H]^+$ [94]; with NH_3 the ions $[M+H-nCO]^+$ (n=0 to 3) are observed [94]. The methane chemical ionization spectrum [54, 94] shows $[M+H-nCO]^+$ (n=0 to 3), $[M-H-nCO]^+$ (n=0, 2, 3), $[M-nCO]^{\dot{+}}$ (n=0 to 3) and $[C_8H_8+H]^+$ [94]; the i-C_4H_{10} chemical ionization spectrum shows $[M+H-nCO]^+$, $[M-H-nCO]^+$, and $[M-nCO]^{\dot{+}}$ (n=0, 1) [94].

Photochemical Reactions. UV irradiation of $C_8H_8Fe(CO)_3$ in n-hexane (O_2-free) for 8 min gives $C_8H_8Fe_2(CO)_5$ (see 2.8.2.2 in "Organoiron Compounds" C5, 1981, Table 7, No. 1, p. 114) and only traces of $C_8H_8Fe_2(CO)_6$ (see 2.8.1.1 in "Organoiron Compounds" C5, 1981, Table 4, No. 25, p. 49). Similar irradiation for less than 3 h gives $C_8H_8Fe_2(CO)_5$, C_8H_8, and CO. The yield of $C_8H_8Fe_2(CO)_5$ (>65%) remains essentially unchanged regardless of reaction temperature (−30 to +60 °C) [52]. Irradiation of $C_8H_8Fe(CO)_3$ in tetrahydrofuran at −78 °C permits cleavage of CO; discontinuation of UV irradiation followed by addition of RC≡CR (R=CO_2CH_3, C_6H_5) gives I [62, 64] in ca. 20% yield [64]. Irradiation of $C_8H_8Fe(CO)_3$ in CH_3CO_2H gives a mixture of olefins [83]. Fourier-transform IR spectroscopic evidence, including ^{13}CO labelling and energy-factored CO force-field fitting, is presented to show that photolysis of "tub"-$C_8H_8Fe(CO)_3$ (see Formula III, p. 234) in a CH_4 matrix at 12 K leads to the formation of a "chair"-$C_8H_8Fe(CO)_3$ species (see Formula IV, p. 236) via the $^2LFe(CO)_3$ intermediate ($^2L=\eta^2$-C_8H_8) [113].

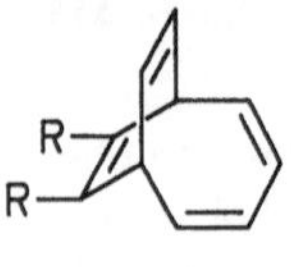

I 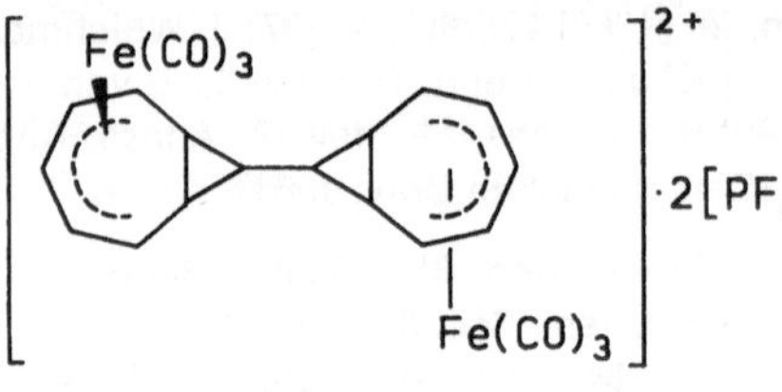 II 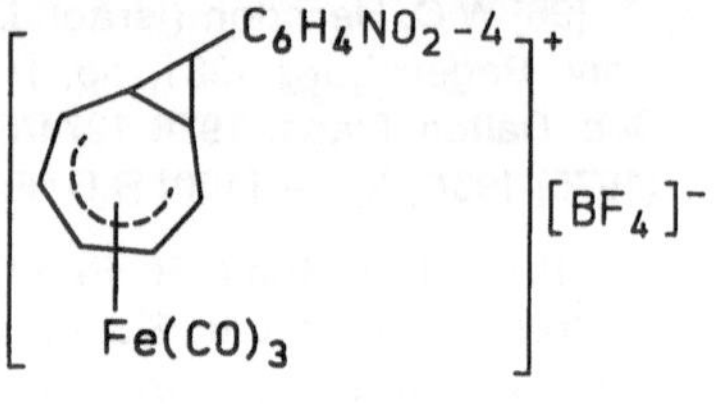 III

Electrochemical Behavior. Cyclic voltammetry demonstrates that $C_8H_8Fe(CO)_3$ undergoes irreversible one-electron oxidation at a Pt electrode [71, 84, 96, 97, 110] in CH_2Cl_2 [96, 97, 110] or CH_3CN [97] at $E_p=0.88$ V [96, 97] (calomel electrode (1 M LiCl) as reference, 0.1 M $[N(C_2H_5)_4]Cl$ as supporting electrolyte [97]). The electrochemical oxidation to $[C_8H_8Fe(CO)_3]^+$ is irreversible even at a scan rate of 200 V·s^{-1} [97]. Although reduction of $C_8H_8Fe(CO)_3$ is not observed down to −1.5 V in CH_2Cl_2 [97], $C_8H_8Fe(CO)_3$ undergoes two one-electron reductions in $(CH_2OCH_3)_2$ [28] or dimethylformamide [72, 93, 103] to **$[C_8H_8Fe(CO)_3]^n$** (n=1−, 2−) [28, 72, 97]. Polarographic reduction in $(CH_2OCH_3)_2$ occurs at $E_{1/2}$= −2.0 V in the first step and $E_{1/2}$=−2.5 V in the second step (10^{-3} M Ag^+/Ag as reference, 10^{-1} M $[N(C_4H_9)_4]ClO_4$ as supporting electrolyte at 22 °C). At 270 K the radical anion in $(CH_2OCH_3)_2$ shows in the ESR spectrum g=2.0079 with ΔH=5.40 G [28, 38]. In dimethylformamide in the presence of proton donors such as $[HN(CH_3)_3]Br$, $E_{1/2}$=−1.13 V. Cyclic voltammetry shows that the reaction of H^+ with $[C_8H_8Fe(CO)_3]^{2-}$ is very rapid. The disappearance of $[C_8H_8Fe(CO)_3]^{\dot{-}}$, which takes place through $[C_8H_8Fe(CO)_3]^{2-}$, is much slower, since the concentration of $[C_8H_8Fe(CO)_3]^{2-}$ in a solution of $[C_8H_8Fe(CO)_3]^{\dot{-}}$ is very small: the dismutation constant K of the radical anion is $K=[C_8H_8Fe(CO)_3]\cdot[C_8H_8Fe(CO)_3^{2-}]/[C_8H_8Fe(CO)_3^{\dot{-}}]^2=$

1.136×10^{-8} so that $[C_8H_8Fe(CO)_3^{2-}]\cong10^{-4}\cdot[C_8H_8Fe(CO)_3^{\dot{-}}]$ [72]. The reduction in dimethylformamide proceeds via two well-separated diffusion-controlled, one-electron, chemically reversible steps with $E_p^1=-1.24$ and $E_p^2=-1.71$ V [28, 72, 93]. Phase-selective alternative current polarography at 298 K in dimethylformamide (0.1 M $[N(C_4H_9)_4]PF_6$, mercury pool at -1.5 V) yields the heterogeneous electron-transfer rate $k_s=0.24$ and $0.15\ cm\cdot s^{-1}$ for the processes $[C_8H_8Fe(CO)_3]^{0/1-}$ and $[C_8H_8Fe(CO)_3]^{1-/2-}$, respectively [93]. The frozen solution of $[C_8H_8Fe(CO)_3]^-$ in dimethylformamide at 163 K shows anisotropy with $g_1=2.016$, $g_2=2.009$, $g_3=1.980$. Extended Hückel calculations show that the odd electron in $[C_8H_8$-$Fe(CO)_3]^-$ resides in the C_8H_8 portion of the molecule [103]. $[C_8H_8Fe(CO)_3]^{2-}$ is said to be a source of other polycyclic hydrocarbon complexes [76, 86] (see below: reaction with potassium). After electrolysis of the solution of $C_8H_8Fe(CO)_3$ in dimethylformamide at -1.3 V an orange solution is obtained which on treatment with H_2O followed by extraction with ether, drying the organic layer ($MgSO_4$), and chromatography on alumina with n-hexane gives $C_8H_{10}Fe(CO)_3$ (C_8H_{10}=cyclooctatriene) in 100% yield. This illustrates the advantage of the electrochemical preparation of $C_8H_{10}Fe(CO)_3$ in a series where the starting materials are costly [72].

Reaction with Elements. No reaction occurs with H_2 in glacial acetic acid or C_2H_5OH [3], with H_2/Pt in glacial acetic acid [3, 8] within 3 h (no H_2 absorption observed, although subsequent reaction takes place with release of CO) [8], with H_2/Pt [3, 7, 10], or Pd, or prereduced Adams catalyst, or H_2/PtO_2 in glacial acetic acid, or C_2H_5OH [3], but with H_2 (1 atm)/Raney Ni in C_2H_5OH at room temperature formation of $C_8H_{12}Fe(CO)_3$ (C_8H_{12}=cycloocta-1,3-diene) within 1.5 h is observed [9]. UV irradiation of a hexane solution of $C_8H_8Fe(CO)_3$ for 5 h with O_2 bubbling continuously through the solution gives C_8H_8 and iron oxides [52]. Contrary to a report that $C_8H_8Fe(CO)_3$ fails to decolorize Br_2 [10], oxidation occurs [1, 2, 7] in CCl_4 [1] at room temperature [2] to give brominated products of C_8H_8 [2], CO [1], and $FeBr_2$ [2]. Oxidation occurs with I_2 [7] yielding decomposition products in toluene [2]; treatment with I_2/KI in H_2O liberates CO [2]. No $C_8H_{12}Fe(CO)_3$ is formed with Na in liquid NH_3 [7, 10]. Reaction with a slight excess of K in tetrahydrofuran at room temperature gives $[C_8H_8Fe(CO)_3]^{\dot{-}}$ in equilibrium with small amounts of $[C_8H_8Fe(CO)_3]^{2-}$ (IR spectrum: three very large ν(CO) bands at 1950 to 1850 cm^{-1}) and $C_8H_8Fe(CO)_3$. Protonation takes place on $[C_8H_8Fe(CO)_3]^{2-}$ to give equal quantities of $C_8H_{10}Fe(CO)_3$ and $C_8H_8Fe(CO)_3$; the latter is produced by a dismutation [72].

Oxidation and Reduction. $C_8H_8Fe(CO)_3$ is oxidized only by strong electrophiles such as $Tl(NO_3)_3$ [99]. Oxidation with H_2O_2 [8], $[NO]PF_6$ in CH_3CN at 0 °C [49], $Cu(NO_3)_2$ in $O(COCH_3)_2$ at 20 °C [49], $AgNO_3$ [7] gives no derivatives, but the major part of $C_8H_8Fe(CO)_3$ decomposes. $AgBF_4$ does not react with $C_8H_8Fe(CO)_3$ [71, 110], but $AgPF_6$ gives $[C_8H_8Fe(CO)_3]PF_6$ [71, 110], identified as II (p. 242, see 2.9.1.3 in "Organoiron Compounds" C5, 1981, p. 132) [84]. $HgSO_4$ gives C_8H_8 (76%) and $C_6H_5CH_2CHO$ [2], and oxidation also occurs with $Hg(O_2CCH_3)_2$ and $FeCl_3$ [7]. $KMnO_4$ causes decomposition [8]. No derivatives are observed with conc. H_2SO_4/$C_2H_5ONO_2$, conc. H_2SO_4/fuming HNO_3 at 0 to 20 °C, conc. $HNO_3/O(COCH_3)_2$ at -20 °C [49], or with perbenzoic acid [8]; but most $C_8H_8Fe(CO)_3$ decomposes under these conditions [8, 49]. Oxidation with ONR_3 liberates C_8H_8 as shown in the following table [66]:

solvent	amine oxide	reaction time (h)	yield C_8H_8 (%)
C_6H_6	$ON(CH_3)_3$	1	78
tetrahydrofuran	$ON(CH_3)_3$	1	80
acetone	$ON(CH_3)_3$	1	68
C_6H_6	$ON(CH_3)_2C_6H_5$	1.5	40

References on pp. 250/3

One-electron oxidation with $[N(C_6H_4Br\text{-}4)_3]PF_6$ in CH_2Cl_2 gives II (see p. 242) [84, 97, 112]; and with $[4\text{-}O_2NC_6H_4N_2]BF_4$ in acetone at −20 °C, salt III (see p. 242) is obtained after 15 min [29, 85].

No reaction is observed with B_2H_6 and H_2O_2/NaOH for 16 h at room temperature [50], but reduction with B_2H_6 in tetrahydrofuran gives $C_8H_8(Fe(CO)_3)_2$ [87]. $LiAlH_4$ does not give $C_8H_{12}Fe(CO)_3$ [7, 10].

Fe(CO)₃Cl

IV

Fe(CO)₃

V

Reaction with Inorganic Acids. Strong acids attack $C_8H_8Fe(CO)_3$ [23, 51, 99], cf. [11, 45, 57, 109]. The proton enters from the exo side of C_8H_8 [51]. Thus, dry HCl gas in ether at 0 °C gives IV [73], and conc. HCl gives V [10]. Reaction with conc. H_2SO_4 gives $[C_8H_9Fe(CO)_3]^+$ [27, 109] isolated as the BF_4^- or $PtCl_6^-$ salts [109] and shown by reaction with conc. D_2SO_4 to have the bicyclic structure V [109]. Reaction of $C_8D_8Fe(CO)_3$ with D_2SO_4 in ether followed by addition of $[NH_4]PF_6$ gives the perdeuterated PF_6 salt of V [111]. Analogous reaction with 40% HBF_4 in $H_2O/O(COCH_3)_2$ [11], with HBF_4 in ether [10], with 70 to 80% aqueous $HClO_4$ [10] or with $HSbCl_6$ in CH_3NO_2 [21] gives $[C_8H_9Fe(CO)_3]X$, originally proposed to have structure VI (R=H) [10] but recognized later as V [11, 12, 20, 56, 109]. Chromatographically purified $C_8H_8Fe(CO)_3$ and excess $HOSO_2F$ in liquid SO_2 at −50 °C give V, isolated as BF_4^- salt [77]. Protonation with $HOSO_2F/SO_2F_2$ (1:3) [45, 56, 60, 75] or with $HOSO_2F/SO_2ClF$ [79] at −120 °C gives VI (R=H) and at −60 °C V [45, 56, 60, 75, 79].

H R X⁻ Fe(CO)₃

VI

OH Fe(CO)₃

VII

Reactions with Inorganic Lewis Acids. In the reaction with $(C_2H_5)_2OBF_3$ in C_6H_6 a yellow crystalline complex precipitates out. But this material is rapidly hydrolyzed to $C_8H_8Fe(CO)_3$ and cannot be identified. Similar reaction of $C_8H_8Fe(CO)_3$ in $CHCl_3$ with $(C_2H_5)_2OBF_3$ in acetic acid followed by hydrolysis gives an isolable product, probably VII [69]. Mixing of $C_8H_8Fe(CO)_3$ with $AlCl_3$ gives a gum formulated as VIII [69]. Reaction with AlX_3 (X=Cl [61, 69, 82], Br [69, 82]) in C_6H_6 at 10 °C [61, 69] for 3.5 h [69] or in CH_2Cl_2 at room temperature for 2 d [82] gives IX in 35% yield [82]. Reaction with a large excess of $AlBr_3$ in CH_2Cl_2 gives X [80, 91]. Reaction with AlX_3 (X=Cl [61, 69, 80, 82, 99], Br [82]) in a CO atmosphere (1 atm [82]) [61, 69, 80, 82, 99] in CH_2Cl_2 at room temperature for 2 d [82] gives a ketone complex [61], X in 65% yield [82] or XI [99], cf. [69, 80]. No reaction is observed with $Al(C_2H_5)_3$ or $SnCl_4$ [69]. Reaction with the catalyst $WCl_6/C_2H_5OH/C_2H_5AlCl_2$ (molar ratio 1.75:3.4:3.5 [102]) [76, 102] in toluene [76, 102] or C_6H_6 [102] in a sealed tube at room temperature for 24 h, followed by addition of more premixed catalyst and further reaction for 17 h gives several products [76, 102]. Chromatography gives XII (see 2.8.1.1 in "Organoiron Compounds" C5, 1981, Table 4, No. 33, p. 50) [76, 102], $W_2Cl_6(OC_2H_5)_6$, $C_8H_{10}Fe(CO)_3$, trans-$C_8H_8(Fe(CO)_3)_2$, coeluting with an unidentified oil (^1H NMR spectrum (acetone-d_6): δ=2.8(m), 3.2(m), 5.5(m), 6.05(d), 6.22(d) ppm; IR spectrum: 1975, 2040 (νCO) cm^{-1}) and other iron carbonyl complexes which cannot be characterized sufficiently for identification. Formation of XII is also observed using Al:W=1.4:1; high Al:W ratios do not appear to favor its formation. The formation of several other compounds is generally favored when using a high O:W ratio in the catalyst [102].

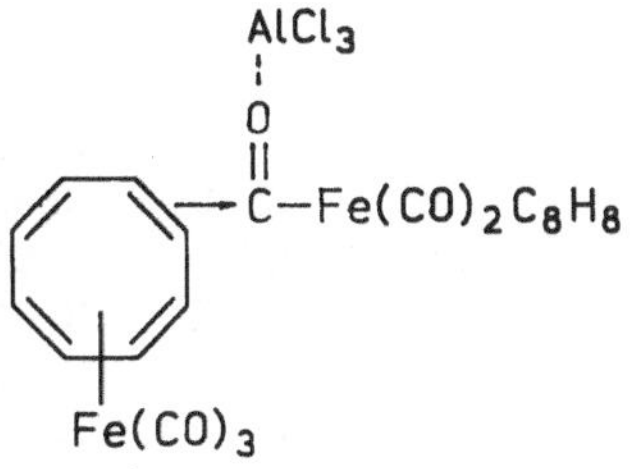

VIII 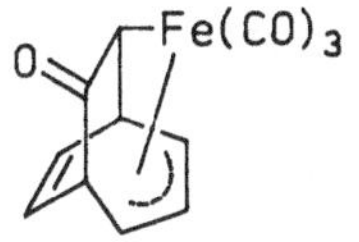 IX 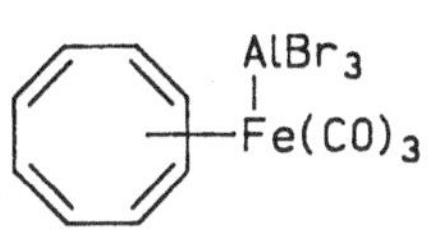 X

No reaction is observed with WCl_6/LiC_4H_9-n or $Mo(NO)_2(P(C_6H_5)_3)_2Cl_2/C_2H_5AlCl_2$ under various conditions [102]. Reaction with $ONCSO_2Cl$ in CH_2Cl_2 at 0 °C followed by dechlorosulfination with C_6H_5SH/pyridine in cold acetone gives XIII [58].

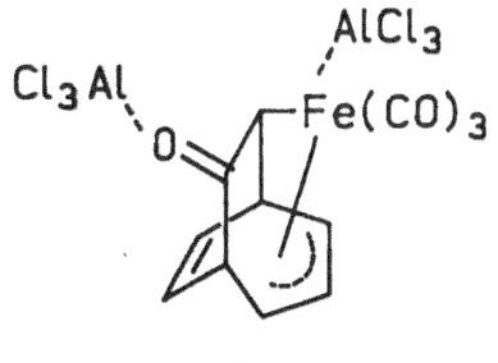

XI 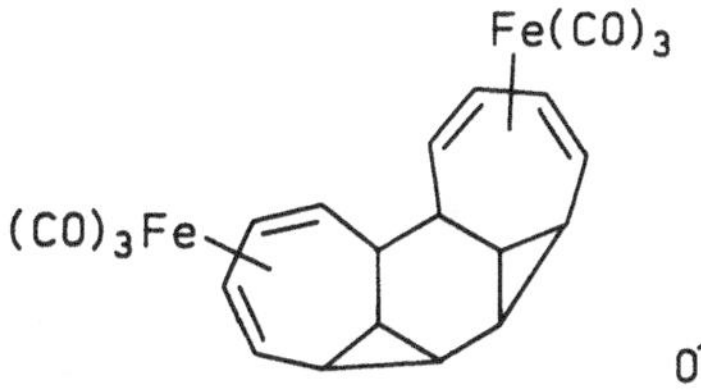 XII 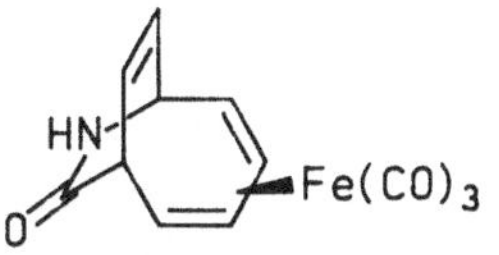 XIII

Reaction with MR_3 (M=P, As, Sb), CO, and CNR. Irradiation of $C_8H_8Fe(CO)_3$ with a large excess of $P(OCH_3)_3$ in tetrahydrofuran for 48 h gives $C_8H_8Fe(CO)_2P(OCH_3)_3$ and $C_8H_8Fe(P(OCH_3)_3)_2CO$ [97]. Similar irradiation with $P(OC_6H_5)_3$ in toluene gives $C_8H_8Fe(CO)_2P(C_6H_5)_3$ [97]. Refluxing with $P(OCH_2)_3CCH_3$ in hexane gives $C_8H_8Fe(CO)_2P(OCH_2)_3CCH_3$ [39]. Refluxing with $P(N(CH_3)_2)_3$ in ethylcyclohexane for 15 h gives $(CO)_3Fe(P(N(CH_3)_2)_3)_2$ [19], but photochemical reaction gives $C_8H_8Fe(CO)_2P(N(CH_3)_2)_3$ [97]. Photochemical reaction of $P(C_6H_5)_3$ with $C_8H_8Fe(CO)_3$ in n-hexane for 22 d gives $C_8H_8Fe(CO)_2P(C_6H_5)_3$ [97]. Irradiation of $C_8H_8Fe(CO)_3$ with ^2D=XIV in hexane gives trans-$(CO)_3Fe^2D_2$ in 70% yield whereas the thermal reaction in boiling methylcyclohexane gives the same product in 78% yield within 5 h [36]. Thermal reaction of $C_8H_8Fe(CO)_3$ with $^2D=P(C_4H_9\text{-}n)_3$ in decalin or methylcyclohexane [37], with $^2D=P(C_6H_5)_3$ in C_6H_6 [7], methylcyclohexane [37], decalin [37, 63], with $^2D=P(C_6H_5)_2C_2H_5$ in decalin [37], or with $^2D=P(C_6H_5)_2CH_2C_6H_5$ in 1,2-dimethylcyclohexane [92] gives compounds of the type trans-$(CO)_3Fe^2D_2$, see 1.1.4.4.3.1 in "Eisen-Organische Verbindungen" B1, 1976, p. 148.

Refluxing $C_8H_8Fe(CO)_3$ in ethylcyclohexane with $^2D=P(C_6H_5)_3$ for 9 h gives trans-$(CO)_3$-Fe^2D_2 and an unstable red oil, probably $C_8H_8Fe(CO)_2P(C_6H_5)_3$ [8]. The pseudo-first order rate constants for some of these thermal reactions $C_8H_8Fe(CO)_3+2\,^2D \rightarrow$ trans-$(CO)_3$-$Fe^2D_2+C_8H_8$ at different temperatures in decalin are shown in the following table [37]:

References on pp. 250/3

^2D	$[^2D] \cdot 10^2$ (M)	t in °C	$k_{obs} \cdot 10^4$ (s^{-1})	^2D	$[^2D] \cdot 10^2$ (M)	t in °C	$k_{obs} \cdot 10^4$ (s^{-1})
$P(C_4H_9\text{-}n)_3$	3.87	90	1.38	$P(C_6H_5)_3$	3.23	127.2	1.00
	7.99	90	3.34		8.36	127.2	2.43
	10.29	90	4.53		13.49	127.2	3.85
	13.59	90	6.13		19.17	127.2	5.48
	16.56	90	7.28		23.70	127.2	6.86
	21.77	90	9.84	$P(C_6H_5)_2C_2H_5$	4.62	115	4.06
	3.21	105	2.95		7.64	115	5.36
	5.49	105	4.68		12.67	115	9.76
	7.76	105	7.29		15.73	115	11.83
	9.65	105	9.17		2.61	127.2	3.90
	14.36	105	13.70		4.03	127.2	6.62
	2.86	110	2.76		6.40	127.2	10.81
	4.82	110	5.02		7.71	127.2	13.47
	5.90	110	6.10		8.67	127.2	15.16
	7.83	110	8.67				
	10.59	110	11.24				
	14.96	110	16.20				

The pseudo-first order rate constants for the thermal reaction with $^2D = P(C_4H_9\text{-}n)_3$ in methylcyclohexane are shown in the following table [37]:

$[^2D] \cdot 10^2$ (M)	t in °C	$k_{obs} \cdot 10^4$ (s^{-1})
4.41	90	1.75
10.51	90	4.25
16.81	90	6.79
26.54	90	10.80
31.37	90	12.54

The data show that k_{obs} values depend both on the nature and on the concentration of ^2D and fit the rate law $k_{obs} = k_2 [^2D]$. The second order rate constants, k_2, are shown in the following table [37]:

^2D	t in °C	$k_2 \cdot 10^3$ ($M^{-1} \cdot s^{-1}$)	solvent
$P(C_4H_9\text{-}n)$	90	4.69	decalin
	90	4.03	methylcyclohexane
	105	9.55	decalin
	110	11.10	decalin
$P(C_6H_5)_3$	127.2	2.85	decalin
$P(C_6H_5)_2C_2H_5$	115	7.42	decalin
	127.2	18.55	

The activation parameters for the reaction with $P(C_4H_9\text{-}n)_3$ in decalin and methylcyclohexane are $\Delta H^{*} = 11.41$ kcal/mol, $\Delta S^{*} = -38.16$ cal $\cdot K^{-1} \cdot mol^{-1}$. The highly negative activation entropy suggests a bimolecular attack of the entering ligand as rate-determining step [37].

References on pp. 250/3

Thermal reaction of diphosphines ^4D gives compounds of the type $(CO)_3Fe^4D$ (see 1.1.4.4.3.2.2 in "Eisen-Organische Verbindungen" B1, 1976, p. 170). Thus, refluxing $^4D=(NCCH_2CH_2)_2P(CH_2)_2P(CH_2CH_2CN)_2$ in C_6H_6 for 4 h gives $(CO)_3Fe^4D$ [14]. Similar reaction of $^4D=(C_6H_5)_2P(CH_2)_2P(C_6H_5)_2$ in boiling C_6H_6 for 3 h gives $(CO)_3Fe^4D$ [14]. Refluxing in toluene for 6 h gives $(CO)_3Fe^4D$ (72% yield) and small amounts of $(C_6H_5)_2P(CH_2)_2(C_6H_5)_2$-$P(Fe(CO)_3P(C_6H_5)_2(CH_2)_2P(C_6H_5)_2)_2$ [17]. Reaction in a sealed tube at 150 °C for 24 h gives $Fe(CO)_5$ and impure $(CO)_3Fe^4D$ [14]. The pseudo-first order rate constant, k_{obs}, and the second order rate constant fitting the rate law $k_{obs}=k_2[^4D]$ for the reaction $C_8H_8Fe(CO)_3+(C_6H_5)_2P(CH_2)_2P(C_6H_5)_2$ (^4D) = cis-$(CO)_3Fe^4D+C_8H_8$ in decalin are shown in the following table at different temperatures:

$[^4D]\cdot 10^2$ (M)	t in °C	$k_{obs}\cdot 10^4$ (s^{-1})	$k_2\cdot 10^3$ ($M^{-1}\cdot s^{-1}$)
1.35	110	1.28	13.10
1.26	117.3	1.92	20.11
1.32	127.2	4.25	30.79

The activation parameters for this reaction are $\Delta H^{\neq}=14.28$ kcal/mol and $\Delta S^{\neq}=-30.34$ cal·$K^{-1}\cdot mol^{-1}$ [37].

Refluxing $C_8H_8Fe(CO)_3$ with $^4D=(C_6H_5)_2P(CH_2)_3P(C_6H_5)_3$ in toluene for 14 h or $(CH_3)_2PCF_2CH_2P(CH_3)_2$ in the dark for 6 h [67], or refluxing $C_8H_8Fe(CO)_3$ with $P(CH_2CH_2P(C_6H_5)_2)_2C_6H_5$ in C_6H_6 for 21 h [44] or heating with XV in 1,2-dimethylcyclohexane at 100 °C for 17 h [92] gives $(CO)_3Fe^4D$. Refluxing $^6D=CH_3C(CH_2P(C_6H_5)_2)_3$ with $C_8H_8Fe(CO)_3$ in cyclohexane for 48 h gives $(CO)_2Fe^6D$, C_8H_8, and CO. Similar reaction in a sealed tube at 100 °C for 48 h gives $(CO)_3Fe(P(C_6H_5)_2CH_2)_3CCH_3$, isolated as $[(CO)_3$-$Fe(P(C_6H_5)_2CH_2)_2C(CH_3)CH_2P(C_6H_5)_2CH_3]I$ [48].

Refluxing $C_8H_8Fe(CO)_3$ with $^2D=M(C_6H_5)_3$ with M = As, Sb in ethylcyclohexane for 15 h gives $C_8H_8Fe(CO)_2M(C_6H_5)_3$ [8]. Refluxing $^4D=(C_6H_5)_2As(CH_2)_2As(C_6H_5)_2$ with $C_8H_8Fe(CO)_3$ in C_6H_6 for 5 h gives $(CO)_3Fe^4D$. Heating these components in C_6H_6 in a sealed tube at 110 to 120 °C for 40 h gives impure $(CO)_3Fe^4D$ [14]. Reaction with CO (high pressure) at 100 °C gives $Fe(CO)_5$ [7], and reaction with CNR gives $(CO)_3Fe(CNR)_2$ [74].

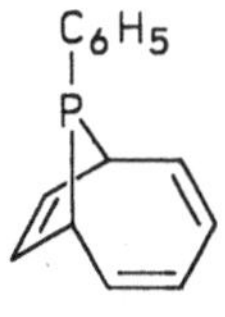

C_6H_5 P

XIV

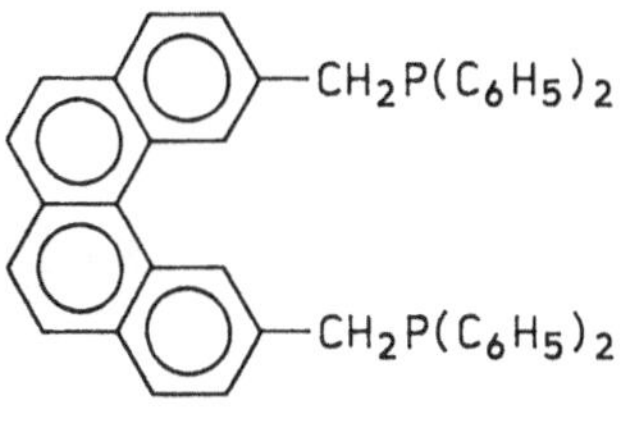

$CH_2P(C_6H_5)_2$ $CH_2P(C_6H_5)_2$

XV

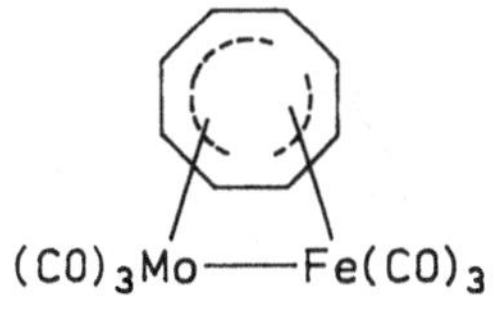

$(CO)_3Mo$——$Fe(CO)_3$

XVI

Reaction with Metal Salts and Metalorganic Compounds. Reaction of $C_8H_8Fe(CO)_3$ with KCN in C_2H_5OH gives C_8H_8 together with a mixture of impure products [65]. $C_8H_8Fe(CO)_3$ is recovered unchanged when reacted with KOC_4H_9-t in $CHCl_3$, NaO_2CCCl_3 in boiling $(CH_2OCH_3)_2$, or $Hg(C_6H_5)CCl_2Br$ in boiling C_6H_6 [89]. No metalation is observed with LiC_4H_9-n or C_6H_5MgBr [7], and no characterizable products are obtained with NaC_5H_5 [8]. Reaction with $NaN(Si(CH_3)_3)_2$ [33, 65, 74, 90, 95] in C_6H_6 at 120 °C [65, 74, 90, 95] in a sealed tube for 14 h [65, 74] gives $Na[C_8H_8Fe(CO)_2CN]$. No reaction is observed with $(CO)_3Cr(NCCH_3)_3$

 References on pp. 250/3

(see 1.1.2.1.2 in "Chrom-Organische Verbindungen" 1971, Table 5, No. 1, p. 27), and $(CO)_3W^2D_3$ ($^2D=(CH_3)_2NCHO$); but reaction with $(CO)_3Mo^2D_3$ ($^2D=O(CH_2CH_2OCH_3)_2$) in tetrahydrofuran at room temperature for 16 h gives XVI (see p. 247) [107]. UV irradiation of $C_8H_8Fe(CO)_3$ with excess $Fe(CO)_5$ gives $C_8H_8(Fe(CO)_3)_2$ [3, 7, 8, 23], and thermal reaction with $Fe_2(CO)_9$ gives the two isomers cis- and trans-$C_8H_8(Fe(CO)_3)_2$ [24]. Refluxing $C_8H_8Fe(CO)_3$ with $Ru_3(CO)_{12}$ (see 3.21.2 in "Ruthenium" Erg.-Bd., 1970, pp. 484/6) for 12 h gives XVII [46]. Reaction of $C_8H_8Fe(CO)_3$ with $[(CO)_2Rh]BF_4$ [108] or $[^4LRh]BF_4$ (^4L = norbornadiene [106, 108], cycloocta-1,5-diene [108]) in acetone gives XVIII with $M=Rh(CO)_2$ or Rh^4L, respectively.

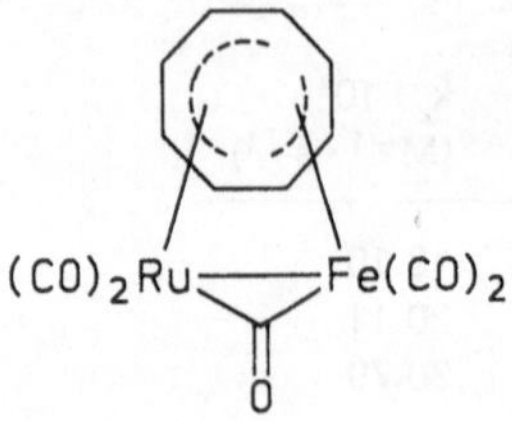

XVII 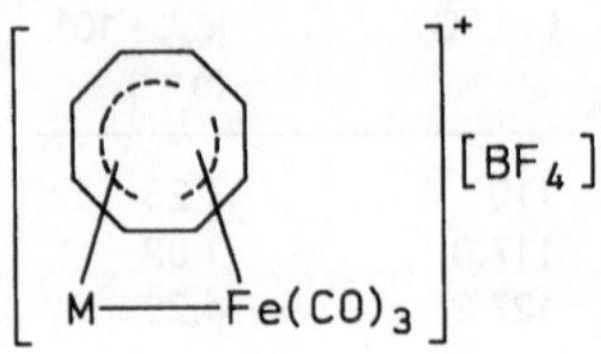 XVIII 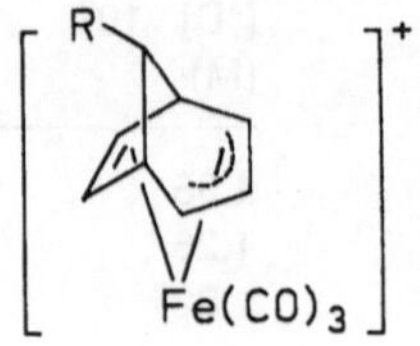 XIX

Reaction with Organic Compounds. The reaction with organic nucleophiles such as thiophene or amines gives only decomposition products as do aromatic compounds like mesitylene or $C_6(CH_3)_6$ [8]. Reaction with ^6D=2,2′:6′,2″-terpyridine in cyclohexane at 150 °C for 3 d gives $(CO)_2Fe^6D$, C_8H_8, and CO [48].

Reaction with Organic Electrophiles. $C_8H_8Fe(CO)_3$ is only attacked by strong electrophiles [99]. The electrophilic substitution reactions of $C_8H_8Fe(CO)_3$ are related to those of $C_4H_4Fe(CO)_3$ (see 1.4.1.4.1.2.1.1 in "Organoiron Compounds" B7, 1981, p. 19). The stability of $[^5LFe(CO)_3]^+$ cations, e.g. XIX, resulting from addition of electrophiles to $C_8H_8Fe(CO)_3$, is presumably the predominant factor responsible for the observed facile electrophilic substitution reactions undergone by this complex. Since the $Fe(CO)_3$ moiety is readily removed from the substituted complex, this reaction sequence in effect allows for the preparation of substituted cyclooctatetraenes that cannot be prepared from C_8H_8 directly.

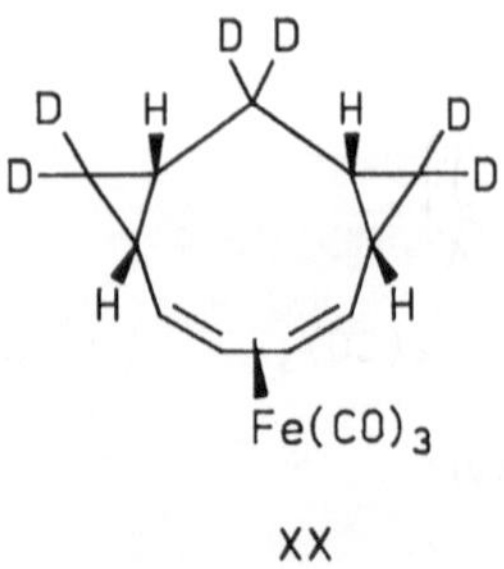

XX 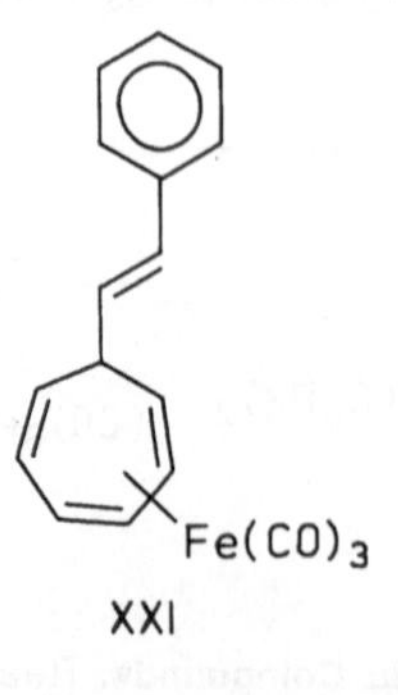 XXI 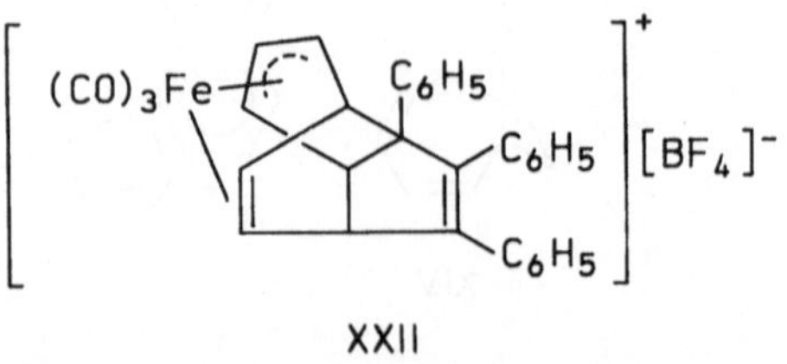 XXII

Refluxing $C_8H_8Fe(CO)_3$ with CH_2I_2/Zn/Cu in the presence of catalytic amounts of I_2 (Simmons-Smith reagent) in ether for 2 d gives XX with H instead of D [68, 87]. Analogous reaction with CD_2I_2 gives XX; no mono- or diadducts are observed [68]. Reaction with $[C(C_6H_5)_3]BF_4$ [25, 70] in CH_2Cl_2 gives VI (see p. 244; $R=C(C_6H_5)_3$, $X=BF_4$) [70]. Reaction with $[C_7H_7]BF_4$ (C_7H_7 = cycloheptatrienyl) gives XXI in 32% yield; the yield is increased (41%) in the presence of pyridine [98, 114]. Reaction of $C_8H_8Fe(CO)_3$ with $[C_6H_5N_2]BF_4$ in acetone at 25 °C for 24 h gives $C_6H_5N_2C_8H_7Fe(CO)_3$ in low yield [85]. For a different reaction with $[4\text{-}O_2NC_6H_4N_2]BF_4$ see "Oxidation and Reduction", p. 244. Reaction with the cyclopropenylium salt, $[C_3(C_6H_5)_3]BF_4$, in CH_2Cl_2 yields a yellow-brown solution from which the salt XXII is isolated [101]. No reaction is observed with $[O(CH_3)_3]PF_6$ or $C_2H_5Br/AlCl_3$ [99]. Contrary to first observations that Friedel-Crafts acylation causes decomposition [7] and gives no isolable products under various conditions [4, 6, 8], $C_8H_8Fe(CO)_3$ undergoes typical Friedel-Crafts reaction [41] to yield a very air-sensitive reaction mixture (IR spectrum ($CHCl_3$): 1709 (νCO) cm^{-1}) [8], containing among other products [41] $CH_3COC_8H_7Fe(CO)_3$ [41, 49] and an isomer mixture [49] of $(CH_3CO)_2C_8H_6Fe(CO)_3$ [41, 49]. Reaction with $CH_3COCl/AlCl_3$ in CH_2Cl_2 [49, 99] at 0 °C for 20 min [49] gives $[CH_3COC_8H_8Fe(CO)_3]^+$, isolated as the PF_6^- salt with $[NH_4]PF_6$ and identified as the cation XIX (see p. 248) with R = $COCH_3$ [99]. Friedel-Crafts reaction with $C_6H_5CH_2COCl/AlCl_3$ gives a very air-sensitive reaction mixture (IR spectrum ($CHCl_3$): 1681, 1721 (νCO) cm^{-1}) with no isolable products [8], but reaction with $C_6H_5COCl/AlCl_3$ in CH_2Cl_2 as above gives $[C_6H_5COC_8H_8Fe(CO)_3]PF_6$ with the cation XIX (R = COC_6H_5) [99]. The reaction of $C_8H_8Fe(CO)_3$ with $(CH_3)_2NCHO/OPCl_3$ [41] in $(CH_3)_2NCHO$ at 4 °C for 50 min [49] gives $OHCC_8H_7Fe(CO)_3$ [41, 49].

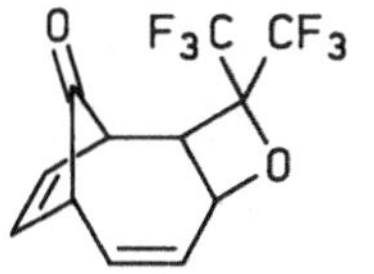

XXIII 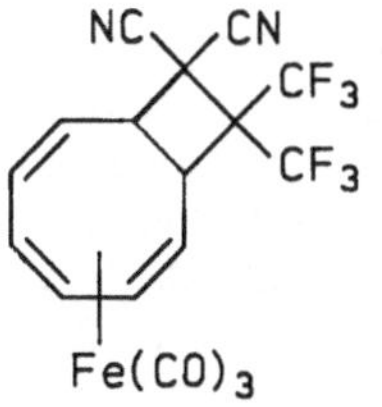 XXIV 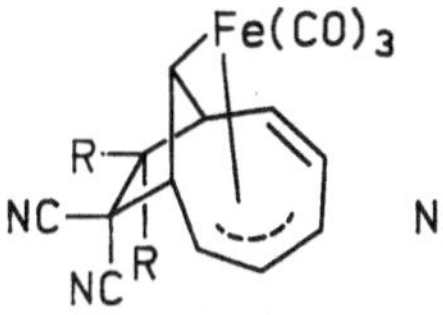 XXV 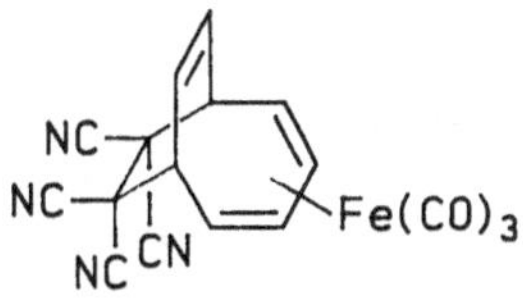 XXVI

Reactions with Alkenes and Analogous Compounds. No Diels-Alder reaction occurs in boiling toluene [7, 10], but $C_8H_8Fe(CO)_3$ is attacked by alkenes with electron withdrawing substituents [99]. Reaction of $C_8H_8Fe(CO)_3$ and $O{=}C(CF_3)_2$ in C_6H_6 at room temperature for 48 h gives trans-$C_8H_8(Fe(CO)_3)_2$ and XXIII [30, 43], cf. [78]. Similar reaction of $C_8H_8Fe(CO)_3$ with $(NC)_2C{=}C(CF_3)_2$ at room temperature in hexane for 10 h gives $C_{14}H_8F_6N_2Fe(CO)_3$ [30, 43] with the proposed structure XXIV [30, 43]; a comparison with the reaction product obtained with $(NC)_2C{=}C(CN)_2$ (see below) shows that structure XXV (R = CF_3) should be preferred for this reaction product [55, 78]. Reaction of $C_8H_8Fe(CO)_3$ with $(NC)_2C{=}C(CN)_2$ [11 to 13, 30, 43, 47, 55, 58, 78] in C_6H_6 at room temperature [11 to 13, 55, 58] for 1 h [11 to 13] or 15 h [55] is said to give a Diels-Alder [27] 1:1 adduct [23, 47], originally formulated as XXVI [11 to 13] and later proposed to have a structure analogous to XXIV [30, 43]. Oxidative degradation led to the proposal of structure XXV (R = CN) [47, 55, 78] for this reaction product, which is confirmed by an X-ray structure determination [58]. The effect of solvent on the rate of $(NC)_2C{=}C(CN)_2$ addition is $k = (8.39 \pm 0.26) \times 10\ M^{-1} \cdot s^{-1}$ for CH_2Cl_2 and $k = (5.83 \pm 0.20) \times 10^2\ M^{-1} \cdot s^{-1}$ for CH_3NO_2. This acceleration of 7.0 on going to the more polar solvent is small and not sufficient to accommodate a polar intermediate. Excluding a free radical mechanism, this reaction is probably concerted [100]. The addition reaction of XXVII (R = CH_3 [78], C_6H_5 [43, 78]) in CH_2Cl_2 at 25 °C gives XXVIII and XXIX [78]. A partial reaction of $C_8H_8Fe(CO)_3$ with maleic anhydride under more severe conditions, namely in the melt or in o-$Cl_2C_6H_4$, is observed (no products mentioned) [3]. Treatment of $C_8H_8Fe(CO)_3$ with dicyclopentadiene at its cracking point yields ferrocene [8]. Refluxing $C_8H_8Fe(CO)_3$ with cyclooctatetraene in heptane for 2 d gives XXX [81]. Irradiation of $C_8H_8Fe(CO)_3$ with equal amounts of C_8H_8 in C_6H_6 gives a yellow product A (m.p. 118 °C) and traces of an isomer B (m.p. 172 °C) [8, 13, 31, 119]. The ratio of isomers is dependent upon the wavelength used for irradiation [119]. Both products of the composition $C_{16}H_{16}Fe(CO)_3$ are also obtained by prolonged irradiation of $Fe(CO)_5$ with excess C_8H_8 [13]. The ^{57}Fe-γ Mössbauer spectrum at 78 °C of a compound of composition $C_{16}H_{16}Fe(CO)_3$ (origin unknown) shows $\delta = 0.21$ [115] and $\Delta = 1.599$ mm/s [115, 116]. A structure analysis showed that isomer A is exo, exo-$C_{10}H_{10}C_6H_6Fe(CO)_3$, and isomer B is $C_6H_6C_2H_2C_2H_2C_6H_6Fe(CO)_3$ (see Nos. 55a and 50, respectively, in Table 12, Section 1.4.1.4.1.2.3.2.1.14, pp. 275 and 276 in "Organoiron Compounds" B8, 1985) [31].

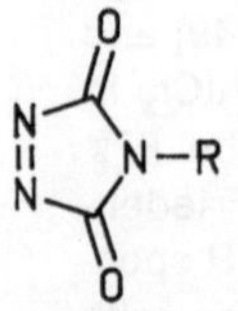

XXVII 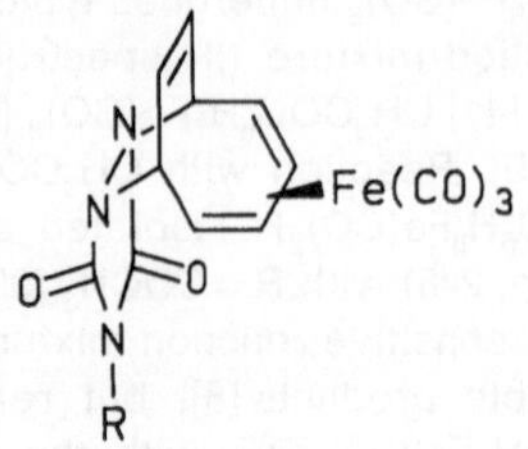 XXVIII 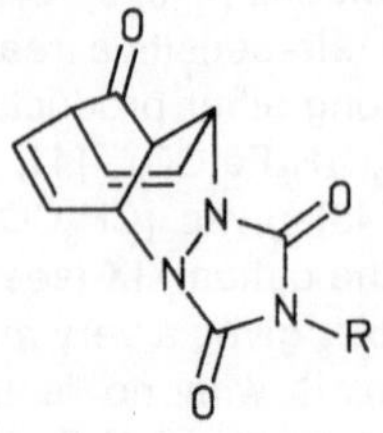 XXIX 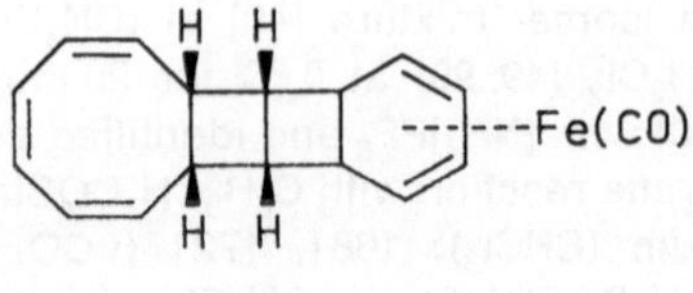 XXX

Reaction with Alkynes. Reaction with $C_6H_5C{\equiv}CH$, $C_6H_5C{\equiv}CCN$, $RC{\equiv}CR$ ($R = CH_3$, CO_2CH_3) gives predominantly trimerizations of the alkynes, formation of organoiron carbonyls, and decomposition products [33]. Thus, reaction with $C_6H_5C{\equiv}CH$ gives XXI ($R = C_6H_5$, $R' = H$) and XXXII (see 2.8.1.1 in "Organoiron Compounds" C5, 1981, Table 4, No. 28, p. 50) as main products, whereas reaction with $CH_3C{\equiv}CCH_3$ gives small amounts of XXXI ($R = R' = CH_3$) [33]. Irradiation of $C_8H_8Fe(CO)_3$ with $CH_3O_2CC{\equiv}CCO_2CH_3$ in tetrahydrofuran at 0 °C gives XXXIV (5%), its $Fe(CO)_3$ complex (15%), and XXXIII ($R' = R = CO_2CH_3$) in ca. 20% yield. Similar photochemical reaction with $C_6H_5C{\equiv}CC_6H_5$ gives XXXIII ($R' = R = C_6H_5$) in ca. 20% yield [64]. Thermal reaction of $C_8H_8Fe(CO)_3$ with $C_6H_5C{\equiv}CC_6H_5$ in boiling mesitylene gives XXXIII ($R' = R = C_6H_5$) and $(C_6H_5)_4C_4Fe_2(CO)_6$ (see 2.4.1.2.2.1 in "Organoiron Compounds" C3, 1980, Table 3, No. 20, p. 31); bubbling CO through the reaction mixture increases the yield of $(C_6H_5)_4C_4Fe_2(CO)_6$ up to 35%. A small pressure of CO (2 to 4 atm) does not further increase the yield. CO stabilizes the $Fe(CO)_3$ moiety in $C_8H_8Fe(CO)_3$ and therefore nearly completely prevents the catalytic trimerization of tolane [33]. Reaction with $C_6H_5C{\equiv}CC_6H_5$ at 140 °C gives $(C_6H_5)_4C_4Fe_2(CO)_6$, XXXIII ($R = R' = C_6H_5$), and $OC_5(C_6H_5)_4$, and at 190 °C $(C_6H_5)_4C_4Fe(CO)_3$ (see 1.4.1.4.1.2.1.5.1 in "Organoiron Compounds" B7, 1981, p. 102) and $C_6(C{\equiv}CC_6H_5)_6$ are formed [18]. Refluxing $C_6H_5C{\equiv}CR$ ($R = CO_2CH_3$, $Si(CH_3)_3$) with $C_8H_8Fe(CO)_3$ in mesitylene gives XXXIII with $R = C_6H_5$ and $R' = CO_2CH_3$ or $Si(CH_3)_3$, respectively [33].

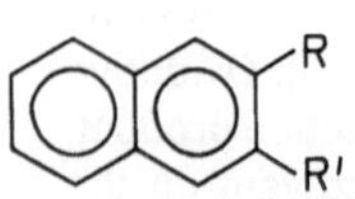

XXXI 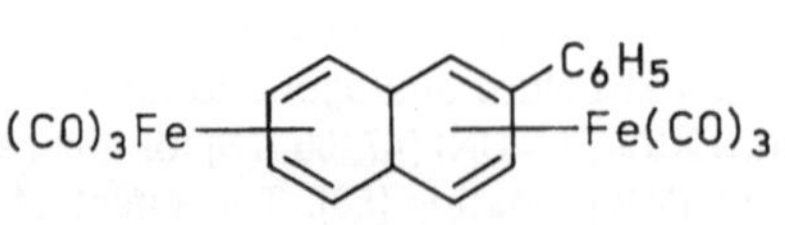 XXXII 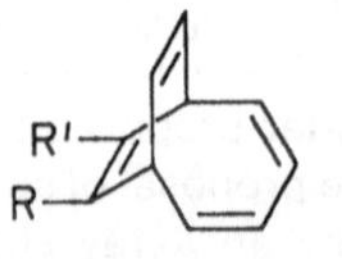 XXXIII 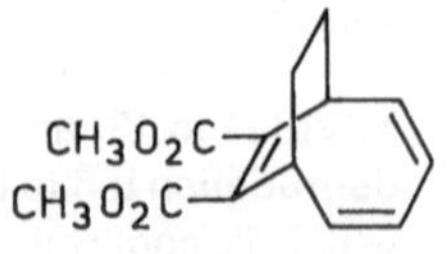 XXXIV

Uses. $C_8H_8Fe(CO)_3$ is proposed as an antiknock agent when added to gasoline, as an additive for fuels, as a drying agent, or as a chemical intermediate for production of metal-containing polymer materials [9].

References:

[1] T.A. Manuel, F.G.A. Stone (Proc. Chem. Soc. **1959** 90). – [2] A. Nakamura, N. Hagihara (Bull. Chem. Soc. Japan **32** [1959] 880/1). – [3] M.D. Rausch, G.N. Schrauzer (Chem. Ind. [London] **1959** 957/8). – [4] R. Burton, G. Wilkinson (unpublished results from [5]). – [5] F.A. Cotton (J. Chem. Soc. **1960** 400/6).

[6] F.G.A. Stone (unpublished results from [5]). – [7] A. Nakamura, N. Hagihara (Mem. Inst. Sci. Ind. Res. Osaka Univ. **17** [1960] 187/90). – [8] T.A. Manuel, F.G.A. Stone (J. Am. Chem. Soc. **82** [1960] 366/72). – [9] K.G. Ihrman, T.H. Coffield, Ethyl Corp. (U.S. 3077489 [1960/63]). – [10] G.N. Schrauzer (J. Am. Chem. Soc. **83** [1961] 2966/7).

[11] A. Davison, W. McFarlane, L. Pratt, G. Wilkinson (J. Chem. Soc. **1962** 4821/9). – [12] A. Davison, W. McFarlane, G. Wilkinson (Chem. Ind. [London] **1962** 820/1). – [13] G.N. Schrauzer, S. Eichler (Angew. Chem. **74** [1962] 585). – [14] F. Zingales, F. Canziani, R. Ugo (Chim. Ind. [Milan] **44** [1962] 1394/6). – [15] R. Schneider (unpublished results from [16]).

[16] H.P. Fritz, H. Keller (Chem. Ber. **95** [1962] 158/73). – [17] T.A. Manuel (Inorg. Chem. **2** [1963] 854/8). – [18] A. Nakamura, N. Hagihara (Nippon Kagaku Zasshi **84** [1963] 339/44). – [19] R.B. King (Inorg. Chem. **2** [1963] 936/44). – [20] J.D. Holmes, R. Pettit (J. Am. Chem. Soc. **85** [1963] 2531/2).

[21] J.L. von Rosenberg (Diss. Univ. Texas 1963; Diss. Abstr. **24** [1964] 3997). – [22] R.T. Bailey, E.R. Lippincott, D. Steele (J. Am. Chem. Soc. **87** [1965] 5346/50). – [23] R.B. King (Organometal. Syn. **1** [1965] 126/7). – [24] C.E. Keller (Diss. Univ. Texas 1966; Diss. Abstr. B **27** [1967] 2651). – [25] C.E. Keller (Diss. Univ. Texas 1966 from [26]).

[26] R. Pettit, L.W. Haynes (Carbonium Ions **5** [1976] 2263/302). – [27] W. McFarlane, G. Wilkinson (Inorg. Chem. **8** [1966] 184/5). – [28] R.E. Dessy, F.E. Stary, R.B. King, M. Waldrop (J. Am. Chem. Soc. **88** [1966] 471/6). – [29] N.G. Connelly, A.R. Lucy, M.W. Whiteley (J. Chem. Soc. Dalton Trans. **1983** 111/5). – [30] M. Green, D.C. Wood (Chem. Commun. **1967** 1062).

[31] G.N. Schrauzer (unpublished results from [32]). – [32] A. Robson, M.R. Truter (J. Chem. Soc. A **1968** 794/801). – [33] U. Krüerke (Angew. Chem. **79** [1967] 55/6). – [34] R.B. King (New Aspects Chem. Metal Carbonyls Deriv. 1st Intern. Symp. Proc., Venice 1968, Abstr. No. E7 from [35]). – [35] R.B. King (Fortschr. Chem. Forsch. **14** [1970] 92/126).

[36] R.B. King, K.H. Pannell (Inorg. Chem. **7** [1968] 273/7). – [37] F. Faraone, F. Zingales, P. Uguagliati, U. Belluco (Inorg. Chem. **7** [1968] 2362/4). – [38] R.E. Dessy, R.L. Pohl (J. Am. Chem. Soc. **90** [1968] 2005/8). – [39] M. Cooke (Diss. Univ. Bristol 1969 from [97]). – [40] M.I. Bruce (Intern. J. Mass Spectrom. Ion Phys. **2** [1969] 349/55).

[41] B.F.G. Johnson, J. Lewis, A.W. Parkins, G.L.P. Randall (Chem. Commun. **1969** 595). – [42] R.B. King (Appl. Spectrosc. **23** [1969] 536/46). – [43] M. Green, D.C. Wood (J. Chem. Soc. A **1969** 1172/5). – [44] R.B. King, P.N. Kapoor, R.N. Kapoor (Inorg. Chem. **10** [1971] 1841/50). – [45] M. Brookhart, E.R. Davis (J. Am. Chem. Soc. **92** [1970] 7622/3).

[46] E.W. Abel, S. Moorhouse (Inorg. Nucl. Chem. Letters **6** [1970] 621/2). – [47] D.J. Ehntholt (Diss. State Univ. New York 1971; Diss. Abstr. Intern. B **32** [1972] 4486). – [48] H. Behrens, H.-D. Feilner, E. Lindner (Z. Anorg. Allgem. Chem. **385** [1971] 321/4). – [49] B.F.G. Johnson, J. Lewis, G.L.P. Randall (J. Chem. Soc. A **1971** 422/9). – [50] C.H. Mauldin, E.R. Biehl, P.C. Reeves (Tetrahedron Letters **1972** 2955/8).

[51] D.F. Hunt, G.C. Farrant, G.T. Rodeheaver (J. Organometal. Chem. **38** [1972] 349/65). – [52] J. Schwartz (J. Chem. Soc. Chem. Commun. **1972** 814/5). – [53] H. Knöchl (Diss. Univ. Erlangen-Nürnberg 1972 from [65]). – [54] D.F. Hunt, J.W. Russell, R.L. Torian (J. Organometal. Chem. **43** [1972] 175/83). – [55] D.J. Ehntholt, R.C. Kerber (J. Organometal. Chem. **38** [1972] 139/45).

[56] M. Brookhart, E.R. Davis, D.L. Harris (J. Am. Chem. Soc. **94** [1972] 7853/8). – [57] R. Aumann (Angew. Chem. **85** [1973] 628/9). – [58] L.A. Paquette, S.V. Ley, M.J. Broadhurst, D. Truesdell, J. Fayos, J. Clardy (Tetrahedron Letters **1973** 2943/6). – [59] R.C. Kerber,

D.J. Ehntholt (J. Am. Chem. Soc. **95** [1973] 2927/34). – [60] R. Aumann (J. Organometal. Chem. **78** [1974] C31/C34).

[61] V. Heil, B.F.G. Johnson, J. Lewis, D.J. Thompson (J. Chem. Soc. Chem. Commun. **1974** 270/1). – [62] R. Pettit (Organotransition-Metal Chem. Proc. 1st Japan.-Am. Semin., Honolulu 1974 [1975], pp. 157/67; C.A. **85** [1976] No. 176360). – [63] B.F.G. Johnson, J. Lewis, M.V. Twigg (J. Chem. Soc. Dalton Trans. **1974** 2546/9). – [64] R.E. Davis, T.A. Dodds, T.H. Hseu, J.C. Wagnon, T. Devon, J. Tancrede, J.S. McKennis, R. Pettit (J. Am. Chem. Soc. **96** [1974] 7562/3). – [65] H. Behrens, M. Moll (Z. Anorg. Allgem. Chem. **416** [1975] 193/202).

[66] Y. Shvo, E. Hazum (J. Chem. Soc. Chem. Commun. **1975** 829/30). – [67] G.R. Langford, M. Akhtar, P.D. Ellis, A.G. MacDiarmid, J.D. Odom (Inorg. Chem. **14** [1975] 2937/41). – [68] D.L. Reger, A. Gabrielli (J. Am. Chem. Soc. **97** [1975] 4421/2). – [69] B.F.G. Johnson, J. Lewis, D.J. Thompson, B. Heil (J. Chem. Soc. Dalton Trans. **1975** 567/71). – [70] B.F.G. Johnson, J. Lewis, J.W. Quail (J. Chem. Soc. Dalton Trans. **1975** 1252/7).

[71] N.G. Connelly, R.L. Kelly (J. Organometal. Chem. **120** [1976] C16/C17). – [72] N. El Murr, M. Riveccié, E. Laviron (Tetrahedron Letters **1976** 3339/40). – [73] A.D. Charles, P. Diversi, B.F.G. Johnson, J. Lewis (J. Organometal. Chem. **116** [1976] C25/C28). – [74] H. Behrens, M. Moll, P. Würstl (Z. Naturforsch. **31b** [1976] 1017/8). – [75] R. Aumann, J. Knecht (Chem. Ber. **109** [1976] 174/9).

[76] H.A. Bockmeulen, R.G. Holloway, A.W. Parkins, B.R. Penfold (Chem. Commun. **1976** 298/9). – [77] G.A. Olah, S.H. Yu, G. Liang (J. Org. Chem. **41** [1976] 2383/6). – [78] H. Olsen (Acta Chem. Scand. B **31** [1977] 635/6). – [79] G.A. Olah, G. Liang, S. Yu (J. Org. Chem. **42** [1977] 4262/3). – [80] K.D. Karlin, B.F.G. Johnson, J. Lewis (J. Organometal. Chem. **160** [1978] C21/C23).

[81] A.P. Humphries, S.A.R. Knox (J. Chem. Soc. Dalton Trans. **1978** 1514/23). – [82] B.F.G. Johnson, K.D. Karlin, J. Lewis (J. Organometal. Chem. **145** [1978] C23/C25). – [83] M. Franck-Neumann, D. Martina, F. Brion (Angew. Chem. **90** [1978] 736/7). – [84] N.G. Connelly, M.D. Kitchen, R.F.D. Stansfield, S.M. Whiting, P. Woodward (J. Organometal. Chem. **155** [1978] C34/C36). – [85] N.G. Connelly, A.R. Lucy, M.W. Whiteley (J. Chem. Soc. Chem. Commun. **1979** 985/6).

[86] G. Deganello (Transition Metal Complexes of Cyclic Polyolefins, Academic, London 1979, p. 224 from [97]). – [87] A.M. Gabrielli (Diss. Univ. South Carolina 1978; Diss. Abstr. Intern. B **39** [1979] 5911). – [88] M.R. Blake, J.L. Garnett, I.K. Gregor, S.B. Wild (J. Organometal. Chem. **178** [1979] C37/C42). – [89] G.A. Taylor (J. Chem. Soc. Perkin Trans. I **1979** 1716/9). – [90] M. Moll, H. Behrens, W. Popp (Z. Anorg. Allgem. Chem. **458** [1979] 202/10).

[91] B.F.G. Johnson, K.D. Karlin, J. Lewis (J. Organometal. Chem. **174** [1979] C29/C31). – [92] R. Holderegger, L.M. Wenanzi (Helv. Chim. Acta **62** [1979] 2154/8). – [93] B. Tulyathan, W.E. Geiger (J. Electroanal. Chem. **109** [1980] 325/31). – [94] M.R. Blake, J.L. Garnett, I.K. Gregor, D. Nelson (J. Organometal. Chem. **188** [1980] 203/10). – [95] H. Behrens, M. Moll, W. Popp, H.-J. Seibold, E. Sepp, P. Würstl (J. Organometal. Chem. **192** [1980] 389/98).

[96] K.L. Amos, N.G. Connelly (J. Organometal. Chem. **194** [1980] C57/C59). – [97] N.G. Connelly, R.L. Kelly, M.D. Kitchen, R.M. Mills, R.F.D. Stansfield, M.W. Whiteley, S.M. Whiting, P. Woodward (J. Chem. Soc. Dalton Trans. **1981** 1317/26). – [98] K. Broadley, N.G. Connelly, R.M. Mills, M.W. Whiteley, P. Woodward (J. Chem. Soc. Chem. Commun. **1981** 19/20). – [99] A.D. Charles, P. Diversi, B.F.G. Johnson, J. Lewis (J. Chem. Soc. Dalton Trans. **1981** 1906/17). – [100] S.K. Chopra, M.J. Hynes, P. McArdle (J. Chem. Soc. Dalton Trans. **1981** 586/9).

[101] K. Broadley, N.G. Connelly, J.A.K. Howard, W. Risse (J. Organometal. Chem. **221** [1981] C29/C32). – [102] H.A. Bockmeulen, A.W. Parkins (J. Chem. Soc. Dalton Trans. **1981** 262/6). – [103] T.A. Albright, W.E. Geiger, J. Moraczewski, B. Tulyathan (J. Am. Chem. Soc. **103** [1981] 4787/94). – [104] J. Weichmann (Dipl.-Arbeit Univ. Regensburg 1981, pp. 1/111). – [105] N.B.H. Henis, W. Lamanna, M.B. Humphrey, M.M. Bursey, M.S. Brookhart (Inorg. Chim. Acta **54** [1981] L11/L12).

[106] A. Salzer, T. Egolf (Proc. 10th Intern. Conf. Organometal. Chem., Toronto 1981, p. 64). – [107] A. Salzer, T. Egolf, L. Linowsky, W. Petter (J. Organometal. Chem. **221** [1981] 339/49). – [108] A. Salzer, T. Egolf, W. von Philipsborn (J. Organometal. Chem. **221** [1981] 351/60). – [109] A. Davison, W. McFarlane, L. Pratt, G. Wilkinson (Chem. Ind. [London] **1961** 553). – [110] N.G. Connelly, R.L. Kelly (J. Organometal. Chem. **120** [1976] C16/C17).

[111] R. Aumann (Chem. Ber. **109** [1976] 168/73). – [112] N.G. Connelly (Abstr. Papers 186th Natl. Meeting Am. Chem. Soc., Washington, D.C., 1983, INOR 41). – [113] R.B. Hitam, R. Narayanaswamy, A.J. Rest (J. Chem. Soc. Dalton Trans. **1983** 1351/6). – [114] K. Broadley, N.G. Connelly, R.M. Mills, M.W. Whiteley, P. Woodward (J. Chem. Soc. Dalton Trans. **1984** 683/8). – [115] R.H. Herber (TID-16124 [1962]; N.S.A. **16** [1962] No. 27037).

[116] G.K. Wertheim, R.H. Herber (unpublished results from [117]). – [117] G.K. Wertheim, R.H. Herber (J. Am. Chem. Soc. **84** [1962] 2274/5). – [118] F.A. Cotton, W.T. Edwards (J. Am. Chem. Soc. **91** [1969] 843/7). – [119] G.N. Schrauzer (unpublished results from [120]). – [120] G.O. Nelson (Diss. Univ. North Carolina 1977; Diss. Abstr. Intern. B **39** [1978] 237/8).

1.4.1.4.1.2.5.3.1.4 The Adduct $C_8H_8Fe(CO)_3AlBr_3$

$C_8H_8Fe(CO)_3AlBr_3$ forms, by IR spectroscopic evidence, in CH_2Cl_2 by addition of $C_8H_8Fe(CO)_3$ to a large excess of $AlBr_3$ in a dilute solution [1]. The IR spectrum in CH_2Cl_2 shows (νCO) bands at 2064 and 2110 cm^{-1}. The proposed structure is shown in Formula I. On hydrolysis, $C_8H_8Fe(CO)_3$ reforms together with a small amount of $C_8H_9Fe(CO)_3Br$ [1, 2].

$AlBr_3$
$Fe(CO)_3$

I

References:

[1] K.D. Karlin, B.F.G. Johnson, J. Lewis (J. Organometal. Chem. **160** [1978] C21/C23). – [2] B.F.G. Johnson, K.D. Karlin, J. Lewis (J. Organometal. Chem. **174** [1979] C29/C31).

1.4.1.4.1.2.5.3.2 ⁴L is a Substituted Cyclooctatetraene

In this section compounds of the general formula $R_nC_8H_{8-n}Fe(CO)_3$ are described where n = 1 (Nos. 1 to 32), 2 (Nos. 36 to 54), 3 (No. 55), and 4 (No. 57). Salts derived from the monosubstituted complexes such as $[R_3MCH_2C_8H_7Fe(CO)_3]PF_6$ where M = N, P, or As are also known (Nos. 33 to 35). Most of these compounds can be prepared by the following methods:

Method I: $Fe(CO)_5$ reacts photochemically or thermally with the free organic ligand ^4L or with a suitable organic substrate in an inert solvent.

a. $Fe(CO)_5$ reacts photochemically with $C_6H_5C_8H_7$ to give No. 31 [4]. Irradiation for 24 h at room temperature gives No. 31 and $C_6H_5C_8H_7$ (4% yield, see Formula I) [6].

References on pp. 277/9

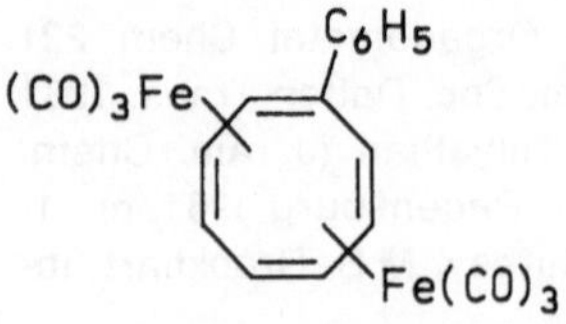

I 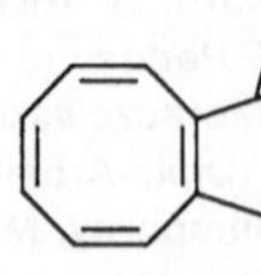 II 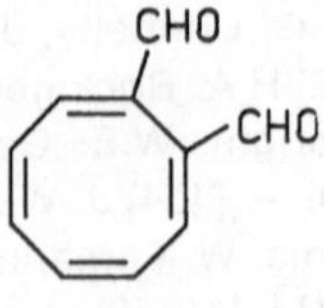 III 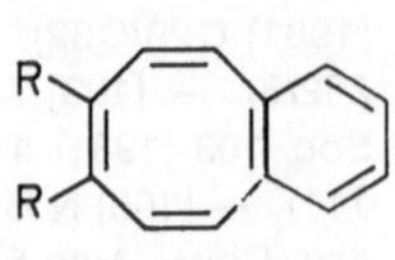 IV

b. $Fe(CO)_5$ and II are refluxed in ethylcyclohexane. Filtration and chromatography on acidic alumina with ether/C_6H_6 gives No. 39. Similar reaction with III in methylcyclohexane for 4 d followed by processing as before also gives No. 39 [12]. $Fe(CO)_5$ is said to give No. 40 in the reaction with IV (R=H) [15, 18, 29], but according to [17], compound No. 40 is not available by this reaction (cf. Method IIIc).

Method II: $Fe_2(CO)_9$ reacts photochemically or thermally with the free organic ligand ^4L or with a suitable organic substrate in an inert solvent.

a. Photochemical reaction of $Fe_2(CO)_9$ with $C_6H_5C_8H_7$ gives No. 31 (cf. Method Ia) [4].

b. $Fe_2(CO)_9$ and RC_8H_7 (R=CH_3O [41], CH_3 [32], CH_3O_2C, C_6H_5 [41]) are refluxed in n-hexane for 0.5 [32] to 1.2 h [41], followed by filtration (No. 3), evaporation, and chromatography of the residue with n-hexane on silica gel (No. 10) [32] or on Florisil with n-hexane (No. 21), or ether/pentane (1:5) for No. 31 [41]. Compound No. 3 is distilled twice by short-path distillation [41]. Similar reaction of $Fe_2(CO)_9$ with BrC_8H_7 in refluxing pentane for 1 h followed by filtration, evaporation, and chromatography of the residue (twice) on alumina with pentane/4% ether gives No. 2 [53]. Reaction of $Fe_2(CO)_9$ with $R_3MC_8H_7$ ($R_3M=(CH_3)_3Si$, $CH_2{=}CHCH_2(CH_3)_2Si$, $(CH_3)_3Ge$, $(CH_3)_3Sn$) in heptane for 12 h followed by chromatography on alumina with hexane gives Nos. 6 [38], cf. [9], and 7 to 9 [38]. Careful chromatography for No. 6 gives additionally I (R=$Si(CH_3)_3$) in ca. 2% yield [38].

c. $Fe_2(CO)_9$ and V react in C_6H_6 at 40 °C. Chromatography on alumina gives VI and No. 21 [33]. Reaction of $Fe_2(CO)_9$ with VII (R=CH_3) at ca. 5 °C for one week, followed by filtration, evaporation, and chromatography of the residue on silica gel with pentane gives No. 38 which cannot be purified and is always contaminated with VII (R=CH_3) and VIII (R=CH_3) [58]. Refluxing $Fe_2(CO)_9$ and IV (R=H) in C_6H_6 for 12 h followed by chromatography on silica gel with petroleum ether as eluent gives No. 40 and $C_{12}H_{10}Fe_2(CO)_6$ (see 2.6.1.2.4 in "Organoiron Compounds" C5, 1981, Table 3, No. 13, p. 29) [17]. Similar reaction of $Fe_2(CO)_9$ with IV (R=CH_3O_2C) gives No. 56 and a dinuclear complex $C_{16}H_{14}O_4Fe_2(CO)_6$ (see 2.6.1.2.4 in "Organoiron Compounds" C5, 1981, Table 30, No. 18, p. 30) [30]. Refluxing $Fe_2(CO)_9$ and $CH_3C_8H_7$ in C_6H_6 for a few minutes yields No. 10 [14].

V 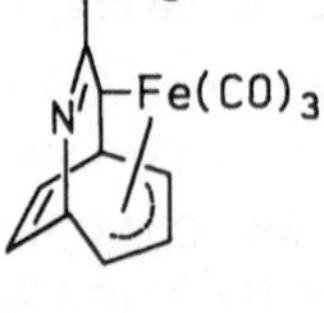 VI 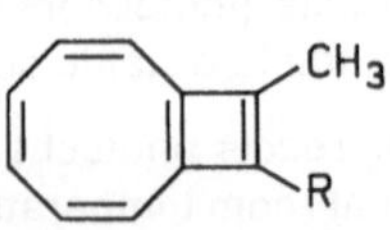 VII 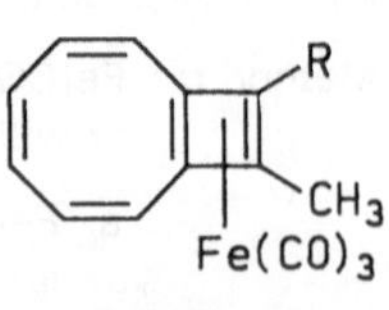 VIII

d. $Fe_2(CO)_9$ and $CH_3SC_8H_7$ in tetrahydrofuran are stirred for 12 h at room temperature. The reaction mixture is evaporated, and the residue is chromatographed on alumina. Elution with hexane develops yellow and orange bands. The former yields I (R=CH_3S) and the latter a small amount of No. 4. Analogous reaction over 60 h to increase the yield of No. 4 does not give No. 4 but $Fe_2(CO)_6(SCH_3)_2$ and $Fe_4(CO)_{12}(SCH_3)_2S$ instead [50]. Refluxing $Fe_2(CO)_9$ with $CH_3OC_8H_7$ in ether for 3 h, filtration, evaporation, and chromatography of the residue on silica gel with petroleum ether as eluent gives a red oil which after bulb-to-bulb distillation and rechromatography gives No. 3. Elution with petroleum ether/ether (5:1) gives IX [41], cf. [59], and elution with ether gives X [41].

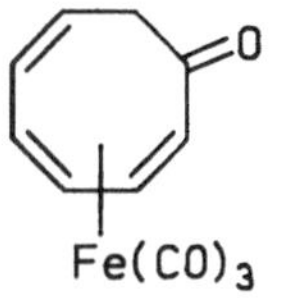

Fe(CO)3 — IX; (CO)3Fe–Fe(CO)3 — X; Cl(CH3), CH3(Cl), (CO)3Fe, Fe(CO)3 — XI; Fe(CO)3, Cl, CH3, Fe(CO)3 — XII

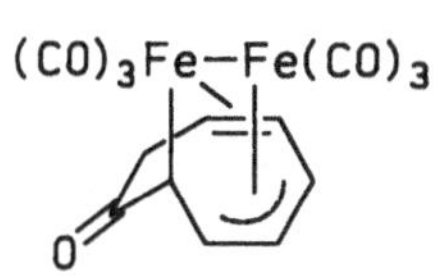

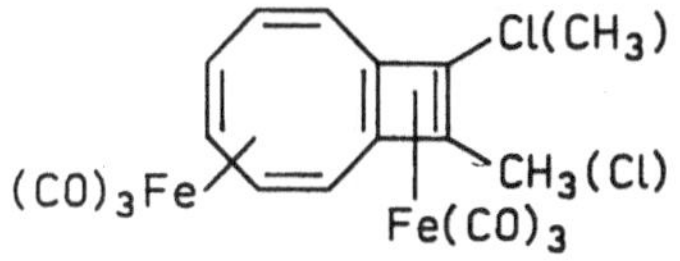

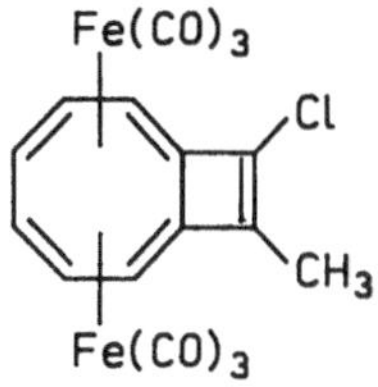

Method III: $Fe_3(CO)_{12}$ reacts with the free organic ligand ^4L or with a suitable organic substrate.

a. Photoreaction of $Fe_3(CO)_{12}$ with $C_6H_5C_8H_7$ gives No. 31 [4].

b. Refluxing $Fe_3(CO)_{12}$ with $CH_3OC_8H_7$ in octane for 2 h followed by chromatography on alumina with light petroleum ether affords I with R=OCH_3 and No. 3 [44]. $Fe_3(CO)_{12}$ in n-heptane is heated at 100 °C, and VII (R=Cl) in n-heptane is added dropwise to this solution. After 5 h the reaction mixture is filtered. The residue is chromatographed on silica gel with pentane to yield VIII (R=Cl), an isomeric mixture of No. 37a and b, and XI. Rechromatography increases the ratio of isomers No. 37 from 2:1 to 3:1 [58]. Similar reaction in toluene at 111 °C for 4 h gives the isomer No. 37 in 38% yield. Reaction in n-heptane at 95 °C for 4 h gives VIII (R=Cl), isomers XI in 60% yield, and XII [57].

c. $Fe_3(CO)_{12}$ and $C_6H_5C_8H_7$ are refluxed in C_6H_6 for 5 h followed by chromatography on alumina with toluene to give No. 31 [31], cf. [27]. Refluxing $Fe_3(CO)_{12}$ with 1,2-$(CH_3O_2C)_2C_8H_6$ [12] in C_6H_6 for 48 h followed by chromatography on silica gel [12, 17] with petroleum ether [17] gives No. 36 in 30% yield [12]. An analogous reaction with IV (R=H) [17] gives No. 40 together with $C_{12}H_{10}Fe_2(CO)_6$ (see 2.6.1.2.4 in "Organoiron Compounds" C5, 1981, Table 3, No. 13, p. 29) [17]. Rechromatography of No. 36 on acid-washed alumina with ether/C_6H_6 gives pure No. 36 in 12% yield [12]. Refluxing $Fe_3(CO)_{12}$ with $C_6H_5CH_2C_8H_7$ in $O(CH_2CH_2OCH_3)_2$ for 30 min, filtration, and evaporation gives No. 15 [5].

d. $Fe_3(CO)_{12}$ reacts with $C_6H_5C{\equiv}CH$ in petroleum ether (b.p. 60 to 70 °C)/C_6H_6 (ratio 0.75:0.1) at reflux temperature for 1.75 h. Chromatography on acidic Al_2O_3 gives $(C_6H_5C_2H)_5COFe(CO)_3$ (see 1.4.1.4.1.2.7, p. 283), $OC_7H_3(C_6H_5)_3Fe(CO)_3$ (see 1.4.1.4.1.2.4.2.4.3, Table 5, No. 25, p. 187), $C_6H_3(C_6H_5)_3COFe_2(CO)_5$ (see 2.6.1.2.1 in "Organoiron Compounds" C5, 1981, p. 7), No. 56, and two isomers of $O{=}C_5(C_6H_5)_2H_2Fe(CO)_3$ (see 1.4.1.4.1.2.2.2.4 in "Organoiron Com-

References on pp. 277/9

pounds" B7, 1981, pp. 163/5). Similar reaction at 75 °C for 0.5 h gives $C_6H_3(C_6H_5)_3COFe_2(CO)_5$ as main product [1].

Method IV: $^4LFe(CO)_3$ compounds with ^4L = norbornadiene or butadiene give ligand exchange reactions with $^4L = RC_8H_7$ (R = Cl, CH_3), and reaction of $C_8H_8Fe(CO)_3$ with electrophiles gives substitution products.

a. $^4LFe(CO)_3$ (^4L = butadiene) is refluxed with $CH_3C_8H_7$ in $O(CH_2CH_2OCH_3)_2$ for 6 h. The reaction mixture is filtered, evaporated, and the residue is chromatographed. Elution with low boiling petroleum ether gives No. 10. Similar reaction of $^4LFe(CO)_3$ (^4L = norbornadiene) with ClC_8H_7 for 6 h gives No. 1 [5].

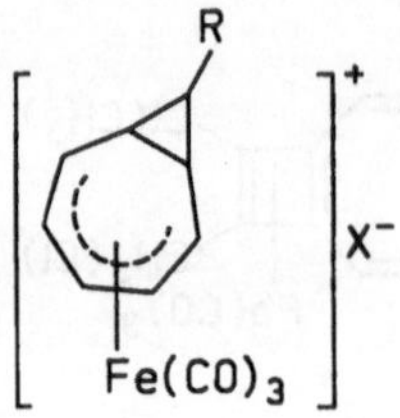

XIII 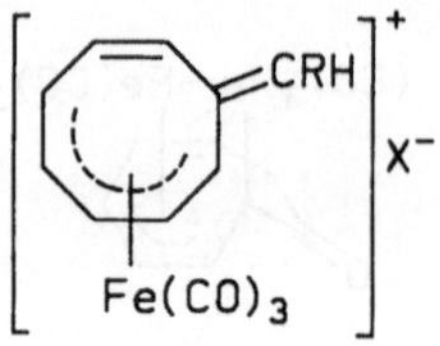 XIV

b. $C_8H_8Fe(CO)_3$ reacts with $CH_3COCl/AlCl_3$ in CH_2Cl_2 at 0 °C for 20 min. Hydrolysis with ice-cold 5% aqueous HCl, extraction with ether, drying of the organic layer ($MgSO_4$), and chromatography on alumina type H with 5% $CH_3CO_2C_2H_5$/$CH_3C_6H_5$ gives No. 28 and a second fraction containing an isomer mixture of $(CH_3CO)_2C_8H_6Fe(CO)_3$ (No. 52) [26], cf. [21]. The aqueous phase obtained after hydrolysis gives $[C_{10}H_{11}OFe(CO)_3]PF_6$ (see Formula XIII, R = CH_3CO, X = PF_6) on treatment with 15% $[NH_4]PF_6/H_2O$ [26]. $C_8H_8Fe(CO)_3$ and $[C_6H_5N_2]BF_4$ at 25 °C for 24 h give on treatment with Al_2O_3 or $NaHCO_3$ No. 5 [47]. Reaction of $C_8H_8Fe(CO)_3$ with $[C(C_6H_5)_3]BF_4$ in CH_2Cl_2 gives No. 30 [42]. $C_8H_8Fe(CO)_3$ reacts with $(CH_3)_2NCHO/OPCl_3$ at 4 °C for 50 min. The reaction mixture is poured into ice-water, extracted with pentane, and the organic layer is dried ($MgSO_4$). Chromatography on alumina with $CH_3C_6H_5$ containing up to 10% $CH_3CO_2C_2H_5$ gives No. 18 [21, 26].

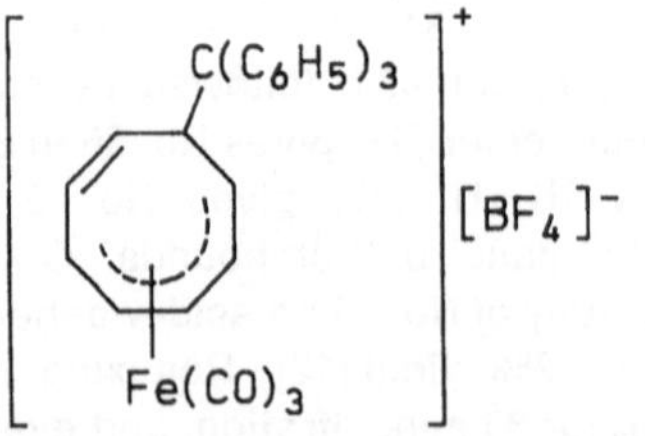

XV 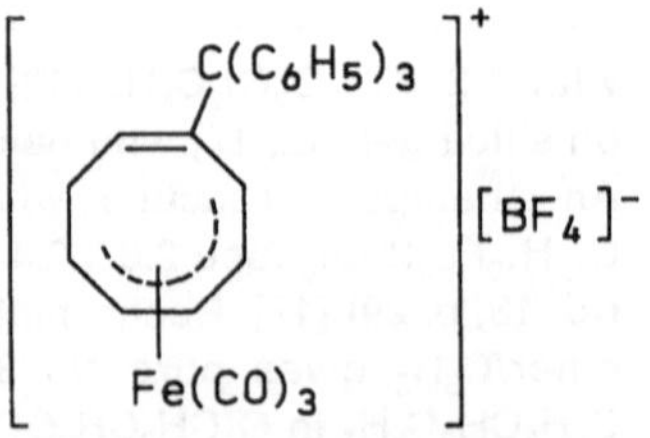 XVI

Method V: Salts $[^5LFe(CO)_3]X$, where ^5L is a cyclooctadienyl derivative, react with nucleophiles.

a. Salt XIV (R = H, X = PF_6) is suspended in CH_2Cl_2 and cooled to 0 °C. An excess of $N(CH_3)_3$ is added. Almost immediately, crystals of No. 33 are deposited [52]. Analogous reaction with $P(C_6H_5)_3$ or $As(C_6H_5)_3$ for 10 min gives Nos. 34

References on pp. 277/9

and 35. The solution is filtered. On addition of pentane, a red oil separates which solidifies on drying in vacuum [52, 54]. Similarly, salt XIV (R=H, X=PF_6) reacts with CH_3OH at room temperature for 20 min. H_2O is added, and the mixture is extracted with ether. The organic layer is dried ($MgSO_4$) and chromatographed on silica gel with 10% $CH_3CO_2C_2H_5/CH_3C_6H_5$ to give No. 12 [21, 26]. Analogous reaction with morpholine in ether followed by processing as before gives No. 13. Similar reaction with NaCN in acetone/H_2O (4:1) at room temperature for 10 min but elution with toluene gives No. 17 [21, 26].

b. Salt XV reacts with $N(C_2H_5)_3$ in CH_2Cl_2. After evaporation the residue of the reaction mixture is dissolved in hexane. Chromatography on silica gel with 10% CH_2Cl_2/hexane gives No. 30 (74% yield). The compound is crystallized from hexane and dried at 80 °C in vacuum for 12 h. Similar reaction of XV with $NaOCH_3/CH_3OH$ gives No. 30, too (59% yield). Neutralization of XVI followed by processing as before gives No. 30, $C_8H_8Fe(CO)_3$, $(C_6H_5)_3CC_8H_7$, and some unidentified substances [42].

c. Deprotonation of XIII (R=$C_6H_4NO_2$-4) with pyridine at −20 °C in acetone gives No. 32. After warming to 30 °C and stirring for 30 min, the solution is evaporated. The residue is dissolved in ether and chromatographed on alumina with hexane/ether (1:1) [55]. Reaction of $[C_{10}H_{11}Fe(CO)_3]PF_6$ (see "Further information" for No. 22, p. 273) with CH_3OH as in Method Va gives No. 23 [21, 26].

Since in the $RC_8H_7Fe(CO)_3$ compounds (Nos. 1 to 32) the $Fe(CO)_3$ moiety is bound to different ring C atoms relative to R and some of the molecules are fluxional or the bonding is not known with certainty, the formulas shown in the table do not give the real structure. In most of the compounds the real position of the $Fe(CO)_3$ is discussed (if known) under "Further information", pp. 269/75. In brief, electron withdrawing groups in $RC_8H_7Fe(CO)_3$ compounds are seen to force the iron residue to the side of the ring opposite the substituent; perhaps because the double bonds adjacent to R experience the greatest level of electron deficiency. Alkyl and heteroatomic groups give a more complex distribution, see, e.g. [18, 50].

Table 9
Compounds of the Type $R_{8-n}C_8H_nFe(CO)_3$.
Further information on numbers preceded by an asterisk is given at the end of the table, pp. 269/77.
For abbreviations and dimensions see p. VIII.

No.	compound	method of preparation (yield in %), properties and remarks	Ref.
compounds of the type $RC_8H_7Fe(CO)_3$ and $[R_3MCH_2C_8H_7Fe(CO)_3]PF_6$:			
*1	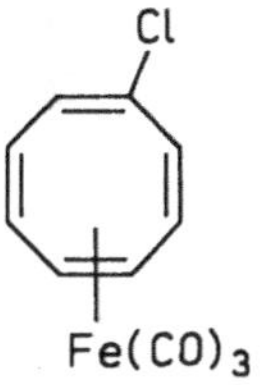 Cl-substituted cyclooctatetraene ring with $Fe(CO)_3$	IVa	[5]

Table 9 [continued]

No.	compound	method of preparation (yield in %), properties and remarks	Ref.
*2	Br (on ring); $Fe(CO)_3$	IIb (46) m.p. 54 to 56°, deep red crystals ^{1}H NMR (benzene-d_6): 3.67 (t, 1H; J=8.0), 4.32 (m, 4H), 5.66 (d, 2H; J=10.0) IR (CCl_4): 1982, 2000, 2057 mass spectrum: $[M-nCO]^+$ (n=0 to 3)	[53]
*3	OCH_3 (ring positions 2 to 8); $Fe(CO)_3$	IIb (45), IId (32), IIIb (79) m.p. 40 to 42°, m.p. 45 to 46°, m.p. 46 to 46.5° (from pentane), subl. 40°/0.1 Torr, b.p. 85 to 90°/0.03 Torr, crystallizes on standing, red-orange plates, large dark red needles ^{1}H NMR ($CDCl_3$): 3.50 (s, CH_3), 5.45 to 5.52 (m, 7H) ^{1}H NMR ($CDCl_3$): 3.54 (s, CH_3), 4.45 to 6.80 (m, H-2 to 8) IR (Nujol): 1985, 2060 IR (cyclohexane): 1974, 1988, 2050 (CO)	[41, 44]
*4	SCH_3; $Fe(CO)_3$	IId (very low yield) m.p. 62 to 63°, red crystals ^{1}H NMR ($CDCl_3$): 2.35 (s, CH_3), 5.1 to 6.7 (m, 7H) IR (hexane): 1976, 1992, 2048 (CO) mass spectrum: $[M]^+$	[50]
*5	$N{=}N{-}C_6H_5$; $Fe(CO)_3$	IVb (low yield)	[47]
*6	$Si(CH_3)_3$; $Fe(CO)_3$	IIb (40) m.p. 49°, deep red crystals ^{1}H NMR (CS_2): 0.17 (s, 9H), 4.43 (d, 2H), 5.35 (m, 2H), 5.81 (m, 3H) IR (hexane): 1977, 2001, 2053 (CO) mass spectrum: $[M]^+$	[9, 37, 38]

References on pp. 277/9

Table 9 [continued]

No.	compound	method of preparation (yield in %), properties and remarks	Ref.
*7	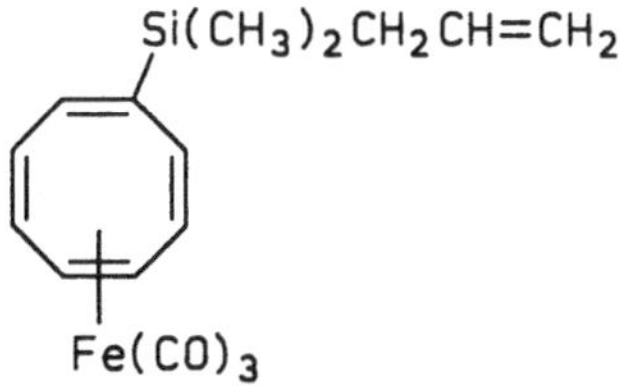 $Si(CH_3)_2CH_2CH{=}CH_2$ ... $Fe(CO)_3$	IIb (60) red oil ^1H NMR (CS_2): 0.25 (s, 6H), 1.55 (d, 2H), 4.36 (d, 2H), 4.75 (m, 1H), 4.85 (s, 1H), 5.2 (t, 2H), 5.7 (m, 4H) IR (hexane): 1968, 1988, 2052 (CO) mass spectrum: $[M]^+$	[38]
*8	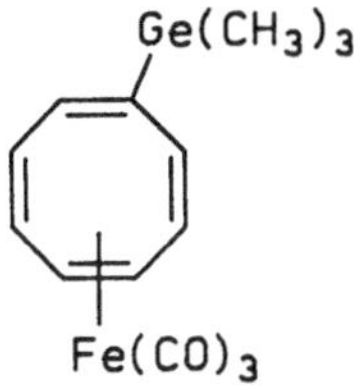 $Ge(CH_3)_3$ ... $Fe(CO)_3$	IIb (60) m.p. 62 to 64°, red crystals ^1H NMR (CS_2): 0.3 (s, 9H), 4.39 (d, 2H), 5.29 (m, 2H), 5.75 (m, 3H) IR (hexane): 1974, 1984, 2044 (CO) mass spectrum: $[M]^+$	[38]
*9	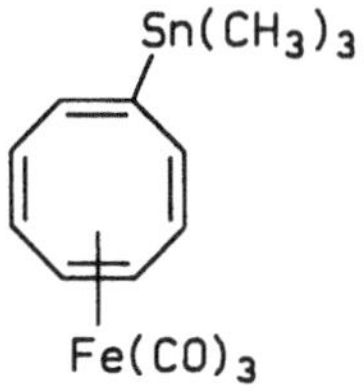 $Sn(CH_3)_3$ ... $Fe(CO)_3$	IIb (10) m.p. 91°, red crystals ^1H NMR (CS_2): 0.25 (s, 9H), 4.23 (d, 2H), 5.3 (m, 2H), 5.9 (m, 3H) IR (hexane): 1971, 1991, 2050 (CO) mass spectrum: $[M]^+$	[38]
*10	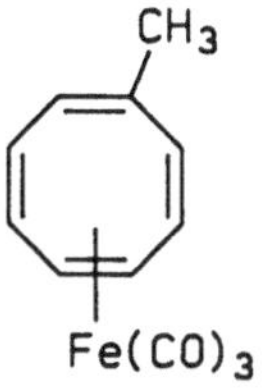 CH_3 ... $Fe(CO)_3$	IIb (49), IIc, IVa (good yield) m.p. 40°, subl. 85°/0.05 Torr, dark red crystals ^1H NMR (CS_2): 1.84 (CH_3), 4.33 (d, 1H; J=10), 4.7 to 5.8 (complex m, 6H) ^1H NMR (at 30°): 1.87 (s, CH_3); 4.5 (d), 5.08 (t), 5.67 (d) (ratio 1.8:2.3:3) ^1H NMR (at −120°): 1.9 (CH_3); 4.38, 5.95 (ratio 1:1) IR: 1967, 1985, 2048 (CO)	[5, 7, 10, 13, 14, 32]
*11	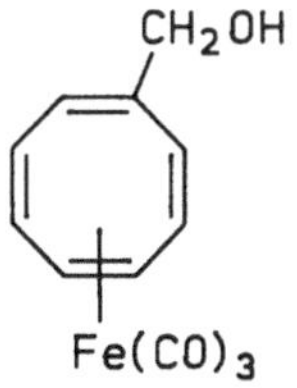 CH_2OH ... $Fe(CO)_3$	see No. 18, p. 273 b.p. 100°/0.05 Torr, red oil ^1H NMR (CS_2): 3.25 (t, OH; J=6), 3.85 (d, CH_2; J=6), 5.20 (m, 7H) IR (CCl_4): 1970, 1990, 2050 (CO) IR (liquid film): 3340 (OH) mass spectrum: $[M-nCO]^+$ (n=0 to 3), $[M-3CO-H_2O]^+$	[21, 26, 36]
*12	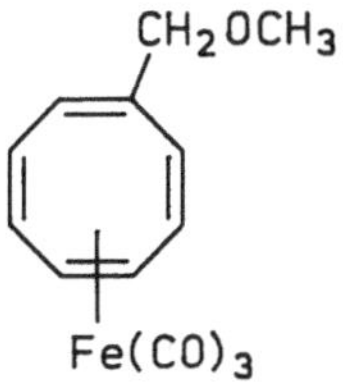 CH_2OCH_3 ... $Fe(CO)_3$	Va (89) b.p. 100°/0.05 Torr, red oil ^1H NMR (CS_2): 3.2 (s, CH_3), 3.7 (s, CH_2), 5.0 (m, 7H) IR (liquid film): 1089 (C-O); 1975, 2050 (CO) mass spectrum: $[M-nCO]^+$ (n=0 to 3), $[M-3CO-CH_2O]^+$, $[CH_3OCH_2C_8H_7]^+$	[21, 26, 36]

References on pp. 277/9

Table 9 [continued]

No.	compound	method of preparation (yield in %), properties and remarks	Ref.
13	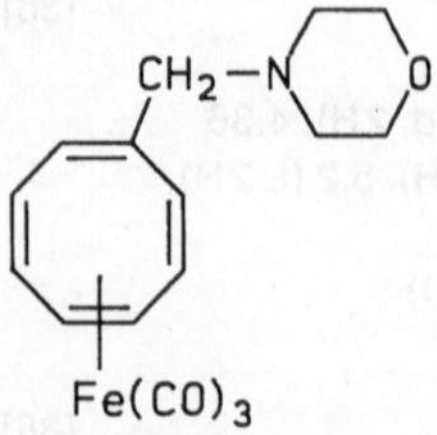 $CH_2-N(CH_2CH_2)_2O$ substituted $C_8H_7\,Fe(CO)_3$	Va b.p. 120°/0.05 Torr, red oil ^{1}H NMR (CS_2): 2.3 (t, CH_2 in $N(CH_2CH_2)_2O$; J=6), 2.8 (s, CH_2), 3.5 (t, CH_2 in $N(CH_2CH_2)_2O$; J=6), 5.1 (m, 7H) IR (CCl_4): 1974, 1993, 2051 (CO) mass spectrum: $[M-nCO]^+$ (n=0 to 3), $[M-3CO-C_2H_2]^+$, $[M-3CO-N(CH_2CH_2)_2O]^+$	[21, 26]
*14	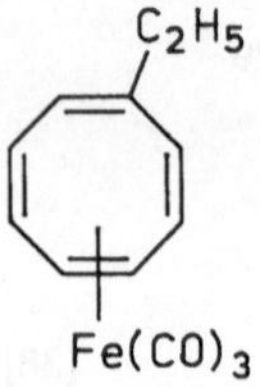 C_2H_5, $Fe(CO)_3$	—	[7, 8]
*15	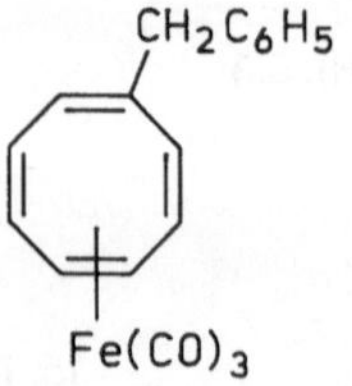 $CH_2C_6H_5$, $Fe(CO)_3$	IIIc	[5]
*16	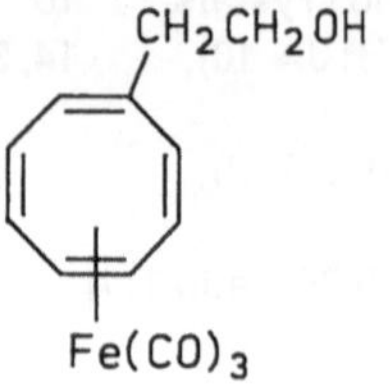 CH_2CH_2OH, $Fe(CO)_3$	—	[54]
17	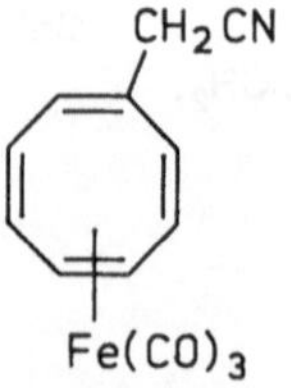 CH_2CN, $Fe(CO)_3$	Va b.p. 50°/0.05 Torr, red oil IR (liquid film): 1100 (C–O); 1990, 2060 (CO); 2220 (CN) mass spectrum: $[M-nCO]^+$ (n=0 to 3), $[M-3CO-HCN]^+$	[21, 26]
*18	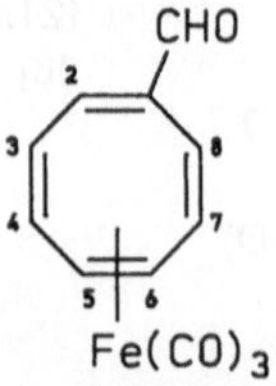 CHO, $Fe(CO)_3$ (ring positions 2 to 8)	IVb (60) m.p. 57 to 58° ^{1}H NMR (CS_2): 4.38 (t, H-5; J=9), 5.34 (m, H-3, 4, 6, 7), 6.09 (d, H-2, 8; J=10) IR (CCl_4): 1680 (C=O); 1988, 2000, 2060 (CO) mass spectrum: $[M-nCO]^+$	[21, 26]

References on pp. 277/9

Table 9 [continued]

No.	compound	method of preparation (yield in %), properties and remarks	Ref.
19	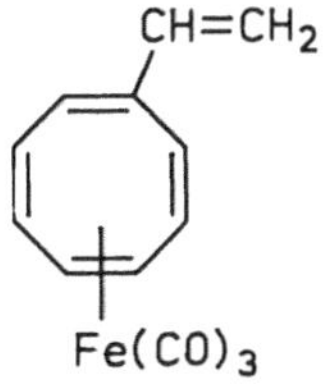 $CH{=}CH_2$-$C_8H_7Fe(CO)_3$	see Nos. 16, p. 272, 18, p. 273, 34, p. 275 red oil ^{1}H NMR ($CDCl_3$): 4.95 to 5.70 (m, 9H), 6.18 (m, 1H; J(H,H) = 10, J(H,H) = 16) IR (pentane): 1981, 1997, 2054 mass spectrum: $[M]^+$	[54]
20a	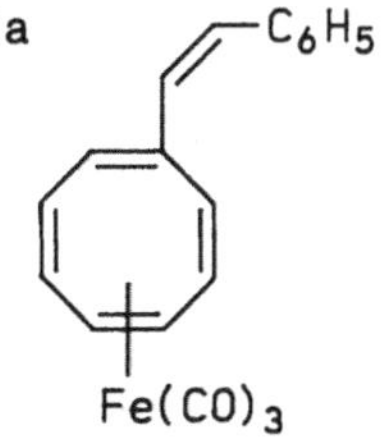 $C_6H_5CH{=}CH$-$C_8H_7Fe(CO)_3$	see No. 34, p. 275 m.p. 68 to 70°, red crystals ^{1}H NMR ($CDCl_3$): 4.70 (d, 2H; J(H,H) = 10), 5.02 (t, 2H; J(H,H) = 10), 5.75 (m, 4H), 6.45 (d, 1H; J(H,H) = 12), 7.22 (s, 5H) IR (CCl_4): 1973, 1990, 2049 mass spectrum: $[M]^+$	[54]
b	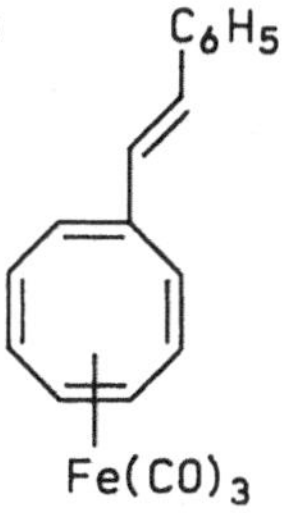 $C_6H_5CH{=}CH$-$C_8H_7Fe(CO)_3$	see No. 34, p. 275 m.p. 87 to 89°, deep red crystals ^{1}H NMR ($CDCl_3$): 5.32 (m, 7H), 6.55 (d, 1H; J(H,H) = 16), 6.82 (d, 1H; J(H,H) = 16), 7.30 (m, 5H) IR (CCl_4): 1973, 1991, 2049 mass spectrum: $[M]^+$	[54]
*21	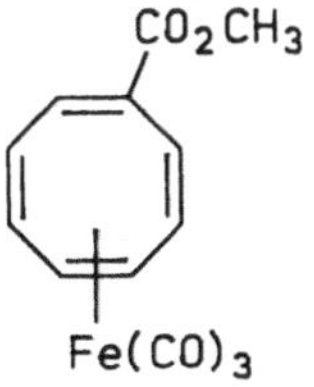 CO_2CH_3-$C_8H_7Fe(CO)_3$	IIb (80), IIc (21) m.p. 44°, m.p. 52.5 to 53°, subl. 48°/0.01 Torr, red or dark red crystals ^{1}H NMR ($CDCl_3$): 4.72 (m), 5.35 (dd), 6.85 (d) IR (hexane): 1720 (C=O); 1974, 2000, 2056 (CO)	[33, 41]
*22	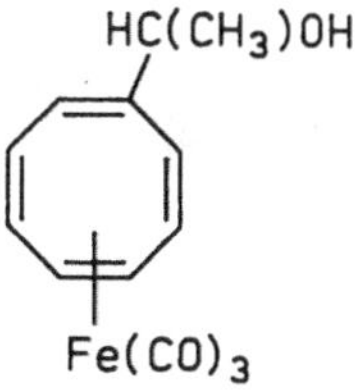 $HC(CH_3)OH$-$C_8H_7Fe(CO)_3$	see Nos. 18, p. 273, and 28, p. 273 b.p. 100°/0.05 Torr, red oil ^{1}H NMR (CS_2): 1.30 (d, CH_3; J = 6), 1.44 (d, OH; J = 3), 3.90 (q of d, CH), 5.0 (m, 7H) IR (CCl_4): 1973, 1993, 2053 (CO) IR (liquid film): 3400 (OH) mass spectrum: $[M-nCO]^+$ (n = 0 to 3), $[M-3CO-H_2O]^+$	[21, 26, 36]
23	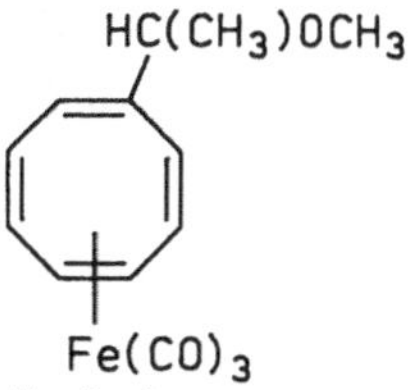 $HC(CH_3)OCH_3$-$C_8H_7Fe(CO)_3$	Vc (83) red oil ^{1}H NMR (CS_2): 1.2 (d, CH_3; J = 6), 3.1 (OCH_3), 3.55 (q, CH; J = 6), 5.1 (m, 7H) mass spectrum: $[M-nCO]^+$ (n = 0 to 3), $[M-3CO-CH_2O]^+$, $[M-3CO-CH_3OH]^+$	[21, 26]

References on pp. 277/9

Table 9 [continued]

No.	compound	method of preparation (yield in %), properties and remarks	Ref.
*24	$HC(C_2H_5)OH$-substituted cyclooctatetraene $Fe(CO)_3$	–	[52]
*25	$HC(C_2H_5)OCH_3$-substituted cyclooctatetraene $Fe(CO)_3$	–	[52]
*26	$HC(C_4H_9\text{-}n)OH$-substituted cyclooctatetraene $Fe(CO)_3$	see No. 18, p. 273 ^{1}H NMR ($CDCl_3$): 0.87 (t, 3H in C_4H_9-n; J=6.5), 1.27 (m, 4H in C_4H_9-n), 1.57 (m, 2H in C_4H_9-n), 2.06 (s, OH), 3.83 (t, 1H, CH; J(CH, C_4H_9)=5.0), 4.50 to 5.90 (m, 7H)	[52]
*27	$HC(C_4H_9\text{-}n)OCH_3$-substituted cyclooctatetraene $Fe(CO)_3$	^{1}H NMR ($CDCl_3$): 0.89 (t, 3H in C_4H_9-n; J=7.0), 1.31 (m, 4H in C_4H_9-n), 1.60 (m, 2H in C_4H_9-n), 3.20 (s, OCH_3), 3.51 (t, CH; J(CH, C_4H_9)=6.5), 4.60 to 5.65 (m, 7H)	[52]
*28	$C(O)CH_3$-substituted cyclooctatetraene $Fe(CO)_3$ (ring positions 2 to 8)	IVb (5) b.p. 100°/0.05 Torr, red oil ^{1}H NMR (CS_2): 2.30 (s, CH_3), 4.46 (t, H-5; J=8), 5.16 (m, H-3, 4, 6, 7) IR (CCl_4): 1663 (C=O); 1983, 1995, 2053 (CO) mass spectrum: $[M-nCO]^+$ (n=0 to 4), $[M-3CO-COCH_3]^+$, $[FeCH_3]^+$	[21, 26, 36]
*29	CN-substituted cyclooctatetraene $Fe(CO)_3$	–	[8, 18 to 20]

References on pp. 277/9

Table 9 [continued]

No.	compound	method of preparation (yield in %), properties and remarks	Ref.
*30	$C(C_6H_5)_3$ … $Fe(CO)_3$	IVb (81), Vb (59 to 74) crystals (from hexane at −20°) ^{1}H NMR ($CDCl_3$): 4.38 and 4.63 (each t, 2.9H, H-4,5,6; J(H-5,6) = 7.7), 5.08 (t, 2.1H, H-3,7; J(H-6,7) = 9), 5.74 (d, 2H, H-2,8; J(H-7,8) = 11.4), 7.28 (m, $3C_6H_5$) ^{13}C NMR ($CDCl_3$): 68.9 ($C(C_6H_5)_3$), 80.8 (C-5), 85.4 (C-4,6), 96.7 (C-3,7), 117.9 (C-2,8), 125.6 (p-C in C_6H_5; 127.6, 131.2 (o-, m-C in C_6H_5), 140.7 (C-1), 145.7 (C-1 in C_6H_5) mass spectrum: $[M-nCO]^+$ (n = 0 to 3), $[M-3CO-C_2H_2]^+$, $[M-3CO-Fe]^+$	[9, 42]
*31	C_6H_5 … $Fe(CO)_3$	Ia (28), IIa, IIb (50), IIIa, IIIc (60) m.p. 65°, subl. 53°/0.01 Torr, subl. 120°/5 Torr, dec. 195°, deep red or dark red crystals (from petroleum ether or CH_3OH), red needles ^{1}H NMR ($CDCl_3$): 5.3 (br, s, 7H), 7.25 (br, s, 5H) IR (Nujol): 654, 698, 757, 772, 835, 880, 900, 970, 1000, 1030, 1075, 1157, 1187, 1250, 1310, 1380, 1415, 1430, 1456, 1470, 1488, 1565, 1581, 1600, 1985, 2055, 2865, 2950, 3020 IR (n-heptane): 1985, 2000, 2060 (CO)	[4, 6, 9, 27, 31, 41, 53]
*32	$C_6H_4NO_2$-4 … $Fe(CO)_3$	Vc (55) m.p. 128 to 131°, red-orange solid ^{1}H NMR ($CDCl_3$): 4.78 (t, H-5; J(H-5,6) = 8), 5.02 (dd, H-4,6; J(H-6,7) = 10), 5.40 (dd, H-3,7; J(H-7,8) = 10), 5.80 (d, H-2,8), 7.46 (d, 2H in C_6H_4; J = 8), 8.16 (d, 2H in C_6H_4) ^{13}C NMR ($CDCl_3$): 87.25 (C-5); 91.98, 97.92, 109.15 (C-2 to 4, 6 to 8); 123.59, 128.02 (4CH in C_6H_4); 128.56, 146.58, 153.01 (C-1, 2C in C_6H_4); 211.38 (CO) IR (CH_2Cl_2): 1989, 2047 (CO) IR (hexane): 1983, 1997, 2055 (CO) mass spectrum: $[M]^+$	[55]
*33	$[CH_2N(CH_3)_3 … Fe(CO)_3]^+ [PF_6]^-$	Va (97) dec. 165°, red crystals ^{1}H NMR (acetone-d_6): 3.12 (s, $3CH_3$), 4.12 (s, CH_2), 5.06 (m, 3H-ring), 5.37 (t, 2H-ring; J = 9.5), 5.82 (d, 2H-ring; J = 11.2) IR (acetone): 1984, 2053	[52]

References on pp. 277/9

Table 9 [continued]

No.	compound	method of preparation (yield in %), properties and remarks	Ref.
*34 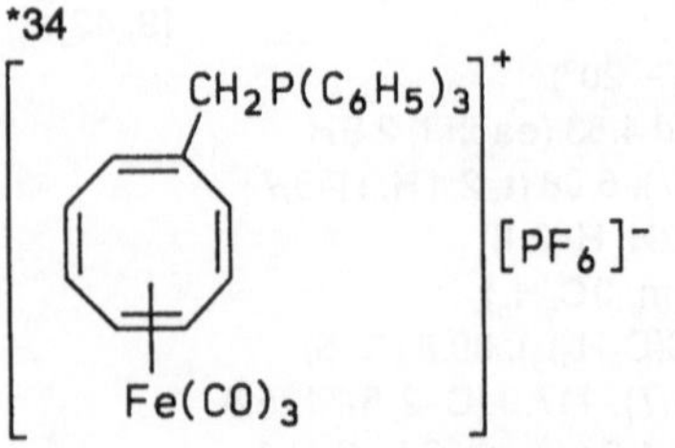	$[CH_2P(C_6H_5)_3$ … $Fe(CO)_3]^+$ $[PF_6]^-$	Va (ca. 95) m.p. 80 to 82°, red solid ^{1}H NMR ($CDCl_3$): 4.03 (d, CH_2; J(H, P) = 13.5), 4.90 to 5.30 (m, 7H), 7.40 to 7.95 (m, C_6H_5) IR ($CHCl_3$): 1989, 2000, 2059	[52, 54]
*35 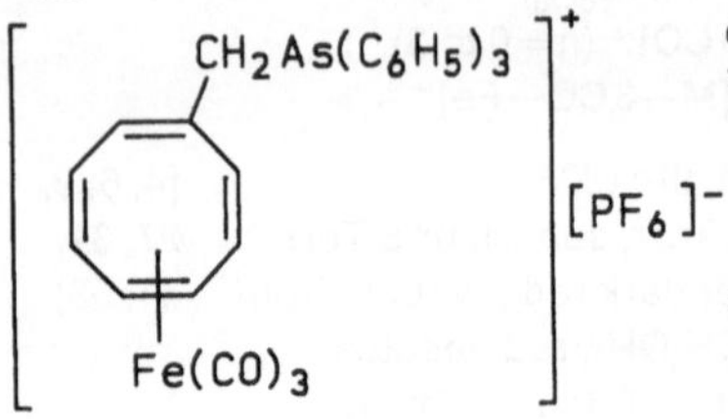	$[CH_2As(C_6H_5)_3$ … $Fe(CO)_3]^+$ $[PF_6]^-$	Va (ca. 95) dec. ca. 155°, red solid ^{1}H NMR (acetone-d_6, free excess $As(C_6H_5)_3$): 4.62 (s, CH_2), 5.03 (m, 4H in C_8H_7), 5.44 (m, 3H in C_8H_7), 7.24 (m, free $As(C_6H_5)_3$), 7.70 (m, C_6H_5) IR (CH_2Cl_2), free excess $As(C_6H_5)_3$): 1989, 2056	[52]

compounds of the type $R_2C_8H_6Fe(CO)_3$ (continued with No. 58, p. 269):

No.	compound	method of preparation (yield in %), properties and remarks	Ref.
*36	CO_2CH_3, CO_2CH_3 (positions 2–7), $Fe(CO)_3$	IIIc (12 to 30) m.p. 114 to 115°, subl. 80°/0.5 Torr, red solid (from petroleum ether) or orange crystalline material ^{1}H NMR ($CDCl_3$): 3.71 (CH_3), 4.09 (complex m, H-3,6), 5.09 (complex m, H-4,5), 7.28 (d, H-2,7; J(H-6,7) = 10) ^{57}Fe-γ (−78°): δ = 0.33, Δ = 1.27 ^{57}Fe-γ (294 K): δ = 0.327, Δ = 1.28 IR (KBr): 1055, 1065, 1250, 1600, 1730, 2000, 2065 IR (CCl_4): 1055, 1065, 1250, 1730, 2000, 2065 UV (C_2H_5OH): $\lambda_{max}(\varepsilon)$ = 270 (16000), 333 (13000) nm mass spectrum: $[M-nCO]^+$ (n = 0 to 3)	[11, 12, 22, 24]
*37a	Cl, CH_3 (positions 2–7), $Fe(CO)_3$	mixture of isomers: IIIb (38 to 60) m.p. ca. 68 to 76°, yellow-red crystals (from pentane) ^{1}H NMR (CS_2): 1.84 (s, CH_3), 2.08 (s, CH_3), 3.72 (complex m, H-5), 4.84 (complex m, H-6, 7), 5.72 (complex m, H-2, 3, 4) IR: 1975, 2050 UV (C_2H_5OH): $\lambda_{max}(\varepsilon)$ = 241 (35000), 320 (6000), 415 (2200) nm	[57, 58]

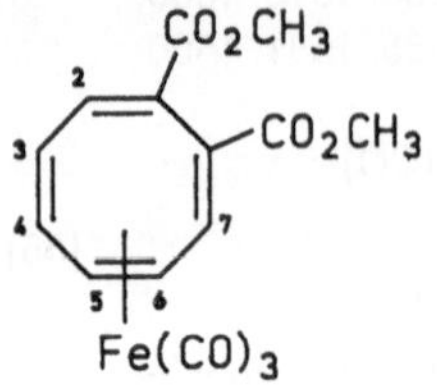

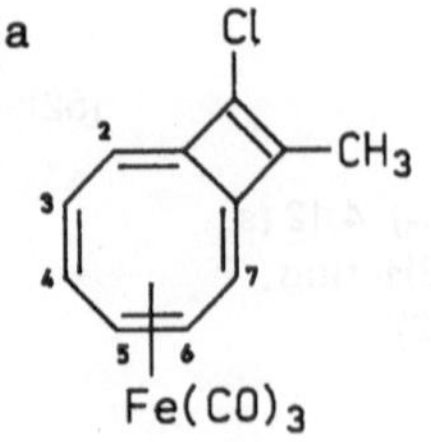

References on pp. 277/9

Table 9 [continued]

No.	compound	method of preparation (yield in %), properties and remarks	Ref.
b	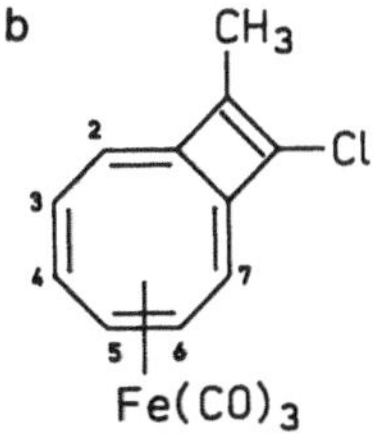 (structure; Fe(CO)3 complex with CH3 and Cl substituents)		
*38	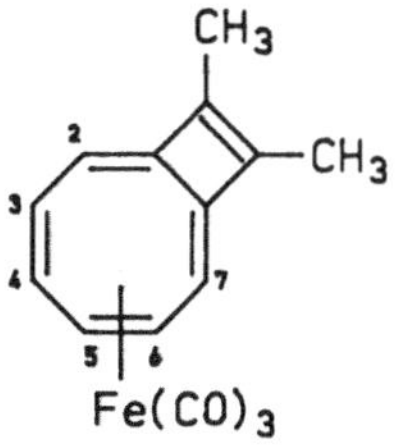 (structure; Fe(CO)3 complex with two CH3 substituents)	IIc ^{1}H NMR (CS_2): 1.80 (q, CH_3; J = 1), 2.05 (q, CH_3), 3.78 (m, H-5), 4.75 (m, H-6, 7), 5.68 (m, H-2, 3, 4)	[57]
*39	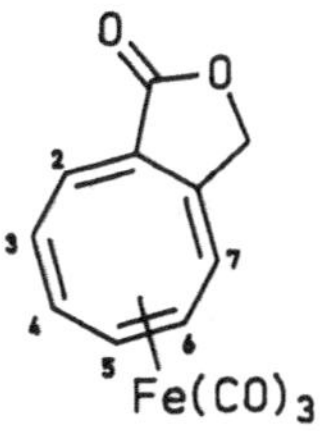 (structure; Fe(CO)3 lactone complex)	Ib (11) m.p. 135 to 137°, red solid ^{1}H NMR ($CDCl_3$): 4.55 (complex, H-4, 7), 4.815 (CH_2), 4.825 (CH_2), 5.1 (d, H-2; J(H-2, 3) = 10.5), 5.55 (H-5, 6), 5.95 (complex m, H-3) ^{57}Fe-γ (294 K): $\delta = 0.33$, $\Delta = 1.25$ ^{57}Fe-γ (294 K, methyltetrahydrofuran): $\delta = 0.33$, $\Delta = 1.17$ IR (KBr): 1050, 1060, 1160, 1750, 1960, 2065 IR (CCl_4): 1050, 1060, 1175, 1760, 1998, 2065, 2800, 2890 UV (C_2H_5OH): $\lambda_{max}(\varepsilon) = 300$ (13100) nm mass spectrum: $[M - nCO]^+$ (n = 0 to 3)	[12, 24]
*40	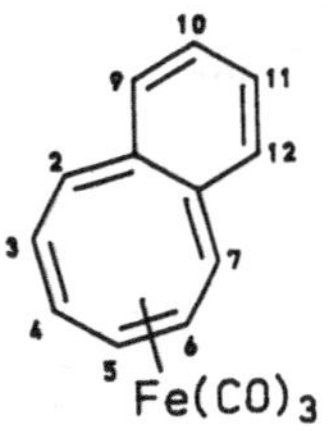 (structure; Fe(CO)3 benzocyclooctatetraene complex)	Ib, IIc (16.3), IIIc (62.3) m.p. 77 to 79° (dec.), yellow-orange prisms (from petroleum ether) ^{1}H NMR (CCl_4): 3.73 (br, s, 1H), 4.60 (br, s, 1H), 4.90 (br, s, 2H), 6.08 (br, s, H-2, 3), 7.13 (s, H-9 to 12) ^{1}H NMR (toluene-d_8, −15°): 3.58 (m, 1H), 4.57 (m, 3H), 6.20 (m, 2H) ^{1}H NMR (toluene-d_8, 93°): 4.10 (s, H-4, 5), 5.40 (H-2, 3, 6, 7) IR (film): 760, 780, 975, 1000, 1025, 1450, 1500, 1950, 1995, 2045 UV (ether): $\lambda_{max}(\varepsilon) = 237.5$ (24100), 270 (20300) mass spectrum: $[M - nCO]^+$	[15, 17, 28 to 30]

References on pp. 277/9

Table 9 [continued]

No.	compound	method of preparation (yield in %), properties and remarks	Ref.
*41	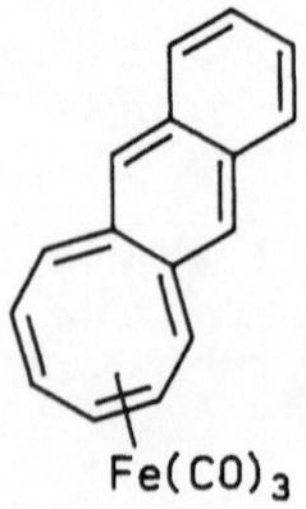	–	[28, 30]
42	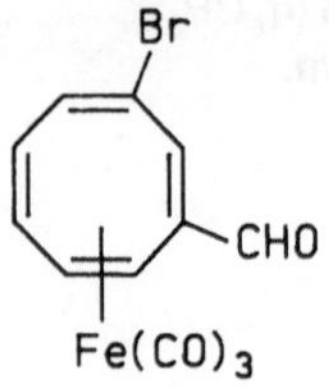	see No. 2, p. 270 m.p. 111 to 112°, red-orange crystals ^{1}H NMR (CS_2): 4.55 (m, 2H), 4.99 (dd, 1H; J = 8.0, J = 10.0), 5.30 (d, 1H; J = 10), 6.30 (d, 1H; J = 10.0), 6.78 (s, 1H), 9.14 (s, 1H) IR (CCl_4): 1687, 1994, 2012, 2065 mass spectrum: $[M-nCO]^+$ (n = 0 to 3)	[53]
*43	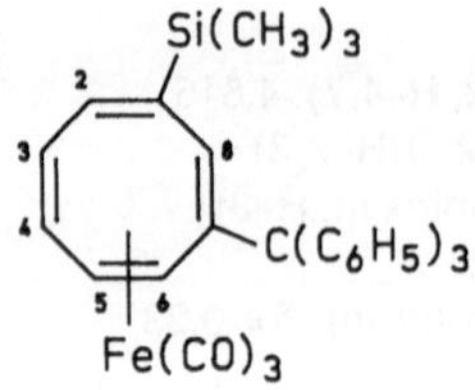	see No. 6, p. 271 m.p. 162 to 165°, red crystals ^{1}H NMR ($CDCl_3$): 0.15 (s, CH_3), 4.43 (dd, H-3), 4.58 (d, H-2; J(H-2,3) = 9.0), 4.68 (s, H-8; J(H-6,8) < 1), 5.57 (dd, H-5), 5.94 (dd, H-4; J(H-4,5) = 10.5, J(H-3,4) = 8.5), 6.43 (d, H-6; J(H-5,6) = 9.5), 7.3 (m, C_6H_5) IR (hexane): 1974, 1990, 2047 (CO) mass spectrum: $[M]^+$	[38]
*44	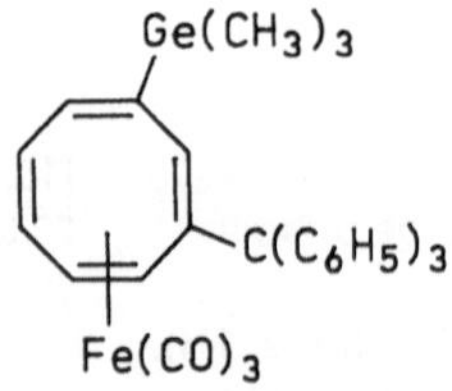	see No. 8, p. 271 m.p. 161 to 163°, red crystals ^{1}H NMR (CS_2): 0.1 (s, CH_3), 4.31 (m, 3H), 5.47 (dd, 1H), 5.9 (dd, 1H), 6.36 (d, 1H), 7.25 (m, C_6H_5) IR (hexane): 1973, 1982, 2041 (CO) mass spectrum: $[M]^+$	[38]
45	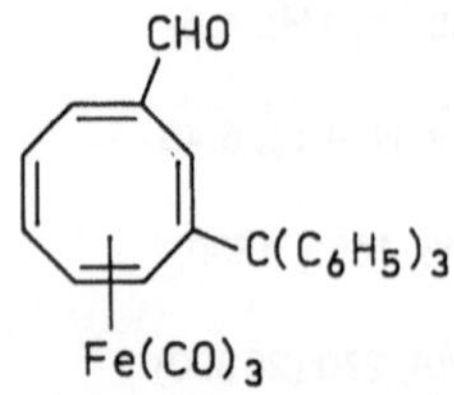	see No. 30, p. 274 m.p. 185° (dec.), red crystals ^{1}H NMR ($CDCl_3$): 4.24 (dd, 1H; J = 10.5, J = 9.0), 4.75 (m, 2H), 5.33 (d, 1H; J = 9.0), 6.24 (d, 1H; J = 10.0), 6.59 (s, 1H), 7.19 (m, C_6H_5), 8.95 (s, 1H) IR (CCl_4): 1672, 1985, 1998, 2057 mass spectrum: $[M-3CO]^+$, $[M-Fe-3CO]^+$	[53]

Table 9 [continued]

No.	compound	method of preparation (yield in %), properties and remarks	Ref.
46	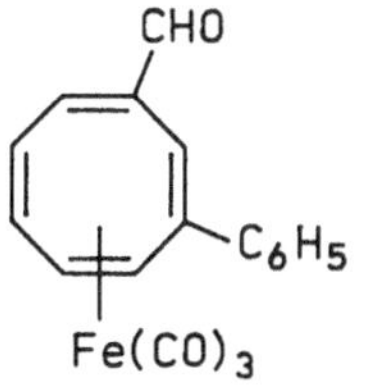 1-CHO, 3-C_6H_5-cyclooctatetraene $Fe(CO)_3$	see No. 31, p. 274 red oil ^{1}H NMR ($CDCl_3$): 4.26 (dd, 1H; J=10.0, J=9.5), 5.34 (d, 1H; J=9.5), 5.70 (d, 1H; J=10.0), 5.92 (s, 1H), 6.13 (dd, 1H; J=9.5, J=9.5), 6.43 (d, 1H; J=9.5), 7.28 (m, C_6H_5), 9.44 (s, 1H) IR (CCl_4): 1679, 1986, 2002, 2058 mass spectrum: $[M-nCO]^+$ (n=0 to 3)	[53]
*47	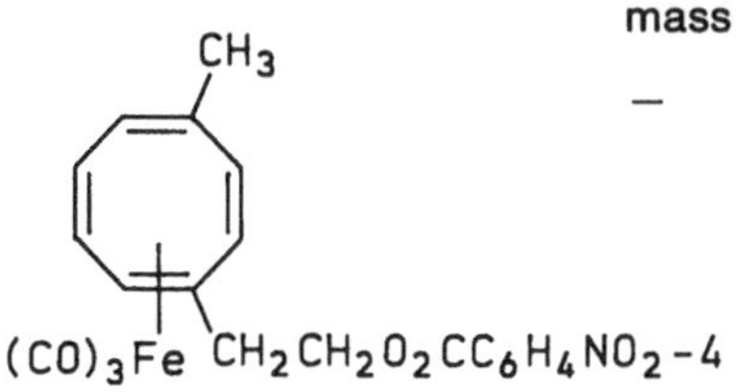 CH_3, $CH_2CH_2O_2CC_6H_4NO_2$-4, $(CO)_3Fe$	–	[39]
*48	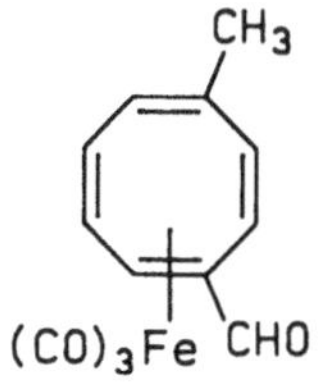 CH_3, CHO, $(CO)_3Fe$	see No. 10, p. 271 deep red oil ^{1}H NMR ($CDCl_3$): 4.39 (d, 1H; J=9.5), 4.82 (d, 1H; J=10.5), 5.63 (dd, 1H; J=10.5, J=9.5), 5.87 (d, 1H; J=10.5), 6.08 (t, 1H; J=10.5), 6.45 (d, 1H; J=10.5), 9.36 (s, 1H) IR (CCl_4): 1682, 1983, 1999, 2057 mass spectrum: $[M-nCO]^+$ (n=0 to 3)	[53]
49	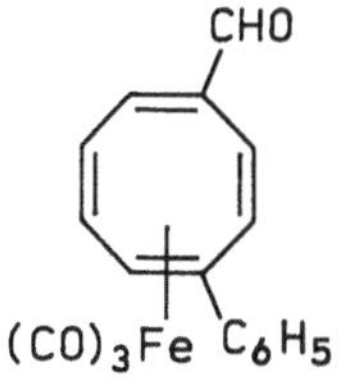 CHO, C_6H_5, $(CO)_3Fe$	see No. 31, p. 274 deep red oil ^{1}H NMR ($CDCl_3$): 4.72 (d, 1H; J=10.0), 5.31 (d, 1H; J=10.0), 5.94 (t, 1H; J=10.0), 6.05 (d, 1H; J=10.0), 6.26 (d, 1H; J=10.0), 7.30 (m, C_6H_5), 9.40 (s, 1H) IR (CCl_4): 1683, 1988, 2002, 2058 mass spectrum: $[M-nCO]^+$ (n=0 to 3)	[53]
*50	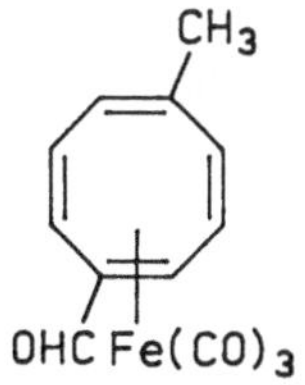 CH_3, OHC, $Fe(CO)_3$	see No. 10, p. 271 red-orange oil ^{1}H NMR ($CDCl_3$): 4.51 (d, 2H; J=9.0), 5.33 (dd, 2H; J=11.5, J=10.0), 6.74 (d, 2H; J=11.5), 9.35 (s, 1H) IR (CCl_4): 1684, 1984, 1998, 2057 mass spectrum: $[M-nCO]^+$ (n=0 to 3)	[53]
51	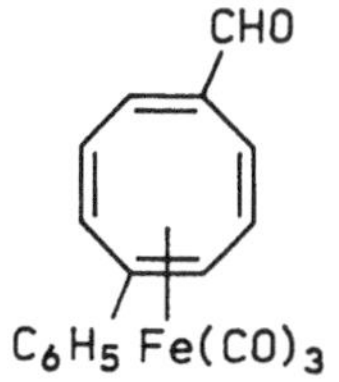 CHO, C_6H_5, $Fe(CO)_3$	see No. 31, p. 274 m.p. 96 to 97°, orange crystals ^{1}H NMR ($CDCl_3$): 4.90 (d, 2H; J=9.0), 5.50 (dd, 2H; J=11.5, J=9.0), 6.84 (d, 2H; J=11.5), 7.32 (m, C_6H_5), 9.34 (s, 1H) IR (CCl_4): 1689, 1986, 2002, 2058 mass spectrum: $[M-nCO]^+$ (n=0 to 3)	[53]

References on pp. 277/9

Table 9 [continued]

No.	compound	method of preparation (yield in %), properties and remarks	Ref.
*52	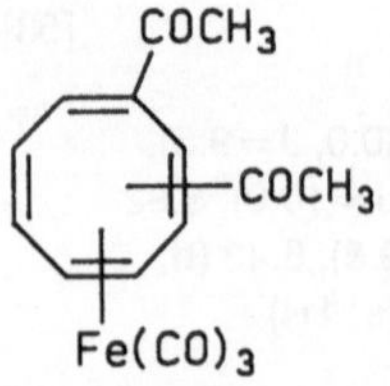 $COCH_3$, $COCH_3$, $Fe(CO)_3$	isomer mixture: IVb (0.9) b.p. 120°/0.05 Torr ^{1}H NMR ($CFCl_3$): 2.26 (s, CH_3), 2.28 (s, CH_3; J=5.4), 2.40 (s, CH_3; J=7.4), 4.53 (d, 12.5H; J=7), 4.63 (d, 12.5H; J=7), 5.04 (t, 12.5H; J=7), 5.94 (s, 14H), 6.06 (s, 6H), 6.42 (d, 22.0H; J=10), 6.82 (d, 6.0H; J=3), 7.02 (s, 6.0H), 7.65 (s, 10.5H) IR (liquid film): 1670, 1990, 2055 mass spectrum: $[M]^+$	[21, 26]
*53	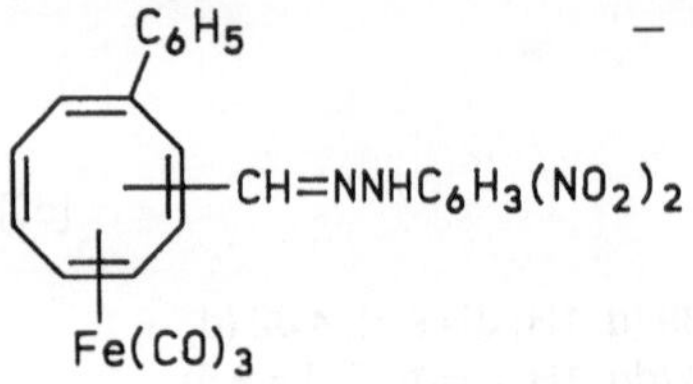 C_6H_5, $CH{=}NNHC_6H_3(NO_2)_2$, $Fe(CO)_3$	—	[31]
*54	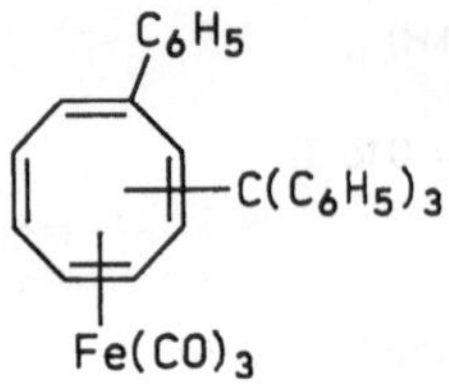 C_6H_5, $C(C_6H_5)_3$, $Fe(CO)_3$	see No. 31, p. 274 m.p. 160 to 162°, red crystals ^{1}H NMR ($CDCl_3$): 4.5 to 5.2 (m, 5H), 5.9 (m, 2H), 7.2 (m, 20H) IR (hexane): 1975, 1985, 2042 (CO) mass spectrum: $[M]^+$	[38]

compound of the type $R'R_2C_8H_5Fe(CO)_3$:

No.	compound	method of preparation (yield in %), properties and remarks	Ref.
55	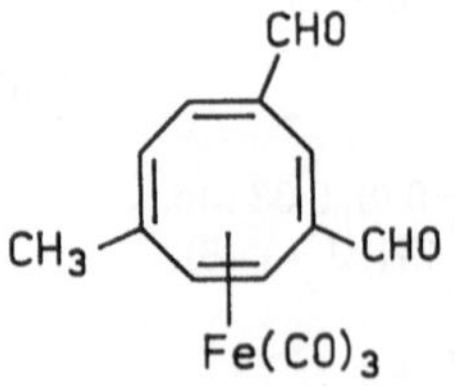 CHO, CH_3, CHO, $Fe(CO)_3$	see No. 10, p. 271 m.p. 95 to 97°, deep red solid ^{1}H NMR ($CDCl_3$): 4.65 (d, 2H; J=10.5), 6.23 (d, 2H; J=10.5), 7.44 (s, 1H), 9.48 (s, 2H) IR ($CHCl_3$): 1689, 1994, 2009, 2064 mass spectrum: $[M-nCO]^+$ (n=0 to 3)	[53]

compounds of the type $R_4C_8H_4Fe(CO)_3$:

No.	compound	method of preparation (yield in %), properties and remarks	Ref.
56	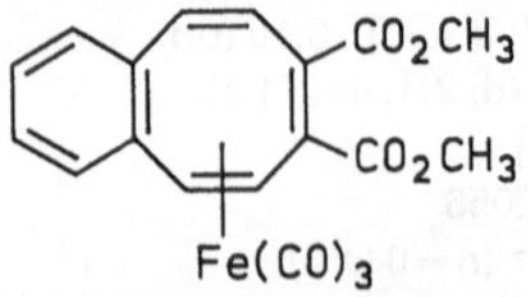 CO_2CH_3, CO_2CH_3, $Fe(CO)_3$ (benzo-fused)	IIc	[30]

References on pp. 277/9

Table 9 [continued]

No.	compound	method of preparation (yield in %), properties and remarks	Ref.
57	$(C_6H_5C_2H)_4Fe(CO)_3$	IIId m.p. 225 to 235° (dec.), deep red or violet crystals (from tetrahydrofuran), only slightly soluble in organic solvents	[1 to 3]

supplements:

compounds of the type $R_2C_8H_6Fe(CO)_3$ (continued from No. 54, p. 268):

No.	compound	method of preparation (yield in %), properties and remarks	Ref.
*58	3 2 H 10 11 12 13 H 4 5 6 7 $Fe(CO)_3$	m.p. 112 to 113°, red crystals ^{1}H NMR (acetone-d_6): 3.95 (m, H-2,7), 5.35 to 5.60 (m, H-9,10,11,12,13,14), 5.95 (s, H-3,4,5,6) ^{13}C NMR (acetone-d_6): 42.37 (C-2,7); 123.15, 125.80 (C-3,4,5,6); 213.72 (CO) IR (CsI): 2042 (CO); 2903, 2934 (C-H); 3011, 3028 (=CH) UV (CH_2Cl_2): $\lambda_{max}(\varepsilon)$ = 257 (14000), 310 (9900) nm mass spectrum: $[M-nCO]^+$ (n = 0 to 3), $[M-3CO-Fe-2H]^+$, $[M-3CO-C_6H_6]^+$	[60, 62]
*59	3 2 10 15 CN 11 12 14 CN 13 4 5 6 7 $Fe(CO)_3$	see No. 58, p. 277 isomer a: m.p. 143 to 144° ^{1}H NMR ($CDCl_3$): 3.0 (m, H-3,8), 3.95 to 4.2 (m, H-4,7), 4.7 to 5.25 (m, H-9,10,11,12,13,14), 6.1 to 6.3 (m, H-5,6) IR (CsI): 1993, 2037 (C=O); 2218 (CN), 2934 (C-H), 3037 (=C-H) UV (CH_2Cl_2): $\lambda_{max}(\varepsilon)$ = 300 (11100) nm mass spectrum: $[M-nCO]^+$ (n = 0 to 3)	[60, 62]
		isomer b: m.p. 159 to 161° ^{1}H NMR ($CDCl_3$): 2.9 to 3.0 (m, H-3,8), 4.0 to 4.2 (m, H-4,7), 5.09 (m, H-9,10,11,12,13,14), 6.1 to 6.28 (m, H-5,6) IR (CsI): 1994, 2046 (C=O); 2223 (CN), 2957 (C-H), 3019 (=C-H) UV (CH_2Cl_2): $\lambda_{max}(\varepsilon)$ = 305 (9900) nm mass spectrum: $[M-nCO]^+$ (n = 0 to 3)	[60, 62]

*Further information:

$ClC_8H_7Fe(CO)_3$ (Table **9**, No. **1**) is proposed as an antiknock additive for gasoline, as an additive for fuels, as a drying agent for metal plating applications, and as a chemical intermediate for metal-containing polymer materials [5].

 References on pp. 277/9

$BrC_8H_7Fe(CO)_3$ (Table **9**, No. **2**). Analysis of the ^1H NMR spectrum indicates that at room temperature in a variety of solvents tautomerism between the structures XVIIa to c (R=Br) exists. Therefore, exclusively electrophilic attack at the position 3 (or 7) is expected if position 8 is rendered sterically inaccessible by the Br substituent. Reaction with $(CH_3)_2NCHO/OPCl_3$ at 0 °C for 8 h followed by hydrolysis (overnight), extraction with ether, drying ($MgSO_4$), evaporation and chromatography of the residue on alumina with pentane/30% $O(C_2H_5)_2$ gives 1-Br(7-OHC)$C_8H_6Fe(CO)_3$ (No. 42) in 25% yield [53]. Reaction with $(NC)_2C{=}C(CN)_2$ in CH_2Cl_2 at room temperature gives XVIII (X=H, Y=Br) [35].

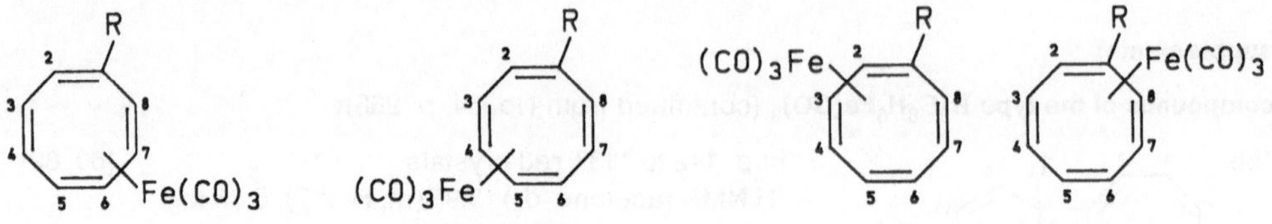

R R R R
(CO)3Fe Fe(CO)3
Fe(CO)3 (CO)3Fe
a b c d

XVII

$CH_3OC_8H_7Fe(CO)_3$ (Table **9**, No. **3**). A comprehensive study of tautomeric equilibria at room temperature proves the existence of at least three different valence tautomers, XVIIa, b, and d ($R{=}OCH_3$) [18, 19]. A solution of the compound in tetrahydrofuran gives on hydrolysis with H_2O/H_2SO_4 (1:1) compound XIX (R=H). Similarly D_2O/D_2SO_4 gives deuterated XIX (R=D) [44]. Reaction with $(NC)_2C{=}C(CN)_2$ in CH_2Cl_2 at room temperature gives XX, showing that bonding to C-6 is preferred [41]. No reaction is reported with XXI ($R{=}CH_3$, C_6H_5) [46].

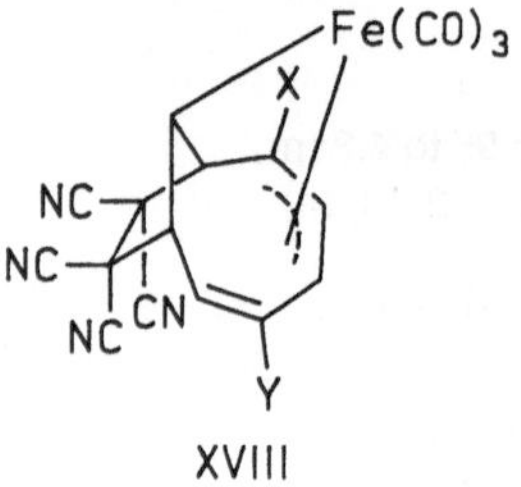

XVIII 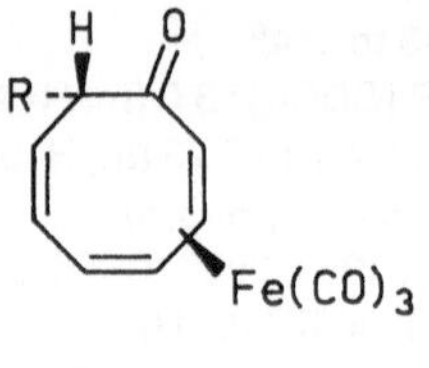 XIX 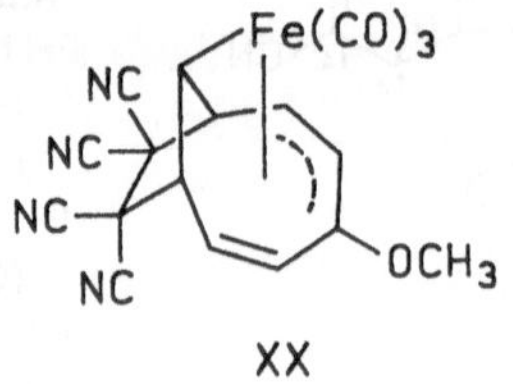 XX 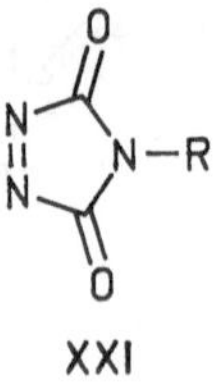 XXI

$CH_3SC_8H_7Fe(CO)_3$ (Table **9**, No. **4**). It is not possible to determine the position of SCH_3 substitution because the ring protons resonate in a narrow range δ=5.1 to 6.7 ppm as a series of overlapping multiplets. But in comparison with similar compounds, it seems likely that bonding to the electron-rich diene unit as in XVIIb occurs [50].

$C_6H_5N_2C_8H_7Fe(CO)_3$ (Table **9**, No. **5**) liberates the organic ligand ^4L with $ON(CH_3)_3 \cdot 2H_2O$ [47].

$RC_8H_7Fe(CO)_3$ (Table **9**, Nos. **6** to **9** with $R{=}Si(CH_3)_3$, $Si(CH_3)_2CH_2CH{=}CH_2$, $Ge(CH_3)_3$, and $Sn(CH_3)_3$) show in the ^1H NMR spectrum a pattern similar to the $RC_8H_7Fe(CO)_3$ compounds with $R{=}CH_3$ (No. 10), C_2H_5 (No. 14), and is reversed from that of CHO (No. 18), CO_2CH_3 (No. 21), $COCH_3$ (No. 28), CN (No. 29) [8, 20]. Compounds No. 6 to 9 have the same general pattern in the ^1H NMR spectrum down to −100 °C [8, 20, 38] and the $Fe(CO)_3$ moiety is bound as shown in XVIId [8, 20, 38, 45]. They are air-stable [38]. Treatment of No. 6 [37, 45] or No. 8 [45] in octane in a sealed tube at 150 to 160 °C for 20 h gives XXII with $R{=}Si(CH_3)_3$ or $Ge(CH_3)_3$ [37, 45], and in the case of No. 6, also $(CH_3)_3SiC_8H_7Fe_2(CO)_5$ (see 2.8.2.2 in

"Organoiron Compounds" C5, 1981, Table 7, No. 3, p. 114) [45]. Reaction of No. 6 or No. 8 with $[C(C_6H_5)_3]BF_4$ in CH_3NO_2 at room temperature for 15 min affords a brownish yellow precipitate of $[RC_8H_6Fe(CO)_3]BF_4$ on addition of ether. This salt is filtered off followed by addition of H_2O and by extraction of the filtrate with ether. The organic solution is dried ($MgSO_4$) and chromatographed on alumina with hexane to give 1-$(CH_3)_3$Si(7-$(C_6H_5)_3$C)C_8H_6-$Fe(CO)_3$ (No. 43) in 15% yield or 1-$(CH_3)_3$Ge(7-$(C_6H_5)_3$C)$C_8H_6Fe(CO)_3$ (No. 44) in 12% yield, respectively [38].

$CH_3C_8H_7Fe(CO)_3$ (Table **9**, No. **10**) shows a similar ^{1}H NMR pattern as those of Nos. 6 to 9 and $C_2H_5C_8H_7Fe(CO)_3$ (No. 14) and a reverse behavior to that of $RC_8H_7Fe(CO)_3$ (R = CHO (No. 18), CO_2CH_3 (No. 21), $COCH_3$ (No. 28), CN (No. 29)) [8]. At room temperature a tautomeric equilibrium between the two valence isomers XVIIb and d exists (ratio 1:4) [13, 18, 19]. The chemical shifts at −120 °C (see Table 9, p. 259) at 4.4 and 5.6 ppm [7, 10, 13] are assigned to H-2,7,8 and H-3,4,5,6, respectively [13]. Deuterated $DH_2CC_8D_6HFe(CO)_3$ (see Formula XXIII), obtained analogous to Method IIc (p. 254) confirms this conclusion [13, 14]. The ^{1}H NMR spectrum of XXIII in $CDCl_3$ with $\delta = 1.85$ (CH_2D) and 4.56 (ring H) ppm is only consistent with XVIId being dominant at room temperature. In $CHCl_2F$ the ring H in XXIII changes from a sharp line at room temperature to two sharp lines at −145 °C. No change is observed in the CH_2D band, apart from some broadening at very low temperature. At the coalescence temperature at −125 °C for the ring H bands, the rate constant for the valency tautomerism in XXIII is 35 s^{-1} and the free energy of activation is $\Delta F^{*} = 7.5$ kcal/mol. Tautomers other than XXIII and XXIV must be present in very small amounts (>90% XXIII and XXIV) [14].

XXII XXIII XXIV XXV

Protonation with concentrated H_2SO_4 gives only one of the six possible homotropylium cations [10]. Protonation with $HOSO_2F/SO_2F_2$ at −120 to −80 °C gives XXV (66%) and XXVI (33%) [23, 32] and occurs initially at C-3 [32]. Analogous reaction at ca. −60 °C gives XXVII (66%) and XXVIII (33%) [23]. A similar way as for the protonation is proposed for the acylation with CH_3COCl or C_6H_5COCl to give bicyclic intermediates which then undergo either a 1,3 or 1,2 shift to produce bicyclic $[^5LFe(CO)_3]^+$ cations. As XXIX (R = COC_6H_5) is a major product from benzoylation and no XXX (R = COC_6H_5) is observed, it is suggested that most of the reaction proceeds from attack on XVIId [51]. Reaction with $AlCl_3$ in C_6H_6 followed by hydrolysis gives XXXI (R = CH_3) [43]. Reaction with $(CH_3)_2NCHO/OPCl_3$ at 0 °C followed by hydrolysis with ice-water, extraction with ether, drying the organic layer with $MgSO_4$, and evaporation gives a mixture of compounds, which is separated by chromatography on silica gel. Elution with C_6H_6/n-C_5H_{12} (1:1) gives a clean separation into five bands. The first band is unreacted starting material (45%). The four product bands are removed and purified by further chromatography on silica gel. Thus, the second band gives a red oil (mass spectrometry: $[M]^+ = 284$) which gives a very complicated ^{1}H NMR spectrum and cannot be identified. A third band gives 1-CH_3(6-OHC)$C_8H_6Fe(CO)_3$ (No. 48) in 20% yield, the fourth band gives 1-CH_3(5-OHC)-$C_8H_6Fe(CO)_3$ (No. 50) in 9% yield, and a final band gives 4-CH_3(1,7-$(OHC)_2$)$C_8H_7Fe(CO)_3$ (No. 55) in 4% yield [53]. Reaction with $[C_7H_7]BF_4$ in acetone at −23 °C gives XXXII [49, 63].

Electrophilic addition of $(NC)_2C{=}C(CN)_2$ in CH_2Cl_2 at room temperature for 1 h gives XVIII ($X{=}CH_3$, $Y{=}H$) [35, 41] and XXXIII [41]. $CH_3C_8H_7Fe(CO)_3$ is proposed as an antiknock additive for gasoline, as an additive to fuels, for metal plating applications, as a drying agent, and as a chemical intermediate for metal-containing polymer materials [5].

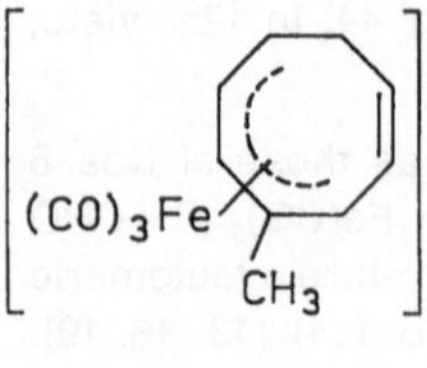

XXVI 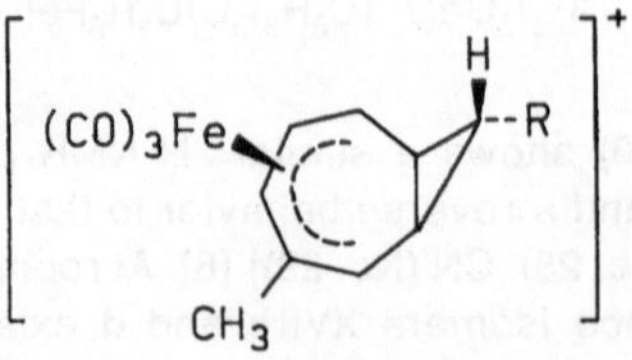 XXVII 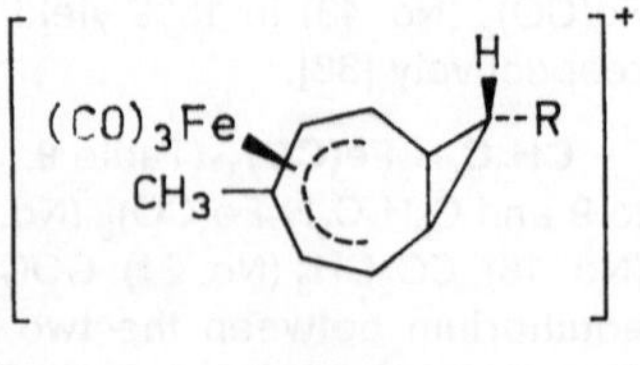 XXVIII

$HOCH_2C_8H_7Fe(CO)_3$ (Table **9**, No. **11**). A comprehensive study of the tautomeric equilibrium of No. 11 at room temperature shows at least three different valence tautomers, XVIIa, b, d (ratio 1:1:2 [18]) with significant populations [18, 19]. The symmetry and multiplicity of the peaks in the ^1H NMR spectrum observed in the spin-decoupling experiments with No. 11 at −55 °C indicate that the rearranging process is such that, on the average, the ligand possesses a planar symmetry. Therefore, at least one mechanism involving a double bond shift must be operating at this temperature [7]. The ^1H NMR spectrum of $HOHDCC_8H_7Fe(CO)_3$ (for preparation see No. 18 ($R{=}CHO$)) is as expected [36]. Protonation of No. 11 with 65% HPF_6/H_2O [21, 26, 54] in ether at 20 °C [26, 54] or reaction with $[C(C_6H_5)_3]BF_4$ [52] gives unstable $[C_9H_9Fe(CO)_3]X$ ($X{=}PF_6$ or BF_4), formulated as shown in XIV ($R{=}H$, p. 256) [52]. The deuterated derivative $HOHDCC_8H_7Fe(CO)_3$ also reacts with 65% HPF_6/H_2O in ether. Until precipitation is complete, the salt is suspended in CH_3OH to give $CH_3OHDCC_8H_7Fe(CO)_3$ (No. 12) after the usual work-up [36].

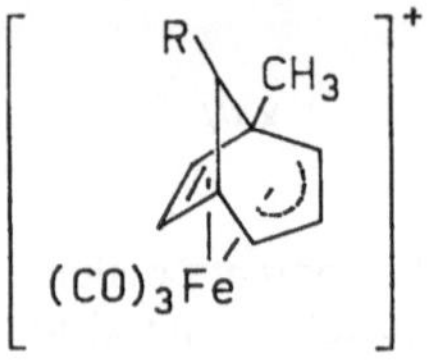

XXIX 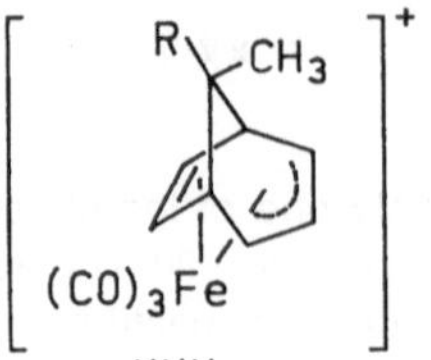 XXX 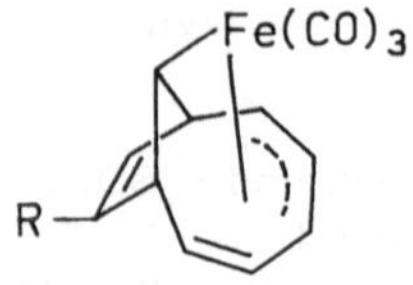 XXXI 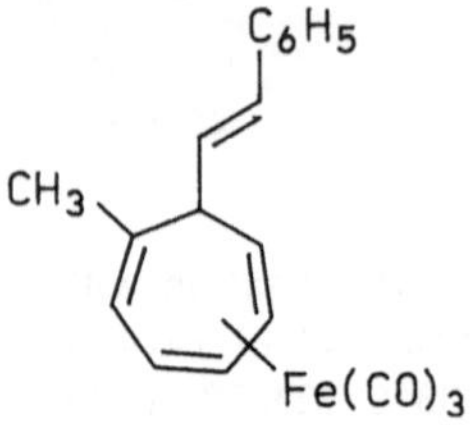 XXXII

$CH_3OCH_2C_8H_7Fe(CO)_3$ (Table **9**, No. **12**). The deuterated complex $CH_3OCHDC_8H_7Fe(CO)_3$ is prepared analogously to Method Va (see pp. 256/7) starting with $[C_9H_8DFe(CO)_3]PF_6$, and exhibiting the expected ^1H NMR spectrum [36]. Oxidation with $[NH_4]_2[Ce(NO_3)_6]$ in C_2H_5OH liberates the organic ligand ^4L in 90% yield [26].

$C_2H_5C_8H_7Fe(CO)_3$ (Table **9**, No. **14**) shows in the ^1H NMR spectrum a pattern similar to Nos. 6 to 10, and a reverse behavior compared with Nos. 18, 21, 28, 29 [7, 8].

$C_6H_5CH_2C_8H_7Fe(CO)_3$ (Table **9**, No. **15**) is proposed as an antiknock additive for gasoline, as an additive for fuels, for metal plating applications, as a drying agent, and as a chemical intermediate for metal-containing polymer materials [5].

$HOCH_2CH_2C_8H_7Fe(CO)_3$ (Table **9**, No. **16**). Dehydration on silica gel gives $H_2C{=}CHC_8H_7Fe(CO)_3$ (No. 19) [54].

$OHCC_8H_7Fe(CO)_3$ (Table **9**, No. **18**). A comprehensive study of the tautomeric equilibrium of No. 18 at room temperature shows that the $Fe(CO)_3$ group is bonded to double bonds opposite to CHO [19]. The ratio of isomers XVIIa:b is 1:1 [18]. This pattern is similar to those of Nos. 21, 28, 29 but reversed from that of Nos. 6 to 10 and 14 [8, 20]. The chemical shifts at 4.38 and 6.09 ppm show a temperature dependent fine structure, best interpreted in terms of a fluxional system [21]. Oxidation with $[NH_4]_2[Ce(NO_3)_6]$ in C_2H_5OH [26] or in C_2H_5OH/H_2O (10:1) [21] liberates the organic ligand ^4L (80% yield) [21, 26]. Reduction with $NaBH_4$ in C_2H_5OH at 0 °C for 5 min, followed by hydrolysis with ice-water, extraction with ether, and chromatography on silica gel with $CH_3CO_2C_2H_5$ gives $HOCH_2C_8H_7Fe(CO)_3$ (No. 11) in 93% yield [21, 26]. Reduction in C_2H_5OH at 0 °C by adding powdered $NaBD_4$ to the solution results in a suspension which on work-up as above gives $HOCHDC_8H_7Fe(CO)_3$ in 89% yield [36]. Reaction with CH_3MgX gives $HO(CH_3)CHC_8H_7Fe(CO)_3$ (No. 22) [21, 26]. Similar reaction with n-C_4H_9MgBr in ether at −78 °C for 1 h, followed by hydrolysis with H_2O at room temperature, separation of the organic layer which is dried ($MgSO_4$) and evaporated, and chromatography on silica gel with C_6H_6 gives $HO(n\text{-}C_4H_9)CHC_8H_7Fe(CO)_3$ (No. 26) in 80% yield [52]. Reaction with $CH_2{=}P(C_6H_5)_3$ gives $CH_2{=}CHC_8H_7Fe(CO)_3$ (No. 19) [54].

$CH_3O_2CC_8H_7Fe(CO)_3$ (Table **9**, No. **21**). A comprehensive study of the tautomeric equilibrium of No. 21 at room temperature shows that the $Fe(CO)_3$ is bonded to double bonds opposite to CO_2CH_3 [19]. The ratio of isomers XVIIa:b is 1:9 [18]. The doublet in the ^{1}H NMR spectrum is observed as the signal at lowest field with the multiplet at higher field. This pattern is similar to those of Nos. 18, 28, 29 but reversed from that of Nos. 6 to 10 and 14 [8, 20]. Reaction with $(NC)_2C{=}C(CN)_2$ in CH_2Cl_2 for 1 h gives a mixture of isomers XVIII (X=H, Y=CO_2CH_3) and XXXIV in the ratio 3.7:1. Analogous reaction in C_6H_6 gives XVIII (X=H, Y=CO_2CH_3) and XXXIV in the ratio 3:1 [41].

$HO(CH_3)CHC_8H_7Fe(CO)_3$ (Table **9**, No. **22**) reacts with 65% HPF_6/H_2O in ether at 20 °C to give $[C_{10}H_{11}Fe(CO)_3]PF_6$ (90% yield) [21, 26].

$R'ORCHC_8H_7Fe(CO)_3$ (Table **9**, Nos. **24** to **27** with R′=H, CH_3 and R=C_2H_5, C_4H_9-n). The compounds are formulated as valence tautomers XVIIb (p. 270) with R=CHROR′. Treatment of compounds No. 24 or 26 with $[C(C_6H_5)_3]BF_4$ or HPF_6 gives the salt XIV (p. 256) with R=C_2H_5 or C_4H_6-n and X=BF_4 or PF_6. Reaction of this salt with excess CH_3OH gives Nos. 25 and 27, respectively [52].

$CH_3COC_8H_7Fe(CO)_3$ (Table **9**, No. **28**). A comprehensive study of the tautomeric equilibrium of No. 28 at room temperature shows that the $Fe(CO)_3$ group is bound to double bonds opposite to $COCH_3$ [19]. The ratio of isomers XVIIa:b is 1:1 [18]. The doublet in the ^{1}H NMR spectrum is observed as the signal at lowest field with the multiplet at higher field. This pattern is similar to those of Nos. 18, 21, and 29 but reversed from that of Nos. 6 to 10 and 14 [8, 20]. Reduction with $NaBH_4$ as described for No. 18 gives No. 22 in 85% yield [26], cf. [21].

References on pp. 277/9

$NCC_8H_7Fe(CO)_3$ (Table **9**, No. **29**). A comprehensive study of the tautomeric equilibrium of No. 29 at room temperature shows that the $Fe(CO)_3$ group is bound to double bonds opposite to CN [19]. The ratio of isomer XVIIa:b is 1:9 [18]. The doublet in the ^1H NMR spectrum is observed as the signal at lowest field with the multiplet at higher field. This pattern is similar to those of Nos. 18, 21, and 28, but reversed from that of Nos. 6 to 10 and 14 [8, 20].

$(C_6H_5)_3CC_8H_7Fe(CO)_3$ (Table **9**, No. **30**) is a fluxional molecule [7, 42] and maintains its fluxional behavior down to −80 °C, and is formulated as structure XVIIa [42] or b (p. 270) [37, 45]. The shift of the $Fe(CO)_3$ group between structures XVIIa to c is discussed in [53]. Treatment of the compound at 160 °C in vacuum for 12 h [42] or heating in octane in a sealed tube at 150 to 160 °C for 20 h [37, 45] gives XXII ($R=C(C_6H_5)_3$) [37, 42, 45]. Oxidation with $[NH_4]_2[Ce(NO_3)_6]$ in C_2H_5OH liberates the organic ligand ^4L, and protonation with CF_3CO_2H in CD_2Cl_2 at 0 °C gives XVI (p. 256) [42]. The compound No. 30 reacts with $(CH_3)_2NCHO/OPCl_3$ at 0 °C within 1 h. The reaction mixture is poured onto ice and left overnight. Usual work-up and chromatography on silica gel gives two bands. The first is unreacted No. 30 (53%) while the second is 1-OHC(7-$(C_6H_5)_3C)C_8H_6Fe(CO)_3$ (No. 45, 18% yield) [53]. Reaction with $(NC)_2C{=}C(CN)_2$ in CH_2Cl_2 or with $(NC)_2C{=}C(CF_3)_2$ in hexane at room temperature gives 1:1 adducts of composition $C_{33}H_{22}N_4Fe(CO)_3$ and $C_{33}H_{22}F_6N_2Fe(CO)_3$ with unknown structure (low solubility) [35], but they will be described in the section devoted to σ,π-allyl bound compounds for reasons of analogy (see 1.4.1.4.3.2 in "Organoiron Compounds" B10, 1985, Nos. 149, 150, p. 154).

$C_6H_5C_8H_7Fe(CO)_3$ (Table **9**, No. **31**). The ^1H NMR spectrum of this compound is temperature dependent [7, 10, 20] and, therefore, the compound is believed to undergo a fluxional process with structures XVIIa to d (p. 270) which averages the same proton pairs but produces a different order of chemical shifts with the doublet H-2 and H-8 at low field [10, 19, 20, 53]. For a figure of the UV spectrum see [6]. The compound is soluble in organic solvents and in conc. H_2SO_4 [4]. Only one of the six possible homotropylium cations is formed in conc. H_2SO_4 [10]. Treating No. 31 in octane at 150 °C in a sealed tube for 20 h gives XXII ($R=C_6H_5$) [45]. Treating with $AlCl_3$ in C_6H_6 gives a complex mixture, no pure complexes isolated [43]. Treatment with $[C(C_6H_5)_3]BF_4$ in CH_3NO_2 at room temperature for 15 min and processing as usual (see No. 6) gives No. 54 in 11% yield [38]. Formylation with $(CH_3)_2NCHO/OPCl_3$ as described in Method IVb (p. 256) gives 1-OHC$(C_6H_5)C_8H_6$Fe-$(CO)_3$(?) in 60% yield as an orange oil (IR (liquid film): 1680 (CHO); 1960, 2063 (CO)) [31].

But the very similar reaction of No. 31 with $(CH_3)_2NCHO/OPCl_3$ at 0 °C for 1 h followed by hydrolysis on ice gives a mixture of three compounds. The reaction mixture is left overnight and then extracted with ether, and the organic layer is dried ($MgSO_4$) and evaporated. Chromatography on silica gel with C_6H_6/25% pentane gives four bands, the first band comprising unreacted No. 31 (51%). Further purification by chromatography on silica gel gives No. 46 (5% yield) as the second band, No. 49 (8.5% yield) as the third band, and finally No. 51 (9% yield) as the last band [53].

Reaction of No. 31 with $(NC)_2C{=}C(CN)_2$ in CH_2Cl_2 at room temperature for 2 h is said to give XVIII ($X=H$, $Y=C_6H_5$, p. 270) [35]. A reinvestigation of this reaction (CH_2Cl_2, 15 min) followed by evaporation and oxidation of the residue in C_2H_5OH at room temperature with $[NH_4]_2[Ce(NO_3)_6]$ gave XXXV and XXXVI. When No. 31 is reacted with $(NC)_2C{=}C(CN)_2$ in C_6H_6 for 50 min, the solid isolated by filtration is XVIII ($X=H$, $Y=C_6H_5$), and from the mother liquor additional XVIII ($X=H$, $Y=C_6H_5$), XXXVII, and XVIII ($X=C_6H_5$, $Y=H$) are isolated [41].

$4\text{-}O_2NC_6H_4C_8H_7Fe(CO)_3$ (Table **9**, No. **32**) is fluxional with positions C-2 and C-8, C-3 and C-7, and C-4 and C-6 rendered equivalent by metal migration. The ^1H NMR spectrum shows that the aryl substituent is endo to Fe as shown in Table 9, p. 263. The compound is only sparingly soluble in hexane [55].

$[R_3MCH_2C_8H_7Fe(CO)_3]PF_6$ (Table **9**, Nos. **33** to **35** with $MR_3=N(CH_3)_3$, $P(C_6H_5)_3$, $As(C_6H_5)_3$). The cations have structures as shown in Formula XVIIb (p. 270). In contrast to the stable salts Nos. 33 and 34, the solution of salt No. 35 shows in the IR spectrum bands at 1989, 2056, 2067, 2114 cm^{-1}. When excess of $As(C_6H_5)_3$ (4:1) is added to the solution, the two bands at 2067 and 2114 cm^{-1} disappear. This suggests that in solution No. 35 is in equilibrium with free arsine and the original carbonium ion XIV (R=H, p. 256). Addition of $N(CH_3)_3$ to the solutions of No. 35 leads to quantitative conversion into No. 33 [52]. Reaction of No. 34 with CO (80 atm) in acetone in an autoclave at 80 °C for 20 h gives $[C_8H_7CH_2P(C_6H_5)_3]PF_6$ in 85% yield; compound No. 34, suspended in ether, is dissolved when reacted with LiC_4H_9 in hexane. After 1 h at −78 °C a deep red solution is formed. After warming to room temperature, CH_2O is bubbled through the solution for 5 min. After filtration and evaporation, the residue is chromatographed on silica gel with pentane to give No. 19 in 82% yield. Similar reaction of No. 34 with LiC_4H_9 at −20 °C, followed by addition of a twofold excess of cyclohexanone and then stirring for 30 min at room temperature, gives $OP(C_6H_5)_3$ and chromatography of the residue as above gives $C_8H_7CH_2C_6H_{11}$ in 30% yield. Analogous reaction with LiC_4H_9/benzophenone (24 h at room temperature) gives $C_8H_7CH{=}C(C_6H_5)_2$ in 70% yield. Reaction with $NaH/OHCC_6H_5$ in CH_2Cl_2 at room temperature for 2 h, followed by hydrolysis, separation of the organic layer which is dried ($MgSO_4$), evaporation and chromatography on silica gel with pentane gives isomer No. 20a in 43% (first band) and isomer No. 20b in 23% yields (second band) [54].

1,7-$(CH_3O_2C)_2C_8H_6Fe(CO)_3$ (Table **9**, No. **36**). The ^{1}H NMR spectrum recorded over a temperature range of −60 to +120 °C does not show any appreciable change in appearance [11, 12]. Thus, the compound does not undergo rapid intramolecular rearrangement in solution [12], and structure XXXVIII is favored [12, 25].

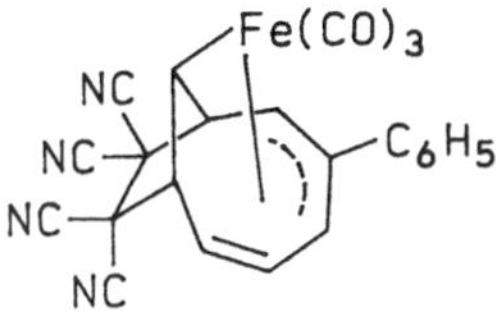

$Cl(CH_3C_2)C_8H_6Fe(CO)_3$ (Table **9**, No. **37**) is monomeric in $CHCl_3$ (osmometry) and shows no temperature dependence in the ^{1}H NMR spectrum at 35 to 80 °C and, thus, possesses structure XXXIX (R=Cl) [58]. Treatment of the isomeric mixture in ClC_6H_5 in a sealed tube at 130 °C gives VIII (R=Cl, p. 254). There is no reaction observed with VII ($R=CH_3$, p. 254) under these conditions [57]. For a detailed theoretical investigation of the minimum energy pathway for the isomerization of No. 37 to VIII (R=Cl) see [56].

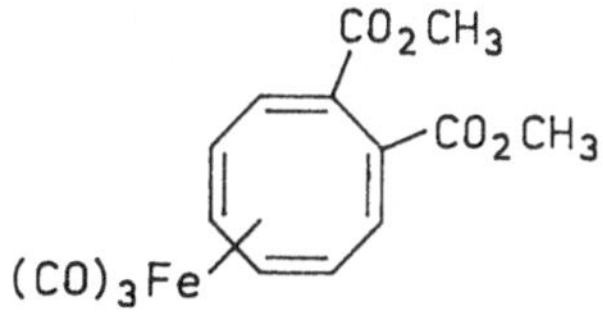

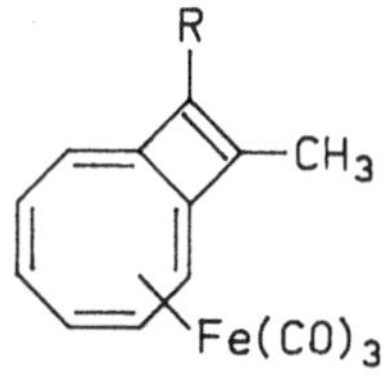

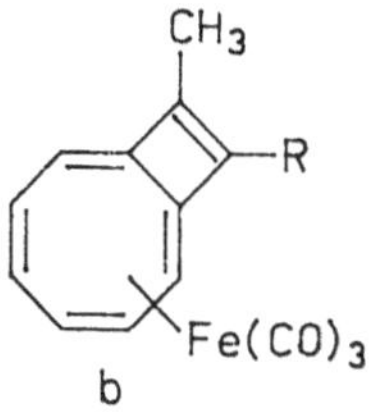

$(CH_3)_2C_2C_8H_6Fe(CO)_3$ (Table **9**, No. **38**) has structure XXXIX. Thermal isomerization of the mixture of No. 38, VIII with $R=CH_3$ (p. 254) and VII ($R=CH_3$, p. 254) (ratio 20:3:1) obtained by Method IIc (p. 254) in ClC_6H_5 at 90 °C gives VIII ($R=CH_3$) [57].

$C_2H_2O_2C_8H_6Fe(CO)_3$ (Table **9**, No. **39**) shows an unchanged ^{1}H NMR spectrum in the temperature range −60 to +120 °C [12] and, thus, structure XL is in agreement with the ^{1}H NMR spectrum [12, 25].

$C_4H_4C_8H_6Fe(CO)_3$ (Table **9**, No. **40**). Leaving $C_{12}H_{10}Fe_2(CO)_6$ (see Methods IIc (p. 254) and IIIc (p. 255)) in ethereal solution at room temperature in the dark for 1 month leads to

References on pp. 277/9

60% decomposition. Insoluble material is separated and small amounts of No. 40 are isolated [17]. Contrary to the statement in [17] shift isomerism of No. 40 is sufficiently rapid to label it as a fluxional molecule [28, 30]. Its structure is shown in XLI a (D=H) [15, 17, 28 to 30, 41].

The coalescence temperature in the ^{1}H NMR spectrum is observed at ca. 55 °C [28]. Examination of XLI a (obtained in analogy to Method I b, p. 254) affords an activation energy of $E_a = 18.6$ kcal/mol [28, 29] for interconversion of the two nuclei XLI a and b; XLI c considered as intermediate. At 333 K, $k = 735\ s^{-1}$ and $\Delta G^{+} = 15.2$ kcal/mol [28]. Protonation gives XLII [15].

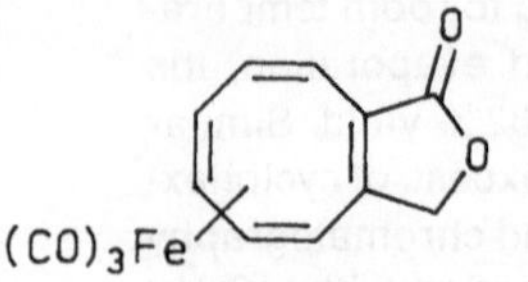

XL 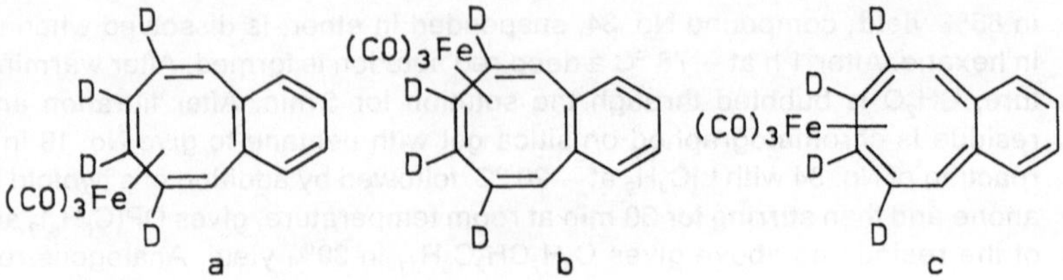 — XLI (a, b, c)

Reaction with $(NC)_2C{=}C(CN)_2$ in C_6H_6 at room temperature gives a mixture of adducts, one of which identified as XLIII. Attempted recrystallization from acetone leads to decomposition, and consequently this mixture is directly oxidized with $[NH_4]_2[Ce(NO_3)_6]$ in 95% C_2H_5OH to give XLIV to XLVI [41].

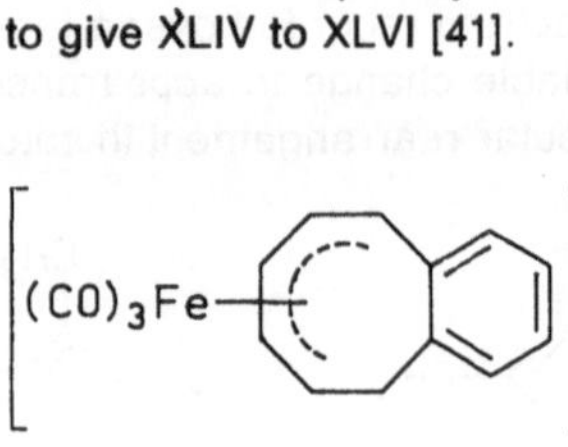

XLII 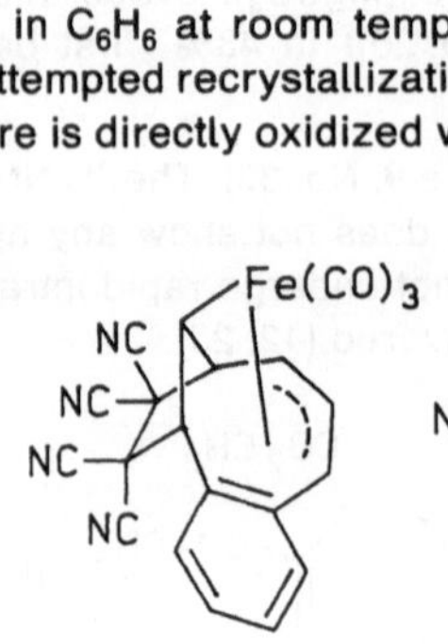 — XLIII 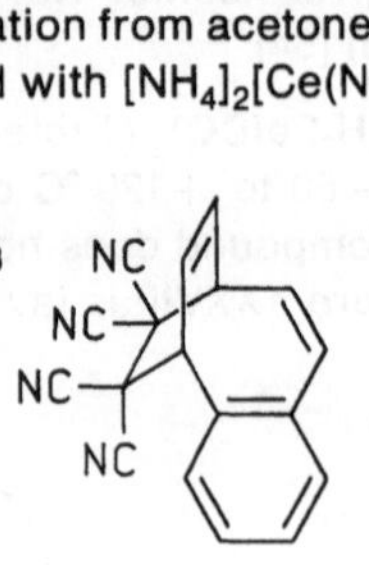 — XLIV 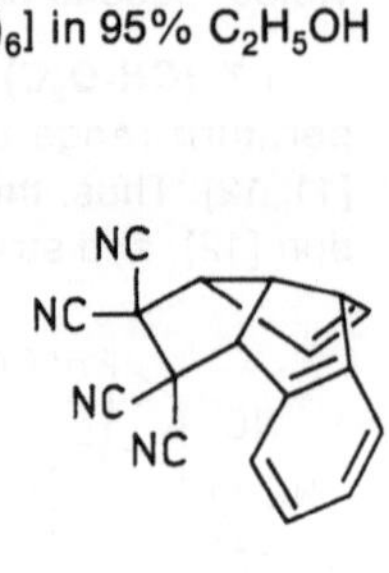 — XLV

$C_8H_6C_8H_6Fe(CO)_3$ (Table **9**, No. **41**). The reaction of iron carbonyls with $^{4}L = C_8H_6C_8H_6$ gives No. 41 and $C_{16}H_{12}Fe_2(CO)_6$ (see 2.6.1.2.4 in "Organoiron Compounds" C5, 1981, Table 3, No. 20, p. 30) [30]. It is a fluxional molecule analogous to No. 40 (cf. XLI) [28, 30].

1-R(7-$(C_6H_5)_3C$)$C_8H_6Fe(CO)_3$ (Table **9**, Nos. **43** and **44** with $R = Si(CH_3)_3$ or $Ge(CH_3)_3$). The ^{1}H NMR spectra of these compounds do not vary with temperature [38]; the structure assigned is shown in XLVII [37, 38, 45]. Treatment of compound No. 43 in octane at 150 °C in a Carius tube for 20 h gives XLVIII [37, 45].

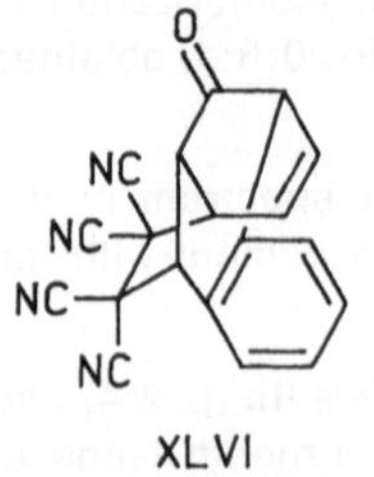

XLVI 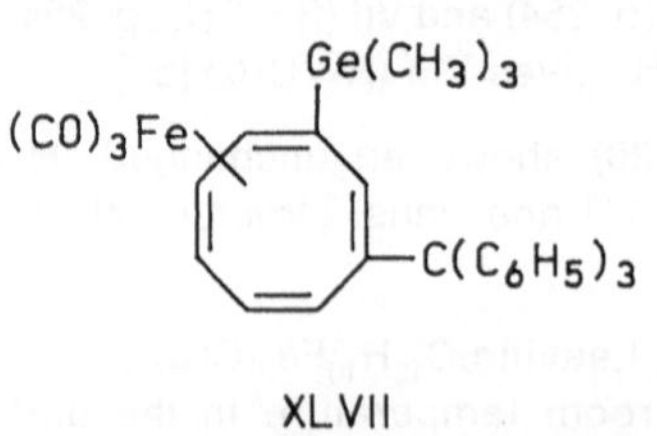 — XLVII 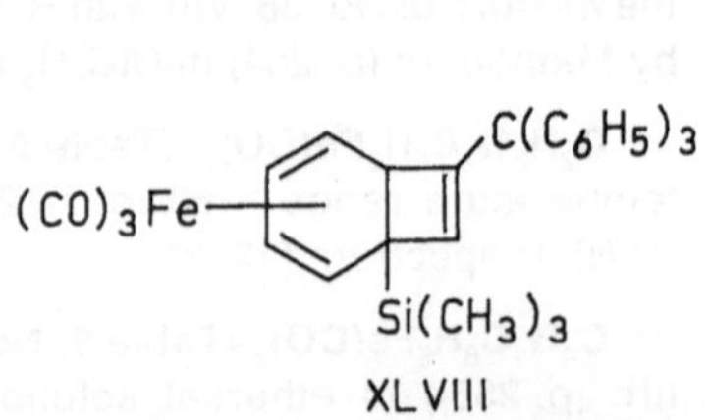 — XLVIII

References on pp. 277/9

1-CH_3[6-(4-$O_2NC_6H_4CO_2CH_2CH_2$)]$C_8H_6Fe(CO)_3$ (Table **9**, No. **47**) is prepared from 1-CH_3(4-$HOCH_2CH_2$)C_8H_6, and an X-ray structure has been performed. No further details are given [39].

1-CH_3(6-OHC)$C_8H_6Fe(CO)_3$ and **1-CH_3(5-OHC)$C_8H_6Fe(CO)_3$** (Table **9**, Nos. **48** and **50**) react with n-C_4H_9MgX. Compound No. 48 gives IL (R=H, R'=CH_3), and No. 50 gives IL with R=CH_3, R'=H [48].

$(CH_3CO)_2C_8H_6Fe(CO)_3$ (Table **9**, No. **52**) is considered to be present in at least three isomeric forms [26].

$(O_2N)_2C_6H_3$NHN=CH(1-C_6H_5)$C_8H_6Fe(CO)_3$ (Table **9**, No. **53**) is readily formed by the reaction of 1-OHC(C_6H_5)$C_8H_6Fe(CO)_3$ (see "Further information" for No. 31, p. 274) with $H_2NNHC_6H_3(NO_2)_2$ [31].

$(C_6H_5)_3$C(1-C_6H_5)$C_8H_6Fe(CO)_3$ (Table **9**, No. **54**) may have a different structure than Nos. 43 or 44. No signal assignable to a single H located between the substituents could be found. The ^{1}H NMR spectrum does not vary with temperature. The compound is air-stable and moderately soluble in hydrocarbons and ether [38].

IL — $[R,R'-C_8H_5(=CHC_4H_9)Fe(CO)_3]^+X^-$; L ; LI — Fe(CO)3

$C_6H_6C_8H_6Fe(CO)_3$ (Table **9**, No. **58**) is prepared by irradiation of L and $Fe(CO)_5$ in pentane (22%). Dinuclear complexes are also observed in this reaction [60, 62]. X-ray structure determination confirms that the $Fe(CO)_3$ group is on the same side of the C_8 ring as the C_6 ring, as shown in the table [61]. Addition of NCC≡CCN gives the two stereoisomeric products No. 59a (56%) and b (28%), easily separable by chromatography [60, 62].

$(NC)_2C_8H_6C_8H_6Fe(CO)_3$ (Table **9**, No. **59**). A mixture of the stereoisomers No. 59a and b is heated at 180 °C to give LI (62% yield) [60, 62] and the phthalonitrile [60].

References:

[1] K.W. Hübel, Soc. Eur. Res. Associates S.A. (Ger. Offen. 1169932 [1958/64]). — [2] W. Hübel, E.H. Braye, I. Caplier, R. Vannieuwenhoven (J. Inorg. Nucl. Chem. **10** [1959] 250/68). — [3] W. Hübel, E.H. Braye, A. Clauss, E. Weiss, U. Krüerke, D.A. Brown, G.S.D. King, C. Hoogzand (J. Inorg. Nucl. Chem. **9** [1959] 204/10). — [4] A. Nakamura, N. Hagihara (Mem. Inst. Sci. Ind. Res. Osaka Univ. **17** [1960] 187/90). — [5] K.G. Ihrman, T.H. Coffield, Ethyl Corp. (U.S. 3077489 [1960/63]).

[6] A. Nakamura, N. Hagihara (Nippon Kagaku Zasshi **82** [1961] 1387/9). — [7] C.E. Keller, B.A. Shoulders, R. Pettit (J. Am. Chem. Soc. **88** [1966] 4760/1). — [8] C.E. Keller (Diss. Univ. Texas 1966 from [38]). — [9] C.E. Keller (Diss. Univ. Texas 1966 from [45]). — [10] C.E. Keller (Diss. Univ. Texas 1966; Diss. Abstr. B **27** [1967] 2651).

[11] R.H. Herber (Symp. Faraday Soc. No. 1 [1967] 86/96). — [12] R. Grubbs, R. Breslow, R. Herber, S.J. Lippard (J. Am. Chem. Soc. **89** [1967] 6864/70). — [13] F.A.L. Anet, H.D. Kaesz, A. Maasböl, S. Winstein (J. Am. Chem. Soc. **89** [1967] 2489/91). — [14] F.A.L. Anet (J. Am. Chem. Soc. **89** [1967] 2491/2). — [15] W. Merk, R. Pettit (unpublished results from [16]).

[16] W. Merk, R. Pettit (J. Am. Chem. Soc. **90** [1968] 814/6). – [17] J.A. Elix, M.V. Sargent (J. Am. Chem. Soc. **91** [1969] 4734/9). – [18] L.A. Bock (Diss. Univ. California 1969 from [41]). – [19] L.A. Bock (Diss. Univ. California 1969; Diss. Abstr. Intern. B **30** [1970] 4967). – [20] L.A. Bock (Diss. Univ. California 1969 from [38]).

[21] B.F.G. Johnson, J. Lewis, A.W. Parkins, G.L.P. Randall (Chem. Commun. **1969** 595). – [22] R.H. Herber (Chem. Anal. [N.Y.] **26** [1969] 315/40). – [23] M. Brookhart, E.A. Davis (Tetrahedron Letters **1971** 4349/52). – [24] R.H. Herber (Introd. Mössbauer Spectrosc. **1971** 138/54). – [25] H.W. Whitlock, C. Reich, W.D. Woessner (J. Am. Chem. Soc. **93** [1971] 2483/92).

[26] B.F.G. Johnson, J. Lewis, G.L.P. Randall (J. Chem. Soc. A **1971** 422/9). – [27] B.F.G. Johnson, J. Lewis, P. McArdle, G.L.P. Randall (Chem. Commun. **1971** 177/8). – [28] H.W. Whitlock, H. Stucki (J. Am. Chem. Soc. **94** [1972] 8594/6). – [29] H. Stucki (Diss. Univ. Wisconsin 1972 from [41]). – [30] H. Stucki (Diss. Univ. Wisconsin 1972; Diss. Abstr. Intern. B **32** [1972] 6936/7).

[31] B.F.G. Johnson, J. Lewis, P. McArdle, G.L.P. Randall (J. Chem. Soc. Dalton Trans. **1972** 2076/83). – [32] M. Brookhart, E.R. Davis, D.L. Harris (J. Am. Chem. Soc. **94** [1972] 7853/8). – [33] Y. Becker, A. Eisenstadt, Y. Shvo (J. Chem. Soc. Chem. Commun. **1972** 1156). – [34] B.F.G. Johnson (unpublished results from [36]). – [35] M. Green, S. Heathcock, D.C. Wood (J. Chem. Soc. Dalton Trans. **1973** 1564/9).

[36] J.E. Alsop, R. Davis (J. Chem. Soc. Dalton Trans. **1973** 1686/91). – [37] M. Cooke, J.A.K. Howard, C.R. Russ, F.G.A. Stone, P. Woodward (J. Organometal. Chem. **78** [1974] C43/C46). – [38] M. Cooke, C.R. Russ, F.G.A. Stone (J. Chem. Soc. Dalton Trans. **1975** 256/9). – [39] J. Clardy (unpublished results from [40]). – [40] L.A. Paquette (Tetrahedron Letters **31** [1975] 2855/83).

[41] L.A. Paquette, S.V. Ley, S. Maiorana, D.F. Schneider, M.J. Broadhurst, R.A. Boggs (J. Am. Chem. Soc. **97** [1975] 4658/67). – [42] B.F.G. Johnson, J. Lewis, J.W. Quail (J. Chem. Soc. Dalton Trans. **1975** 1252/7). – [43] B.F.G. Johnson, J. Lewis, D.J. Thompson, B. Heil (J. Chem. Soc. Dalton Trans. **1975** 567/71). – [44] B.F.G. Johnson, J. Lewis, D. Wege (J. Chem. Soc. Dalton Trans. **1976** 1874/80). – [45] M. Cooke, J.A.K. Howard, C.R. Russ, F.G.A. Stone, P. Woodward (J. Chem. Soc. Dalton Trans. **1976** 70/5).

[46] H. Olsen (Acta Chem. Scand. B **31** [1977] 635/6). – [47] N.G. Connelly, A.R. Lucy, M.W. Whiteley (J. Chem. Soc. Chem. Commun. **1979** 985/6). – [48] P. Hackett (unpublished results from [52]). – [49] K. Broadley, N.G. Connelly, R.M. Mills, M.W. Whiteley, P. Woodward (J. Chem. Soc. Chem. Commun. **1981** 19/20). – [50] S.C. Carleton, F.G. Kennedy, S.A.R. Knox (J. Chem. Soc. Dalton Trans. **1981** 2230/4).

[51] A.D. Charles, P. Diversi, B.F.G. Johnson, J. Lewis (J. Chem. Soc. Dalton Trans. **1981** 1906/17). – [52] P. Hackett, B.F.G. Johnson, J. Lewis (J. Chem. Soc. Dalton Trans. **1982** 1253/9). – [53] P. Hackett, B.F.G. Johnson, J. Lewis (J. Chem. Soc. Dalton Trans. **1982** 939/42). – [54] P. Hackett, B.F.G. Johnson, J. Lewis, G. Jaouen (J. Chem. Soc. Dalton Trans. **1982** 1247/51). – [55] N.G. Connelly, A.R. Lucy, M.W. Whiteley (J. Chem. Soc. Dalton Trans. **1983** 111/5).

[56] T.A. Albright, P. Hofmann, R. Hoffmann, C.P. Lillya, P.A. Dobosh (J. Am. Chem. Soc. **105** [1983] 3396/411). – [57] M. Magon, G. Schröder (Liebigs Ann. Chem. **1978** 1379/89). – [58] G. Schröder, S.R. Ramadas, P. Nikoloff (Chem. Ber. **105** [1972] 1072/83). – [59] M.S. Brookhart, G.W. Koszalka, G.O. Nelson, G. Scholes, R.A. Watson (J. Am. Chem. Soc. **98** [1976] 8155/61). – [60] D. Kawka, P. Mues, E. Vogel (Angew. Chem. **95** [1983] 1006/7).

[61] P. Mues, J. Knebel, J. Lex (unpublished results from [62]). – [62] D. Kawka, P. Mues, E. Vogel (Angew. Chem. Suppl. **1983** 1371/8). – [63] K. Broadley, N.G. Connelly, R.M. Mills, M.W. Whiteley, P. Woodward (J. Chem. Soc. Dalton Trans. **1984** 683/8).

1.4.1.4.1.2.6 ^{4}L is a Nine-Membered Ring System

$C_9H_{12}Fe(CO)_3$ (Formula I) is prepared by the reaction of $Fe_2(CO)_9$ and cyclonona-1,3,6-diene in C_6H_6 at 70 °C for 1.5 h. Removal of the volatiles, followed by chromatography of the residue on neutral alumina (activity III) and cooling to −80 °C, gives pure $C_9H_{12}Fe(CO)_3$ (25% yield). Further elution with C_6H_6 gives II (see 2.5.2.3.3.1.4 in "Organoiron Compounds" C4, 1981, Table 20, No. 5, p. 111). The compound $C_9H_{12}Fe(CO)_3$ is also obtained from $C_9H_{12}Fe(CO)_4$ (C_9H_{12}=cyclonona-1,3,6-diene) in C_6H_6 at 70 °C. The orange crystals (from hexane at −30 °C) of $C_9H_{12}Fe(CO)_3$ melt at 45 °C. The compound shows in the ^{13}C NMR spectrum in benzene-d_6 chemical shifts at δ=24.9, 27.0, 32.2 (C-7 to 9); 60.2, 62.1, 88.2, 92.8 (C-1 to 4); 126.4, 130.5 (C-5,6), and 211.7 (CO) ppm. The IR spectrum in n-hexane shows ν(CO) bands at 1976, 1986, and 2048 cm^{-1}. Refluxing $C_9H_{12}Fe(CO)_3$ in heptane for 1 h gives III. Protonation with HBF_4 gives [1-5η-cyclo-$C_9H_{13}Fe(CO)_3$]BF_4 with protonation of the noncoordinated double bond. Reduction with $NaBH_4$ gives small amounts of an unstable pale yellow product which decomposes on attempted purification by chromatography [13].

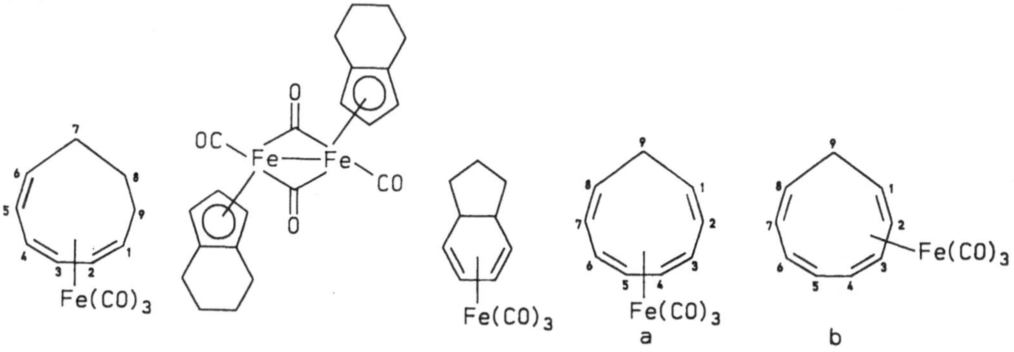

$C_9H_{10}Fe(CO)_3$ (Formula IVa) is prepared by the reaction of $Fe_2(CO)_9$ with cis-bicyclo-[6.1.0]nona-2,4,6-triene by a similar processing as described for its isomer $C_9H_{10}Fe(CO)_3$ (IVb, following compound) at room temperature for 40 h. Repeated chromatography of the first fraction after elimination of V (R=H) gives a reproducible ca. 9:1 ratio of IVa:b [1, 8]. Similar reaction of $Fe_2(CO)_9$ with cis-bicyclo[6.1.0]nona-2,4,6-triene in CH_3OH at room temperature for 2 d followed by chromatography on neutral alumina (activity III) with hexane gives IVa and VI in the first band and pure VII in the second band. Rechromatography on silica gel (activity III) with hexane gives a complete separation of IVa and VI [13].

Isomer IVa is an orange-yellow oil (contaminated with IVb) [1] and shows in the ^1H NMR spectrum (benzene-d_6) chemical shifts at δ=ca. 1.6 to 2.5 (m, H-9,9′), ca. 3.2 to 3.7 (m, H-3,6), ca. 4.3 to 4.8 (m, H-4,5), and ca. 4.4 to 5.1 (m, H-1,2,7,8) [8]. The ^{13}C NMR spectrum (benzene-d_6) shows δ=27.4 (C-9); 60.6, 89.2 (C-3 to 6); 121.3, 122.2 (C-1,2,7,8); 211.1 (CO) ppm [13]. The IR spectrum shows the usual three bands [8] in n-hexane at 1977, 1992, 2050 (CO) cm^{-1} [13], and the mass spectrum shows the ions $[M-nCO]^+$ (n=0 to 3) [8]. Pure IV or the mixture of $C_9H_{10}Fe(CO)_4$, $C_9H_{10}Fe(CO)_3$, IVa and b undergoes very slow rearrangement to V (R=H) at ca. 100 °C [1, 8].

References on p. 282

$C_9H_{10}Fe(CO)_3$ (Formula IVb) is prepared by several methods. Irradiation of $Fe(CO)_5$ with cis-bicyclo[6.1.0]nona-2,4,6-triene [2, 9] in C_6H_6 at 40 °C for 4 d, followed by evaporation and chromatography of the residue on basic alumina (activity II) with hexane, gives minor amounts of IVb and V (R=H), but chiefly complex VI [2]. A second fraction eluted with hexane consists predominantly of IVb but also contains V (R=H), elution with hexane/3% C_6H_6 gives VII. Further purification is achieved by repeated fractional chromatography. The yields of IVb, V (R=H), VI, and VII are 35, 6, 15, and 1%, respectively [2].

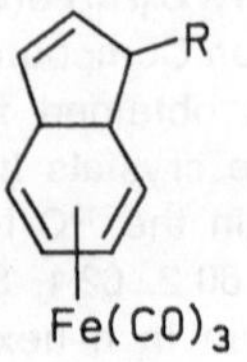

R
$Fe(CO)_3$ $Fe(CO)_3$ $(CO)_3Fe—Fe(CO)_3$

V  VI 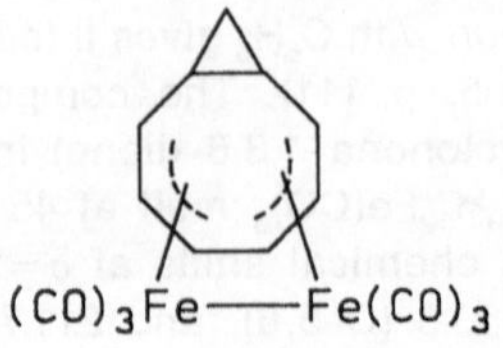 VII 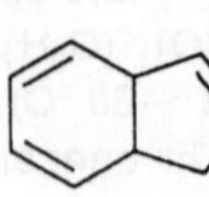 VIII

$Fe_2(CO)_9$ reacts with cis-bicyclo[6.1.0]nona-2,4,6-triene in ether (also petroleum ether, C_6H_6 etc. [1]) at room temperature [1, 2, 8, 9] for 40 h [2, 8] or 4 d [1]. The reaction mixture is filtered, the filtrate is evaporated, and storage of the residual dark viscous oil at 0 °C for several days produces a crystalline mass which is separated to give VII [2]. The remaining oil is chromatographed on neutral alumina (activity I [2] or II [1]) with hexane [2] or petroleum ether [1] to give a mixture of IVb [2] and V (R=H) [1, 2] as a yellow oil and VII (20% C_6H_6/hexane as eluent) [2]. Rechromatography gives $C_9H_{10}Fe(CO)_4$, $C_9H_{10}Fe(CO)_3$, IVa, b, and VII [1]. Chromatography of the mixture of IVb and V (R=H) on basic alumina (activity II) with hexane gives V (R=H). Later fractions are enriched in the minor component IVb, which by further fractional chromatography is obtained in good purity. Complexes IVb and V (R=H) can also be collected via preparative gas-liquid chromatography (silicon gum rubber UC-W98, 110 °C). While V (R=H) elutes intact, IVb partially rearranges to V (R=H) [2]. The yields of IVb, V (R=H), VI, and VII are 12, 46, 15, and 14%, respectively [2]. Repeated chromatography of the first fraction after elimination of V (R=H) gives a reproducible ca. 9:1 ratio of IVa:b [8]. Increasing the temperature results in a decrease of all compounds in favor of V (R=H) and a number of new complexes are formed [1]. Refluxing $Fe_3(CO)_{12}$ with VIII in C_6H_6 or xylene gives V (R=H) in good yield and other unidentified complexes [1].

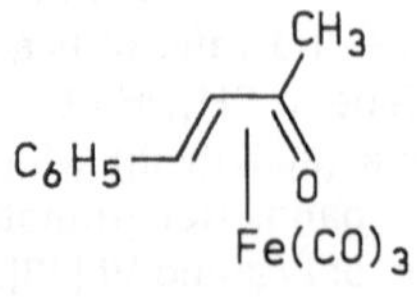

CH_3
C_6H_5 O
$Fe(CO)_3$ $Fe(CO)_3$ $[Fe(CO)_3]^+$

IX 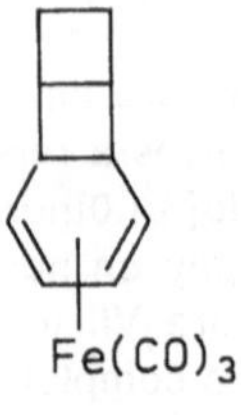 X 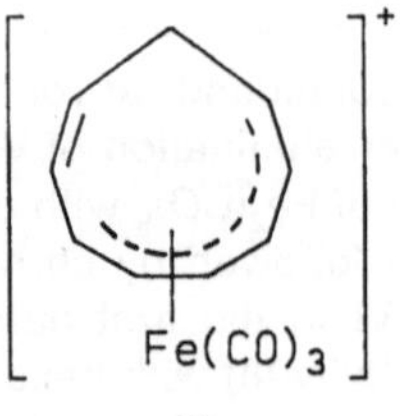 XI

Reaction of IX with cis-bicyclo[6.1.0]nona-2,4,6-triene [3, 10, 11] in C_6H_6 [10, 11] at 55 [3] to 65 °C [11] for 45 [11] to 48 h [3], followed by filtration and chromatography of the concentrated solution on alumina (activity II [11]) with hexane, gives a 5.4:2.4:1 ratio of X, V (R=H), and IVb (78% [11] or ca. 50% overall yield [3] and 9% yield of IVb [1, 10]) [3, 11]. Separation is achieved by medium-pressure liquid chromatography (silica gel, 4.9 atm) [11]. The molar ratio of the starting products is 3.5:1. With increasing IX a decrease

References on p. 282

in the yield of X is observed along with decreased trapping of cis[4]-cyclononatetraene [3]. Refluxing VII in toluene may involve IV as intermediate [4].

Isomer IVb is an orange-yellow [1] or yellow solid [2], m.p. ca. 30 [1] or 36 to 37 °C [2]. The ^{1}H NMR spectrum in benzene-d_6 shows $\delta=1.62$ (H-9′), 2.29 (H-9), 3.32 (H-1,4), 4.42 (H-2,3), and 5.04 to 5.98 (H-5 to 8) ppm [1, 8], and in cyclohexane-d_{12} $\delta=1.68$ (H-9′; J(H-9,9′)=14.5), 2.60 (H-9; J(H-1,9)=11.0), 3.61 (H-1; J(H-1,9′)=5.2), 3.77 (t with additional fine structure, H-4), 4.79 to 5.07 (H-2,3), 5.41 to 6.00 (H-5 to 8; J(H-8,9′)=9.5, J(H-8,9)=6.2) [2]. The IR spectrum (cyclohexane) shows bands at 1990, 2001, 2042 (CO) cm^{-1} [1] and the mass spectrum shows the ions $[M-nCO]^+$ (n=0 to 3) [2]. The organic ligand has all cis configuration [2, 10].

Isomer IVb is stable for days at room temperature, but at 101 °C in octane in a sealed tube undergoes electrocyclic ring closure to V (R=H) [1, 2]. The first-order rate constant for this process is $2.4\times10^{-4}\,s^{-1}$ corresponding to $\Delta G^{*}=28.4$ kcal/mol. Oxidation with $[NH_4]_2[Ce(NO_3)_6]$ in acetone at 0 °C gives cis-8,9-dihydroindene (83% yield) and small amounts of higher molecular weight products, but no trace of the trans isomer or other C_9 isomers. Similar oxidation at −10 °C gives cis[4]-cyclononatetraene. Protonation with $HOSO_2F/ClSO_2F$ (1:2) in CD_2Cl_2 at −120 °C occurs at C-6 to yield XI, see p. 279 [2]. Attempted reaction with C_4H_9Li or $NaOCH_3/CH_3OH$ failed [10].

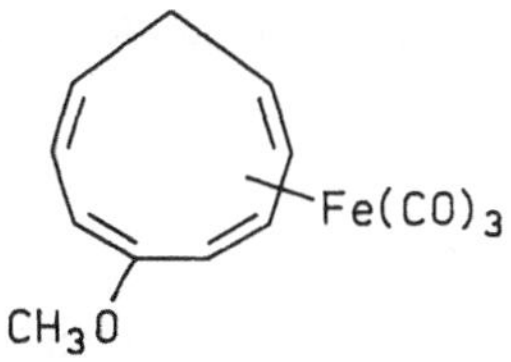

XII 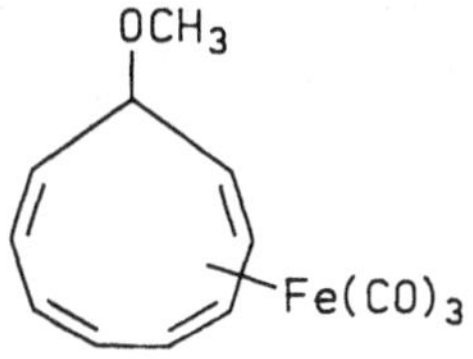 XIII 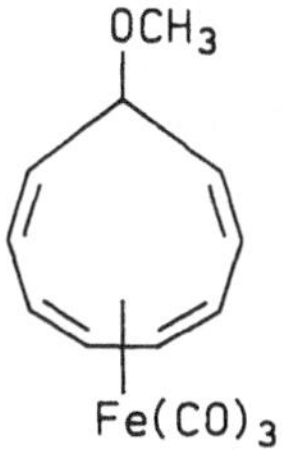 XIV 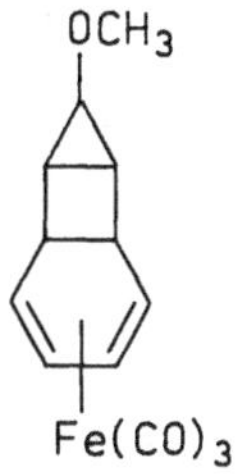 XV

5-$CH_3OC_9H_9Fe(CO)_3$ (see Formula XII) and **9-$CH_3OC_9H_9Fe(CO)_3$** (see Formulas XIII and XIV). Reaction of IX with 9-methoxybicyclo[6.1.0]nonatriene gives XV, V ($R=OCH_3$), and 5-$CH_3OC_9H_9Fe(CO)_3$ (see Formula XII) in 1% yield. Photolysis of $Fe(CO)_5$ with 9-methoxybicyclo[6.1.0]nonatriene gives V ($R=OCH_3$), XV, and XII (12% yield), as well as XIII (18%) and XIV (2%) [2], cf. [6]. All three compounds XII to XIV have cis configuration in the ring. Attempted protonation of XIII in CF_3CO_2H at 0 °C, in HBF_4 at 0 °C, and in $HOSO_2F$ at −78 °C gives several unidentified species, rather than the expected $[^5LFe(CO)_3]^+$ cation (^5L=cyclononatetraenyl) [2].

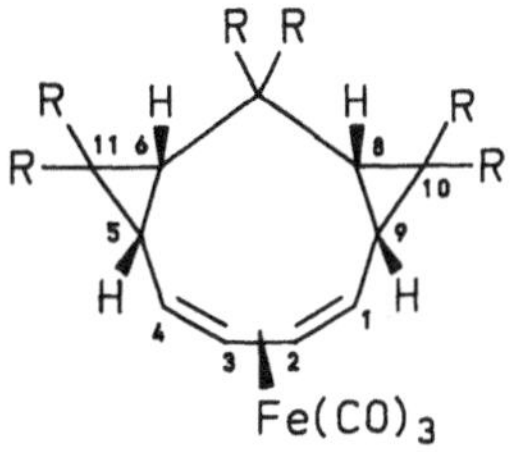

XVI

$C_{11}H_{14}Fe(CO)_3$ (see Formula XVI with R=H) is obtained by refluxing $C_8H_8Fe(CO)_3$ with excess Simmons-Smith reagent (CH_2I_2/Zn/Cu and catalytic amounts I_2) [5, 12, 14] in ether

 References on p. 282

for 2 d [5] or 18 h [14]. Evaporation of the reaction mixture followed by chromatography on alumina [5] or 75% Woelm, 25% Alcoa alumina [14] with hexane [5, 14] gives XVI (R=H) in 25 [5] or 42% yield [14]. Analogous treatment with CD_2I_2 instead of CH_2I_2 gives XVI (R=D); no mono- or diadducts are observed [5, 14]. The compound is a yellow solid [5, 14] showing in the ^1H NMR spectrum (benzene-d_6) chemical shifts at $\delta=-0.11$ (m, H-10endo, 11endo), 0.46 (m, H-7endo); 0.85 (4H), 1.47 (2H) (m, H-5,6,8,9,10exo, 11exo); 2.01 (br, d, H-7exo), 3.57 (dt, H-1,4), 4.67 (dt, H-2,3) ppm [14]. The deuterated compound XVI (R=D) shows in the ^1H NMR spectrum (CS_2) $\delta=1.30$ (H-5,6,8,9; second order AB pattern), 3.78 (m, H-1,4), 5.35 (m, H-2,3) ppm. The proton decoupled ^2H NMR spectrum (CS_2, measured from external acetone-d_6) shows $\delta=1.00$, 1.39, 1.98, and 3.01 ppm, relative intensity 2:1:2:1 [14], cf. [5]. The ^{13}C NMR spectrum of $C_{11}H_{14}Fe(CO)_3$ in CS_2 shows chemical shifts at $\delta=18.6$, 19.1 (C-6,8,10,11); 24.8 (C-5,9), 29.5 (C-7), 66.0 (C-1,4), 91.9 (C-2,3), 211.4 (CO) ppm [14], cf. [5]. The IR spectrum (hexane) shows bands at 1960, 1970, 2035 [5, 14]. From these NMR spectra it is concluded that the cyclopropyl groups are on the same side of the ring, presumably anti to the metal [14]. Oxidation with $[NH_4]_2[Ce(NO_3)_6]$ in CH_3OH at 0 °C for 1 h gives the free ligand ^4L in small amounts [14].

References:

[1] G. Deganello, H. Maltz, J. Kozarich (J. Organometal. Chem. **60** [1973] 323/33). – [2] E.J. Reardon, M. Brookhart (J. Am. Chem. Soc. **95** [1973] 4311/6). – [3] G. Scholes, C.R. Graham, M. Brookhart (J. Am. Chem. Soc. **96** [1974] 5665/7). – [4] G. Deganello, L. Toniolo (J. Organometal. Chem. **74** [1974] 255/62). – [5] D.L. Reger, A. Gabrielli (J. Am. Chem. Soc. **97** [1975] 4421/2).

[6] G. Deganello (unpublished results from [7]). – [7] G. Deganello, P. Uguagliati, L. Calligaro, P.L. Sandrini, F. Zingales (Inorg. Chim. Acta **13** [1975] 247/89). – [8] M. Airoldi, G. Deganello, J. Kozarich (Inorg. Chim. Acta **20** [1976] L5/L6). – [9] M.S. Brookhart, G.W. Koszalka, G.O. Nelson, G. Scholes, R.A. Watson (J. Am. Chem. Soc. **98** [1976] 8155/61). – [10] C.R. Graham (Diss. Univ. North Carolina 1976; Diss. Abstr. Intern. B **38** [1977] 688/9).

[11] C.R. Graham, G. Scholes, M. Brookhart (J. Am. Chem. Soc. **99** [1977] 1180/8). – [12] A.M. Gabrielli (Diss. Univ. South Carolina 1978; Diss. Abstr. Intern. B **39** [1979] 5911). – [13] A. Salzer, W. von Philipsborn (J. Organometal. Chem. **161** [1978] 39/49). – [14] D.L. Reger, A. Gabrielli (J. Organometal. Chem. **187** [1980] 243/52).

1.4.1.4.1.2.7 ^4L is a Ring System Greater than C_9

Most of the following compounds are rather ill-defined except $C_{10}H_{12}Fe(CO)_3$ (I). They were either isolated around 1960 or are cited only in patents.

Fe(CO)3 Fe(CO)3

I II

$C_{10}H_{12}Fe(CO)_3$ (see Formula I) is prepared from cis-bicyclo[6.2.0] deca-2,4,6-triene, which is known to contain at least 5% cyclooctatetraene, with $Fe_2(CO)_9$ in ether at room temperature for 90 h. The mixture of products is separated by chromatography on alumina (activity II) with hexane. The large yellow band containing II is rechromatographed several times on alumina (activity II) until the yellow band separates into two distinct bands. The first band contains only II, but the second band shows a very complex ^1H NMR spectrum indicating a mixture of products. This second band is evaporated to dryness and crystals of $(C_8H_9Fe(CO)_3)_2$ (see 2.8.1.1 in "Organoiron Compounds" C5, 1981, Table 4, No. 36, p. 51) are recovered by crystallization from petroleum ether (b.p. 30 to 60 °C) at −5 °C. A few crystals of I are obtained by reducing the mother liquor from the above crystallization to dryness in vacuum and subliming the resulting solid at 40 °C under atmospheric pressure. The IR spectrum (CS_2) shows strong ν(CO) bands at 1990, 1995, 2050 cm^{-1}.

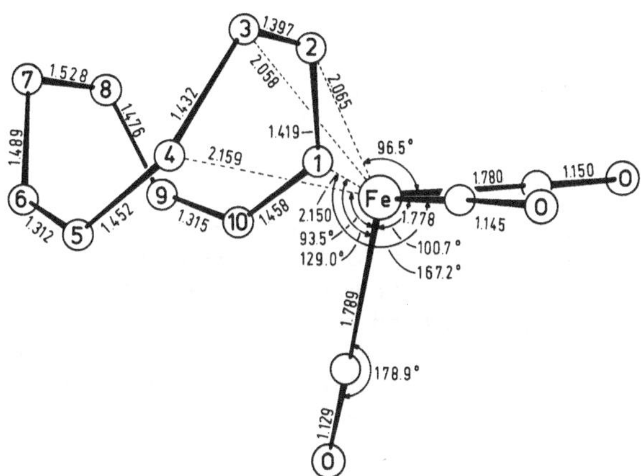

Fig. 27. Molecular structure of $C_{10}H_{12}Fe(CO)_3$ showing how the conformation of the C_{10} ring defines four planes [10].

$C_{10}H_{12}Fe(CO)_3$ crystallizes in the triclinic space group $P\bar{1}-C_i^1$, S_2^1 with the unit cell parameters a = 10.401 (2), b = 9.779 (2), c = 6.554 (2) Å; α = 98.69 (2)°, β = 111.76 (2)°, γ = 82.65 (2)°; Z = 2 molecules per unit cell. The main bond distances and angles are shown in **Fig. 27**. As shown the ring atoms define four planes P_1 to P_4: P_1 with C-1 to 4, P_2 with C-1, 4, 5, 10, P_3 with C-5, 6, 9, 10, and P_4 with C-6 to 9. The dihedral angles between these planes are P_1/P_2 = 41.5°, P_2/P_3 = 50.6°, and P_3/P_4 = 53.8° [10], cf. [12].

$(C_6H_5C_2H)_5COFe(CO)_3$ is prepared by refluxing $Fe_3(CO)_{12}$ with $C_6H_5C{\equiv}CH$ in petroleum ether (b.p. 60 to 70 °C)/C_6H_6 (0.75:0.1) for 1.75 h followed by chromatography on acidic alumina. The first band gives a compound of this composition. Similar reaction at 75 °C for 0.5 h followed by evaporation, dissolving the residue in CS_2, and chromatography as before gives $(C_6H_5C_2H)_5COFe(CO)_3$ as main product [1, 2, 5], cf. [3].

The thin orange-yellow needles melt at 242 °C (dec.). The compound is monomeric and shows two bands in the IR spectrum [2]. For a figure of the K-edge absorption see [6]. The compound is described as a cyclization product with one ketonic CO group; no attempts have been made to localize the aryl groups [1 to 3, 5]. It is very soluble in tetrahydrofuran, and also soluble in C_6H_6, acetone, and ether [2]. The first polarographic wave gives $E^1_{1/2}$ = −1.72 V (from three-electrode potentiostatic geometry, 2×10^{-3} M in substrate, 1×10^{-1} M $[N(C_4H_9)_4]ClO_4$ in $(CH_2OCH_3)_2$ with reference to 1×10^{-3} M $AgClO_4/Ag$ and 1×10^{-1} M $[N(C_4H_9)_4]ClO_4$ in $(CH_2OCH_3)_2$ at 22 °C). No evidence of even a transiently stable radical anion is found [8].

References on p. 284

$(4\text{-}BrC_6H_4C_2H)_5COFe(CO)_3$ is prepared by the reaction of $4\text{-}BrC_6H_4C{\equiv}CH$ and $Fe_3(CO)_{12}$ in light petroleum ether at 65 °C. Chromatography as for the foregoing compound gives a mixture of compounds. Elution with C_6H_6 and some ether gives this product in a small amount. The brown product decomposes at 240 to 260 °C and shows in the IR spectrum (KBr) bands at 1658, 1704 (C=O), and 1972, 2024 (CO) cm^{-1} [7], cf. [3]. The compound is described as a cyclization product with one ketonic CO group; no attempts have been made to localize the aryl groups [7].

Fe(CO)3

III

(E,E,E)-$C_{12}H_{18}Fe(CO)_3$ (see Formula III) can be prepared from $Fe(CO)_5$ and (E,E,E)-dodeca-1,5,9-triene in a sealed tube at 130 °C. After 18 h the reaction mixture is evaporated, and the residue is chromatographed on neutral Al_2O_3 (activity II) with C_6H_6 to give III (9.9% yield) [4]. The compound is also obtained by the reaction of $Fe_2(CO)_9$ and cyclo-$C_{12}H_{18}$ in a sealed tube in ether at 150 to 180 °C for 50 h. Filtration, evaporation, and chromatography gives the pure compound [9]. The deep yellow oil [9] or orange-red viscous liquid boils at 95 to 100 °C/high vacuum [4], is monomeric in C_6H_6 (by cryoscopy), and has in C_6H_6 a dipole moment of $\mu_{25°}=2.85\pm0.04$ D [4, 11]. The ^1H NMR spectrum ($CDCl_3$) shows chemical shifts at $\delta=1.0$ to 2.5 (m) and 3.5 to 6.5 (m) ppm in the ratio 2:1 [9]. The IR spectrum (cyclohexane) shows bands at 970 (C=C) [4], 1971 and 2015 (CO) cm^{-1} [9] or at 1973, 1981, 2047 (CO) cm^{-1} [11]. The compound is fairly air-stable [4, 9] and quite soluble in organic solvents such as C_6H_6, ether, or petroleum ether [4]. Reduction with H_2/PtO_2 in C_2H_5OH for 2 d gives $C_{12}H_{20}Fe(CO)_3$ (next compound) [4].

$C_{12}H_{20}Fe(CO)_3$ is prepared from (E,E,E)-$C_{12}H_{18}Fe(CO)_3$ (see foregoing compound). It is an extremely viscous orange-red liquid, boiling at 115 to 120 °C/high vacuum [4].

References:

[1] K.W. Hübel, Soc. Eur. Res. Associates S.A. (Belg. 567743 [1958]). – [2] K.W. Hübel, Soc. Eur. Res. Associates S.A. (Ger. Offen. 1169932 [1958/64]). – [3] W. Hübel, E.H. Braye, A. Clauss, E. Weiss, U. Krüerke, D.A. Brown, G.S.D. King, C. Hoogzand (J. Inorg. Nucl. Chem. **9** [1959] 204/10). – [4] W. Fröhlich (Diss. München Univ. 1961, pp. 1/118). – [5] W. Hübel, Union Carbide Corp. (Brit. 885514 [1961]).

[6] A. Schneider (Z. Physik. Chem. [Frankfurt] **31** [1962] 249/73). – [7] E.H. Braye, W. Hübel (J. Organometal. Chem. **3** [1965] 25/37). – [8] R.E. Dessy, R.L. Pohl (J. Am. Chem. Soc. **90** [1968] 1995/2001). – [9] S.S. Ullah, S.E. Kabir (J. Bangladesh Acad. Sci. **4** [1980] 47/50). – [10] F.A. Cotton, J.M. Troup (J. Organometal. Chem. **212** [1981] 411/8).

[11] R.D. Fischer (Diss. München Univ. 1961). – [12] F.A. Cotton, J.M. Troup (J. Am. Chem. Soc. **95** [1973] 3798/9).

Table of Conversion Factors

Following the notation in Landolt-Börnstein [7], values which have been fixed by convention are indicated by a bold-face last digit. The conversion factor between calorie and Joule that is given here is based on the thermochemical calorie, cal_{thch}, and is defined as 4.1840 J/cal. However, for the conversion of the "Internationale Tafelkalorie", cal_{IT}, into Joule, the factor 4.1868 J/cal is to be used [1, p. 147]. For the conversion factor for the British thermal unit, the Steam Table Btu, Btu_{ST}, is used [1, p. 95].

Force	N	dyn	kp
1 N (Newton)	1	10^{5}	0.1019716
1 dyn	10^{-5}	1	1.019716×10^{-6}
1 kp	9.80665	9.80665×10^{5}	1

Pressure	Pa	bar	kp/m^2	at	atm	Torr	lb/in^2
1 Pa (Pascal) = 1 N/m^2	1	10^{-5}	1.019716×10^{-1}	1.019716×10^{-5}	0.986923×10^{-5}	0.750062×10^{-2}	145.0378×10^{-6}
1 bar = 10^{6} dyn/cm^2	10^{5}	1	10.19716×10^{3}	1.019716	0.986923	750.062	14.50378
1 kp/m^2 = 1 mm H_2O	9.80665	0.980665×10^{-4}	1	10^{-4}	0.967841×10^{-4}	0.735559×10^{-1}	1.422335×10^{-3}
1 at = 1 kp/cm^2	0.980665×10^{5}	0.980665	10^{4}	1	0.967841	735.559	14.22335
1 atm = 760 Torr	1.01325×10^{5}	1.01325	1.033227×10^{4}	1.033227	1	760	14.69595
1 Torr = 1 mm Hg	133.3224	1.333224×10^{-3}	13.59510	1.359510×10^{-3}	1.315789×10^{-3}	1	19.33678×10^{-3}
1 lb/in^2 = 1 psi	6.89476×10^{3}	68.9476×10^{-3}	703.069	70.3069×10^{-3}	68.0460×10^{-3}	51.7149	1

Work, Energy, Heat	J	kWh	kcal	Btu	MeV
1 J (Joule) = 1 Ws = 1 Nm = 10^7 erg	1	2.778×10^{-7}	2.39006×10^{-4}	9.4781×10^{-4}	6.242×10^{12}
1 kWh	$\mathbf{3.6} \times 10^6$	1	860.4	3412.14	2.247×10^{19}
1 kcal	4184.**0**	1.1622×10^{-3}	1	3.96566	2.6117×10^{16}
1 Btu (British thermal unit)	1055.06	2.93071×10^{-4}	0.25164	1	6.5858×10^{15}
1 MeV	1.602×10^{-13}	4.450×10^{-20}	3.8289×10^{-17}	1.51840×10^{-16}	1

1 eV/mol ≙ 23.0578 kcal/mol = 96.473 kJ/mol

Power	kW	PS	kp m/s	kcal/s
1 kW = 10^{10} erg/s	1	1.35962	101.972	0.239006
1 PS	0.73550	1	7**5**	0.17579
1 kp m/s	$9.8066\mathbf{5} \times 10^{-3}$	0.01333	1	2.34384×10^{-3}
1 kcal/s	4.18**40**	5.6886	426.650	1

References:

[1] A. Sacklowski, Die neuen SI-Einheiten, Goldmann, München 1979. (Conversion tables in an appendix.)
[2] International Union of Pure and Applied Chemistry, Manual of Symbols and Terminology for Physicochemical Quantities and Units, Pergamon, London 1979; Pure Appl. Chem. **51** [1979] 1/41.
[3] The International System of Units (SI), National Bureau of Standards Spec. Publ. No. 330 [1972].
[4] H. Ebert, Physikalisches Taschenbuch, 5th Ed., Vieweg, Wiesbaden 1976.
[5] Kraftwerk Union Information, Technical and Economic Data on Power Engineering, Mülheim/Ruhr 1978.
[6] E. Padelt, H. Laporte, Einheiten und Größenarten der Naturwissenschaften, 3rd Ed., VEB Fachbuchverlag, Leipzig 1976.
[7] Landolt-Börnstein, 6th Ed., Vol. II, Pt. 1, 1971, pp. 1/14.
[8] ISO Standards Handbook 2, Units of Measurement, 2nd Ed., Geneva 1982.

Key to the Gmelin System of Elements and Compounds

System Number	Symbol	Element
1		Noble Gases
2	H	Hydrogen
3	O	Oxygen
4	N	Nitrogen
5	F	Fluorine
6	**Cl**	**Chlorine**
7	Br	Bromine
8	I	Iodine
	At	Astatine
9	S	Sulfur
10	Se	Selenium
11	Te	Tellurium
12	Po	Polonium
13	B	Boron
14	C	Carbon
15	Si	Silicon
16	P	Phosphorus
17	As	Arsenic
18	Sb	Antimony
19	Bi	Bismuth
20	Li	Lithium
21	Na	Sodium
22	K	Potassium
23	NH_4	Ammonium
24	Rb	Rubidium
25	Cs	Caesium
	Fr	Francium
26	Be	Beryllium
27	Mg	Magnesium
28	Ca	Calcium
29	Sr	Strontium
30	Ba	Barium
31	Ra	Radium
32	**Zn**	**Zinc**
33	Cd	Cadmium
34	Hg	Mercury
35	Al	Aluminium
36	Ga	Gallium

HCl

$ZnCl_2$

System Number	Symbol	Element
37	In	Indium
38	Tl	Thallium
39	Sc, Y La—Lu	Rare Earth Elements
40	Ac	Actinium
41	Ti	Titanium
42	Zr	Zirconium
43	Hf	Hafnium
44	Th	Thorium
45	Ge	Germanium
46	Sn	Tin
47	Pb	Lead
48	V	Vanadium
49	Nb	Niobium
50	Ta	Tantalum
51	Pa	Protactinium
52	**Cr**	**Chromium**
53	Mo	Molybdenum
54	W	Tungsten
55	U	Uranium
56	Mn	Manganese
57	Ni	Nickel
58	Co	Cobalt
59	Fe	Iron
60	Cu	Copper
61	Ag	Silver
62	Au	Gold
63	Ru	Ruthenium
64	Rh	Rhodium
65	Pd	Palladium
66	Os	Osmium
67	Ir	Iridium
68	Pt	Platinum
69	Tc	Technetium[1]
70	Re	Rhenium
71	Np,Pu. . .	Transuranium Elements

$CrCl_2$

$ZnCrO_4$

Material presented under each Gmelin System Number includes all information concerning the element(s) listed for that number plus the compounds with elements of lower System Number.

For example, zinc (System Number 32) as well as all zinc compounds with elements numbered from 1 to 31 are classified under number 32.

[1] A Gmelin volume titled "Masurium" was published with this System Number in 1941.

A Periodic Table of the Elements with the Gmelin System Numbers is given on the Inside Front Cover